NEW DIRECTIONS in ORGANIC and BIOLOGICAL CHEMISTRY

Series Editor: C.W. Rees, FRS
Imperial College of Science, Technology and Medicine
London, UK

New and Forthcoming Titles

Chirality and the Biological Activity of Drugs
Roger J. Crossley

Enzyme-Assisted Organic Synthesis
Manfred Schneider and Stefano Servi

C-Glycoside Synthesis
Maarten H.D. Postema

Organozinc Reagents in Organic Synthesis
Ender Erdik

Activated Metals in Organic Synthesis
Pedro Cintas

Capillary Electrophoresis: Theory and Practice
Patrick Camilleri

Cyclization Reactions
C. Thebtaranonth and Y. Thebtaranonth

Mannich Bases: Chemistry and Uses
Maurilio Tramontini and Luigi Angiolini

Vicarious Nucleophilic Substitution and Related Processes in Organic Synthesis
Mieczyslaw Makosza

Radical Cations and Anions
M. Chanon, S. Fukuzumi, and F. Chanon

Chlorosulfonic Acid: A Versatile Reagent
R.J. Cremlyn and J.P. Bassin

Aromatic Fluorination
James H. Clark and Tony W. Bastock

Selectivity in Lewis Acid Promoted Reactions
M. Santelli and J.-M. Pons

Dianion Chemistry
Charles M. Thompson

Asymmetric Methodology in Organic Synthesis
David J. Ager and Michael B. East

Synthesis Using Vilsmeier Reagents
C.M. Marson and P.R. Giles

The Anomeric Effect
Eusebio Juaristi

Chiral Sulfur Reagents
M. Mikołajczyk, J. Drabowicz, and P. Kiełbasiński

C–Glycoside Synthesis

Synthesis

Maarten H.D. Postema, Ph.D.

CRC Press
Taylor & Francis Group
Boca Raton London New York

CRC Press is an imprint of the
Taylor & Francis Group, an **informa** business

CRC Press
Taylor & Francis Group
6000 Broken Sound Parkway NW, Suite 300
Boca Raton, FL 33487-2742

© 1995 by Taylor & Francis Group, LLC
CRC Press is an imprint of Taylor & Francis Group, an Informa business

First issued in paperback 2019

No claim to original U.S. Government works

ISBN 13: 978-0-367-44920-9 (pbk)
ISBN 13: 978-0-8493-9150-7 (hbk)

**Visit the Taylor & Francis Web site at
http://www.taylorandfrancis.com**

**and the CRC Press Web site at
http://www.crcpress.com**

Library of Congress Cataloging-in-Publication Data

Postema, Maarten H.D.
    C-Glycoside synthesis / Maarten H.D. Postema.
        p.   cm. — (New directions in organic and biological
    chemistry)
    Includes bibliographical references and index.
    ISBN 0-8493-9150-4
    1. Glycosides—Synthesis. I. Title. II. Series.
QP702.G59P67 1995
547.7'83—dc20
                                                                94-41409
                                                                    CIP

Library of Congress Card Number 94-41409

For Michael Hogben, my Family, and Mara;
without their help this would not have been possible.

Preface

The growth of the field of C-glycoside chemistry has been enormous over recent years. This is attested by comparing the knowledge accumulated in the early 1970's[1] to what is summarized in this text. There have been several reviews over recent years. Some have been overviews[2, 3] while others have tried to concentrate on one specific aspect of C-glycoside preparation[4] or application.[5] No text[6] exclusive to the synthesis of this compound class has yet appeared.

The organization of the text is straightforward. The type of reaction used to assemble the C-glycoside is the differentiating factor that separates Chapters 1 through 7. The author has tried to stress which methods are particularly well suited for either α- or β-C-glycoside formation. One of the aims of this text is to help workers in the field quickly select which method would be best for synthesizing a particular type of C-glycoside. Where appropriate, the use of C-glycosides as synthons in natural product synthesis has also been mentioned.

Chapters 1 and 2 deal with the use of electrophilic sugars in C-glycoside preparation. This is a very popular method since it uses the natural electrophilicity of the anomeric center to advantage for C-glycosidation. Although there is conceptual overlap in these sections, the organization tries to limit this drawback.

Chapter 3 concerns itself with the umpolung concept, as applied to the anomeric center and the use of these species in C-glycoside preparation. The work here is fairly recent and this methodology constitutes a novel approach for making C-glycosides. Once thought to be limited to 2-deoxy sugars, this technology is now versatile enough to be used routinely.

Chapter 4 deals with Wittig approaches to C-glycoside synthesis and this mature topic has been used by many workers over recent years. Newer chemistry is also covered, which includes the use of sulfur based ylides to make C-glycosides as well as the chemistry of exomethylenic C-glycosidic compounds.

The work outlined in the fifth chapter deals with the application of palladium chemistry to C-glycoside formation. This is a fairly recent approach and both π-allyl complex chemistry and Heck type couplings comprise the bulk of this chapter.

Sigmatropic rearrangements and cycloaddition chemistry as applied to the formation of C-glycosides are the topics covered in Chapter 6. Sigmatropic rearrangements have been applied to both C-furanoside and C-pyranoside synthesis, while cycloaddition chemistry has been limited mainly to the pyranoside series. Both of these approaches are fairly modern in origin. Some cycloaddition approaches are discussed in Chapter 11.

The use of free radical methods, a fairly young field, at least in synthetic applications, has found a powerful position in C-glycoside preparation. This methodology can be used to selectively prepare both α- and β-C-pyranosides.

The next four chapters deal with the synthesis of particular C-glycosides. Chapter 8 addresses the preparation of C-disaccharides and from the length (34 pages) it is evident that this is a very popular research area. One reason for the interest is that C-disaccharides (or similar structures) may hold special promise as enzyme inhibitors since the glycosidic linkage is no longer prone to enzymatic or hydrolytic cleavage.

The synthesis of naturally alkyl and aryl C-glycosides is dealt with in Chapters 9 and 10 respectively, while Chapter 11 serves as an overview (not comprehensive by any means) to C-nucleoside preparation. The final chapter examines some of the applications of C-glycosides, especially biological ones.

The preparation of this manuscript has been by no means a simple task and it would not have been possible without the help of certain individuals. I would like to thank Professor Ole Hindsgaul for the initial inspiration to write on this topic and also for allowing me to perform literature searches on his system, Dr. Todd Lowary for his advice and input for certain chapters and Dr. Frank Barresi for help in computer searching. A whole lot of thanks goes to Rudy and Jeannine Postema who allowed me to stay at their home (rent-free) to complete this manuscript and Karen Postema for her help in final manuscript preparations. Without their encouragement, this book would never have come to fruition. I would also like to extend my gratitude to Mara for her encouragement to "Get to work!" when I felt least like doing so. Finally, I would like to thank Mr. Navin Sullivan at CRC Press for the invitation to write this text, my assistant editor Michelle Veno and my assistant managing editor Gerry Jaffe for their input and helpful suggestions aimed at improving this text. A debt of thanks is owed to two of my past teachers Master Paul Desmarais for teaching me what I needed to know and to Stanley Choi for introducing me to a thing called chemistry.

It is hoped that the reader of this text will find it useful and enlightening and will not hesitate to make suggestions and comments directed at the improvement of this work.

References

1. Hanessian, S.; Pernet, A.G. *Adv. Carbohydr. Chem.* **1976**, *33*, 111.
2. Postema, M.H.D. *Tetrahedron* **1992**, *48*, 8545.
3. Herscovici, J.; Antonakis, K. in *Studies in Natural Product Chemistry*; Vol. 10; Rahman, A.U.; Ed.; Elsevier: Amsterdam, 1992, 337.
4. Jaramillo, C.; Knapp, S. *Synthesis* **1994**, 1.
5. Hacksell, U.; Daves, G.D., Jr. *Progress Med. Chem.* **1985**, *22*, 1.
6. An entire issue of Carbohydrate Research has been dedicated to this topic: See *Carbohydr. Res.* **1987**, *171*.

The Author

Maarten H. D. Postema was born in the Hague, Netherlands in 1965 and came to Canada when he was two years old. Having established an early interest in Science, he obtained his B.Sc. degree with Honors in Chemistry from Concordia University in Montreal. This was followed by graduate work at the University of Alberta in Edmonton. His areas of interest include organic synthesis with emphasis on free radical chemistry, C-glycoside chemistry, and synthetic methodology utilizing the chiral pool approach. Dr. Postema is a member of the Canadian Institute of Chemistry and the American Chemical Society. Besides reading the literature and discussing chemistry, his hobbies include practicing Karate and scuba diving as well as building model ships and planes. He is currently a research associate at the Scripps Research Institute in La Jolla, California.

Table of Contents

Chemical Abbreviations

AIBN	azobisisobutyronitrile
Ac	acetyl
Ar	aryl
9-BBN	9-borabicyclo[3.3.1]nonane
Bn	benzyl
BPM	4-phenyl-benzyl
n-BuLi	*n*-butyllithium
Bz	benzoyl
CAN	ceric ammonium nitrate
cat.	catalytic
CBn	*p*-chlorobenzyl
CSA	camphorsulfonic acid
DBU	1,8-diazabicyclo[4.3.0]non-5-ene
DDQ	2,3-dichloro-5,6-dicyano-1,4-benzoquinone
DCC	dicyclohexylcarboimide
DEAD	diethylazidodicarboxylate
DHP	dihydropyran
DIBAL(H)	diisobutylaluminum hydride
DMAP	4-dimethylaminopyridine
DME	dimethoxyethane
DMF	*N,N*-dimethylformamide
DMSO	dimethylsulfoxide
DPS or TBPS	diphenyl-*t*-butylsilyl
Et	ethyl
Imid.	imidazole
hν	light
LAH	lithium aluminum hydride
LDA	lithium diisopropylamide
LiHMDS	lithium hexamethyldisilazane
LN	lithium naphthalenide
m	meta
m-CPBA	*m*-chloroperbenzoic acid
Me	methyl
MOM	methoxymethyl
Ms	methanesulfonyl
n	normal
NBS	*N*-bromosuccinimide
NCS	*N*-chlorosuccinimide
NIS	*N*-iodosuccinimide
NMNO	*N*-methylmorpholine-*N*-oxide
N-PSP	*N*-phenylselenophthalimide
o	ortho
p	para
PCC	pyridinium chlorochromate
PDC	pyridinium dichromate
Phth	phthalimido
Pmb	*p*-methoxybenzyl
PNB	*p*-nitrobenzoate
PPTS	pyridinium *p*-toluenesulfonate

pyr.	pyridine
Swern [O]	(COCl)$_2$, DMSO, then Et$_3$N
t	tertiary
TBAF	tetrabutylammonium fluoride
TBS or TBDMS	*t*-butyldimethylsilyl
Tf	trifluoromethanesulfonyl
TfOH	trifluoromethanesulfonic acid
THF	tetrahydrofuran
TFA	trifluoroacetic acid
TFAA	trifluoroacetic anhydride
TIPS	triisopropylsilyl
TMEDA	*N,N,N′,N′*-tetramethylenediamine
TMS	trimethylsilyl
TsOH	*p*-toluenesulfonic acid
TTN	thallium trinitrate
Ts	tosyl

t

C-Glycoside Synthesis

Chapter 1

ELECTROPHILIC SUGARS IN *C*-GLYCOSIDE SYNTHESIS I

I. INTRODUCTION

The most common method for carbon-carbon bond formation at the anomeric carbon involves nucleophilic attack on this naturally electrophilic center. A wide variety of electrophilic sugars have been employed, such as reducing sugars (or lactols), alkyl glycosides, anomeric esters, anomeric trichloroacetimidates, and glycosyl halides. The carbon nucleophiles that have been used include silyl enol ethers, olefins, allyl-, propargylsilanes, cyanides, homoenolates, and organometallics such as Grignard reagents, organolithiums, cuprates, and aluminates. This chapter will be divided into categories based on the nature of the electrophilic sugar and its reactions with various nucleophiles. The advantages of certain methods and applications in total synthesis will also be discussed.

II. LACTOLS

A. SILICON AND AROMATIC BASED NUCLEOPHILES
1. Enol Ethers

Allevi and co-workers[1] have used the reaction of silyl enol ethers with the gluco derivative 1 (eq. 1) to synthesize *C*-glycosides.

eq. 1

Table 1: Reaction of 1 with Various Nucleophiles.

Entry	Nucleophile	Product R =	Lewis Acid	Yield	α:β
a	$MeC(OSiMe_3)=CH_2$	CH_2COMe	$ZnCl_2$	45	α
b	$t\text{-}Bu(OSiMe_3)=CH_2$	$CH_2CO\text{-}t\text{-}Bu$	$BF_3{\cdot}OEt_2$	50	α
c	$PhC(OSiMe_3)=CH_2$	CH_2COPh	$BF_3{\cdot}OEt_2$	75	α
d	$p\text{-}ClC_6H_4(OSiMe_3)=CH_2$	$CH_2COC_6H_4Cl\text{-}p$	$BF_3{\cdot}OEt_2$	82	α
e	$\beta\text{-}C_{10}H_7C(OSiMe_3)=CH_2$	$CH_2COC_{10}H_7\text{-}\beta$	$BF_3{\cdot}OEt_2$	72	α
f	$Me_3SiCH_2CH=CH_2$	$CH_2CH=CH_2$	$BF_3{\cdot}OEt_2$	85	4:1
g	$MeOPh$	$p\text{-}MeOC_6H_4$	$BF_3{\cdot}OEt_2$	50	β
h	$m\text{-}(MeO)_2C_6H_4$	$o,p\text{-}(MeO)_2C_6H_3$	$BF_3{\cdot}OEt_2$	78	β

Reproduced with permission from ref. 1, pg. 1246, Copyright 1987, courtesy   Royal Society of Chemistry, Thomas Graham House, Science Park, Milton Road, Cambridge, CB4 4WF, UK.

1

When **1** is first treated with TFAA and then exposed to Lewis acidic conditions an intermediate oxonium ion is formed (Fig. 1) which undergoes condensation with an appropriate nucleophile (Table 1) to give the corresponding *C*-glycoside compounds in good yield. The products obtained are almost always the α-products, except in the case of entries 7 and 8. For entry 7 the weak nucleophile reacts, but in low yield only.

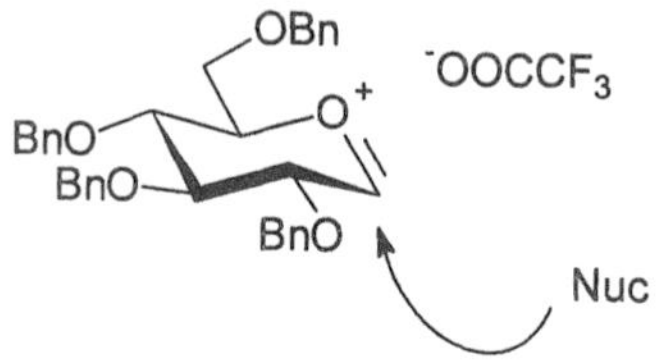

Figure 1

The above examples show that attack of the nucleophile on the intermediate pyranoxonium trifluoroacetate is predominantly from the bottom face under control of the anomeric effect. This is a general trend and this chapter will illustrate how many of the *C*-glycosides made from electrophilic sugars are indeed controlled by this factor, there are some exceptions though.

Equation 2 shows the reaction of a simple enol acetate with lactol **3** to give a pseudo *C*-glycosidic compound.[2] The lactol was treated with stannic chloride and NCS to generate a cyclic oxycarbenium ion that underwent condensation with the enol acetate to lead to the product *C*-glycoside **4** as a mixture of unassigned *syn* and *anti* isomers (40:60).

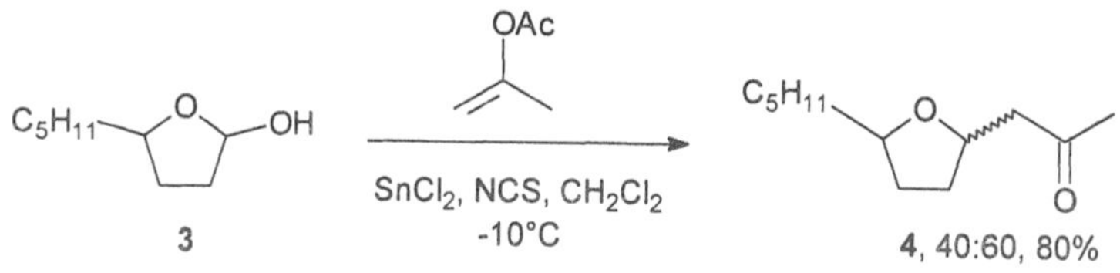

eq. 2

2.   Aromatic Nucleophiles
a.   *Rearrangement*
Kometani and collaborators[3] have taken advantage of an O→C rearrangement

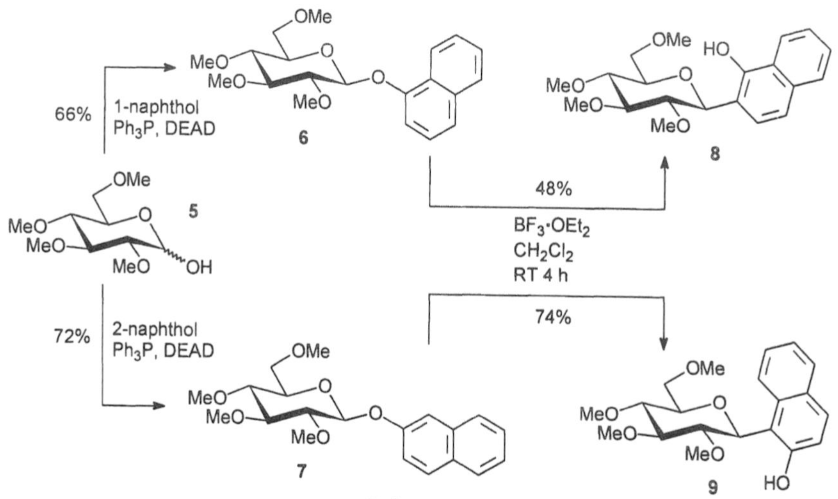

Scheme 1

reaction to generate aromatic *C*-glycosides. The starting material is an *O*-phenyl glycoside that is conveniently prepared by Mitsunobu's method[4] to give for example, **6** and **7**. When these compounds are exposed to the action of boron trifluoride etherate rearrangement occurs to give the thermodynamically more stable β-*C*-glycosides **8** and **9** in 48% and 74% yield, respectively.

Toshima and co-workers[5] have used both protected and unprotected sugars in their synthesis of aryl *C*-glycosides by a Lewis acid catalyzed process. For example, the protected sugars **10a** to **10d** when treated with 2-naphthol in the presence of trimethylsilyltriflate-silver perchlorate complex gave excellent yields of the corresponding aryl β-*C*-glycosides **11a** to **11d**. The β:α ratios were fairly high usually 99:1; the lowest ratio obtained was 15:1 (Scheme 2).

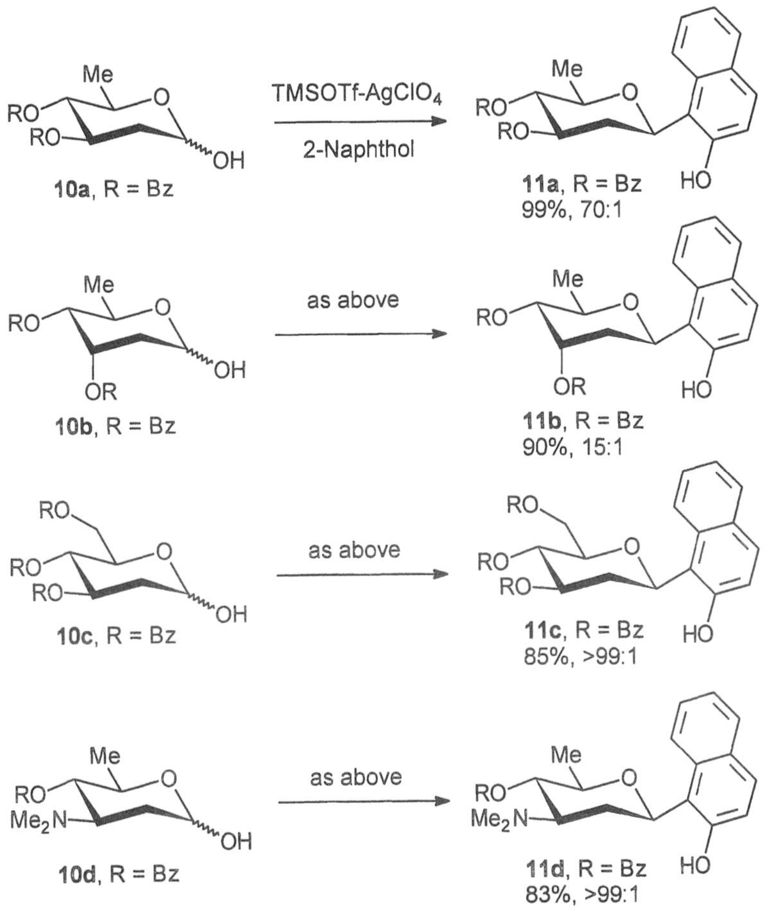

Scheme 2

The unprotected sugars **12a** and **12b** gave good yields of **13a** and **13b** with excellent anomeric selectivity (Scheme 3). Again the β-isomer was the major isomer formed in the reaction and this is best rationalized by assuming that the intermediate undergoes ring opening to the intermediate **14** which recyclizes to the thermodynamically more stable anomer. The reaction is useful since it provides quick access to the 2-deoxy aryl *C*-glycosides which are found in many naturally occurring aryl *C*-glycosides. It is also noteworthy that a 3-amino sugar can also be used in these reactions, although the use of dichloromethane was crucial to the success of this reaction.

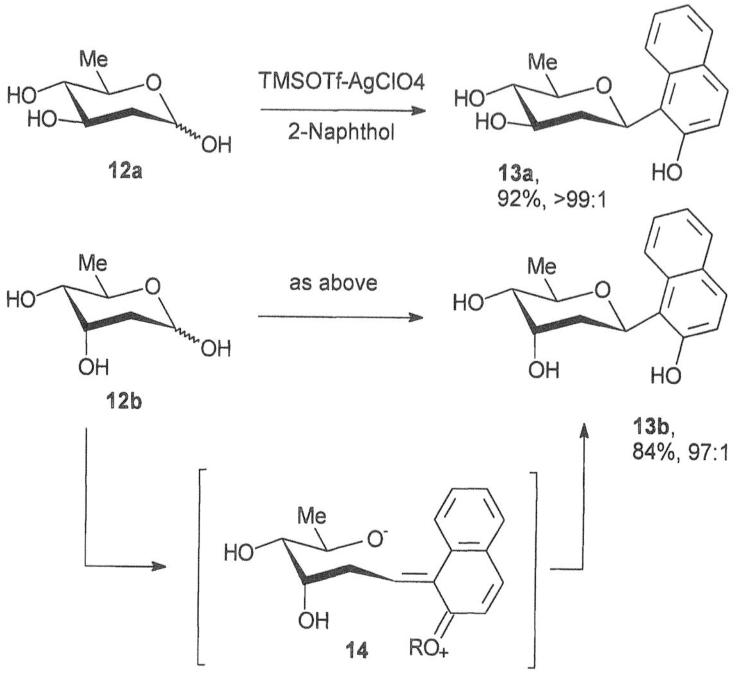
Me
HO
HO
O
OH
12a
TMSOTf-AgClO4
2-Naphthol
13a,
92%, >99:1
HO
12b
as above
OH
13b,
84%, 97:1
14
RO+

Scheme 3

b. Phenolates

Cornia *et al.*[6] have found that the condensation of bromomagnesium phenolates leads to product *C*-glycosides directly. Treatment of the lactol **15** with various bromomagnesium phenolates **16a-d** gave the aromatic *C*-glycosides **17a-d** and **18a-d** in the yields and ratios shown in Scheme 4. The mixture of anomers was of no consequence since treatment of the minor isomers **18a-d** with EtMgBr in dichloroethane resulted in epimerization to the more stable β-anomers **17a-d**.

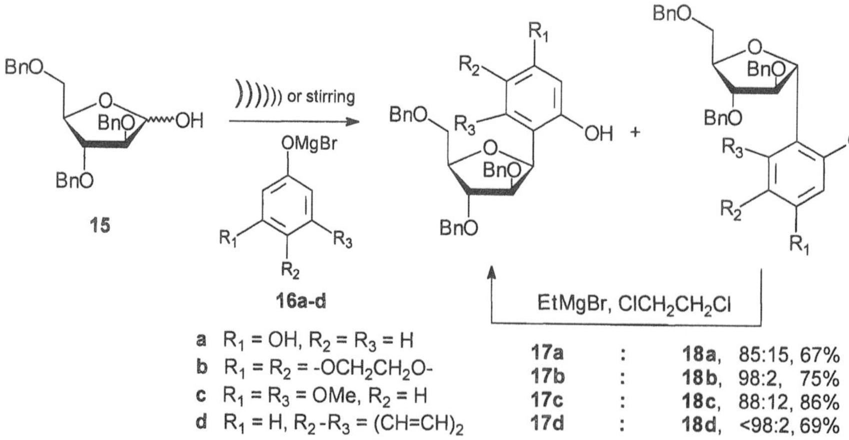
BnO
BnO
OH
15
))))) or stirring
OMgBr
R1
R2
R3
16a-d
+
EtMgBr, ClCH2CH2Cl
a R1 = OH, R2 = R3 = H
b R1 = R2 = -OCH2CH2O-
c R1 = R3 = OMe, R2 = H
d R1 = H, R2-R3 = (CH=CH)2
17a : 18a, 85:15, 67%
17b : 18b, 98:2, 75%
17c : 18c, 88:12, 86%
17d : 18d, <98:2, 69%

Scheme 4

3. Stabilized Nucleophiles

Yamaguchi and collaborators[7] have found the stabilized anion derived from β-oxoglutarate condenses well with the lactol **19** to give good yields of β-*C*-glycosides. Reaction of **20** with anion **21** then yielded intermediate **22** which condensed to **23** by the action of calcium acetate in refluxing methanol to give the aromatic *C*-glycosides **23a** to **23c**, Scheme 5.

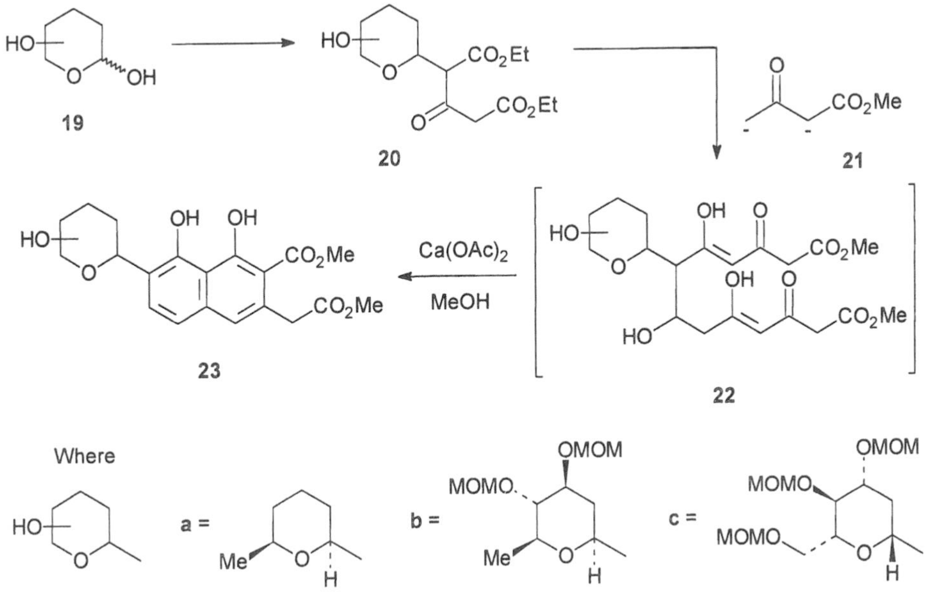

Scheme 5

B. ORGANOMETALLIC NUCLEOPHILES

Tsuchihashi and co-workers[8] have found that access to the open chain diol **27** or a *C*-glycoside like structure is dependent upon the reaction conditions. If the lactol **24** is treated with a lactol then the metallated species presumably exist in equilibrium between forms **25** and **26**, Scheme 6.

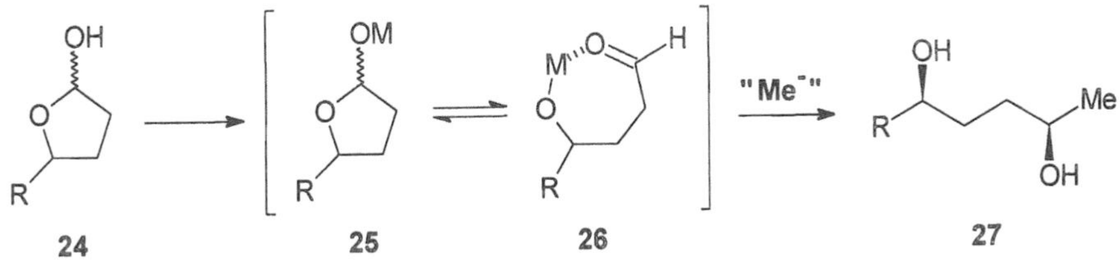

Scheme 6

The chelated form **26** is supported by the fact that when a non-participating solvent and a large nucleophile is used in the reaction good selectivities are obtained. Since the chiral center that is responsible for the induction is far away, the reaction must proceed via **26**. The reaction is governed by the factors that control addition onto faces of ring-like structures, such as the chelated aldehyde **26**, Scheme 7.

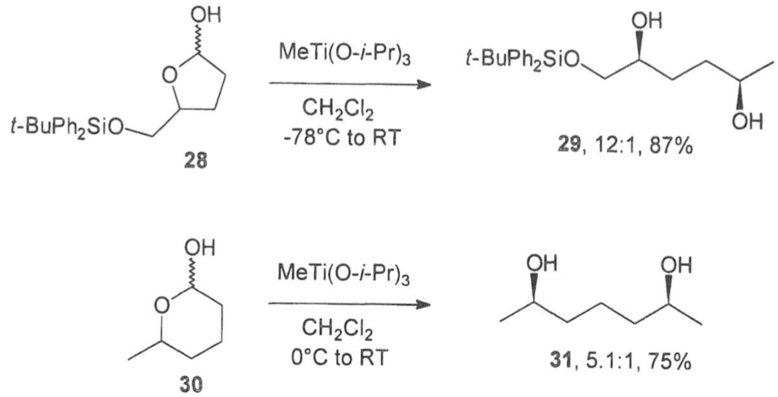

Scheme 7

When a Lewis acid is used[9] in the reaction, either before or after the addition of the nucleophile, then good yields and selectivities of *C*-glycosides are obtained, Scheme 8.

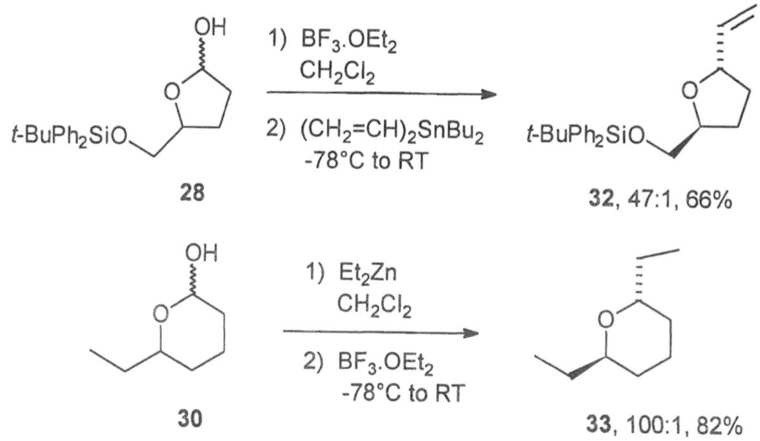

Scheme 8

III. *O*-GLYCOSIDES

A.   SILICON BASED NUCLEOPHILES
1.   Allylsilanes

Work by Hosomi[10] entailed the use of allylsilanes nucleophiles to make allyl *C*-glycosides. This type of compound has been extensively studied and it has been found that for pyranosides the α-anomer usually predominates. The reaction was carried out using trimethylsilyltriflate as a Lewis acid and carried out in acetonitrile as solvent. Various allyl groups were utilized and the α:β selectivity was found to be at least 6:1 usually in the range of 10:1 (eq. 3, Table 2). Again, the anomeric effect dictates the approach of the nucleophile from the bottom face.

eq. 3

Table 2: Products for Equation 3.

Entry	Allylsilane	R_1	R_2	Yield	α:β Ratio
1	$Me_3SiCH_2CH=CH_2$	H	H	86	10:1
2	$Me_3SiCH_2C(Me)=CH_2$	H	Me	87	6:1
3	$Z-Me_3SiCH_2CH=CHMe$	Me	H	68	-
4	$Me_3SiCH_2C(Br)=CH_2$	H	Br	71	1:0

Reproduced from ref. 10, Copyright 1984, pg. 2384 with kind permission from Elsevier Science Ltd, The Boulevard, Langford Lane, Kidlington, OX5 1GB, UK.

Similar chemistry was carried out on mannopyranoside **36** which gave superior α:β ratios (Table 3) and this expected purely on the basis of steric effects.

eq. 4

Table 3: Products for Equation 4.

Entry	Allylsilane	R_1	Yield	α:β Ratio
1	$Me_3SiCH_2CH=CH_2$	H	87	1:0
2	$Me_3SiCH_2C(Me)=CH_2$	Me	81	8:1
3	$Me_3SiCH_2C(Br)=CH_2$	Br	35	1:0

Reproduced from ref. 10, Copyright 1984, pg. 2384 with kind permission from Elsevier Science Ltd, The Boulevard, Langford Lane, Kidlington, OX5 1GB, UK.

Bennek and Gray[11] have shown that free methyl glycosides react with allyltrimethylsialne to give the corresponding α-C-glycoside. The reaction is conducted by treating **38** with bis(trimethylsilyl)trifluoroacetamide which serves to silylate the hydroxyl groups *in situ*. The allylation is then carried out and treatment of the intermediate persilylated C-glycoside with water causes hydrolysis of the silyl groups to **39** (eq. 5). The net effect is a one pot allylation on an unprotected methyl glycoside. The reaction was also applied to suitable ribo and fructo derivatives.

eq. 5

Work by Nicotra and co-workers[12] also utilized the allylation reaction as an access to C-glycosides. The fructofuranoside **40** was exposed to standard allylation conditions and gave the α-anomer **41** in an 80:20 ratio in 94% combined yield. This C-glycoside was then transformed into **43** by double bond migration, ozonolysis, and reduction, Scheme 9. Compound **43** may prove to be useful in the study of carbohydrate metabolism.

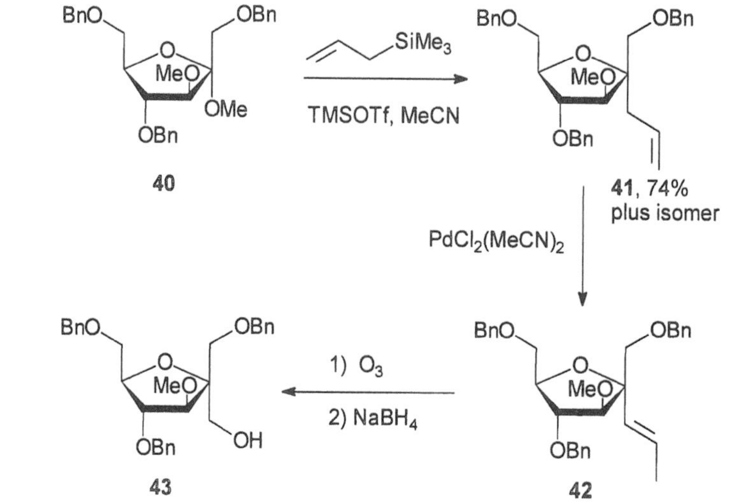

Scheme 9

The allylation reaction sometimes gives open chain compounds as shown in Scheme 10.[13]

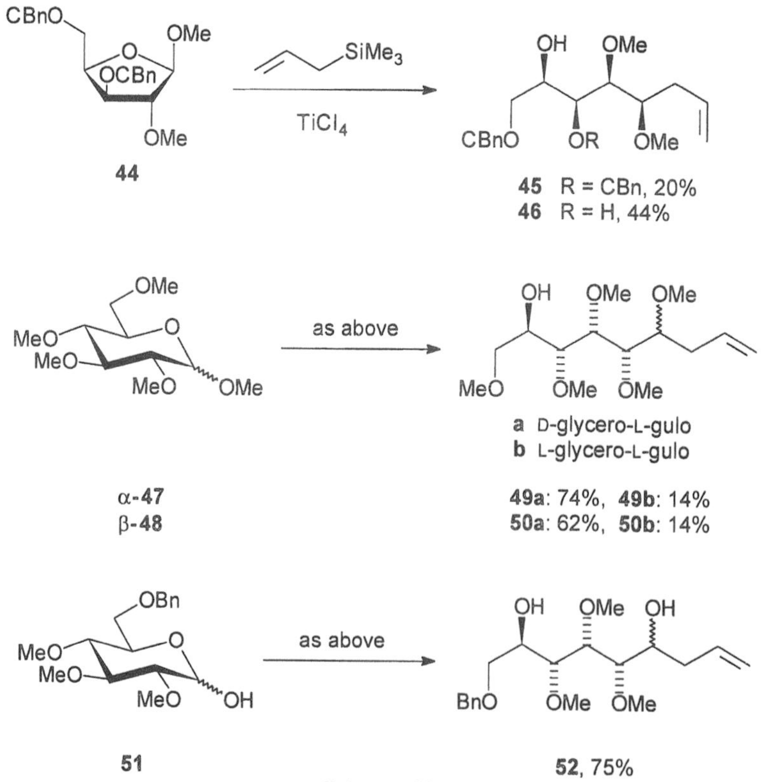

Scheme 10

The workers propose that the reason for the propensity of formed open chain compounds is due chiefly to the Lewis acid (TiCl$_4$) and the chelating effect of the O-5 and O-6 oxygen atoms as shown in Scheme 11. When the O-5 oxygen group is replaced with hydrogen as in **55** only C-furanoside **56** is formed.

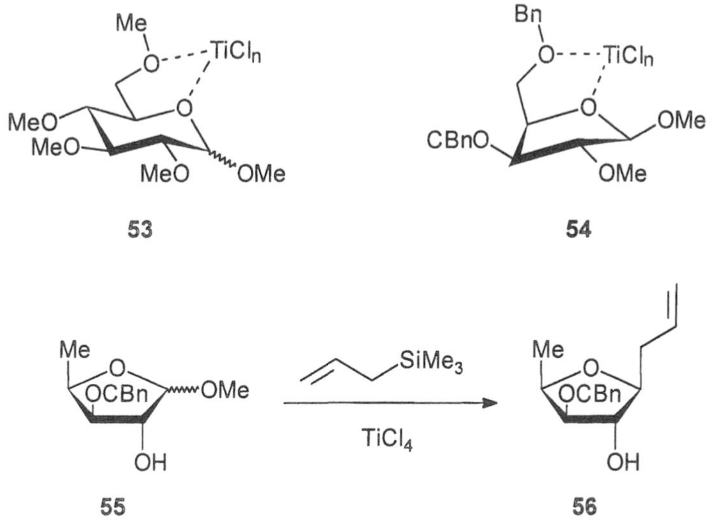

Scheme 11

Bednarski and Bertozzi[14] have extended the utility of the allylation reaction to include 2-azido sugars. This compound has been difficult to synthesize since the amino group is incompatible with the usual reaction conditions. Azidonitration of tri-O-benzyl glycal gave the required starting material **57** which was treated with a Lewis acid and allyltrimethylsilane to give the aldehydo C-glycosides **59**, after ozonolysis. Similar reaction of **57b** with propargylsilane gave the allene derivative **60** which was cleaved (O$_3$) to the shorter α-homologue **61**, Scheme 12.

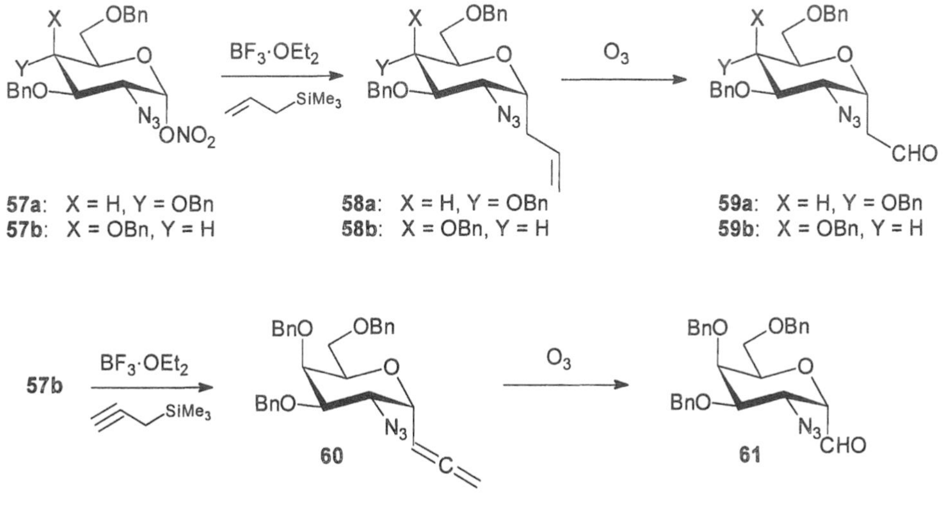

57a: X = H, Y = OBn
57b: X = OBn, Y = H

58a: X = H, Y = OBn
58b: X = OBn, Y = H

59a: X = H, Y = OBn
59b: X = OBn, Y = H

Scheme 12

2.   Aromatic Nucleophiles

Toshima and co-workers[15] applied the same reaction they described with lactols and unprotected sugars with the corresponding methyl glycosides to give good yields of aryl β-*C*-glycosides, Scheme 13.

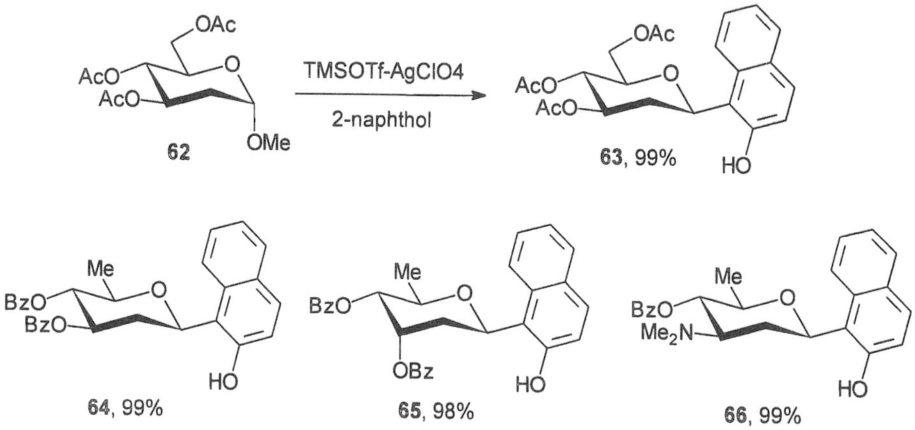

Scheme 13

3.   Stabilized Nucleophiles

Chapleur and co-workers[16] have used lactols to synthesize an intermediate species (**68**) that then underwent displacement with stabilized nucleophiles to give *C*-furanosides **69a-d**. Although some of the yields are low, recovered starting material (15-35%) was isolated in every case.

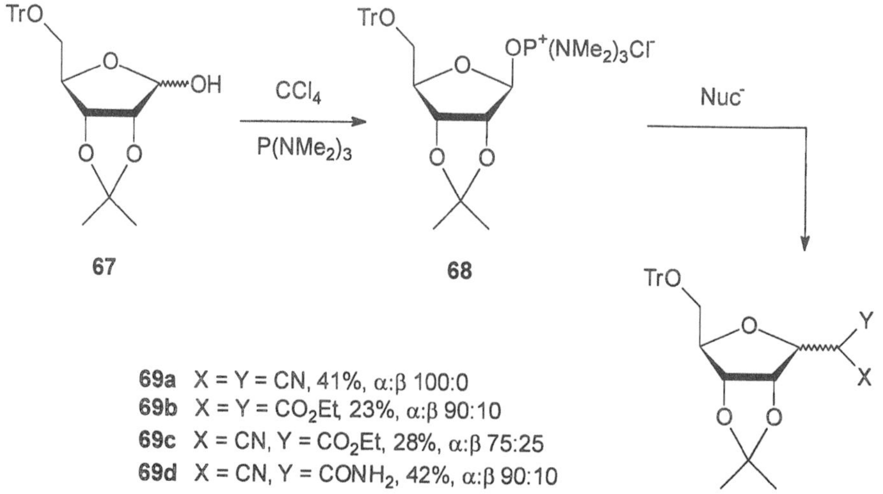

69a X = Y = CN, 41%, α:β 100:0
69b X = Y = CO$_2$Et, 23%, α:β 90:10
69c X = CN, Y = CO$_2$Et, 28%, α:β 75:25
69d X = CN, Y = CONH$_2$, 42%, α:β 90:10

Scheme 14

IV. *S*-GLYCOSIDES

A. THIOPYRIDYL GLYCOSIDES

1. Intermolecular Condensations

Williams and Stewart[17] pioneered the use of thiopyridyl glycosides as precursors to *C*-glycosides. The reaction conditions required for *C*-glycosidation are very mild and entail the use of silver triflate as a Lewis acid. Several nucleophiles were used in the condensation and some of the results are shown in Scheme 15. The yields are good and the anomeric selectivity is often excellent.

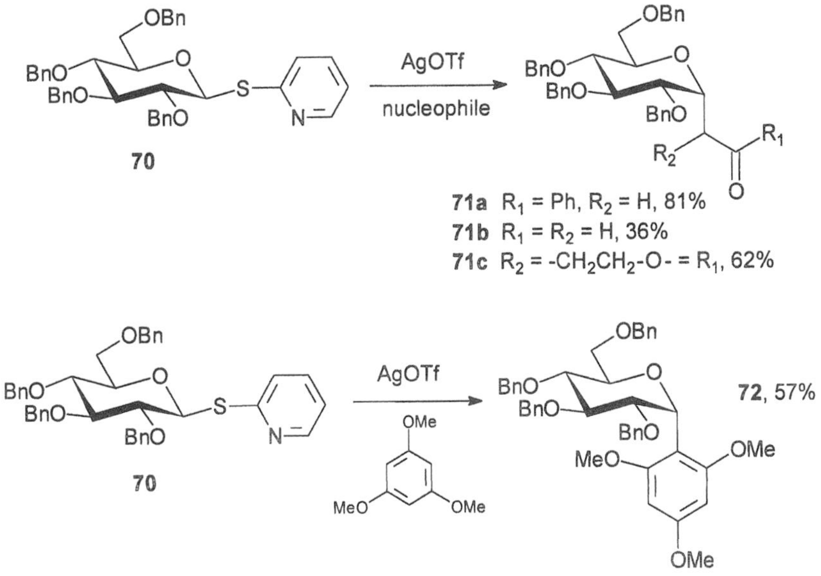

71a R_1 = Ph, R_2 = H, 81%
71b R_1 = R_2 = H, 36%
71c R_2 = -CH$_2$CH$_2$-O- = R_1, 62%

72, 57%

Scheme 15

2. Intramolecular Condensations

Craig and Munasinghe[18] have used the thiopyridyl group to form *C*-glycosides in an intramolecular fashion. Both the α- and β-anomers of **73** and **75** gave the same *cis*-fused lactones **74** and **76** (respectively) in good yield, Scheme 16.

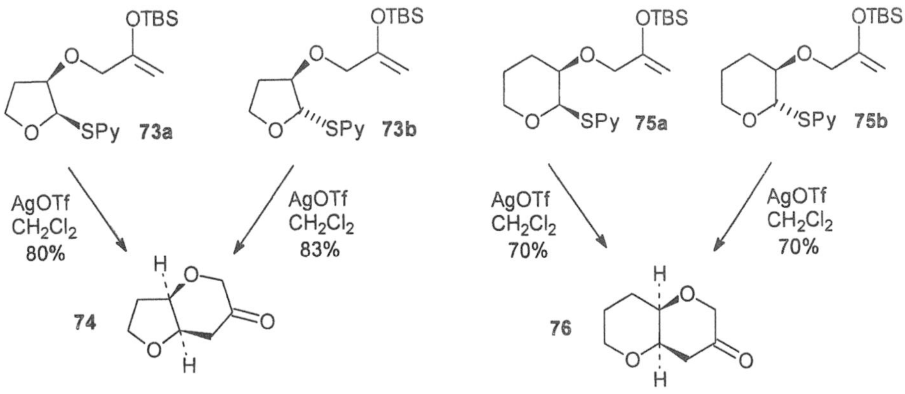

Scheme 16
Reprinted from ref. 18, pg. 665, Copyright 1992, with kind permission
from Elsevier Science Ltd, The Boulevard, Langford Lane, Kidlington 0X5 1GB, UK.

The same workers[19] then applied similar technology to the preparation of the *cis*-fused *C*-glycoside **80**, which was made from methyl glycoside **77** as outlined in Scheme 17. Alkylation, deketalization and benzylation gave **78** which was converted to the mixture of pyridyl thioglycosides and silyl enol ether formation furnished **79**. Exposure to silver triflate then gave the *C*-glycoside **80** as a single isomer at the ester bearing carbon in 50% yield.

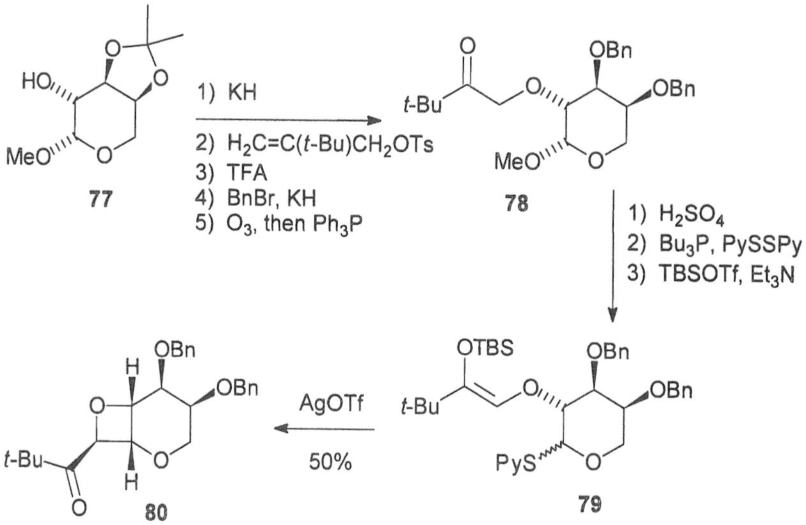

Scheme 17

B.   PHENYL THIOGLYCOSIDES

Work by Keck and collaborators[20] has shown that the allylation of **81** with the methallyltin derivative gave the β-anomer **82** almost exclusively. In the furanoside series the α-isomer **84a** was favored when *O*-5 was protected as its benzyloxy ether, but the ratio shifted, still in favor of the α-anomer, when the *t*-butyldiphenylsilyl protecting group was used on *O*-5, Scheme 18.

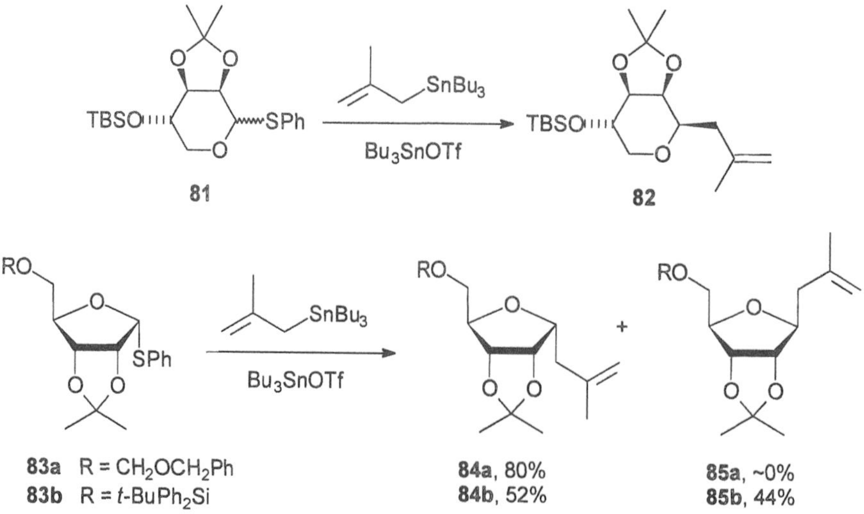

83a   R = CH₂OCH₂Ph   **84a**, 80%   **85a**, ~0%
83b   R = *t*-BuPh₂Si   **84b**, 52%   **85b**, 44%

Scheme 18

Kozikowski[21] has studied the reaction between diorganozinc compounds and phenyl thioglycosides. Reaction of **86** with various nucleophiles gave good yields of *C*-glyosides with the α-anomer predominating in all the cases (eq. 6, Table 4).

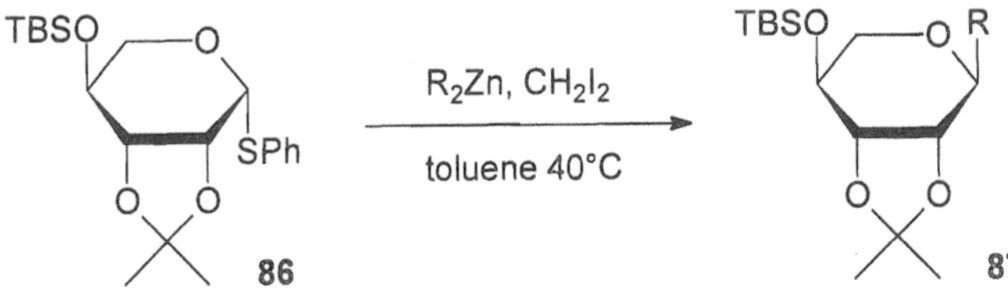

eq. 6

Table 4: Yields and Products for Equation 6.

Entry	R group	Yield	α:β Ratio
a	CH_3	72	94:6
b	CH_3CH_2	78	94:6
c	$CH_3CH_2CH_2$	76	94:6
d	$CH_3(CH_2)_2CH_2$	75	93:7
e	$C_2H_5CH(CH_3)$	72	64:36
f	$(CH_3)_2CH$	73	82:18
g	$CH_2=CH-CH_2$	35	60:40

Reproduced with permission from ref. 21, pg. 112, Copyright 1987, with kind permission from Elsevier Science, Ltd., The Boulevard, Langford Lane, Kidlington, OX5 1GB, UK.

The reaction generally proceeds with inversion of configuration at the anomeric center. Several other sugars were also examined. The gluco derivative **88** gave **89** as the major product while **90** gave a slightly poorer ratio of anomers. Finally, glycal **92** gave **93** as a mixture of anomers, Scheme 19.

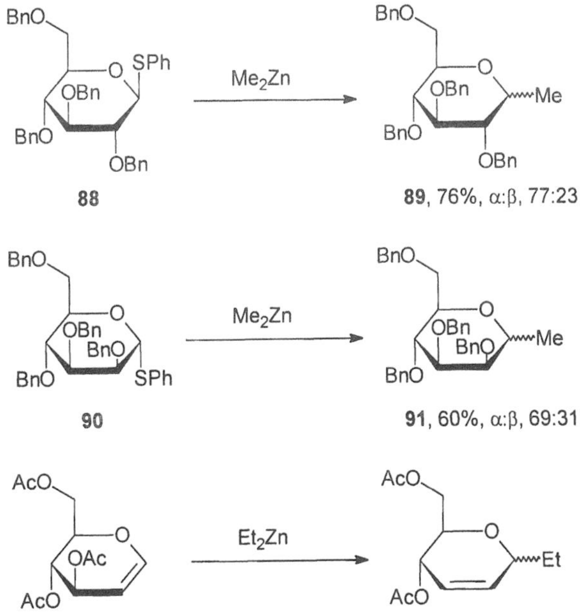

Scheme 19

Reproduced with permission from ref. 21, pg. 112, Copyright 1987, with kind permission from Elsevier Science, Ltd., The Boulevard, Langford Lane, Kidlington, OX5 1GB, UK.

C. ANOMERIC SULFONES

1. With Enol Ethers

Work by Ley[22] has exploited the synthetic utility of sulfone function in forming carbon-carbon bonds on the carbon atom adjacent to oxygen. Treatment of various sulfones with aluminum trichloride followed by the appropriate enol ether gave good yields of adducts, Scheme 20.

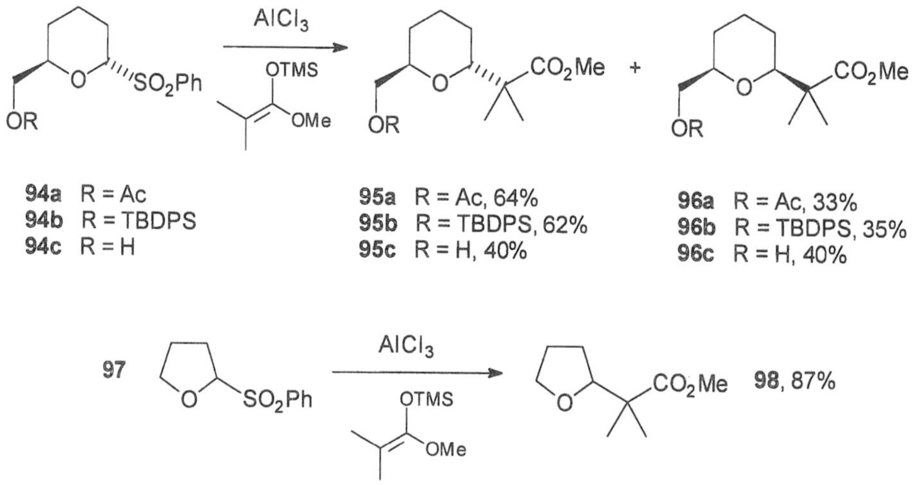
$AlCl_3$
OTMS
OMe
SO_2Ph
OR
CO_2Me
94a R = Ac
94b R = TBDPS
94c R = H
95a R = Ac, 64%
95b R = TBDPS, 62%
95c R = H, 40%
96a R = Ac, 33%
96b R = TBDPS, 35%
96c R = H, 40%
97
98, 87%

Scheme 20

This methodology has also been extended[23] to include reactions with allylsilane, trimethylsilyl cyanide, and trimethyl aluminum to give the corresponding pseudo *C*-glycosides, Scheme 21.

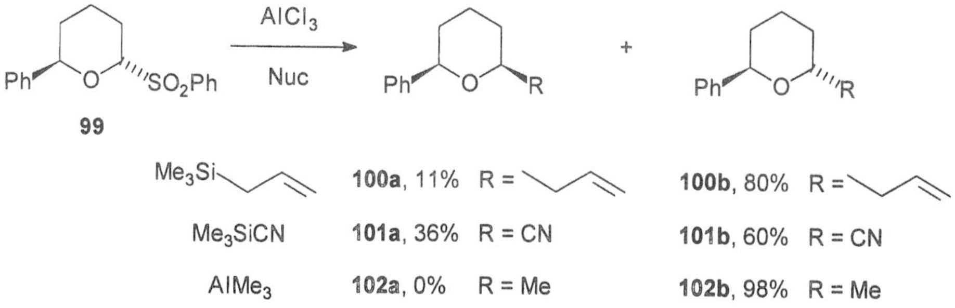
$AlCl_3$
Nuc
Ph
SO_2Ph
R
99
Me_3Si
100a, 11% R =
100b, 80% R =
Me_3SiCN
101a, 36% R = CN
101b, 60% R = CN
$AlMe_3$
102a, 0% R = Me
102b, 98% R = Me

Scheme 21

Grignard reagents[23] can also be used as the nucleophiles, provided zinc bromide is included in the reaction mixture. Vinyl and aryl organometallics are suitable reagents, but the reaction fails with alkyl Grignards, Scheme 22.

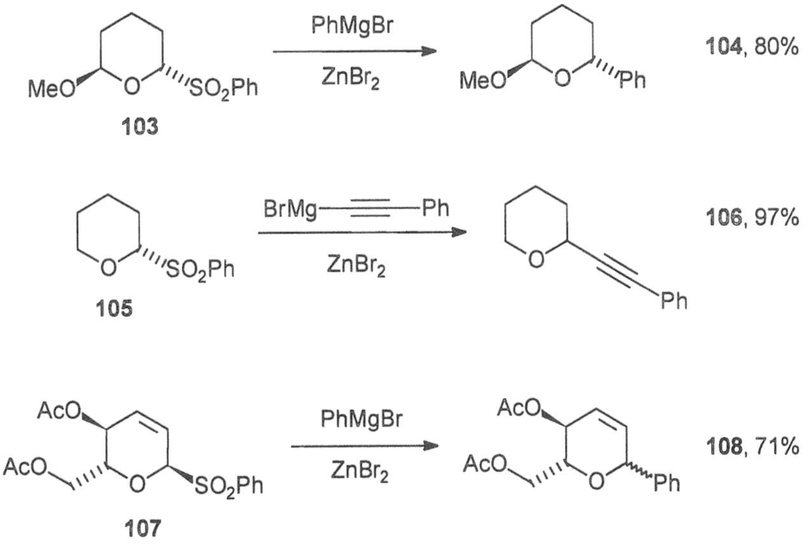

Scheme 22

V. ANOMERIC *O*-TRICHLOROACETIMIDATES

A. SILYL ENOL ETHERS

The development of this methodology is due to Schmidt,[24] and his group has found that reaction of **109** various nucleophiles, as shown in Table 5, leads to *C*-glycosides in good yields. Again, the α-anomer usually predominates. Silyl enol ethers, allylsilanes and trimethylsilyl cyanide can all be used as nucleophiles to form the corresponding *C*-glycosides. The reactions also worked well with enol ethers derived from cyclic ketones (not shown).[25]

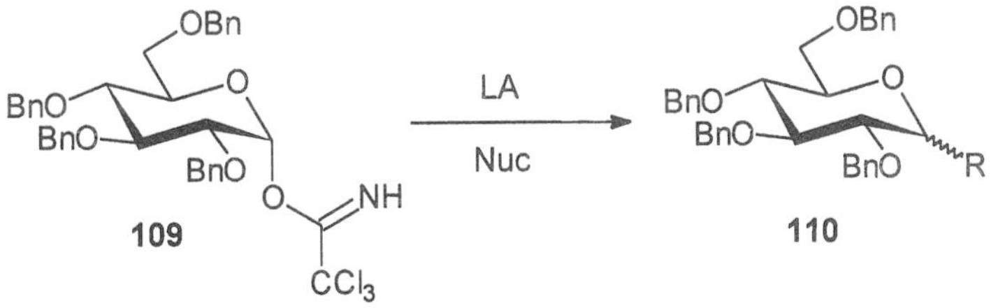

Table 5: Products for Equation 7.

Entry	Lewis acid	R group	Yield, α:β ratio
a	TMSOTf	CN	87, 0:100
b	$BF_3 \cdot OEt_2$	$EtOCOCHCH_2CO_2Et$	70, 1:5
c	"	CH_2COPh	76, 1:1
d	$ZnCl_2$	CH_2COPh	73, 1:0
e	"	$CH_2CO\text{-}t\text{-}Bu$	69, 1:0
f	"	CH_2COMe	68, 1:0
g	"	$CH_2CH=CH_2$	75, 1:0

Reproduced with permission from ref. 24, pg. 406, Copyright 1983, VCH
Verlagsgesellschaft mbH, P.O. Box 101161, D-69451, Weinheim, FRG.

B.   AROMATICS

Compound **109** has also proven useful for the synthesis of aryl *C*-glycosides.[26] Scheme 23 shows some of the reactions of **109** with various aromatics and some of the subsequent chemistry performed on the adduct *C*-glycosides.   The furan **111** can be derivatized in the usual way and both the thiophene *C*-glycoside **115** and indolic *C*-glycoside **116** are available by this methodology.

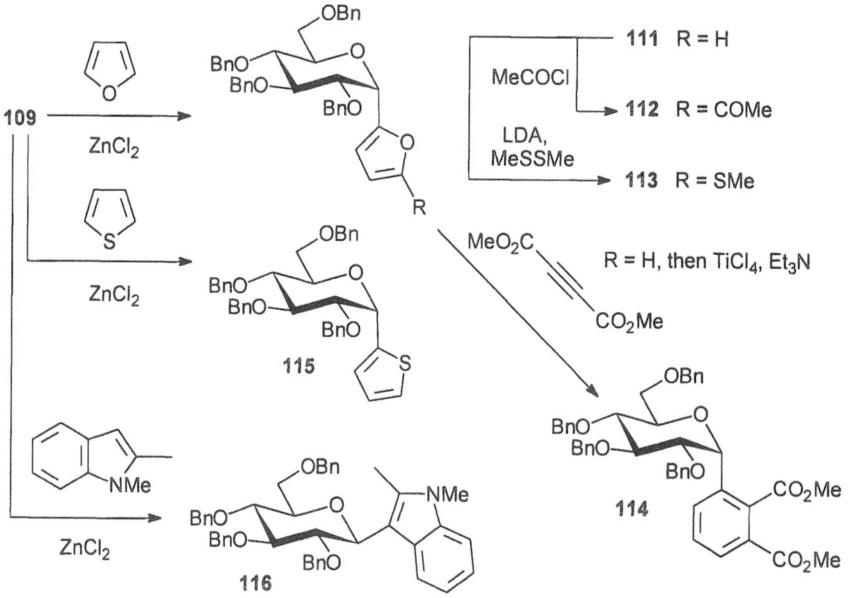

Scheme 23

Work by the same group[27] has also shown that aryl β-*C*-glycosides are available from **109** by reaction with certain phenols using TMSOTf as a catalyst, Scheme 24. The yields for the formation of the adducts range from 60 to 70%. The use of less reactive aromatics led only to the formation of the corresponding *O*-glycosides.

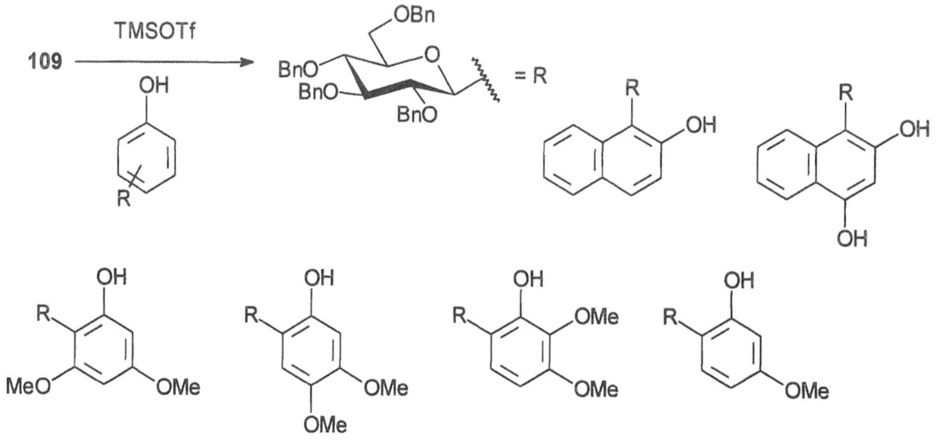

Scheme 24

Furanosides are also suitable substrates for *C*-glycosidation by this methodology, Scheme 25.[28]

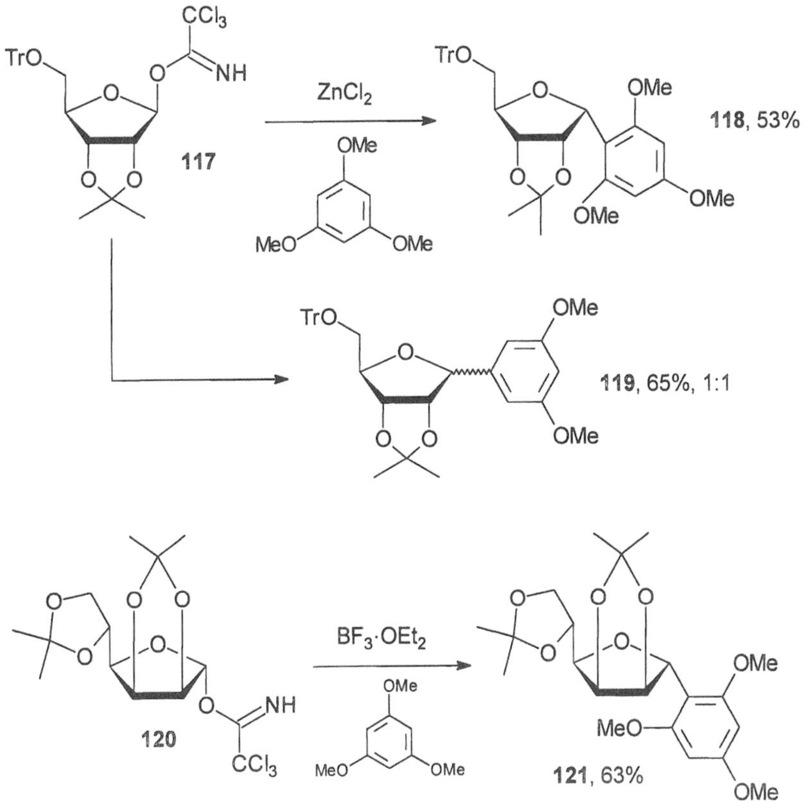

Scheme 25

VI. ANOMERIC ACETATES

There exists a significant amount of literature on the use of anomeric acetates as precursors, some dating back to 1945. This type of reaction—the condensation of a suitable nucleophile with an anomeric acetate—is not a novel synthetic transformation and the development of this field is in part due to the pioneering work of Hurd and Bonner.[29] As can be seen from the material presented thus far, there is conceptually a lot of repetition. Naturally, new substrates and variations in the nucleophile differentiate each set of examples, but overall there is a tremendous amount of overlap and this will become even more pronounced as this chapter unfolds.

A. SILICON BASED NUCLEOPHILES
1. Pyranosides

Noyori and Murata[30] published the reaction of silyl enol ethers with 2-acetoxytetrahydropyrans as a powerful method for forming carbon-carbon bonds. Several nucleophiles were studied and the salient results are shown in Scheme 26. The yields are good and the reaction tolerates substitution on the enol ether. In the case of **124** and **126** the *cis*-products are formed exclusively. Presumably the reaction proceeds via an S_N2 type of process.

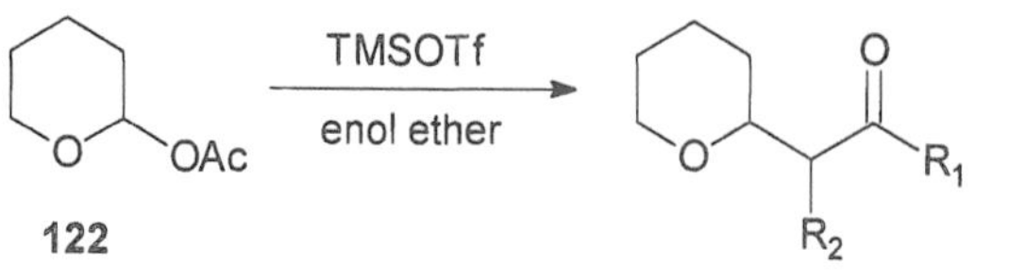

122

123a R_1 = Ph, R_2 = H, 81%
123b R_1 = *t*-Bu, R_2 = H, 96%
123c R_1 = H, R_2 = Me, 78%, 70:30
123d R_1 = Ph, R_2 = Me, 91%, 62:38
123e R_1 = *t*-BuS, R_2 = Me, 79%, 78:22

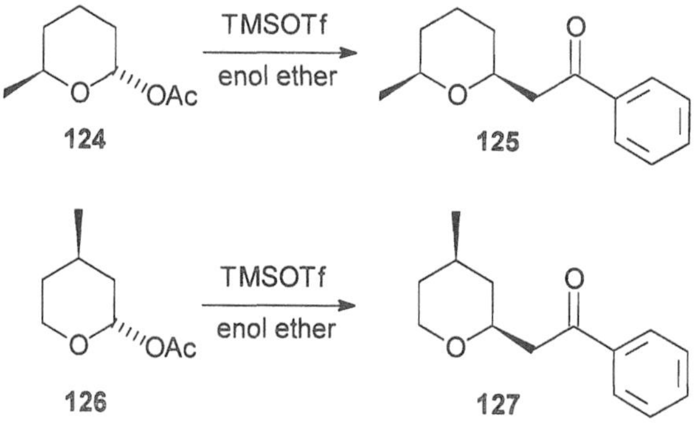

124 → 125

126 → 127

Scheme 26

Kishi, Lewis, and Cha[31] have developed methodology for the convenient access to α-*C*-glycosides from anomeric *p*-nitrobenzoates, which are themselves available from the free sugar. Treatment of the anomeric benzoate **128** with boron trifluoride etherate in acetonitrile at 0°C followed by addition of allyltrimethylsilane led preferentially to the α-allyl glycoside **129** (10:1).

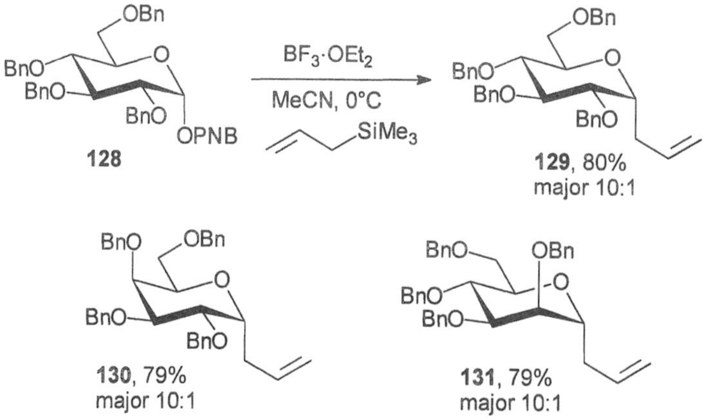

128

129, 80% major 10:1

130, 79% major 10:1

131, 79% major 10:1

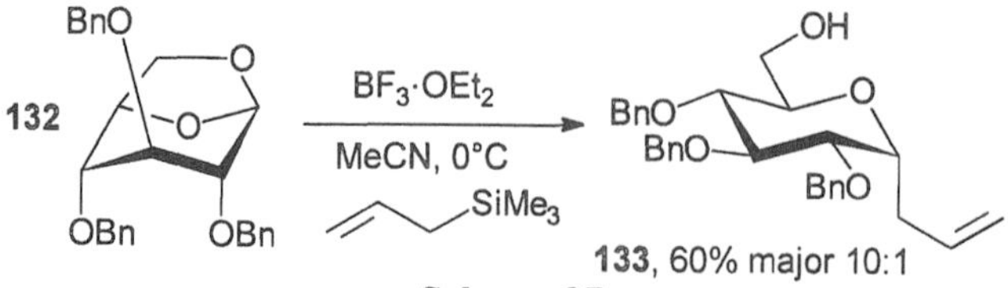

132

133, 60% major 10:1

Scheme 27

Exposure of the corresponding manno and galacto compounds led to the production of **131** and **130** in comparable yield and anomeric ratio. Interestingly enough, the anhydrosugar **132** also reacted under the reaction conditions to give a 60% yield of α-*C*-glycoside **133** in which the 6-hydroxyl group has now been liberated, Scheme 27.

Kozikowski and Sorgi[32] have published similar chemistry with the lyxo derivatives **134** and **136**. Scheme 28 shows the reaction of the protected lyxo compounds and in the case of **134** the α-anomer predominates. With a participating group at *C*-2, as in **136**, the major product formed is the β-isomer. It would seem that this reaction is controlled by steric factors.

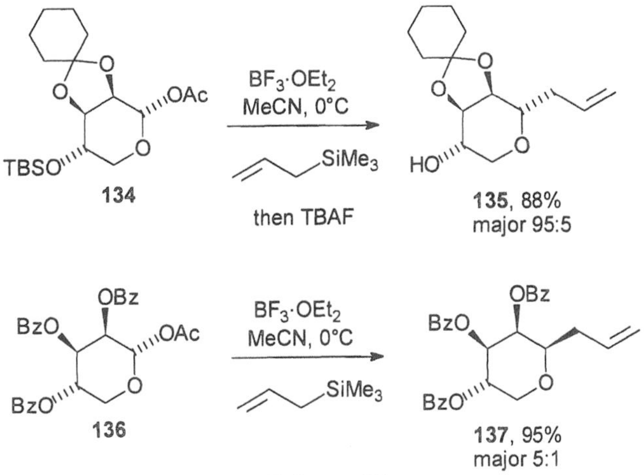

Scheme 28

The *C*-glycosides **138-141** were made from the corresponding peracetates by reaction with allyltrimethylsilane in acetonitrile in the presence of boron trifluoride etherate.[33]

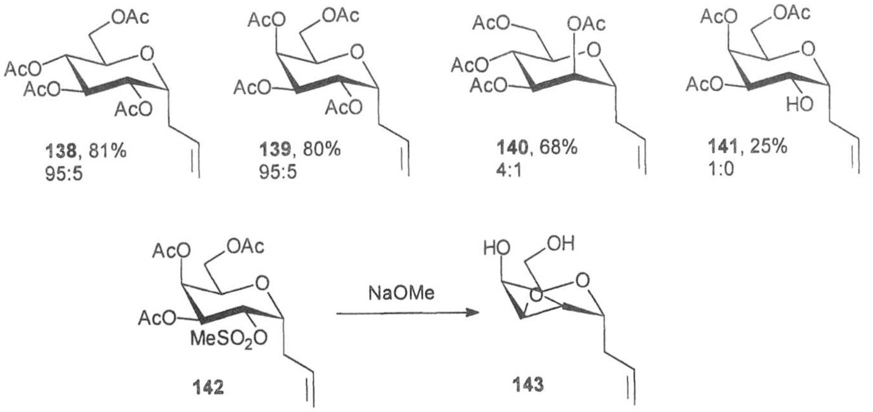

Scheme 29

The derivative **142** was also made from the corresponding tetraacetate by similar reaction, and exposure to sodium methoxide gave epoxide **143** which may be a suicide inhibitor of β-galactosidase from E. coli. Disaccharide **144** was also amenable to the reaction conditions and gave the *C*-glycoside **145**, but in low yield, Scheme 30. The workers reported that attempted *C*-allylation reactions with numerous derivatives of 2-deoxy-2-acetimido glucose failed.[33]

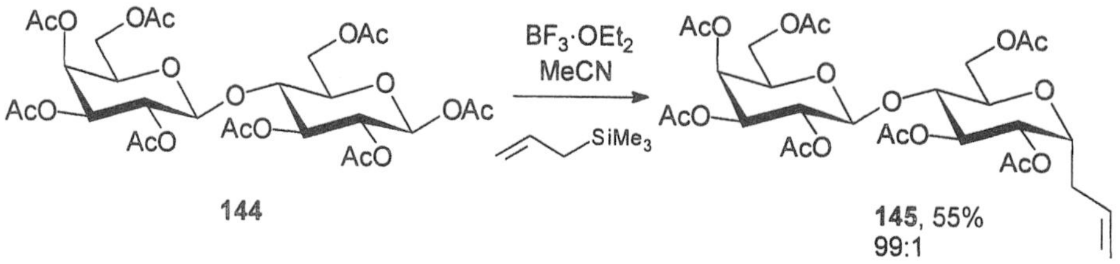

144

145, 55%
99:1

Scheme 30

Horton and Miyake[34] used a slightly more complex allylsilane in their *C*-glycosidation of **146**, eq. 8.

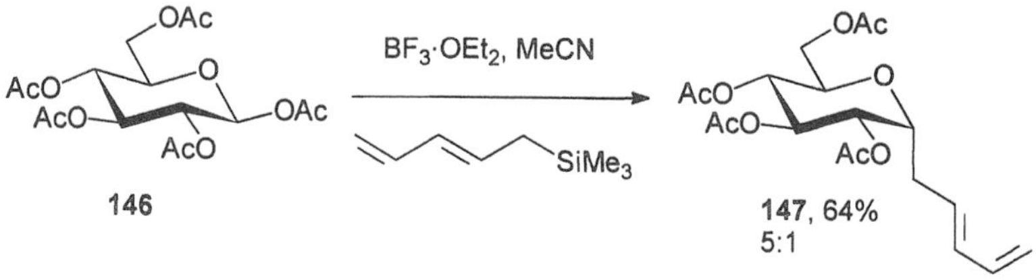

146

147, 64%
5:1

eq. 8

The 2-amino sugar **148** reacted with the same allylsilane to give the corresponding axial *C*-glycoside **149**, eq. 9.[35]

148

149

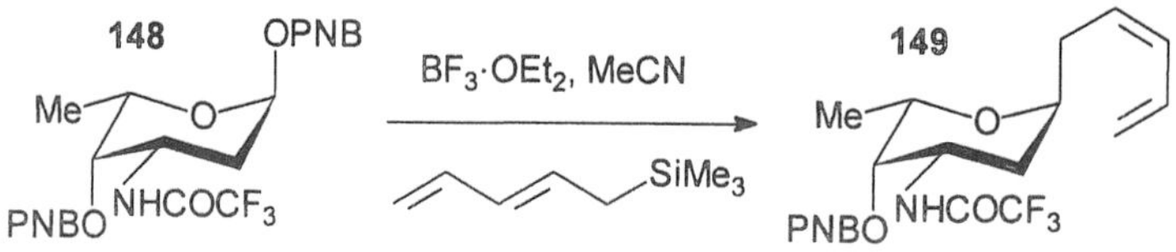

eq. 9

Gurjar *et al.*[36] used the allyl *C*-glycoside as a starting point for the synthesis of *C*-α-D-glucosyl amino acids.

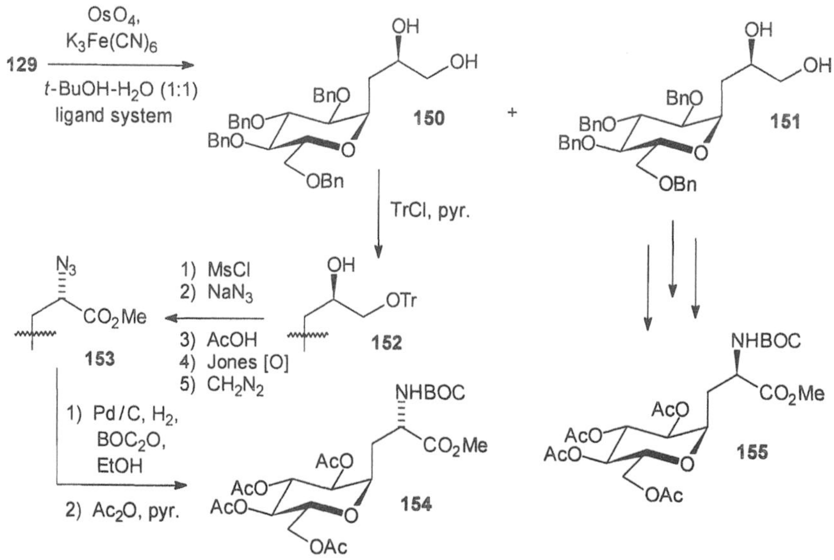

Scheme 31

Hydroxylation of **129** gave a mixture of isomers in ratios that were dependent on the ligand system used in the hydroxylation reaction. Protective group manipulations of **150** gave **152** in which the secondary hydroxyl was converted to the azide ester **153** by mesylation, displacement and deprotection/oxidation. Hydrogenation and protection then gave the target **154**. Compound **151** was converted into **155** in a similar fashion, Scheme 29.

Homoenolates have also found application in *C*-glycoside synthesis.[37] Acetate **156** reacted with **157** under Lewis acidic conditions to give a 74% yield of isomers with the α-anomer **158** predominating by a factor 10:1.

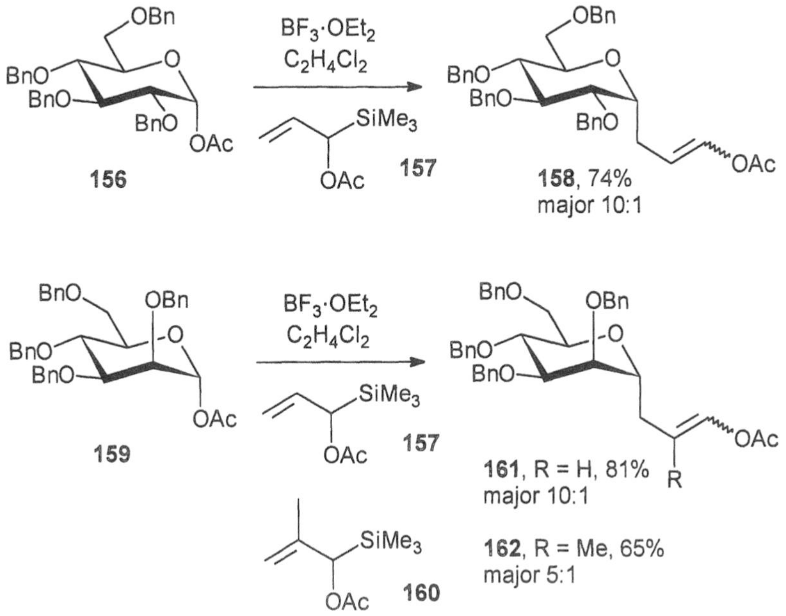

Scheme 32

The manno derivative **159** reacted with **157** to give a 10:1 ratio of anomers in 81% yield and also with **160** to give a 65% yield of anomers (α:β, 5:1).

2. Furanosides

Early work focused in the condensation of silyl enol ether **164** with acetate **163** to give the *C*-glycoside **165** (eq. 10).[38]

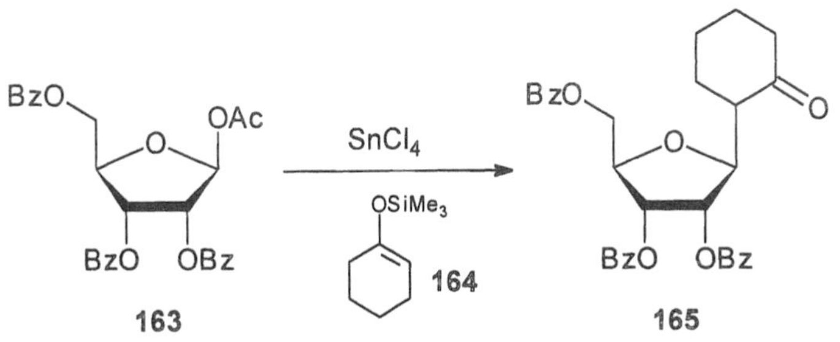

Work has also focused on reactions of silicon based nucleophiles of the corresponding *C*-furanosides. For example reaction of **166** with various nucleophiles under trityl perchlorate catalysis gave a good yield of the corresponding *C*-glycosides in which the α-anomer predominated in all the cases, Scheme 33.[39]

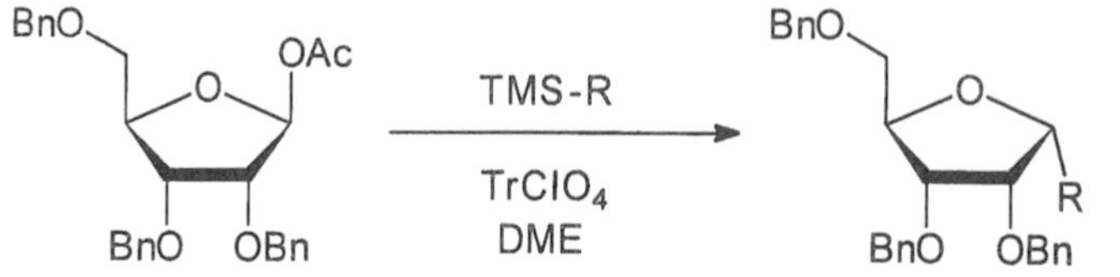

167a   R = CH$_2$COCMe$_3$, 93%, 99:1
167b   R = CH$_2$COPh, 97%, 100:0
167c   R = CH$_2$CH=CH$_2$, 90%, 100:0
167d   R = CN, 93% (in Et$_2$O) 93:7

Scheme 33

The workers also succeeded in converting polystyrene-bound triphenylmethanol into its perchlorate salt, thereby creating a polymer bound catalyst. Flow reactions with 166, the silyl enol ether of pinacolone, and the immobilized catalyst gave an 86% yield of the *C*-glycoside 167a with an α:β ratio of 24:1. The methodology proved to be very convenient since the catalyst can be recycled. This methodology is also well suited to large scale operations.

The corresponding tri-*O*-benzoate showed similar selectivity. The α-anomer 168 was the major product (7:1) under optimized conditions (eq. 11).[32]

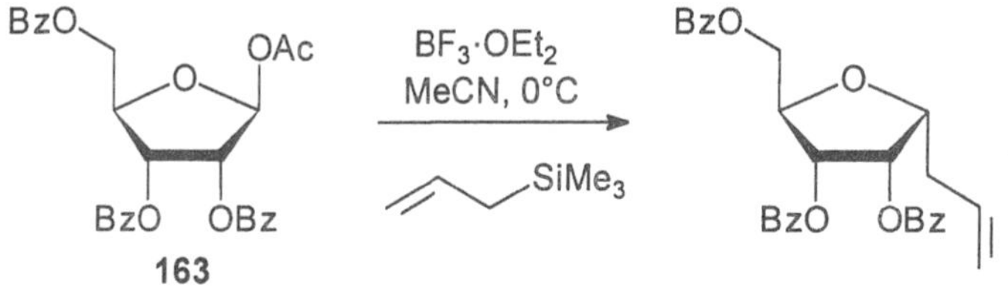

168, 93%
major 7:1   eq. 11

Scandium perchlorate has recently been found to be an effective catalyst for the *C*-glycosidation reaction. Compound 166 reacts with, for example allyltrimethylsilane, to give predominantly 167c (>99:1) in 80% yield (eq. 12).[40]

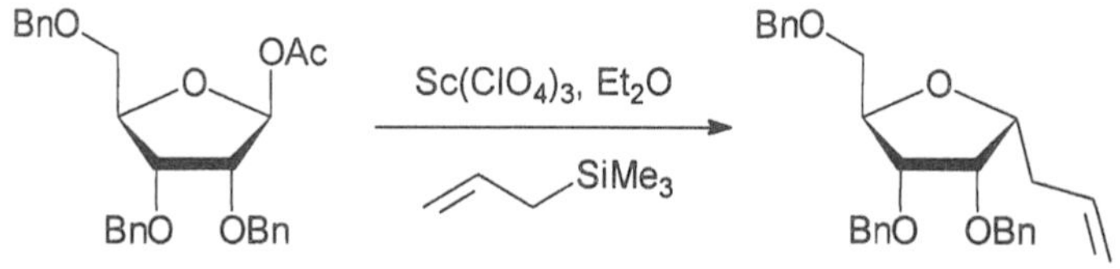

eq. 12

167c, 80%

Solvent effects have also been studied for the allylation of ribofuranosyl acetates and chlorides and it has been found that less polar solvents do influence the stereoselectivity of the allylation reaction.[41]

Exposure of 163 to stannic chloride in dichloromethane presumably gives the intermediate 170, which when treated with the enol ethers 169e to 169g gave compounds 172e to 172g, respectively. The presence of an α-hetero substituent in the enol ether dictates the site of attack on intermediate **170**. If no α-heteroatom is present the enol ether preferentially attacks the *C*-2 benzoyl carbon to produce ketals **171**. Apparently, when no α-heteroatom is present silyl-stannyl exchange occurs to produce the α-trichlorostannyl carbonyl compound, which prefers to react selectively on the benzoyl group carbon. Introduction of the α-heteroatom seems to prevent this exchange causing reaction to occur at the anomeric carbon atom.[42]

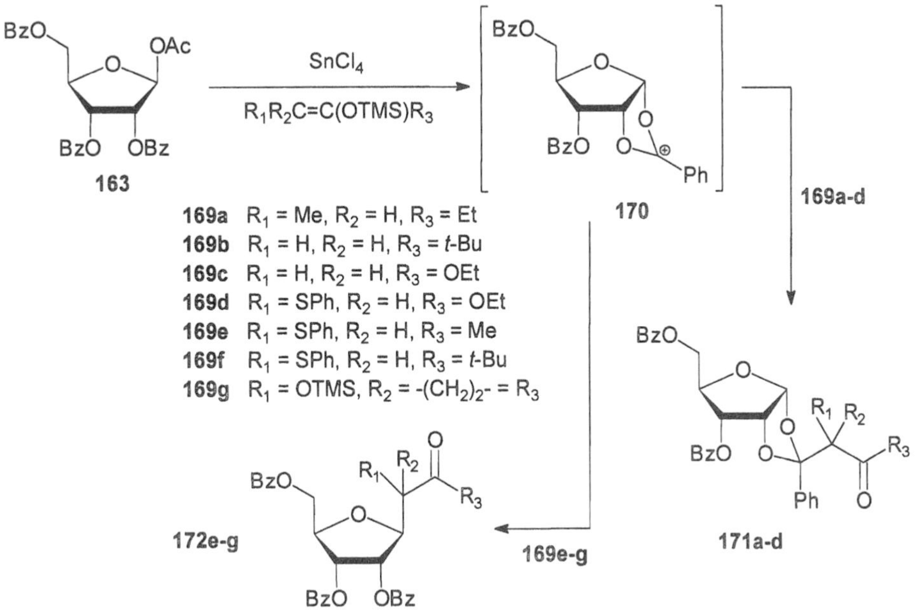
BzO
OAc
BzO OBz
163
SnCl4
R1R2C=C(OTMS)R3
170
Ph
169a-d
169a R1 = Me, R2 = H, R3 = Et
169b R1 = H, R2 = H, R3 = t-Bu
169c R1 = H, R2 = H, R3 = OEt
169d R1 = SPh, R2 = H, R3 = OEt
169e R1 = SPh, R2 = H, R3 = Me
169f R1 = SPh, R2 = H, R3 = t-Bu
169g R1 = OTMS, R2 = -(CH2)2- = R3
172e-g
169e-g
171a-d

Scheme 34

Narasaka and co-workers[43] have utilized neighboring group participation to influence the product ratio in the *C*-glycosidation reaction of the 2-deoxy ribose **173**.

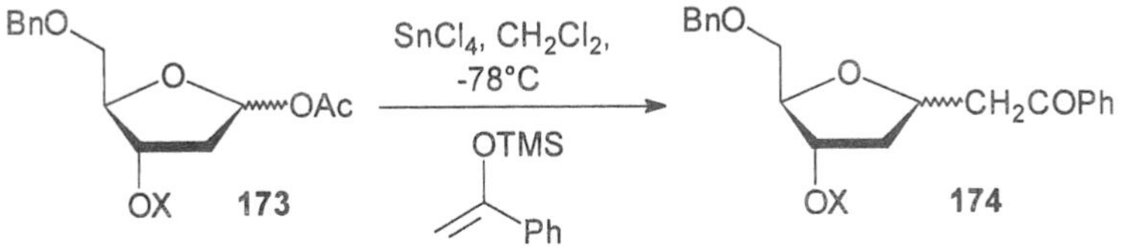
BnO
OAc
OX 173
SnCl4, CH2Cl2,
-78°C
OTMS
Ph
BnO
CH2COPh
OX 174

eq. 13

When X = Bn the ratio was found to be in favor of the α-anomer. The ethylthiomethyl group caused a shift in selectivity towards the β-anomer (42:58). With the corresponding sulfoxide, the ratio shifted to favor the β-anomer by a factor of 68:32 (Table 6).

Table 6: Product Ratios for Equation 13.

Entry	X =	Yield	α:β Ratio
a	CH_2Ph	92%	82:18
b	CH_2SMe	56%	82:18
c	$CH_2S(O)Me$	53%	52:48
d	CH_2CH_2SMe	46%	42:58
e	$CH_2CH_2S(O)Me$	92%	32:68

This effect was even more pronounced using silyl enol ether **176** in which **177** was formed in 86% yield with an α:β ratio of 9:91 (eq. 14).

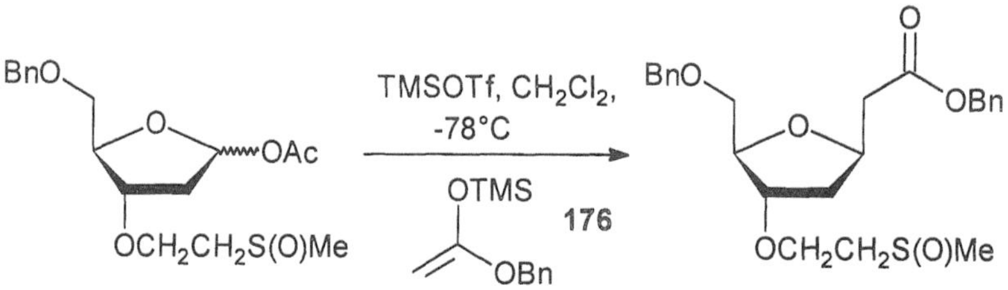

eq. 14

The sulfoxide oxygen is now shielding the α-face of the molecule by participating in oxonium ion stabilization thereby forcing attack from the more accessible top β-face.

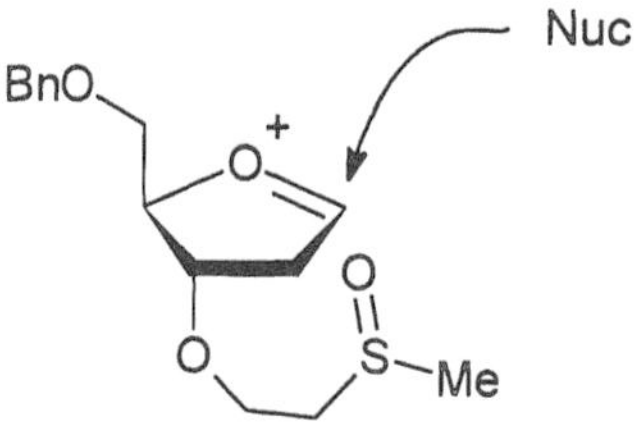

Figure 2

Chemistry describing the synthesis of anomeric cyanides from anomeric acetates using TMSCN and various Lewis acids has been described in both the furanose and pyranose series.[44]

B.   AROMATIC NUCLEOPHILES

Hurd and Bonner[45] reacted **146** with aluminum trichloride and benzene and obtained only the diphenyl derivative **178**. This is in contrast to reaction with the corresponding anomeric chloride that gave the bis aryl *C*-glycoside product (see later).

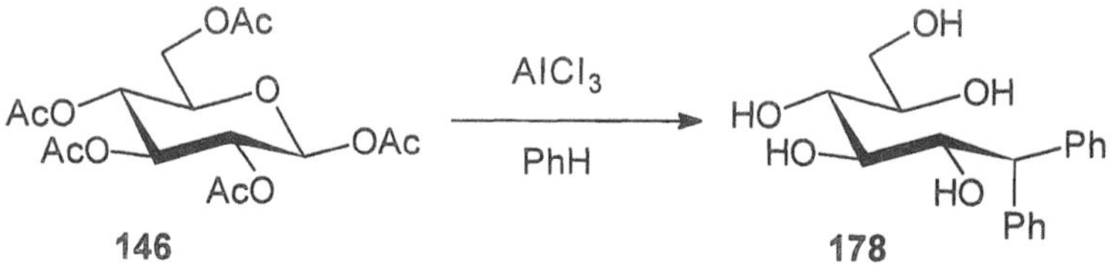

eq. 15

Early work by Šorm and co-workers[46] focused on the preparation of **180** an intermediate used in their showdomycin synthesis. Reaction of **166** with trimethoxybenzene under stannic chloride catalysis gave **179** which was exposed to ozone to give the target molecule, Scheme 35. Other oxygenated aromatics *C*-furanosides have also been prepared in a similar manner by catalysis with stannic chloride.[47]

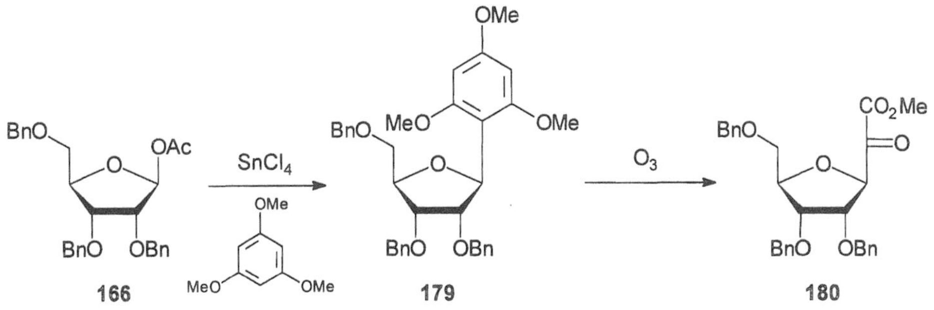

Scheme 35

Reaction of **163** with trimethoxybenzene in the presence of TMSOTf gave a 60% yield of the β-*C*-glycoside **182** along with 5% of the dimeric species **181**, Scheme 36.[48]

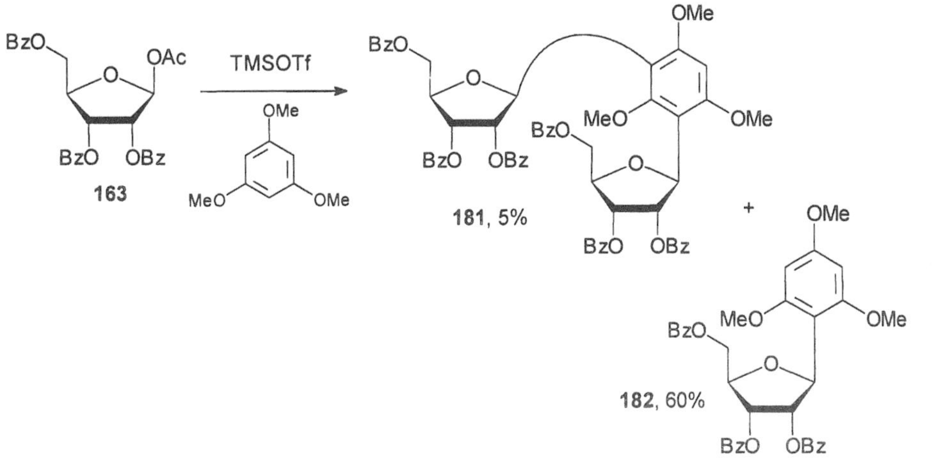

Scheme 36

Cai and Qiu[49] have used aromatic acetates (and trifluoroacetates) which bear electron withdrawing groups that serve to improve the leaving group ability of the acyl function thereby permitting milder conditions under which *C*-glycosidation can occur.

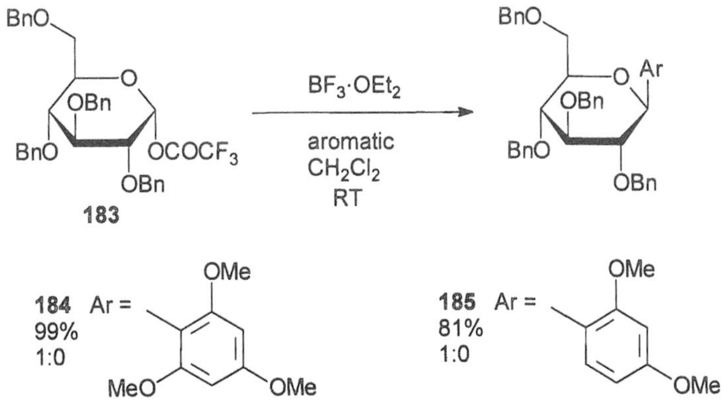

Scheme 37

Scheme 37 shows some of the aryl *C*-glycosides made using the trifluoroacetate group as the anomeric activator. The yields are good as are the selectivities.

The dinitrobenzoate was also used as an anomeric activator and reaction with electron rich aromatic gave the β-anomers exclusively regardless whether or not boron trifluoride etherate or aluminum trichloride was used as the Lewis acid. The use of the disilyl aromatic **188** led to anomeric mixtures, but now in favor of the α-anomer, **189α**. The authors explain this difference in selectivity by invoking steric arguments in which the aryl derivative is too large and forces the reaction under influence of the anomeric effect and not thermodynamic control (i.e. the stability of the product *C*-glycoside).[50]

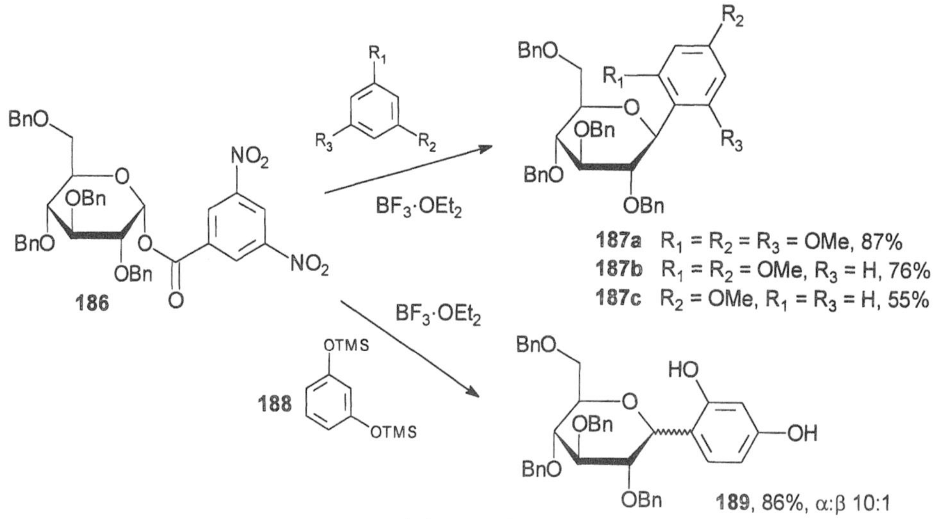

187a $R_1 = R_2 = R_3 = OMe$, 87%
187b $R_1 = R_2 = OMe$, $R_3 = H$, 76%
187c $R_2 = OMe$, $R_1 = R_3 = H$, 55%

189, 86%, α:β 10:1

Scheme 38
Reprinted from ref. 50, pg. 852 by courtesy Marcel Dekker Inc.

The same workers have also used 1-*O*-trimethylsilyl derivatives as well as *p*-nitrobenzoates for the synthesis of *C*-glycosides.[51]

Yagen and co-workers[52] used a Lewis acid catalyzed glycosidation of Δ^6 THC **190** with acetate **191** and obtained a 20% yield of **192**. They carried out this work to determine if the *C*-glucuronide was indeed formed during metabolism of THC.

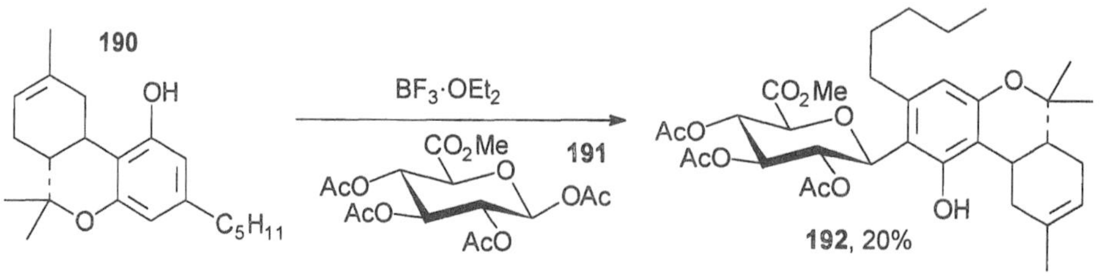

Scheme 39

Suzuki[53] has demonstrated the advantage of using the Hafnium based catalyst for the formation of aryl *C*-glycosides by an O→C rearrangement from pyranosyl acetates. Scheme 40 shows the reactions and in all the cases the use of other Lewis acids such as tin tetrachloride or boron trifluoride etherate led to inferior yields of products and in some cases *O*-glycosides were isolated from the reaction mixture.

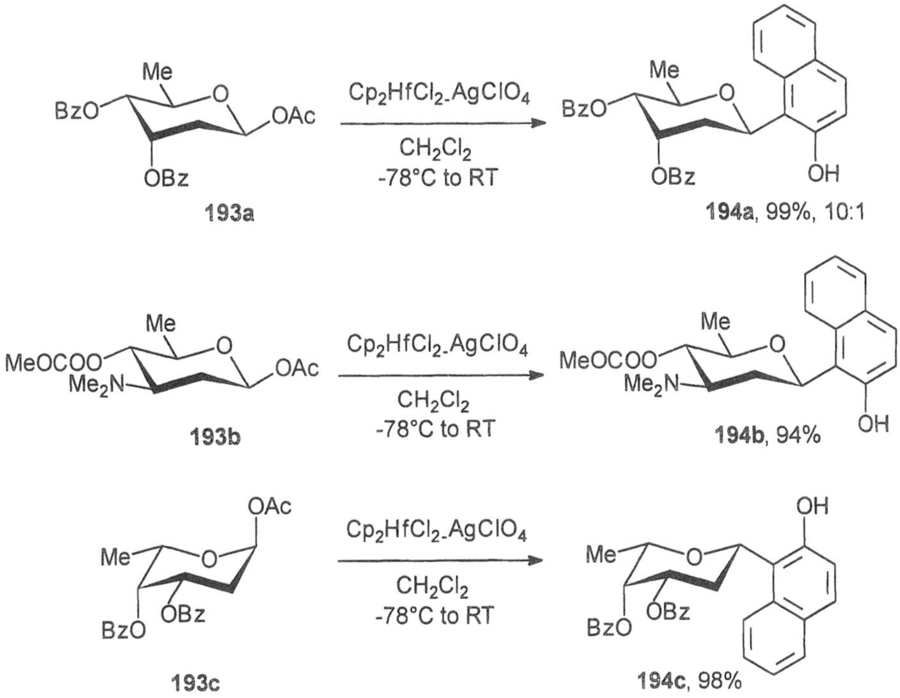

193a

194a, 99%, 10:1

193b

194b, 94%

193c

194c, 98%

Scheme 40

Heteroaromatics have also been employed in the key coupling reaction as shown in Scheme 38. Either the anomeric acetate or the *p*-nitrobenzoate could be used in this reaction and the yields of products obtained were in the 50 to 60% range.[54]

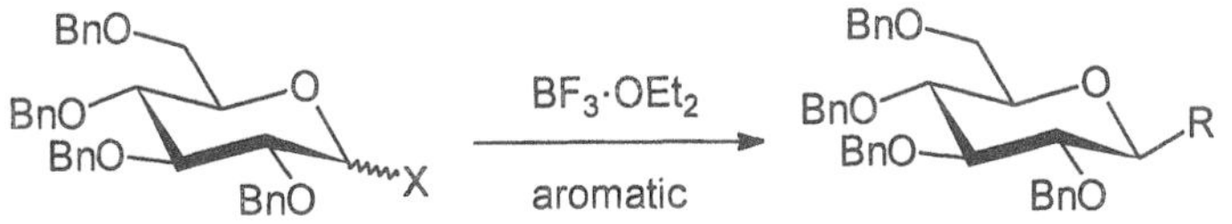

195a X = OAc
195b X = OCO-$C_6H_4NO_2$-*p*

196 R = 2-furyl
197 R = 2-thienyl
198 R = ferrocenyl
199 R = 4-methoxyphenyl
200 R = 2,4,6-trimethoxyphenyl

Scheme 41

C. ORGANOMETALLICS

Oguni and co-workers[55] have used Reformatsky reagents with anomeric acetates to yield *C*-glycosides. The reaction is carried out in dichloromethane in the presence of a small amount of titanium tetrachloride (Scheme 42).

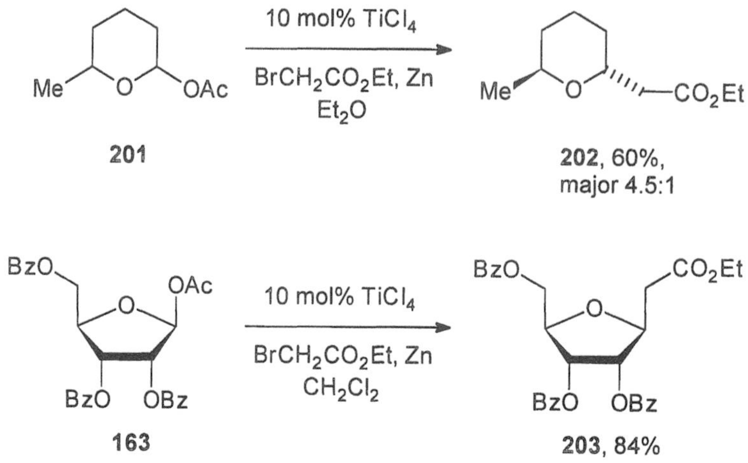

Scheme 42

It has been found that the Grignard reagent, under the influence of zinc chloride, reacts with the anomeric ester **204** to yield the *C*-glycoside **205** in fair yield (eq. 16).[56]

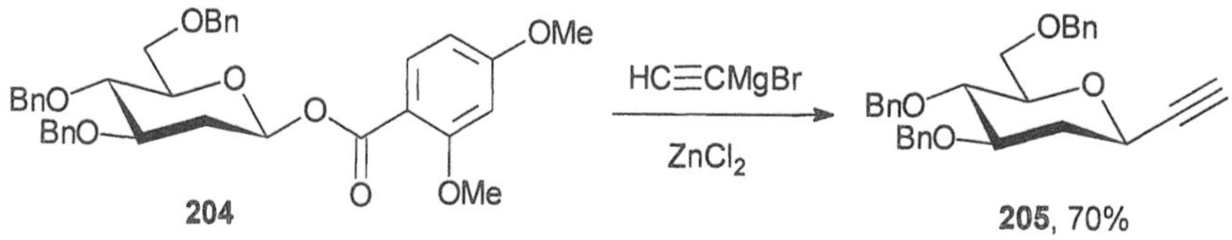

VII. GLYCOSYL HALIDES

Glycosyl halides have been used fairly extensively in *C*-glycoside preparation. Anomeric chlorides, bromides, and fluorides have all received significant synthetic attention.

A.   GLYCOSYL CHLORIDES
1.   Silicon Based Nucleophiles and Aromatics

Hurd and Bonner[57] were the first to carry out Lewis acid mediated condensations of aromatics with pyranosyl halides. Their first attempts led them to a low yield of **207** as a mixture of isomers, after acetylation, since the acetate groups were incompatible with the reaction conditions and led to the formation of acetophenone and other tar by-products.

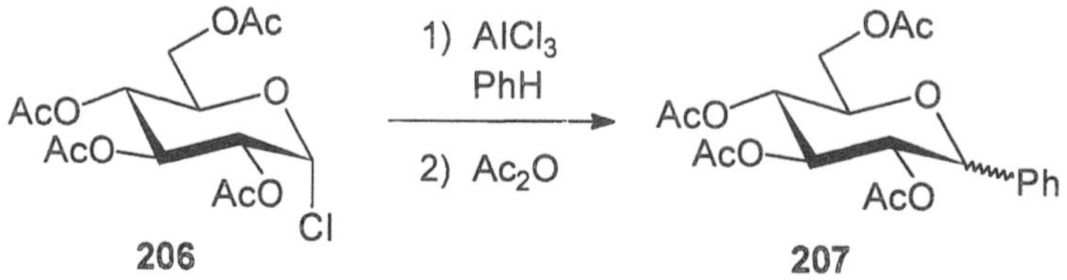

Allevi and co-workers[58] have condensed enol silyl ethers with the α-glucopyranosyl chloride **208** and obtained good yields and selectivities of α-*C*-glycosides. The anomeric position was activated through the use of silver(I) triflate in dichloromethane; Table 7 shows the results.

eq. 18

Table 7: Products from Reactions of Anomeric Chloride 208.

Entry	Enol Ether	R_1	R_2	Yield	$\alpha:\beta$
a	$EtOCH=C(OSiMe_3)OEt$	OEt	CO_2Et	75	α
b	$H_2C=C(OSiMe_3)Ph$	Ph	H	88	α
c	$H_2C=C(OSiMe_3)C_6H_4Cl-p$	C_6H_4Cl-p	H	80	α
d	$H_2C=C(OSiMe_3)Bu-t$	Bu-t	H	83	α
e	$H_2C=C(OSiMe_3)Me$	Me	H	85	α

Reproduced with permission from ref. 58, pg. 101, Copyright 1987, courtesy   Royal Society of Chemistry, Thomas Graham House, Science Park, Milton Road, Cambridge, CB4 4WF, UK.

Prompted by the difficulty in condensing silyl enol ether of β-ketoesters with activated sugars, Allevi and co-workers[59] turned to the less nucleophilic enamines and discovered that this was an efficient way to assemble these types of C-glycosides. Equation 19 and Table 8 illustrate this point. It is noteworthy that only the α-anomer was obtained in these reactions.

eq. 19

Table 8: Reactions of 209 with Enamines.

Entry	Nucleophile	R	R_1	Yield (α)
1	$Me(C_4H_8N)C=CHCO_2Me$	CO_2Me	COMe	85
2	$Et(C_4H_8N)C=CHCO_2Me$	CO_2Me	COEt	80
3	$MeOCH_2(C_4H_8N)C=CHCO_2Me$	CO_2Me	$COCH_2OMe$	85
4	$Ph(C_5H_{10}N)C=CH_2$	H	COPh	85
5	$Ph(Me_2N)C=CH_2$	H	COPh	85
6	p-$ClC_6H_4(C_5H_{10}N)C=CH_2$	H	COC_6H_4Cl-p	83
7	Bu-$t(C_5H_{10}N)C=CH_2$	H	COBu-t	40

Reproduced with permission from ref. 59, pg. 58, Copyright 1988, courtesy   Royal Society of Chemistry, Thomas Graham House, Science Park, Milton Road, Cambridge, CB4 4WF, UK.

Russian workers[60] have used the addition of PhSCl to glycals as an entry to the general structure **213** that when treated with a Lewis acid followed by an appropriate nucleophile furnishes β-C-glycosides.

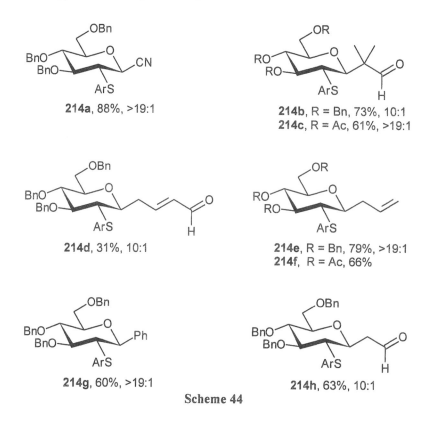

Scheme 43

This strategy is complementary to the previous approaches depicted in this chapter since the majority led exclusively or predominantly to the α-*C*-glycosides. Scheme 44 shows some of the β-*C*-glycosides made by this methodology. Both benzyl and acetate protecting groups are suitable for this type of sequence and Grignard reagents (**214g**) can also be applied to this type of reaction (not shown).

214a, 88%, >19:1

214b, R = Bn, 73%, 10:1
214c, R = Ac, 61%, >19:1

214d, 31%, 10:1

214e, R = Bn, 79%, >19:1
214f, R = Ac, 66%

214g, 60%, >19:1

214h, 63%, 10:1

Scheme 44

2.   Organometallic Reagents

The reaction of either β- or tetraacetyl-α-D-glucopyranosyl chloride with phenyl-magnesium bromide gave comparable yields and ratios of anomeric *C*-glycosides.[61] Reactions with aryllithiums have also been reported by the same group.[62]

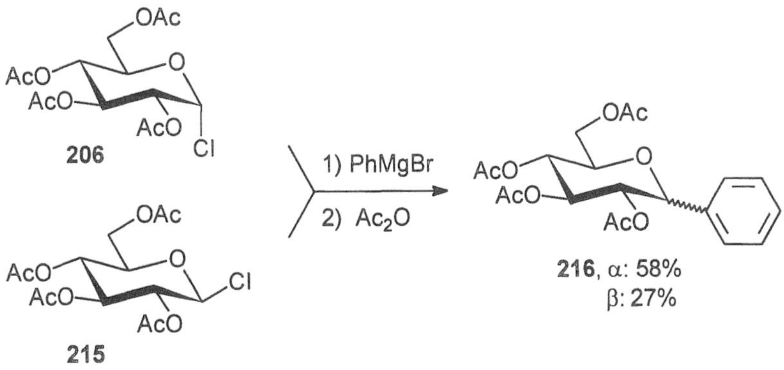

Scheme 45

3. Stabilized Nucleophiles

Hanessian[63] has condensed sodio malonates with furanosyl chlorides to yield *C*-glycosides in fair yield. The major product is the β-anomer **218** obtained in 46% yield and the α-anomer **219** which was formed in 31% yield after hydrogenolysis of the benzyl groups to facilitate separation.

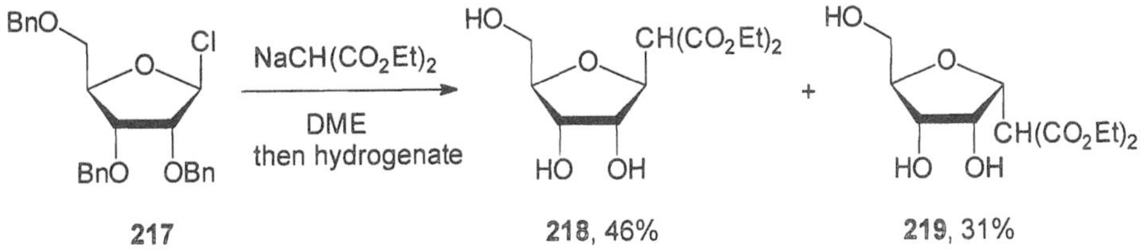

Scheme 46

A similar reaction[64] has been applied to the 2-deoxy amino sugar **220**. Reaction with the potassium salt gave a 21% yield of the β-*C*-glycoside **221**. Although the yield is low, the starting material is readily available (eq. 20).

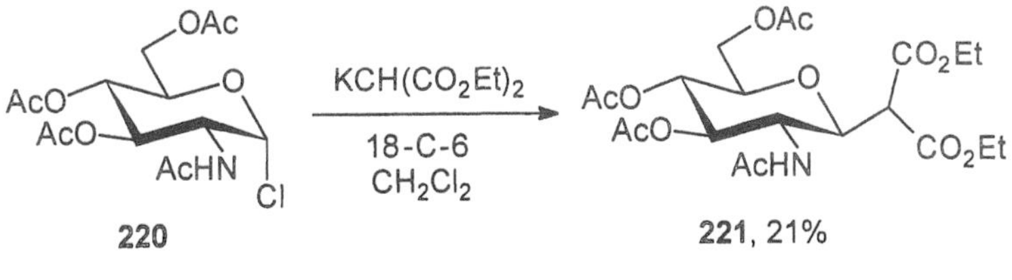

B. GLYCOSYL BROMIDES
1. Amino *C*-glycosides

Glycosyl bromides have also seen use in the synthesis of *C*-glycosides since they are fairly reactive species as well as readily available intermediates.

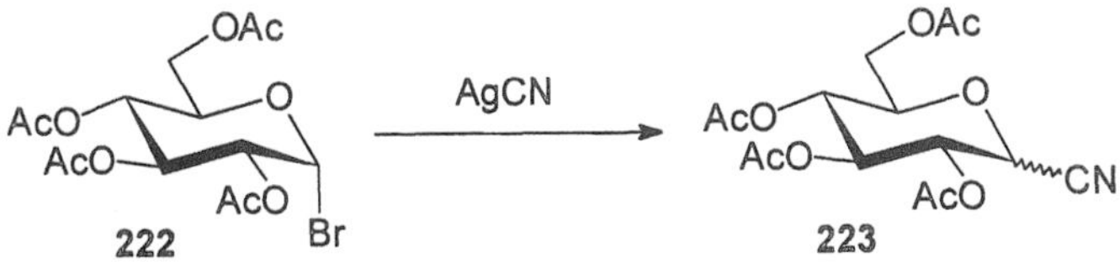

Buerger[65] was the first to condense cyanide ion with a glucopyranosyl bromide and obtained a mixture of anomeric cyanides (eq. 21).

This type of reaction has since been used by others.[66] Myers and Lee[67] carried out similar reactions with the galacto, manno, fuco and 2-deoxy-2-amino gluco sugars. Scheme 47 illustrates the reactions.

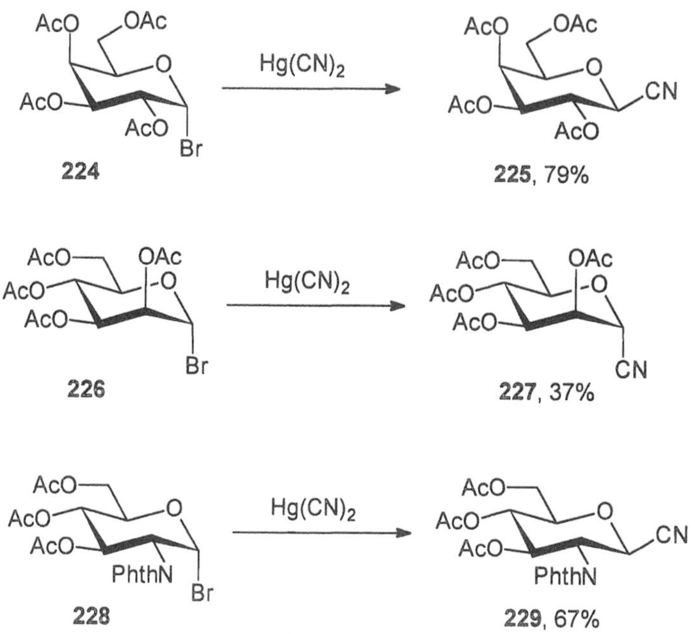

Scheme 47

When the tetraacetate **230a** was condensed with enol ether **231a** only compound

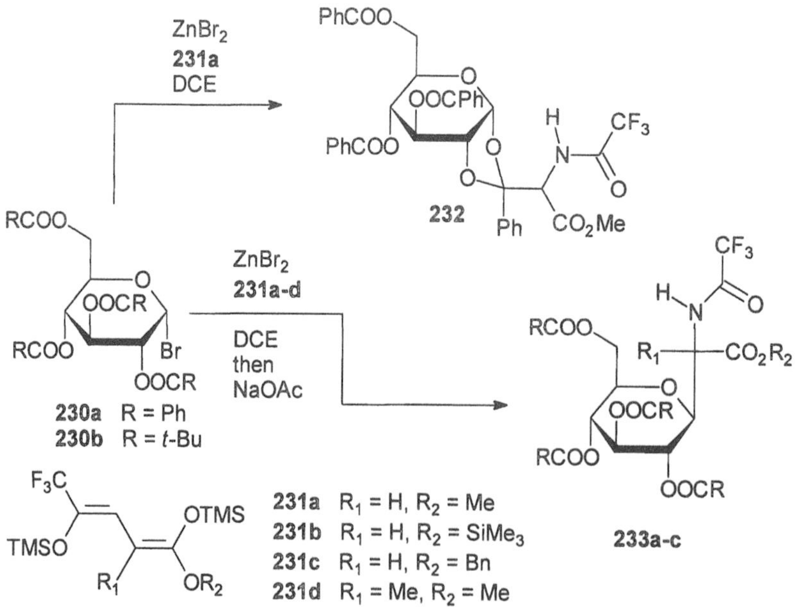

231a R_1 = H, R_2 = Me
231b R_1 = H, R_2 = SiMe$_3$
231c R_1 = H, R_2 = Bn
231d R_1 = Me, R_2 = Me

230a R = Ph
230b R = *t*-Bu

233a-c

Scheme 48

232 was isolated, a change from phenyl to *t*-butyl now caused the condensation to proceed on the anomeric carbon and the *C*-glycosides **233a-c** were obtained in good yield.[68]

Although not a Lewis acid catalyzed process, eq. 22 shows the condensation of anion **234** (derived from deprotonation by lithium dicyclohexylamide) with α-acetobromoglucose gave **235** (after hydrolysis and acetylation) as a mixture of isomers in a 1:1 ratio.[69]

eq. 22

2. Aromatics

Early work focused on the condensation of 1,4,5-trimethoxybenzene with the bromide **236** under zinc oxide catalysis to yield the β-*C*-ribofuranoside **237** which was an early intermediate in a synthesis of showdomycin (eq. 23).[70]

eq. 23

The cannabis derivative **190** was allowed to react with the bromide **238** under mercuric cyanide catalysis to yield the β-*O*-glycoside and the β-*C*-glycoside **239** (eq. 24).[71]

eq. 24

239, 20% plus O-glycoside, 7%.

3. Organometallics

Hurd and Bonner[72] were the first to condense Grignard reagents with pyranosyl bromides to obtain *C*-glycosides. Several aromatic Grignard reagents were examined but only the results from phenylmagnesium bromide will be presented. Reaction of the aforementioned with α-acetoglucosyl bromide gave a mixture of α- and β-anomers after acetylation (not shown).

Reaction with the corresponding organocadmium reagents with the gluco and manno bromides gave a 20% yield of the tetraacetyl-β-D-glucopyranosylbenzene and 30% of the tetraacetyl-α-D-mannopyranosylbenzene (also not shown).[73]

The reaction of tolylmagnesium bromide with acetobromoglucose has since been re-investigated and it was found that the reaction gives a good yield of two compounds **241** and **242**. The benzenoid *C*-glycoside is formed by a chelation controlled delivery of as shown in Scheme 49 since when the 6-acetoxy is removed (e.g. xylose) then only benzyl *C*-glycoside is isolated. The group at *C*-2 also plays a role in the delivery. It is interesting to note the selectivity for this reaction is excellent, and it is believed that the participating group at *C*-2 is responsible for this fact.[74]

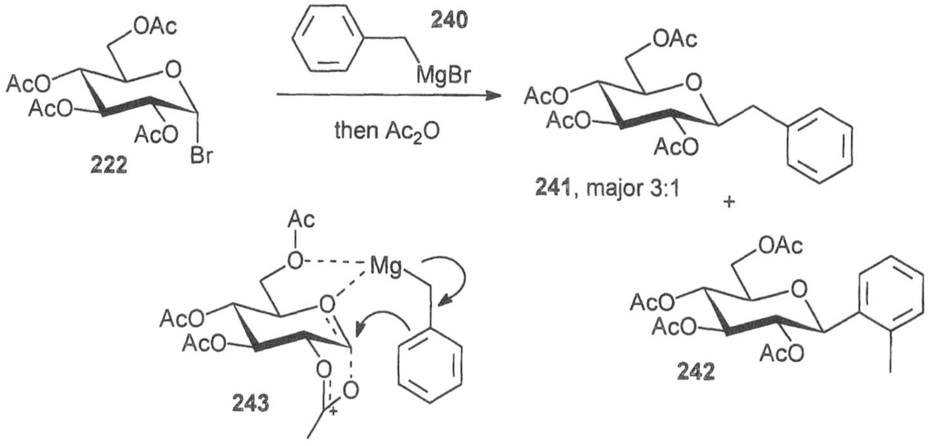

Scheme 49

During the synthesis of quantamycin (**246**), a computer generated antibiotic, an intermediate *C*-glycoside was required. Treatment of **244** with vinylmagnesium bromide gave compound **245** which was then elaborated into the target **246**.[75]

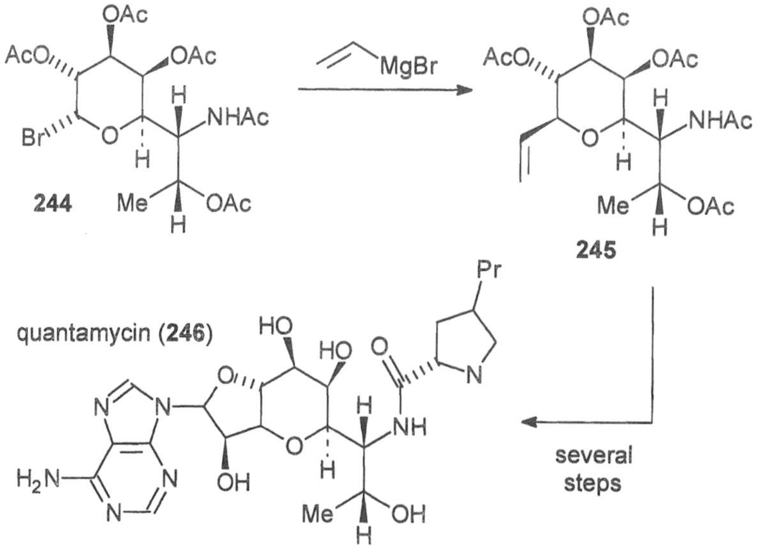

Scheme 50

During model studies towards the ezomycins Hanessian and co-workers[76] used the condensation of vinyl magnesium bromide with the bromide **247** to obtain a good yield of the net product of inversion **248**. Epoxidation was followed by acid induced cyclization to **250**. This molecule matches the CD ring system of ezomycin A$_2$.

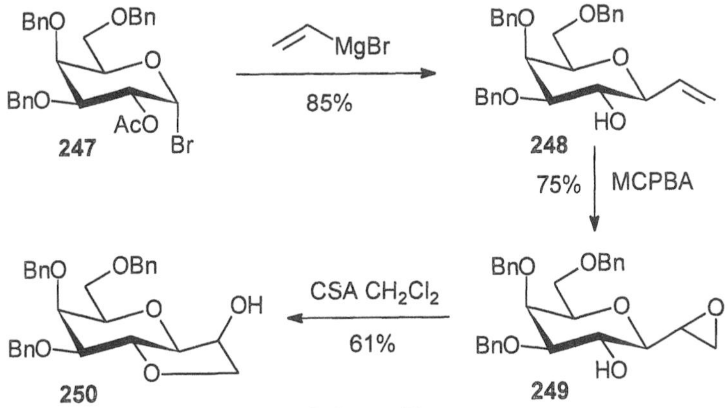

Scheme 51

The allyl derivative **251** was prepared similarly to **248** and oxidative cleavage followed by methylation provided **252**. Lactonization was followed by exomethylene formation to furnish model compound **254** (Scheme 52).

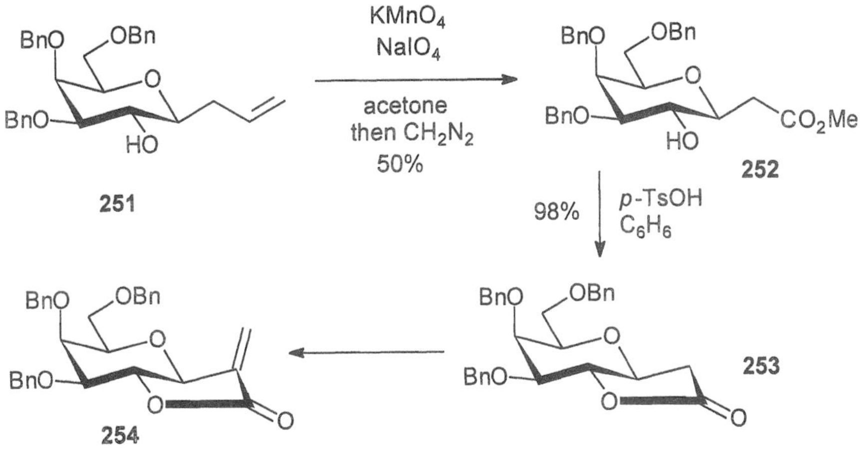

Scheme 52

Bihovsky and co-workers[77] have studied the reaction of various cuprates with the permethylglucosyl bromide **255**. The best results were obtained when **255** was condensed with lithium dimethylcuprate in diethyl ether (eq. 25).

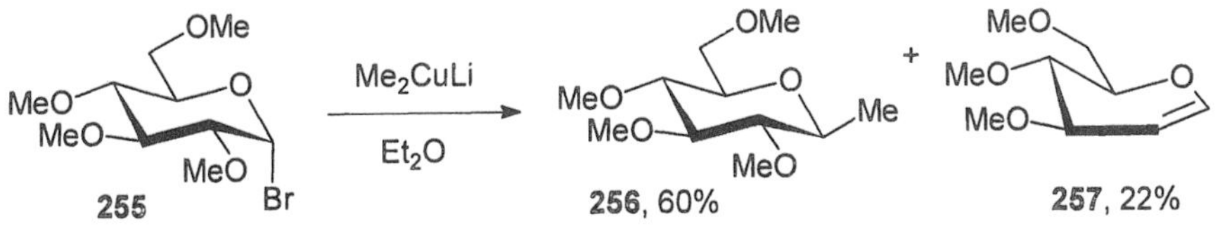

The β-*C*-glycoside **256** was formed in 60% yield accompanied by some glycal **257** (22%), eq. 25.

Reaction of the organotin acetylide with bromide **258** under zinc chloride catalysis yielded the product of retention **259** as the sole *C*-glycoside product, eq. 26.[78]

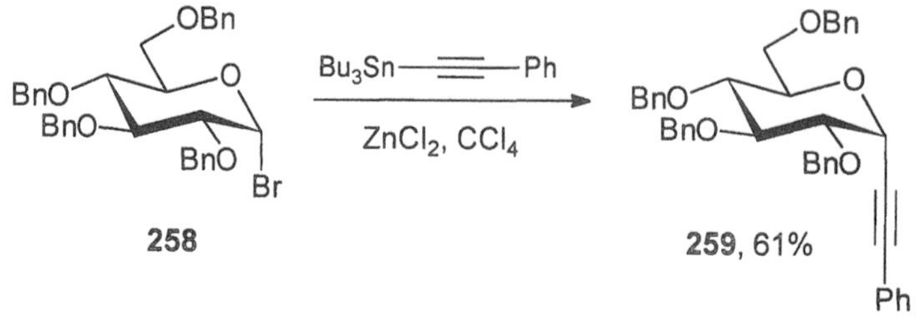

258 **259**, 61% eq. 26

In the furanose series the anomeric ratio was dependent upon the nature of the group on the terminus of the acetylene. When R = *n*-hexyl only the α-isomer is formed. If R = Ph the β-anomer now predominates in a 2.8:1 ratio, whereas with a methoxymethyl group a 1:1 mixture is obtained. The authors offer no explanation for these observations.

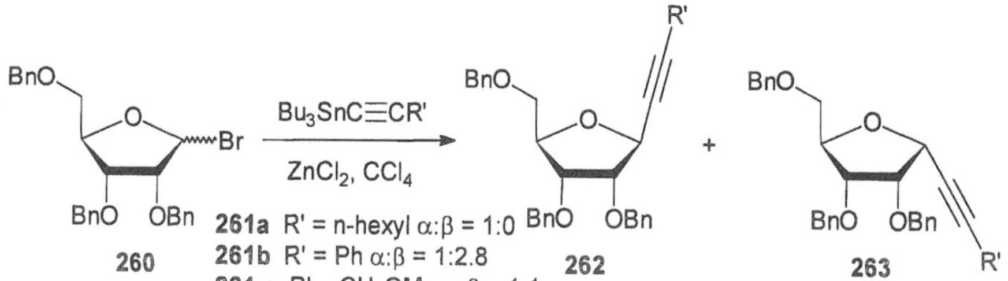

260 **262** **263** eq. 27

261a R' = n-hexyl α:β = 1:0
261b R' = Ph α:β = 1:2.8
261 c R' = CH_2OMe α:β = 1:1

Hanessian and Pernet[79] condensed the sodium salt of dibenzyl malonate with α-acetobromoglucose and obtained an 80% yield of the β-*C*-glycoside **264**. Compound **264** was converted to acid **265** by hydrogenolysis and decarboxylation, Scheme 53. Reaction with diethylmalonate gave only a 20% yield of the corresponding β-*C*-glycoside.

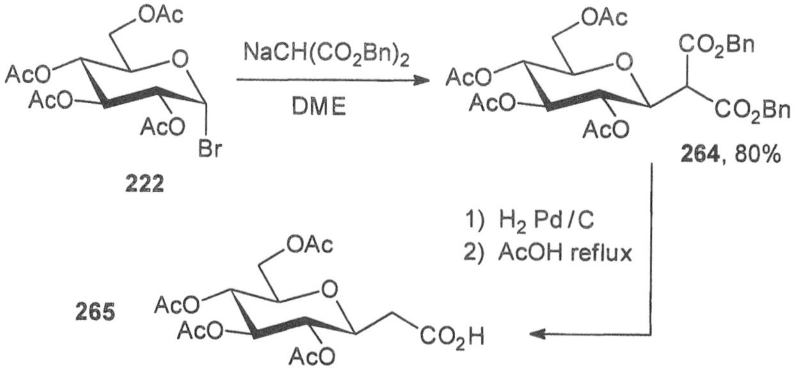

222 **264**, 80% **265**

Scheme 53

C. GLYCOSYL FLUORIDES

Over recent years several workers have begun to utilize glycosyl fluorides as precursors to *C*-glycosides. They have been condensed with silyl enol ethers, electron rich aromatics as well as aluminum reagents.

1. Silicon Based Nucleophiles

Nicolaou[80] has condensed several silicon based nucleophiles with the anomeric fluorides **266** to obtain *C*-glycosides with good to adequate anomeric selectivity.

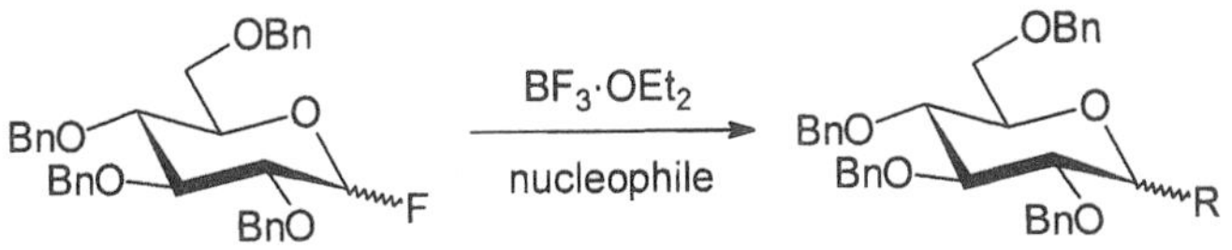

eq. 28

Table 9: Reaction of 266 with Silicon Based Nucleophiles.

Entry	Nucleophile	R	Yield	α:β Ratio
1	$Me_3SiCH_2CH=CH_2$	$CH_2CH=CH_2$	95%	>20:1
2	Me_3SiCN	CN	96%	*ca.* 3:1
3	Me_3SiCH_2CN	CH_2CN	85%	*ca.* 3:1
4	$H_2C=C(OSiMe_3)Ph$	CH_2COPh	95%	*ca.* 2:1

Reproduced with permission from ref. 80, pg. 1154, Copyright 1984, courtesy Royal Society of Chemistry, Thomas Graham House, Science Park, Milton Road, Cambridge, CB4 4WF, UK.

The workers have also applied similar reactions to the disaccharide **268** to obtain the *C*-glycosides **269a-b**.

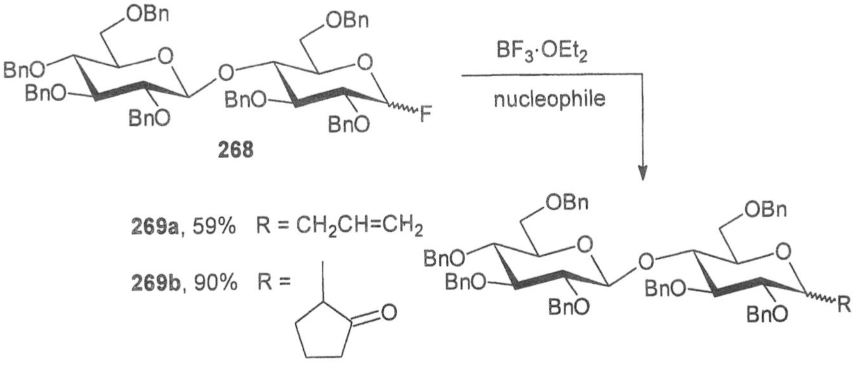

269a, 59% R = $CH_2CH=CH_2$

269b, 90% R =

Scheme 54

Similar reactions have been applied to *C*-furanoside preparation (eq. 29).

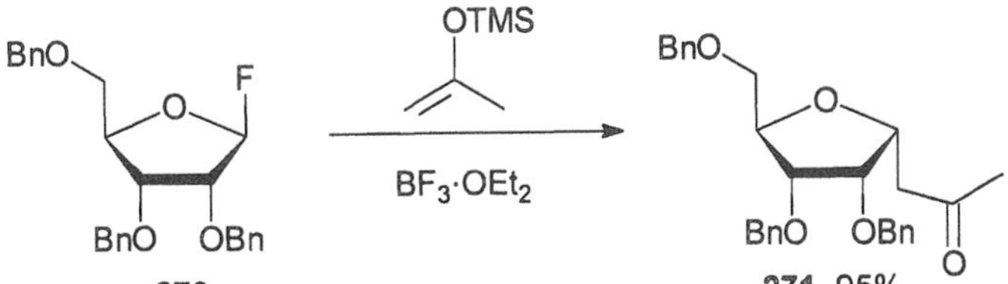

eq. 29

270

271, 95%
major >20:1

The best ratios were obtained when a minimum amount of catalyst was employed. With 0.05 equivalents a >20:1 ratio of α:β was obtained in 95% yield (eq. 29).[81]

Reaction of **270** with allyltrimethylsilane gave the α-anomer **272** in 93% yield. Reaction with trimethylsilylcyanide gave nitrile **273** and isonitrile **274** when the reaction was performed in diethyl ether. Using dichloromethane as solvent led only to the α-*C*-glycoside **273α**. The gluco derivative **275** gave a mixture of α- and β-anomers in 72% and 22% respectively, Scheme 55.[82]

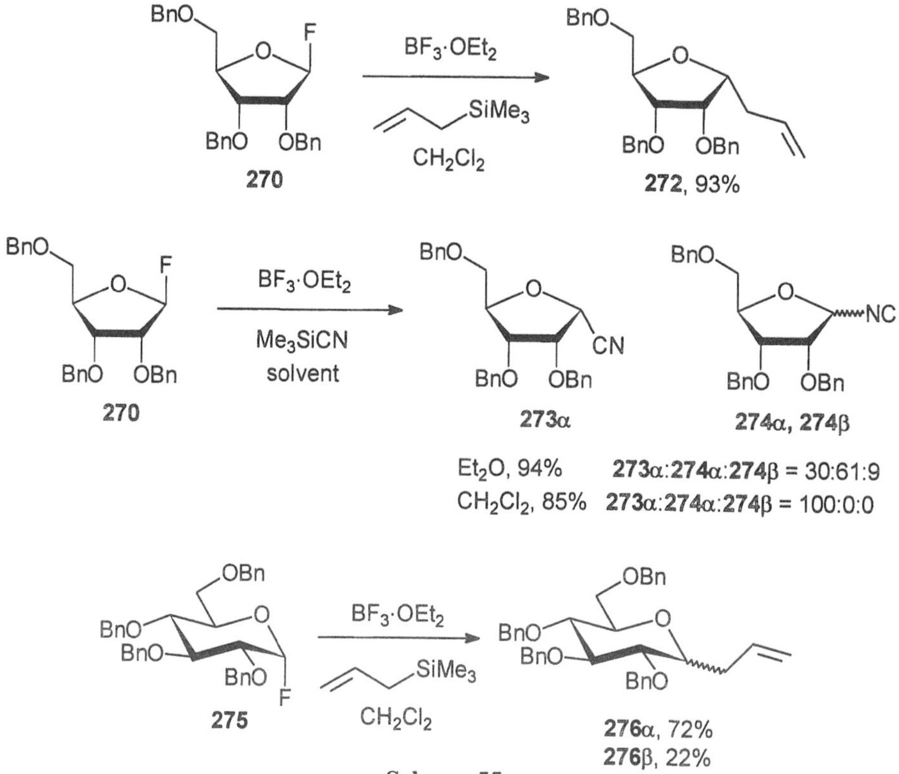
BnO
F
BF$_3$·OEt$_2$
SiMe$_3$
CH$_2$Cl$_2$
BnO OBn
270
272, 93%
BF$_3$·OEt$_2$
Me$_3$SiCN
solvent
CN
NC
273α
274α, 274β
Et$_2$O, 94% 273α:274α:274β = 30:61:9
CH$_2$Cl$_2$, 85% 273α:274α:274β = 100:0:0
OBn
275
276α, 72%
276β, 22%

Scheme 55

2. Aromatics

Suzuki and co-workers[83] used the catalyst Cp_2HfCl_2-$AgClO_4$ in their coupling of aromatics with glycosyl fluorides to obtain aryl *C*-glycosides, eq. 30. The fluorides were coupled with phenol and *p*-methoxyphenol to give good yields and acceptable selectivities of *C*-glycosides. The use of other Lewis acids tended to give substantial amounts of *O*-glycosides.

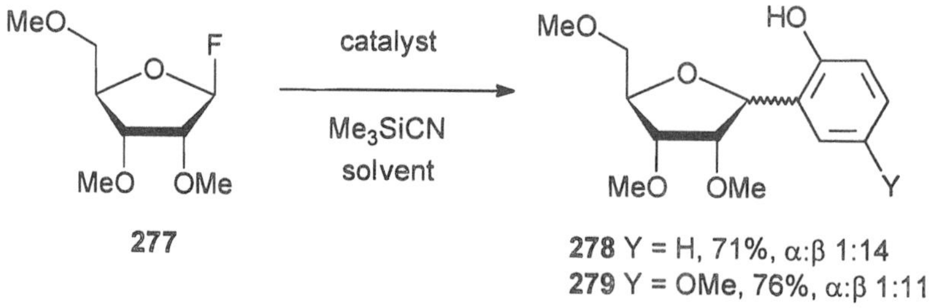
MeO
F
catalyst
Me$_3$SiCN
solvent
HO
MeO OMe
Y
277
278 Y = H, 71%, α:β 1:14
279 Y = OMe, 76%, α:β 1:11

eq. 30

Coupling in pyranose series with trimethoxybenzene gave the β-anomer, but it was found to decompose under the reaction conditions. Reaction with the rhamnosyl fluoride **280** gave more encouraging results. Shorter reaction times tended to favor the β-anomer **281** almost exclusively. Several naphthalenic *C*-glycosides (**281a-b**) were synthesized by this method.[84]

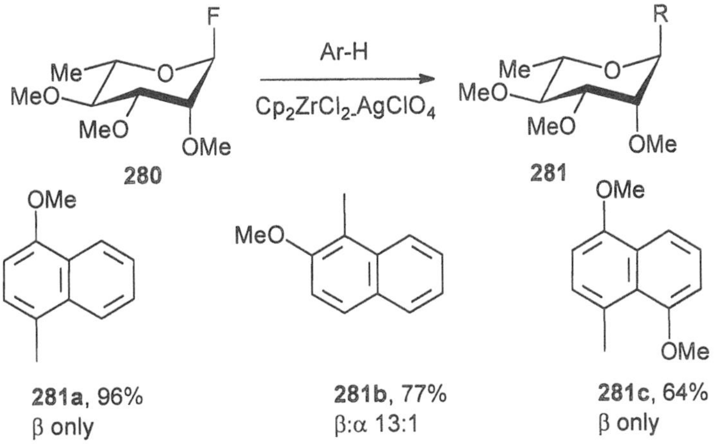

Scheme 56

The coupling reaction was also extended to 2-deoxy sugars with a more complex aromatic substrate.[85] This reaction proceeded in 86% yield and is a model study towards the vineomycins.

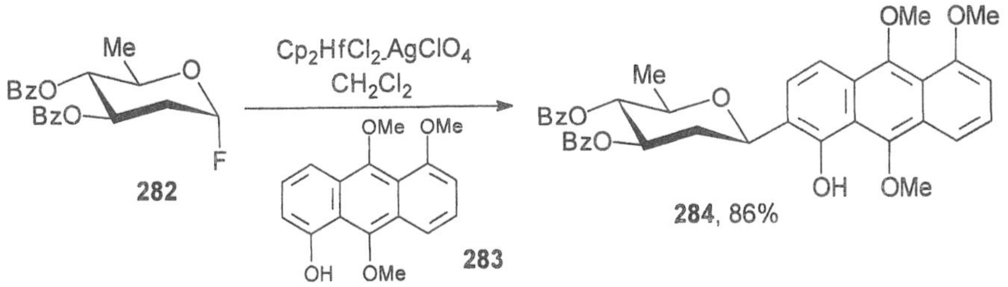

3. Aluminum Based Reagents

Posner and Haines[86] have used alkyl aluminums to prepare *C*-glycosides from anomeric fluorides. The best selectivity was obtained with the mannofuranosyl fluoride **285**. The yield of *C*-glycoside **286** is good and the selectivity for other examples (not shown) was also found to be >20:1.

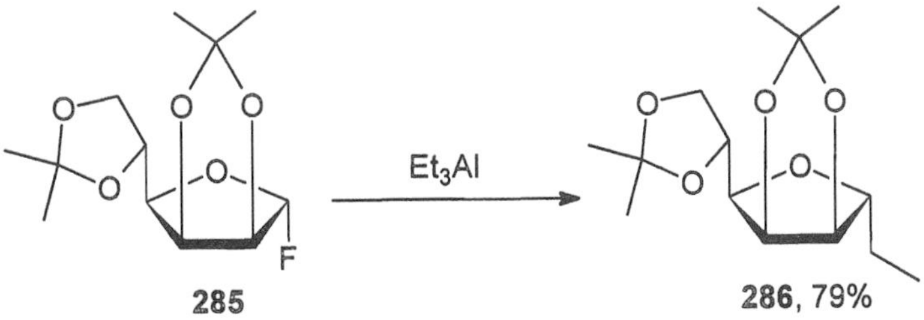

Kende and Fujii[87] have found that the aluminate 287 could be coupled with fluorides 266; two products 288a and 288b were obtained in 56% and 25% yield respectively. The low preponderance of β-isomer was not of consequence since anomerization of the pivaloyl derivative 289a with triflic acid gave the β-anomer 289b almost exclusively. This strategy was applied by the workers in their synthesis of ambruticin (see Chapter 9).

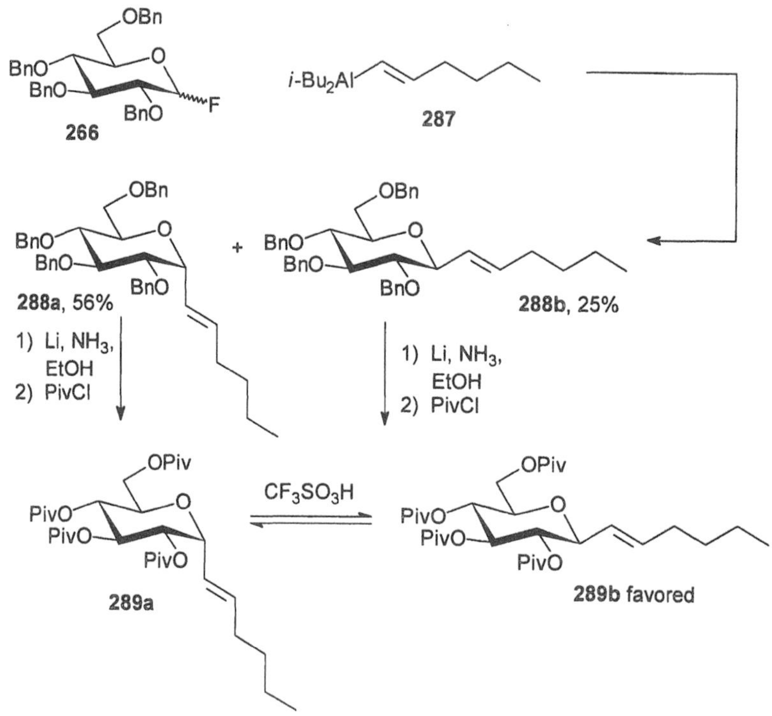

Scheme 57

Nicolaou has also employed aluminum based reagents in *C*-glycoside synthesis.[80]

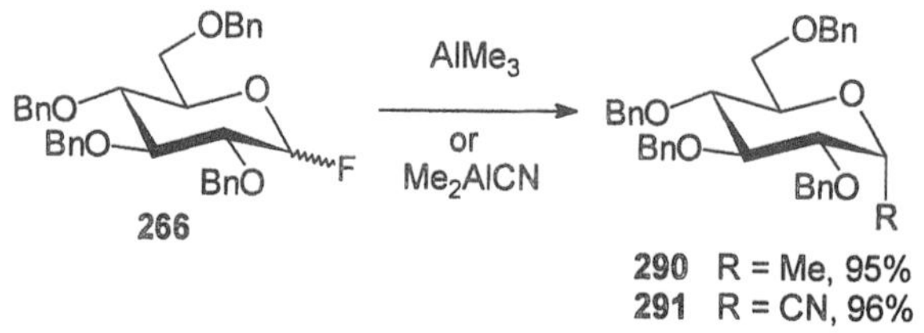

290 R = Me, 95%
291 R = CN, 96%

eq. 33

VIII. REFERENCES

1.   Allevi, P.; Anastasia, M.; Ciuffreda, P.; Fiecchi, A.; Scala, A. *J. Chem. Soc., Chem. Commun.* **1987**, 1245.
2.   Masuyama, Y.; Kobayashi, Y.; Kurusu, Y. *J. Chem. Soc., Chem. Commun.* **1994**, 1123.
3.   Kometani, T.; Kondo, H.; Fujimori, Y. *Synthesis* **1988**, 1005.
4.   For a dated but excellent review of the Mitsunobu reaction see: Mitsunobu, O. *Synthesis* **1981**, 1.
5.   Toshima, K.; Matsuo, G.; Ishizuka, T.; Nakata, M.; Kinoshita, M. *J. Chem. Soc., Chem. Commun.* **1992**, 1641.
6.   Cornia, M.; Casiraghi, G.; Zetta, L. *Tetrahedron* **1990**, *46*, 3071.

7. Yamaguchi, M.; Horiguchi, A.; Ikeura, C.; Minami, T. *J. Chem. Soc., Chem. Commun.* **1992**, 435.
8. Tomooka, K.; Okinaga, T.; Suzuki, K.; Tsuchihashi, G.-i. *Tetrahedron Lett.* **1987**, *28*, 6335.
9. Tomooka, K.; Matsuzawa, K.; Suzuki, K.; Tsuchihashi, G.-i. *Tetrahedron Lett.* **1987**, *28*, 6339.
10. Hosomi, A.; Sakata, Y.; Sakurai, H. *Tetrahedron Lett.* **1984**, *25*, 2383.
11. Bennek, J.A.; Gray, G.R. *J. Org. Chem.* **1987**, *52*, 892.
12. Nicotra, F.; Panza, L.; Russo, G. *J. Org. Chem.* **1987**, *52*, 5627.
13. Martin, O.R.; Rao, S.P.; Yang, T.-F.; Fotia, F. *Synlett* **1991**, 702.
14. Bertozzi, C.R.; Bednarski, M. *Tetrahedron Lett.* **1992**, *33*, 3109. See also: Kobertz, W.R.; Bertozzi, C.R.; Bednarski, M. *Tetrahedron Lett.* **1992**, *33*, 7373 for a similar reaction with non-deoxy sugars.
15. Toshima, K.; Matsuo, G.; Tatsuta, K. *Tetrahedron Lett.* **1992**, *33*, 2175.
16. Germain, F.; Chapleur, Y.; Castro, B. *Tetrahedron* **1982**, *38*, 3593.
17. Williams, R.M.; Stewart, A.O. *Tetrahedron Lett.* **1983**, *24*, 2715 and *ibid. J. Am. Chem. Soc.* **1985**, *107*, 4289.
18. Craig, D.; Munasinghe, V.R.N. *Tetrahedron Lett.* **1992**, *33*, 663.
19. Craig, D.; Munasinghe, V.R.N. *J. Chem. Soc., Chem. Commun.* **1993**, 901.
20. Keck, G.E.; Enholm, E.J.; Kachensky, D.F. *Tetrahedron Lett.* **1984**, *25*, 1867.
21. Kozikowski, A.P.; Konoike, T.; Ritter, A. *Carbohydr. Res.* **1987**, *171*, 109.
22. Brown, D.S.; Ley, S.V.; Bruno, M. *Heterocycles* **1989**, *28*, 773.
23. Brown, D.S.; Bruno, M.; Davenport, R.J.; Ley, S.V. *Tetrahedron* **1989**, *45*, 4293.
24. Schmidt, R.R.; Hoffman, M. *Angew. Chem. Intl. Ed. Engl.* **1983**, *22*, 406.
25. Hoffmann, M.G.; Schmidt, R.R. *Liebigs Ann. Chem.* **1985**, 2403.
26. Schmidt, R.R.; Effenberger, G. *Liebigs Ann. Chem.* **1987**, 825.
27. Mahling, J.-A.; Schmidt, R.R. *Synthesis* **1993**, 325.
28. Schmidt, R.R.; Hoffmann, M. *Tetrahedron Lett.* **1982**, *23*, 409.
29. Hurd, C.D.; Bonner, W.A. *J. Am. Chem. Soc.* **1945**, *67*, 1759.
30. Murata, S.; Noyori, R. *Tetrahedron Lett.* **1982**, *23*, 2601.
31. Lewis, M.D.; Cha, J.K.; Kishi, Y. *J. Am. Chem. Soc.* **1982**, *104*, 4976.
32. Kozikowski, A.P.; Sorgi, K.L. *Tetrahedron Lett.* **1983**, *24*, 1563.
33. Giannis, A.; Sandhoff, K. *Tetrahedron Lett.* **1985**, *26*, 1479.
34. Horton, D.; Miyake, T. *Carbohydr. Res.* **1988**, *184*, 221.
35. Acton, E.M.; Ryan, K.J.; Tracy, M. *Tetrahedron Lett.* **1984**, *25*, 5743.
36. Gurjar, M.K.; Mainkar, A.S.; Syamala, M. *Tetrahedron: Asymmetry* **1993**, *4*, 2343.
37. Panek, J.S.; Sparks, M.A. *J. Org. Chem.* **1989**, *54*, 2034.
38. Kuo, Y.N.; Chen, F.; Ainsworth, C.; Bloomfield, J. *J. Chem. Soc., Chem. Commun.* **1971**, 136.
39. Mukaiyama, T.; Kobayashi, S.; Shoda, S.-i. *Chemistry Lett.* **1984**, 1529 and Mukaiyama, T; Kobayashi, S. *Carbohydr. Res.* **1987**, *171*, 81.
40. Hachiya, I.; Kobayashi, S. *Tetrahedron Lett.* **1994**, *35*, 3319.
41. Wilcox, C.S.; Otoski, R.M. *Tetrahedron Lett.* **1986**, *27*, 1011.
42. Yokoyama, Y.S.; Elmoghayar, M.R.H.; Kuwajima, I. *Tetrahedron Lett.* **1982**, *23*, 2673.
43. Ichikawa, Y.-i.; Kubota, H.; Fujita, K.; Okauchi, T.; Narasaka, K. *Bull. Chem. Soc. Jpn.* **1989**, *62*, 845.
44. See: De las Heras, F.G.; Fernández-Resa, P. *J. Chem. Soc., Perkin Trans. I* **1982**, 903 and García López, M.-T.; De las Heras, F.G.; San Felix, A. *J. Carbohydr. Chem.* **1987**, *6*, 273.
45. Hurd, C.D.; Bonner, W.A. *J. Am.Chem. Soc.* **1945**, *67*, 1759.

46. Kalvoda, L.; Farkaš, J.; Šorm, F. *Tetrahedron Lett.* **1970**, 2297.
47. Kalvoda, L. *Collect. Czech. Chem. Commun.* **1973**, *38*, 1679. See also: Ohrui, H.; Kuzuhara, H.; Emoto, S. *Agr. Biol. Chem.* **1972**, *36*, 1651.
48. Vorbrüggen, H.; Krolikiewicz, K.; Bennua, B. *Chem. Ber.* **1981**, *114*, 1234.
49. Cai, M.-S.; Qiu, D.-X. *Carbohydr. Res.* **1989**, *191*, 125.
50. Cai, M.-S.; Qiu, D.-X. *Synthetic Commun.* **1989**, *19*, 851, Marcel Dekker, Inc., N.Y.
51. Cai, M.-S.; Qiu, D.-X. *Acta. Chim. Sin. (Engl. Ed.)* **1989**, *4*, 372 and Cai, M.-S.; Qiu, D.-X. *Chin. Sci. Bull.* **1990**, *35*, 615.
52. Yagen, B.; Levy, S.; Mechoulam, R.; Ben-Zvi, Z. *J. Am. Chem. Soc.* **1977**, *99*, 6444.
53. Matsumoto, T.; Hosoya, T.; Suzuki, K. *Tetrahedron Lett.* **1990**, *31*, 4629.
54. Grynkiewicz, G.; BeMiller, J.N. *Carbohydr. Res.* **1984**, *131*, 273.
55. Hayashii, M.; Sugiyama, M.; Toba, T.; Oguni, N. *J. Chem. Soc., Chem. Commun.* **1990**, 767.
56. Bolitt, V.; Mioskowski, C.; Falck, J.R. *Tetrahedron Lett.* **1989**, *30*, 6027.
57. Hurd, C.D.; Bonner, W.A. *J. Am. Chem. Soc.* **1945**, *67*, 1664.
58   Allevi, P.; Anastasia, M.; Ciuffreda, P.; Fiecchi, A.; Scala, A. *J. Chem. Soc., Chem. Commun.* **1987**, 101.
59. Allevi, P.; Anastasia, M.; Ciuffreda, P.; Fiecchi, A.; Scala, A. *J. Chem. Soc., Chem. Commun.* **1988**, 57.
60. Smolyakova, I.P.; Smit, W.A.; Zal'chenko, E.A.; Chizov, O.S.; Shashkov, A.S.; Caple, R.; Sharpe, S.; Kuehl, C. *Tetrahedron Lett.* **1993**, *34*, 3047.
61. Hurd, C.D.; Holysz, R.P. *J. Am. Chem. Soc.* **1950**, *72*, 1732.
62. Hurd, C.D.; Holysz, R.P. *J. Am. Chem. Soc.* **1950**, *72*, 1735.
63. Hanessian, S.; Pernet, A.G. *Can J. Chem.* **1974**, *52*, 1280.
64. Kim, K.-I.; Hollingsworth, R.I. *Tetrahedron Lett.* **1994**, *35*, 1031.
65. Buerger, L.R. *J. Am. Chem. Soc.* **1934**, *56*, 2494.
66. See for example: Coxon, B.; Fletcher, Jr., H.G. *J. Am. Chem. Soc.* **1963**, *85*, 2637.
67. Myers, R.W.; Lee, Y.C. *Carbohydr. Res.* **1984**, *132*, 61.
68. Simchen, G.; Pürkner, E. *Synthesis* **1990**, 525.
69. Rosenthal, A.; Brink, A.J. *J .Carbohydr. Nucleotides Nucleosides* **1975**, *2*, 343.
70. Kalvoda, L.; Farkaš, J.; Šorm, F. *Tetrahedron Lett.* **1970**, 2297.
71. Zehavi, U.; Mechoulam, R. *Carbohydr. Res.* **1981**, *98*, 143.
72. Hurd, C.D.; Bonner, W.A. *J. Am. Chem. Soc.* **1945**, *67*, 1972.
73. Hurd, C.D.; Holysz, R.P. *J. Am. Chem. Soc.* **1950**, *72*, 2005.
74. Pangiot, M.J.; Curley, Jr., R.W. *J. Carbohydr. Chem.* **1994**, *13*, 293.
75. Hanessian, S.; Sato, K.; Liak, T.J.; Dixit, D. *J. Am. Chem. Soc.* **1984**, *106*, 6114.
76. Hanessian, S.; Liak, T.J.; Dixit, D.M. *Carbohydr. Res.* **1981**, *88*, C14.
77. Bihovsky, R.; Selick, C.; Giusti, I. *J. Org. Chem.* **1988**, *53*, 4026.
78. Zhai, D.; Zhai, W.; Williams, R.M. *J. Am. Chem. Soc.* **1988**, *110*, 2501.
79. Hanessian, S.; Pernet, A.G. *Can. J. Chem.* **1974**, *52*, 1266.
80. Nicolaou, K.C.; Dolle, R.E.; Chucholowski, A.; Randall, J.L. *J. Chem. Soc., Chem. Commun.* **1984**, 1153.
81. Araki, Y.; Watanabe, K.; Kuan, F.-H.; Itoh, K.; Kobayashi, N.; Ishido, Y. *Carbohydr. Res.* **1984**, *127*, C5.
82. Araki, Y.; Kobayashi, N.; Watanabe, K.; Ishido, I. *J. Carbohydr. Chem.* **1985**, *4*, 565.
83. Matsumoto, T.; Katsuki, M.; Suzuki, K. *Tetrahedron Lett.* **1988**, *29*, 6935.
84. Matsumoto, T.; Katsuki, M.; Suzuki, K. *Tetrahedron Lett.* **1989**, *30*, 833.
85. Matsumoto, T.; Katsuki, M.; Jona, H.; Suzuki, K. *Tetrahedron Lett.* **1989**, *30*, 6185.
86. Posner, G.H.; Haines, S.R. *Tetrahedron Lett.* **1985**, *26*, 1823 see also Drew, K.N.; Gross, P.H. *J. Org. Chem.* **1991**, *56*, 509.
87. Kende, A.S.; Fujii, Y. *Tetrahedron Lett.* **1991**, *32*, 2199.

Chapter 2

ELECTROPHILIC SUGARS IN *C*-GLYCOSIDE SYNTHESIS II

I. INTRODUCTION

The vastness of this area of *C*-glycoside synthesis is exemplified by the need to break-up this section into two separate chapters. This continuation of Chapter 1 will deal with the electrophilic reactions of glycals, enitols, and anhydro sugars. Sugar lactones will also be discussed since they have been used by several workers in the preparation of β-*C*-pyranosides. The same organization, as far as nucleophiles is concerned, will be loosely followed. Finally, and maybe a little bit out of context, a fairly new reaction involving intramolecular arylation by participation of neighboring aromatic rings will be discussed, which is a reaction discovered by Martin. As in the last chapter, there will be a significant amount of conceptual overlap, especially in the use of glycals; therefore information will be summarized and presented as practically as possible.

II. GLYCALS

Glycals have been used extensively as precursors to *C*-glycosides. This section will be divided up into sections addressing reactions with silyl enol ethers and dicarbonyl compounds, allylsilanes, olefinic derivatives, aromatics, cyanides, and finish with organometallic based nucleophiles.

A. ENOL ETHERS AND RELATED SPECIES

In 1981 Fraser-Reid and Dawe[1] published the Lewis acid catalyzed reaction of glucal (1) with the silyl enol ether of acetophenone to obtain a mixture of *C*-glycosides in good yield. Several sets of conditions were examined and the best results were obtained using boron trifluoride etherate in dichloromethane at -40°C. A 4:1 ratio of α:β isomers was obtained in 99% combined yield (eq. 1). Compound 2 also served as a precursor for model studies on the trichothecane natural products.[2]

Grynkiewicz and BeMiller[3] have extended this reaction to the glycals shown in Scheme 1. In these substrates a heteroatom is present at *C*-2 thereby leading to a masked carbonyl compound in the product. The reactions were all stereoselective and led to the α-isomers 4, 6, and 8.

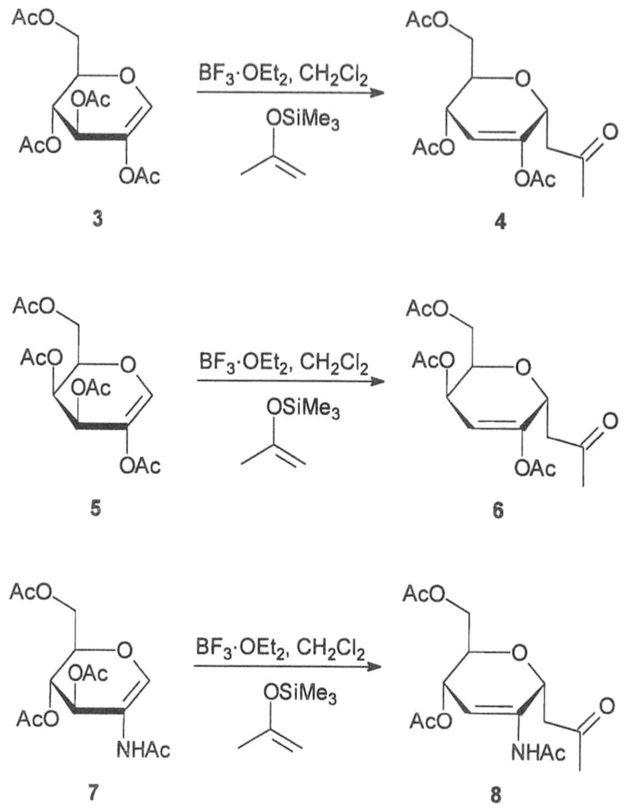

Scheme 1

Yougai and Miwa[4] extended their palladium mediated *C*-glycoside synthesis to the use of a Lewis acid to form an intermediate allyl carbonium ion that was trapped with a suitable dicarbonyl compound to give *C*-glycosides. The two examples shown in Scheme 2 illustrate the use of acetylacetone as the nucleophile.

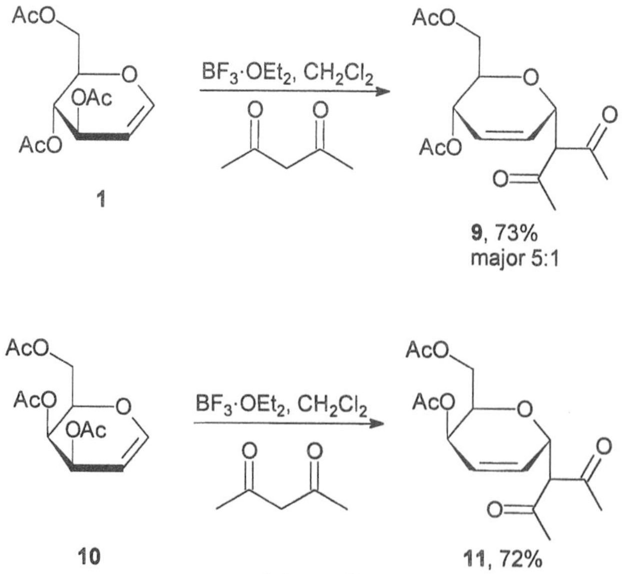

Scheme 2

B.   ALLYLATION

Danishefsky et al.[5] have extended the reactivity of the allylic transposition reaction of glycals to include reaction with allyltrimethylsilane to gain access to allylic C-glycosides. Treatment of glucal (1) with allyltrimethylsilane in the presence of titanium tetrachloride at low temperature gave an 85% yield of 12 and the β-isomer (not shown) in a 16:1 ratio. The corresponding allo (95%, α:β, 6:1) and galacto (93%, α:β, 30:1) derivatives gave ratios in the same direction with the galacto derivative showing the highest selectivity, Scheme 3.

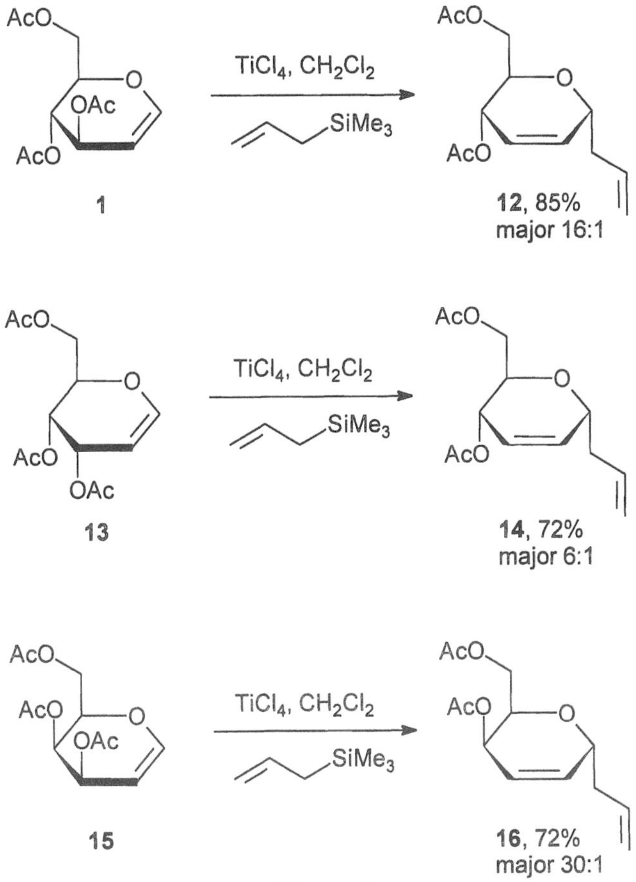

Scheme 3

Isobe and co-workers[6] have extended this type of reaction to include bis(trimethylsilyl)-acetylene as a nucleophile (eq. 2). The α-isomer 17 was the only C-glycosidic product isolated from the reaction mixture.

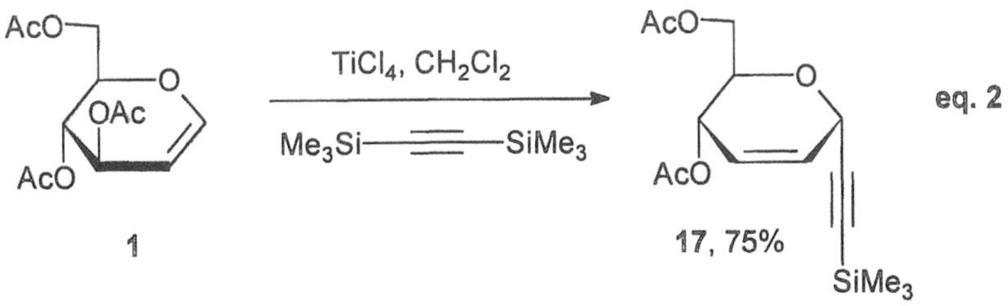

The workers[7] subsequently devised a method to convert the α-isomer **17** to the β-isomer **20**. The triple bond was complexed as its dicobalt hexacarbonyl complex (**18**) (Scheme 4) and epimerization affected by treatment with triflic acid to give a 6:1 ratio of β:α anomers. Better selectivities (19:1) could be obtained with the corresponding saturated derivative. Decomplexation then gave the free acetylene **20**. Other examples were also studied (not shown).

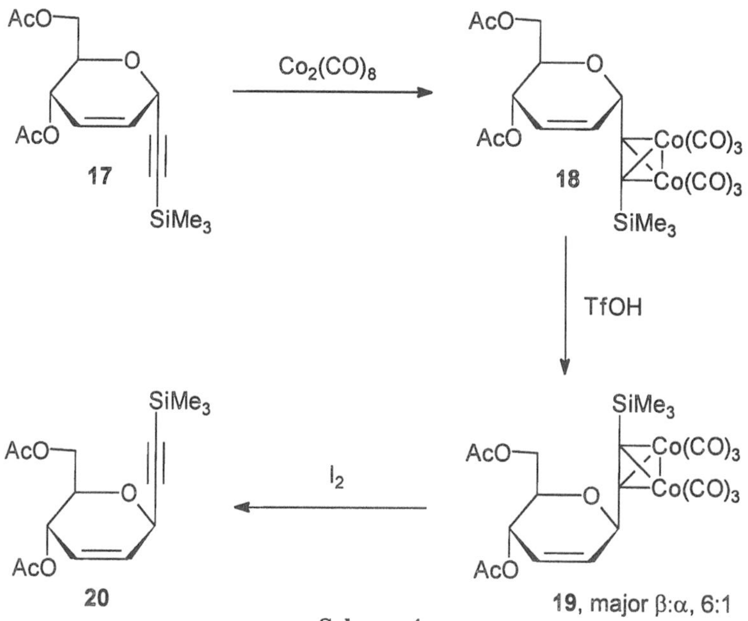

Scheme 4

If bis-silylacetylene **21** is used then two sugar units can be linked together to form a pseudo disaccharide. For example, exposure of **1** to acetylene **21** in the presence of tin tetrachloride furnished adduct **22** in 54% yield (eq. 3).[8]

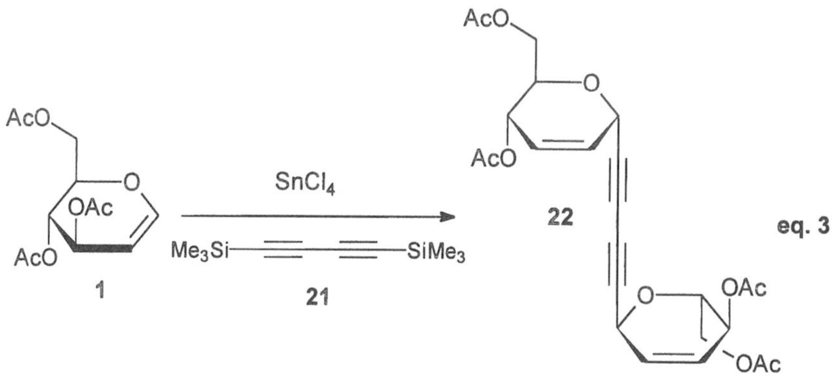

Work from the same group[9] involved conversion of *C*-glycoside **12** into **23** which was subsequently transformed into **24** which corresponds to the *C*-15 to *C*-27 segment of okadaic acid (**25**), Scheme 5.

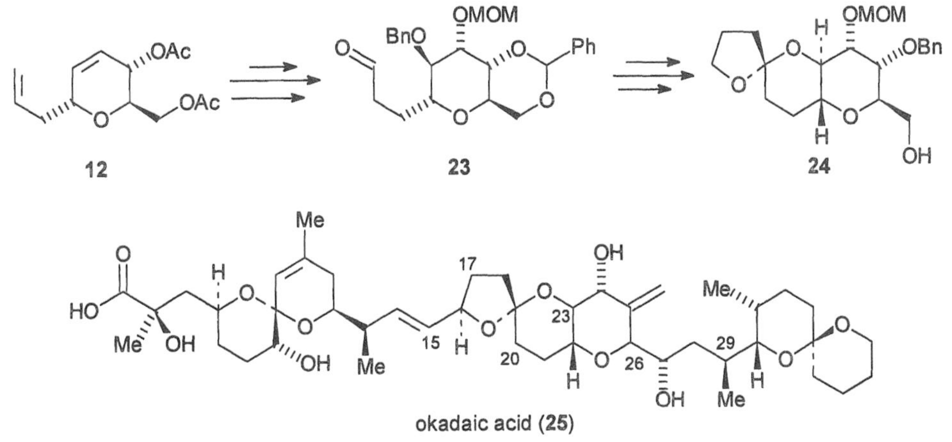

12 **23** **24**

okadaic acid (**25**)

Scheme 5

Danishefsky and collaborators[10] also used the same starting material in model studies directed at the avermectins spiroketal sub-unit. The terminal double was oxidatively cleaved to aldehyde **26** and cycloaddition with Danishefsky's diene gave **28**. The dihydropyran ring was converted to **29** in several steps which was then oxidatively cyclized to spiroketal **30**.

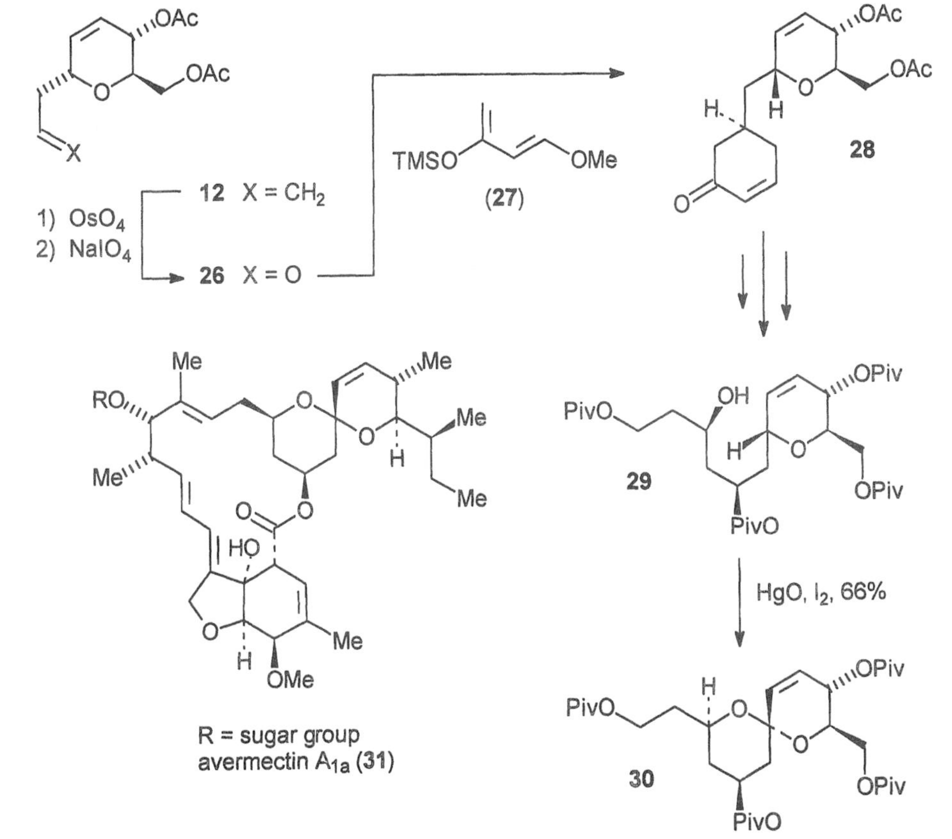

12 X = CH$_2$

1) OsO$_4$
2) NaIO$_4$

26 X = O

(**27**)

28

29

PivO

HgO, I$_2$, 66%

30

R = sugar group
avermectin A$_{1a}$ (**31**)

Scheme 6

Model studies[11] directed towards indanomycin involved allylation with a crotylsilane. This reaction leads to a *C*-glycoside that possesses a new stereogenic center in the carbon side chain. The best selectivity observed (eq. 4) was with silane **34** and the use of boron trifluoride etherate in acetonitrile to furnish **33** as the major isomer (7:1).

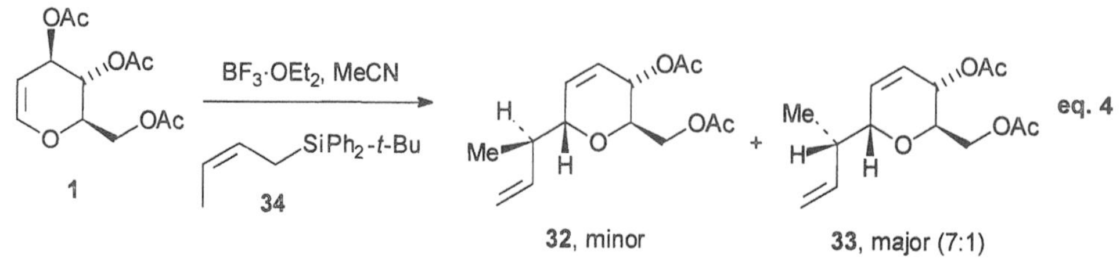

32, minor   **33, major (7:1)**

During the synthesis of the ABC ring system of brevetoxin B, a marine toxin, Nicolaou and co-workers[12] used the allylation of **1** as their starting point. Allylation, followed by chain extension at *C*-6 and oxidation of *O*-4 provided **35**. Treatment of **35** with trimethyl aluminum, silylation, and DIBAL reduction was followed by Sharpless epoxidation to bring the sequence up to **36**. Final steps then gave **37** a compound which corresponds rather nicely to the ABC ring system of the target structure, Scheme 7.

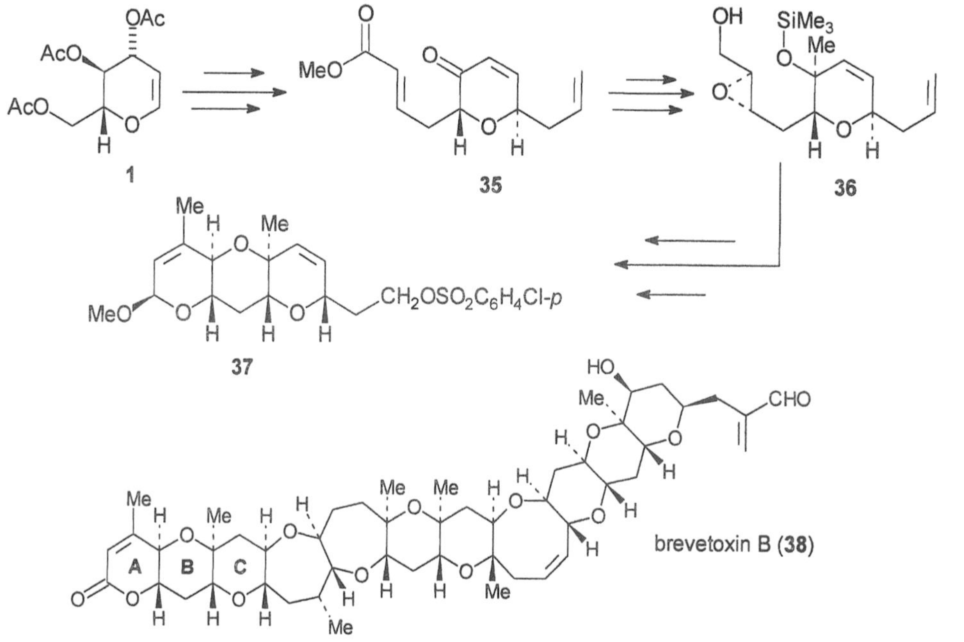

Scheme 7

It was recently found by Toshima[13] that DDQ mediates allylation of glycals. This reaction provides a mild alternative to the usual Lewis acidic conditions required to carry out this transformation. Scheme 8 shows that the yields and selectivities for this reaction fall within the useful range and are as good as those obtained using Lewis acidic methods.

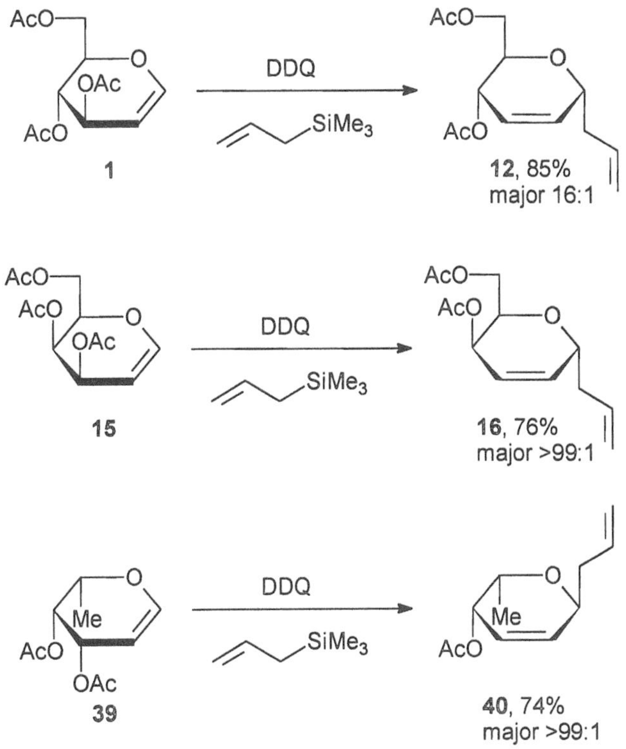

Scheme 8

It has been found that carbinols of the type **42** and **44** undergo rearrangement, under acidic conditions, to give rise to the spirocyclic *C*-glycoside **43** and **45**, respectively (Scheme 9).[14]

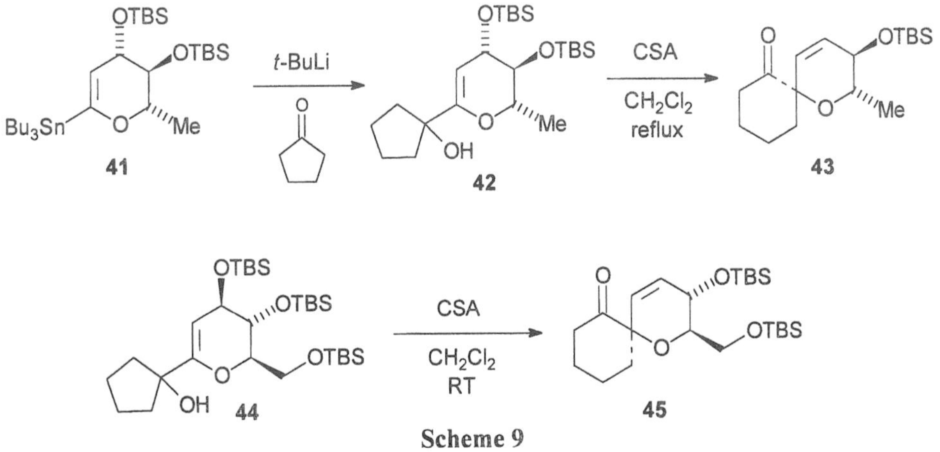

Scheme 9

C.   OLEFINS

Herscovici and co-workers[15] have extended the use of glycals in *C*-glycoside synthesis by developing methodology that enables olefins to be used as the nucleophile. An extensive study has been published and only a few examples will be presented here to illustrate the general concepts. Glucal reacted with methylenecyclohexane to give a 94% yield of the α-anomer **46** exclusively. Reaction of glycal **3** also led to the α-anomer as the sole product of reaction. Compound **47** contains an α,β-unsaturated ketone moiety, and, structurally, this compound is rather distant from the parent sugar and should prove useful in synthesis. Interestingly, use of the cyclobutene gave *C*-glycoside **48** in which the formed carbonium ion had been trapped by chloride before elimination could occur to give the olefin. Olefin **49** also proved useful for *C*-glycoside preparation (**1**→**50**), Scheme 10.

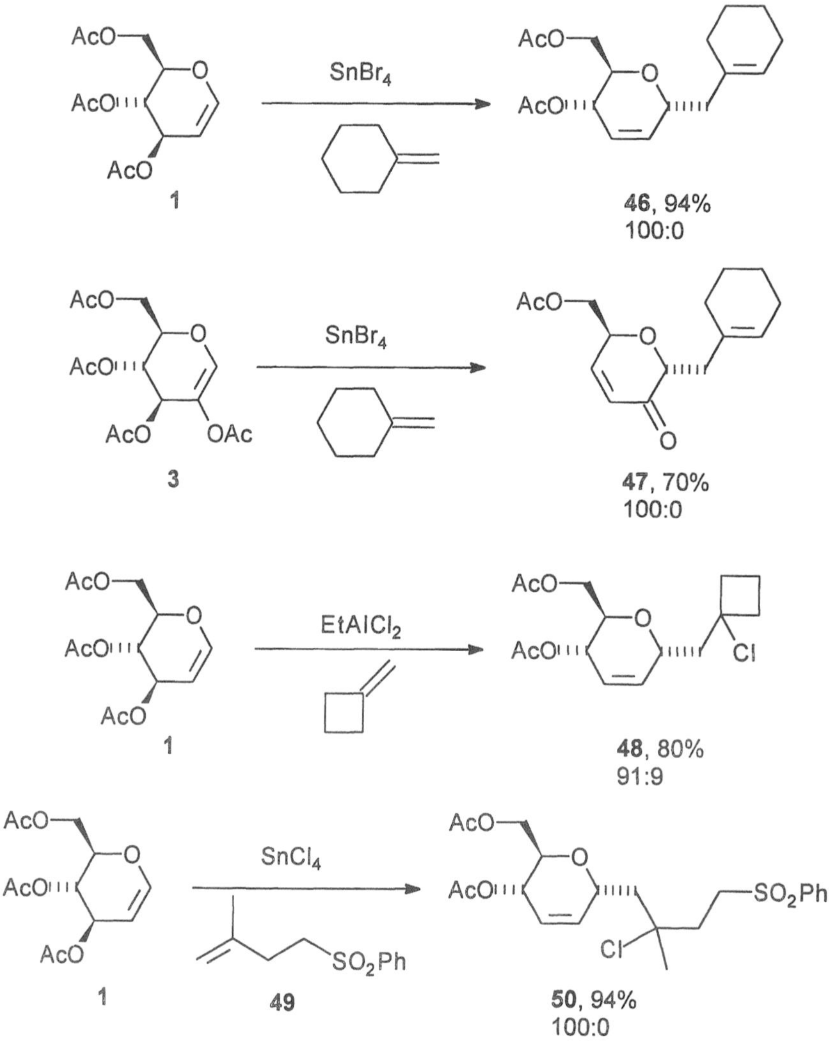

Scheme 10

Reproduced with permission from ref. 15, pg. 1996-1997, Copyright 1990, courtesy of the Royal Society of Chemistry, Thomas Graham House, Science Park, Milton Road, Cambridge, CB4 4WF, UK.

This reaction has also been extended to include homoenolates as nucleophiles.[16] Several glycals were condensed with either 1-[(thexyldimethylsilyl)oxy]-2-methyl-2-propene (**51**) or **52** under Lewis acid conditions to give yields of *C*-glycosides in the yields and anomeric ratios shown. Depending on the nature of the nucleophile (**51** or **52**) ketones or aldehydes could be synthesized by this methodology. This methodology complements allylation and condensation with enol ethers since the carbonyl group is located 3 carbon atoms away from the anomeric carbon atom.

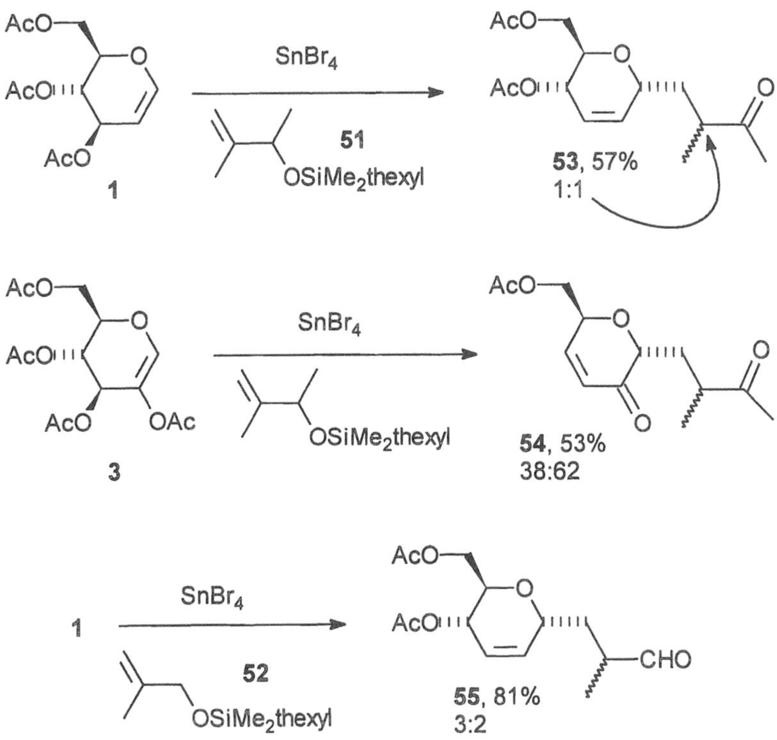

Scheme 11

D. AROMATICS

Casiraghi and co-workers[17] have reported that the Lewis acid catalyzed condensation of anisole with glucal gives the β-isomer **56** and not the α-isomer as previously reported[18] (eq. 5). Their assignment is based on a single X-ray analysis of the product *C*-glycoside.

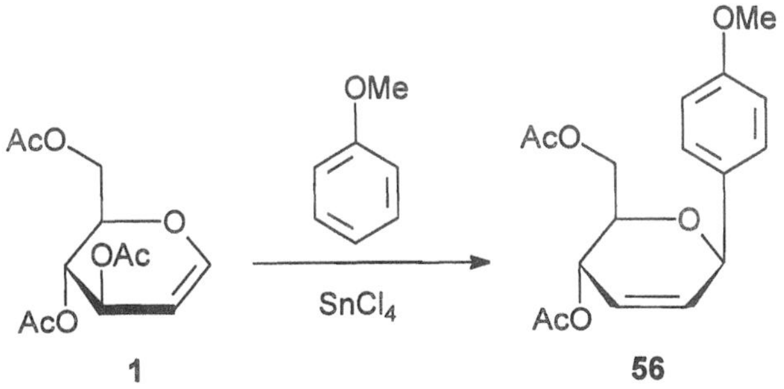

eq. 5

Work from the same group[19] focused on the arylation of glycals by reaction with bromomagnesium phenolates which allowed entry into aryl α-*C*-glycosides. The reactions were conducted by sonicating a mixture of the bromomagnesium phenolate with the glycal in dichloromethane. Under these conditions the α-anomer was either the exclusive or sole product of the reaction. Table 1 and eq. 6 show some of the *C*-glycosides prepared by this procedure.

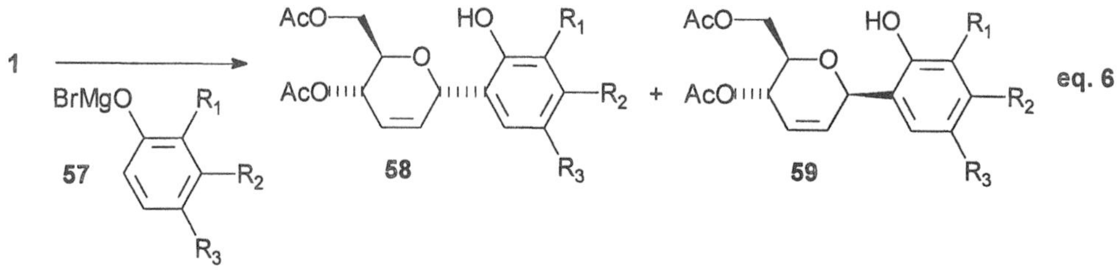

eq. 6

Table 1: Reactions of Various Bromomagnesium Phenolates.

Entry	R_1	R_2	R_3	Yield	$\alpha{:}\beta$ ratio
a	H	H	*t*-Bu	71	100:1
b	H	-O-CH$_2$-O-	-	80	27:1
c	H	H	H	63	26:1
d	H	H	OMe	82	100:1
e	H	OMe	H	69	21:1
f	*t*-Bu	H	OMe	66	60:1

Grynkiewicz and BeMiller[20] have examined the reaction of furan and thiophene with several glycals (Scheme 12). Reaction of glucal 1 with either furan or thiophene gave a 1:1 mixture of regioisomeric products **60** and **61**.

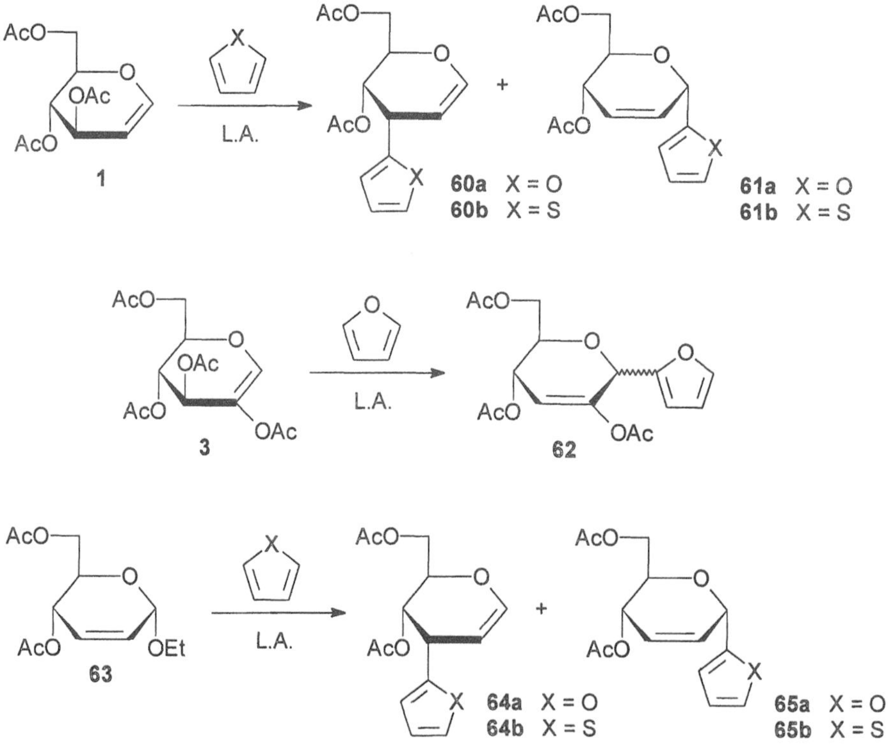

Scheme 12

Glycal **3** on the other hand gave only *C*-glycoside **62**, but as a mixture of anomers. The hex-2-ene-pyranoside **63** on reaction with furan or thiophene gave similar regioisomeric mixtures as with the reaction of glucal.

E.   CYANIDES

Work from the same laboratories[21] concentrated on the reaction of various glycals with trimethylsilylcyanide. Reaction of glucal with TMSCN in the presence of boron trifluoride etherate gave the α-anomer **66** exclusively.   Glycals **3** and **5** under the same reaction conditions gave mixtures of α- and β-*C*-glycosides **67** and **68**. Similar work has also been reported by another group.[22]

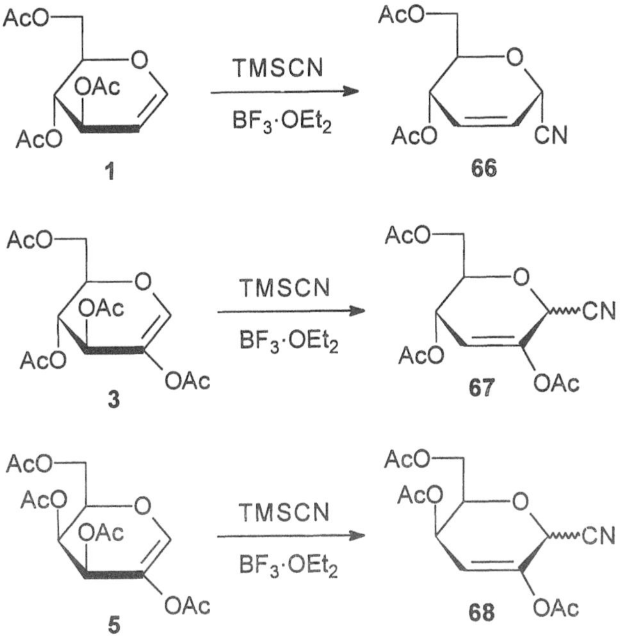

Scheme 13

Reaction of **1** with Et$_2$AlCN led to a mixture of anomers with the β-anomer predominating in a 3:2 ratio.[23] These results are in contrast to those obtained by Fraser-Reid and Tulshian who obtained a 9:1 ratio of α:β anomers.[24] This discrepancy was explained by submitting pure β to the reaction conditions where it was observed that equilibration to a mixture of α- and β-anomers had occurred (eq. 7).

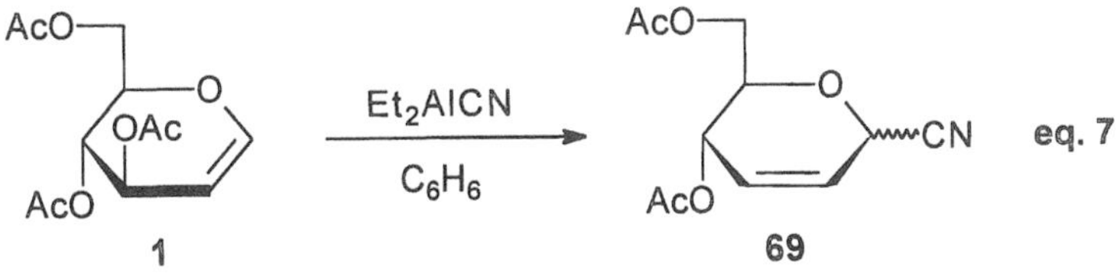

F. ORGANOMETALLICS

Mitsunobu[25] has used an S_N2' type process to introduce alkyl groups on the anomeric carbon with allylic transposition to give *C*-glycosides of the general type **71** (Scheme 14).

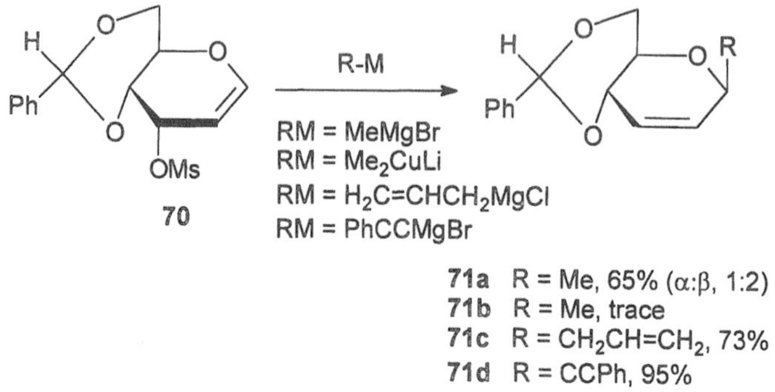
H
Ph
O
O
O
OMs
70
R-M
RM = MeMgBr
RM = Me$_2$CuLi
RM = H$_2$C=CHCH$_2$MgCl
RM = PhCCMgBr
H
Ph
O
O
O
R
71a R = Me, 65% (α:β, 1:2)
71b R = Me, trace
71c R = CH$_2$CH=CH$_2$, 73%
71d R = CCPh, 95%

Scheme 14

Yamamoto and co-workers used aluminum reagents to affect a similar transformation.[26] Scheme 15 shows the examples and in most cases the *trans* isomer was either the exclusive or favored product of reaction.

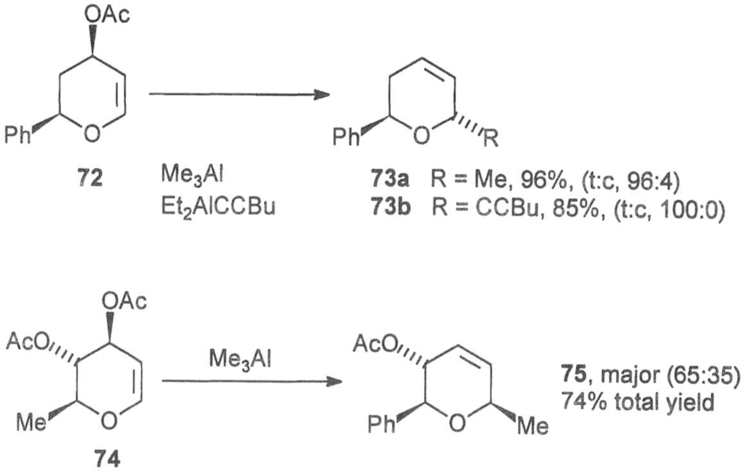
OAc
Ph
O
72
Me$_3$Al
Et$_2$AlCCBu
Ph
O
R
73a R = Me, 96%, (t:c, 96:4)
73b R = CCBu, 85%, (t:c, 100:0)
OAc
AcO
Me
O
74
Me$_3$Al
AcO
Ph
O
Me
75, major (65:35)
74% total yield

Scheme 15

Orsini and Pelizzoni[27] used Reformatsky reagents in reaction with glycals in the presence of a Lewis acid to obtain a mixture of anomers in good yield (eq. 8). Rhamnal and galactal were also studied and also gave mixtures of anomers (not shown).

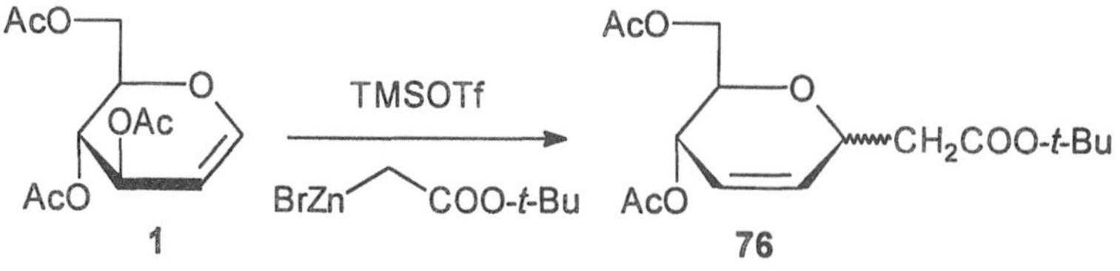
AcO
O
OAc
AcO
1
TMSOTf
BrZn
COO-*t*-Bu
AcO
O
CH$_2$COO-*t*-Bu
AcO
76

eq. 8

III. ENITOLS AND ANHYDRO SUGARS

A. ENITOLS
1. Addition of Cyanide and Enol Ethers

Tatsuta and co-workers[28] have added cyanide ion in a 1,4-manner to the hex-1-ene-3-uloses 77 and 80 followed by hydride reduction to furnish the 2-deoxy-C-glycosides 79 and 82 as the major compounds from reduction. The addition of cyanide was selective in both cases; some of the other isomers were obtained in the reduction, but the amounts were not very substantial.

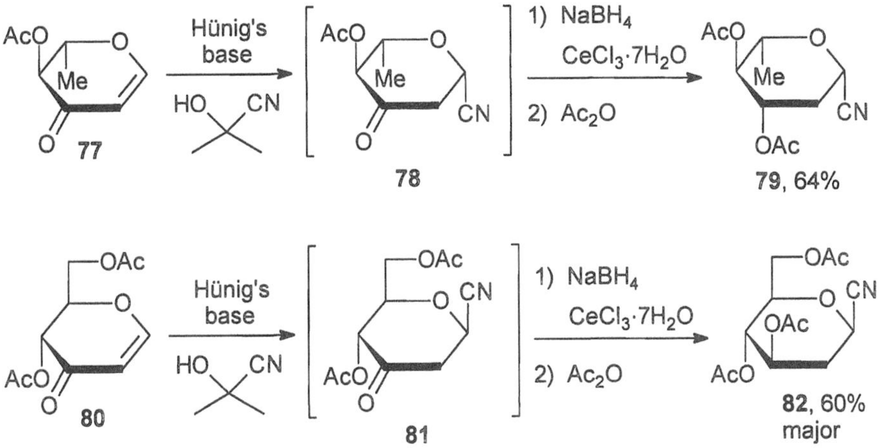

Scheme 16

Reproduced from ref. 28, Copyright 1989, pg. 491, with kind permission from the Chemical Society of Japan, 1-5 Kanda-Surugadai, Chiyoda-ku, Tokyo, 101, Japan.

Glycal 83 was amenable to 1,4-addition by silyl enol ethers. The β-isomer was the major one formed. The product C-glycoside has lost an acetoxy group via allylic transposition and the ketone function had shifted to C-2, although a masked ketone still exists at C-3. Scheme 17 shows the general reaction with the silyl enol ether of cyclohexanone and some of the other C-glycosides made by this methodology.[29]

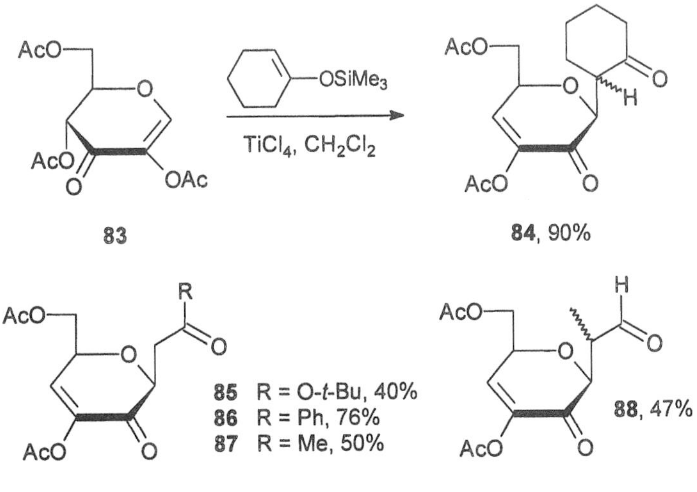

85 R = O-t-Bu, 40%
86 R = Ph, 76%
87 R = Me, 50%

88, 47%

Scheme 17

2.   Cuprates

The next example, involving a 1,5-anhydro-enitol, shows a reversal in product stereochemistry.[30]   Reaction of **80** with the diphenyl cuprate gave the α-*C*-glycoside **89**, after *in situ* acetylation. This is significant since unmasking of the ketone should provide a 2-deoxy-α-*C*-glycoside, a compound that is not that easily accessible.

eq. 9

Taking a lead from the earlier work of Fraser-Reid,[31] Goodwin and co-workers[32] have found that addition of organo(hetero)cuprates of the type **91** (where R = *n*-Bu and *t*-Bu) to enone **90** gave acceptable yields (~65%) of 2-deoxy-3-keto-α-*C*-glycosides **92a** and **92b** (eq. 10).

eq. 10

92a, R = *n*-Bu
92b, R = *t*-Bu

3.   Aromatics

Although the last example of this section does not deal with the preparation of a *C*-glycoside, the goal of this work was to synthesize such a compound. Booma and Balasubramanian[33] treated the glycal **93** with **94** and obtained the *C*-3 arylation product **95** exclusively (eq. 11). This probably arises from a [3,3] rearrangement of the initially formed *O*-glycoside. The driving force for this reaction is the presence of the electron withdrawing aldehyde group.

eq. 11

B.   ANHYDRO SUGARS

During the synthesis of palytoxin Kishi and co-workers[34] used a 1,2-anhydro sugar as the electrophilic partner in coupling with an organocuprate nucleophile (eq. 12). The manno epoxide **96** was treated with Grignard reagent **97** in the presence of Li_2CuCl_4 to afford alcohol **98** exclusively. It is noteworthy that the reagent attacks from the axial direction to give the α-*C*-glycoside selectively.

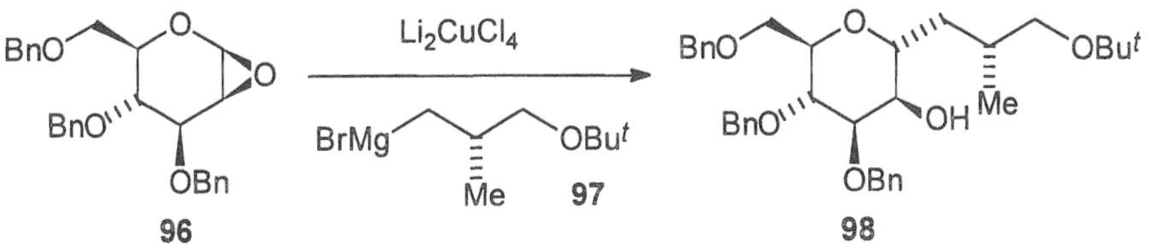

eq. 12

Several years later Bellosta and Czernecki[35] studied this reaction a little more closely and established its viability in *C*-glycoside synthesis. Treatment of Brigg's anhydride with lithium dimethyl or lithium diphenyl cuprate gave the *C*-glycosides **100a** and **100b** (66% and 25%) with the gluco configuration at *C*-3, respectively. This shows that the reaction proceeds by axial addition of the organometallic species onto the epoxide. This was firmly established by treating the manno isomer **101** with the same cuprates to obtain the manno *C*-glycosides in good yield and with excellent stereoselectivity (Scheme 18). Deoxygenation at *C*-2 could also be carried out to gain access to 2-deoxy-α-aryl or alkyl *C*-glycosides.

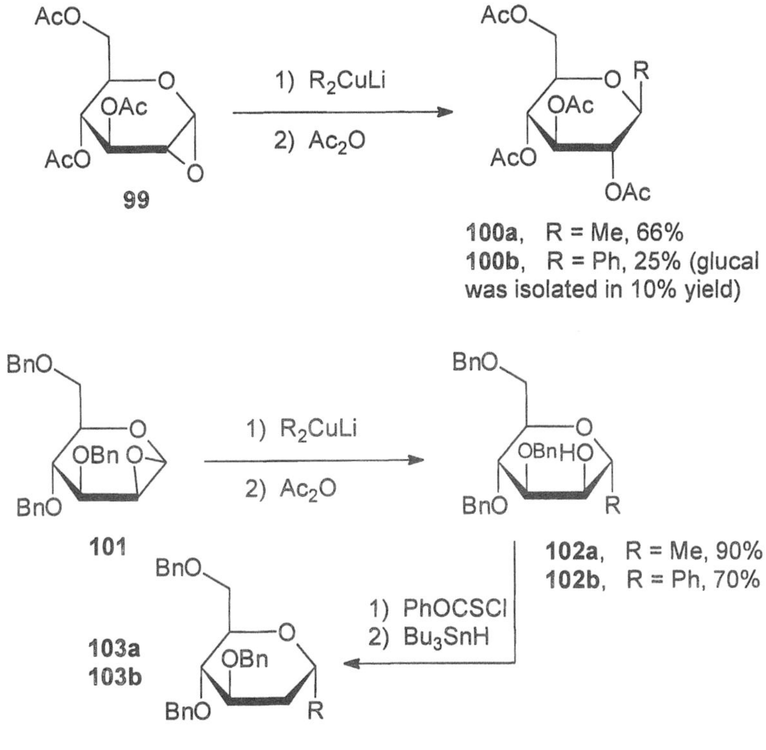

Scheme 18

IV. LACTONES

All of the approaches in this section are very similar in concept. They involve the addition of an organometallic reagent to a sugar lactone to usually give a mixture of lactols that is reduced selectively to the β-*C*-glycoside. This approach (at least with gluco and galacto sugars) is a very reliable method for synthesizing β-*C*-glycosides.

The pioneering work of Kishi and collaborators[36] is illustrated in Scheme 19. The organometallic reagents used were either the allyl Grignard or the lithio ester. The intermediate lactols are not shown, but reduction is selective for both the gluco and galacto sugars. The manno isomer gives a poor ratio after reduction, but this is probably due to the steric effect of the adjacent substituent at *C*-2.

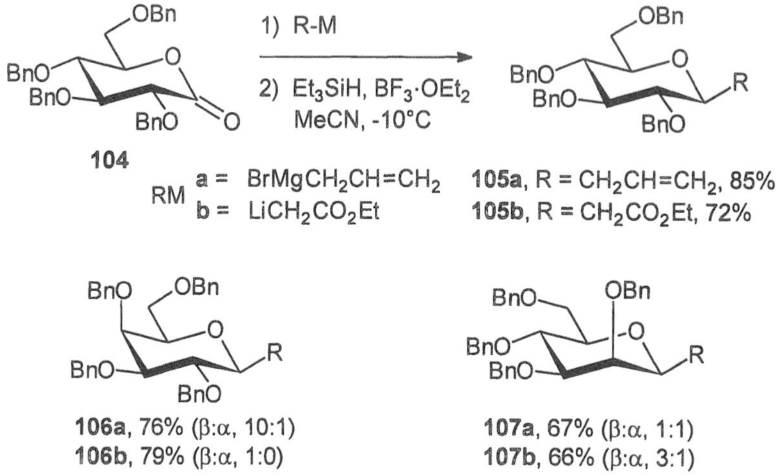
1) R-M
2) Et_3SiH, $BF_3\cdot OEt_2$
MeCN, -10°C
104
RM a = $BrMgCH_2CH{=}CH_2$
b = $LiCH_2CO_2Et$
105a, R = $CH_2CH{=}CH_2$, 85%
105b, R = CH_2CO_2Et, 72%
106a, 76% (β:α, 10:1)
106b, 79% (β:α, 1:0)
107a, 67% (β:α, 1:1)
107b, 66% (β:α, 3:1)

Scheme 19

Sinaÿ and co-workers[37] also used a similar strategy to gain access to either the *E* or *Z* vinyl β-*C*-glycosides. Addition of lithium acetylide **108** to the protected gluconolactone **104** gave a mixture of epimers **109**. Stereoselective reduction with triethylsilane and boron trifluoride etherate then gave β-anomer **110** as the exclusive product. Scheme 20 also shows how the triple bond may be manipulated to give either the *Z*-isomer **111** or the *E*-isomer **112**.

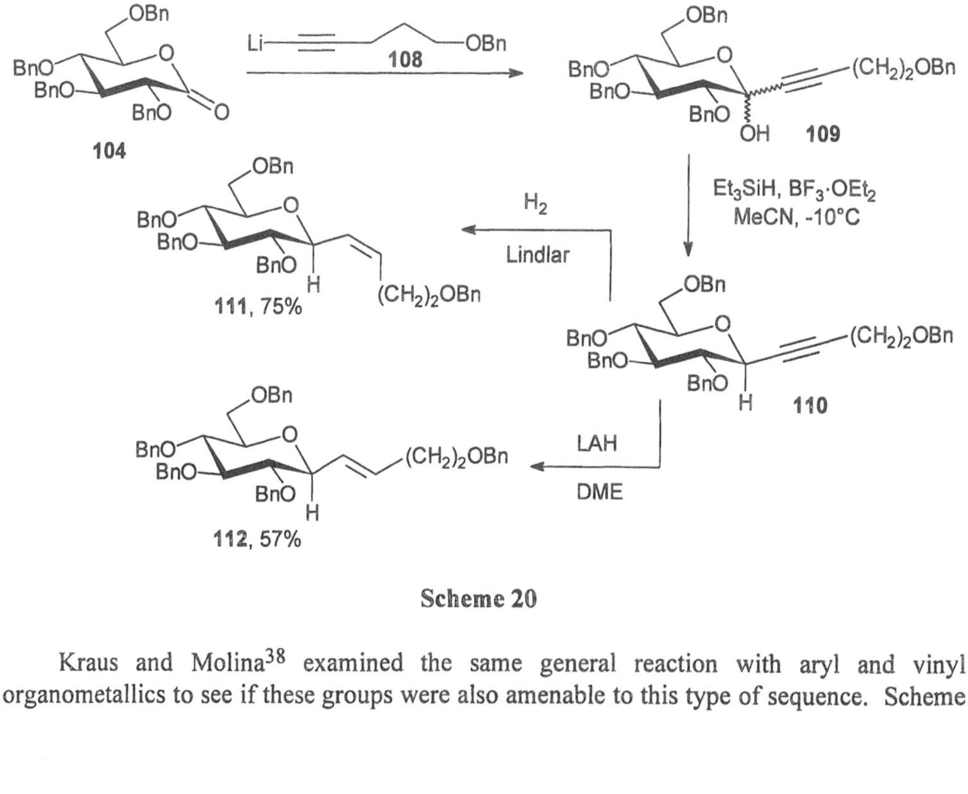
108
104
109
Et_3SiH, $BF_3\cdot OEt_2$
MeCN, -10°C
H_2
Lindlar
111, 75%
110
LAH
DME
112, 57%

Scheme 20

Kraus and Molina[38] examined the same general reaction with aryl and vinyl organometallics to see if these groups were also amenable to this type of sequence. Scheme

21 shows the general reaction and the results. The reaction generally proceeds well and **113g** is an interesting compound since it can provide access, via hydroxylation, to an eight carbon sugar precursor.

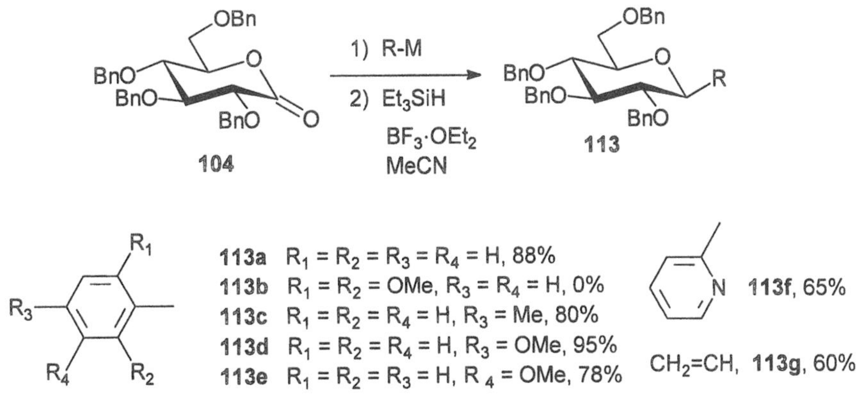

113a $R_1 = R_2 = R_3 = R_4 = H$, 88%
113b $R_1 = R_2 = OMe$, $R_3 = R_4 = H$, 0%
113c $R_1 = R_2 = R_4 = H$, $R_3 = Me$, 80%
113d $R_1 = R_2 = R_4 = H$, $R_3 = OMe$, 95%
113e $R_1 = R_2 = R_3 = H$, $R_4 = OMe$, 78%

113f, 65%

$CH_2=CH$, **113g**, 60%

Scheme 21

Czernecki and Ville[39] extended this reaction to synthesize the 2- and 3-furyl-β-*C*-glycosides as shown below in eq. 13.

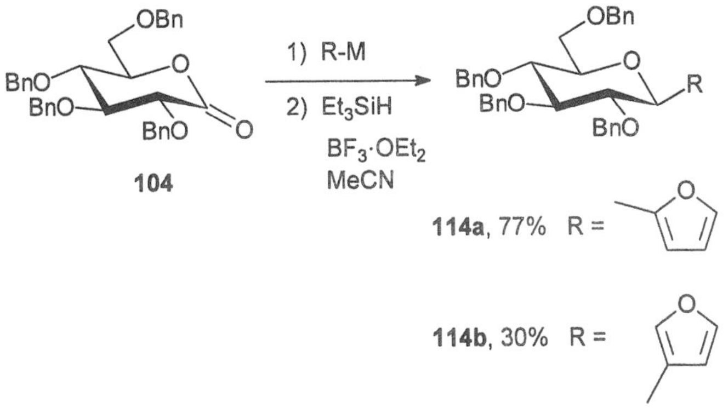

eq. 13

114a, 77% R =

114b, 30% R =

Dondoni[40] has applied a similar reaction and used it to gain access to the β-*C*-glycosyl carboxylate by ozonolysis of the product furan.

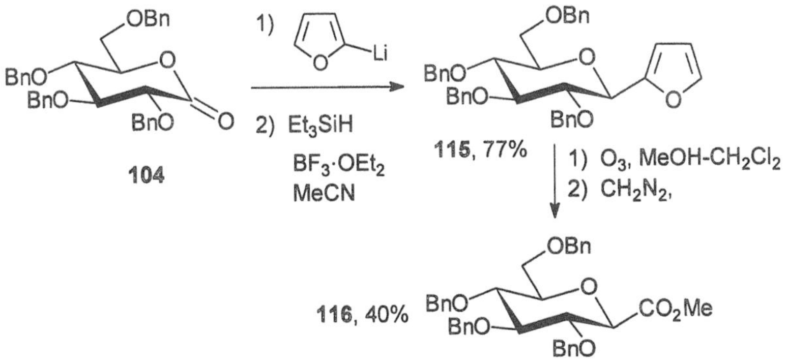

115, 77%

1) O_3, MeOH-CH_2Cl_2
2) CH_2N_2,

116, 40%

Scheme 22

Access to the α-isomer was available by Lewis acid catalyzed arylation of the anomeric acetate followed by ozonolytic cleavage to the acid group.

Diketal **117** was also studied and contrary to previous reports of incompatibility of the acetonide functions to the reducing conditions it was found that the ester **119** could be made in good overall yield, Scheme 23.

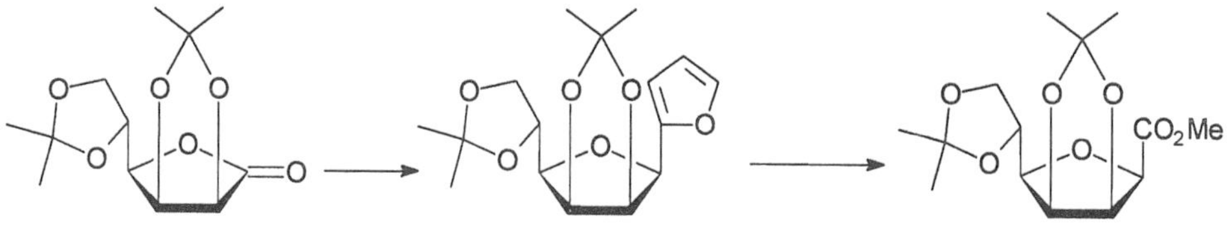

117   **118**   **119**

Scheme 23

Using the thiazole anion gave the β-isomer **120** exclusively.[41] Acetylation followed by rather vigorous conditions for reduction gave a mixture of α and β-anomers. This was not of consequence since reduction and cleavage of the thiazole ring system gave a mixture of aldehydes which were epimerized to the more stable β-configuration **122**, Scheme 24. Similar chemistry was carried to gain access to the β-manno furanoside aldehyde.

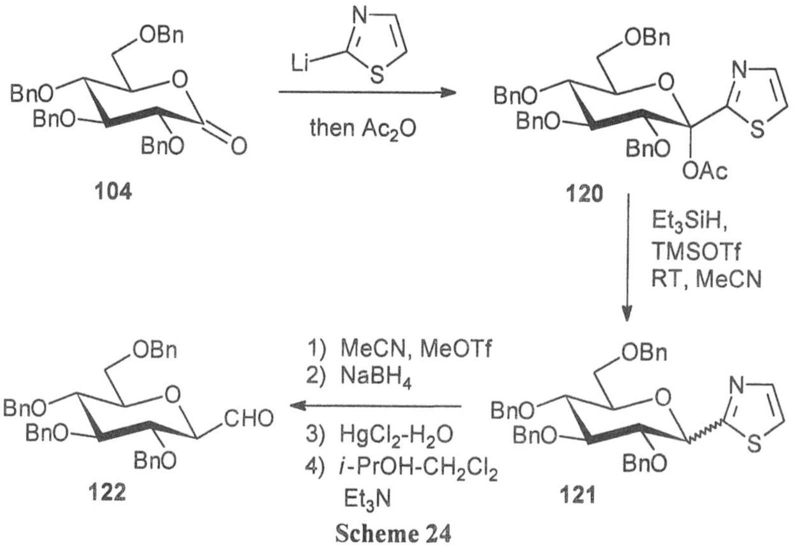

104   **120**

122   **121**

Scheme 24

V. INTRAMOLECULAR ARYLATION

The work in this section can be attributed almost entirely to the efforts of Martin. He has found that an adjacent benzyloxy group is well located to participate in arylation reactions at the anomeric site. This chemistry was well developed in the furanose series although some work dealing with pyranosides will also be described.

Early work focused on the acetate **123** which when treated with tin tetrachloride gave adduct **125** in 46% yield.[42] Similar reaction with **126** was also described leading to the *C*-glycoside like structure **127**. Obviously the adjacent aromatic is attacking the intermediate carbonium ion to give, after rearomatization, the products.

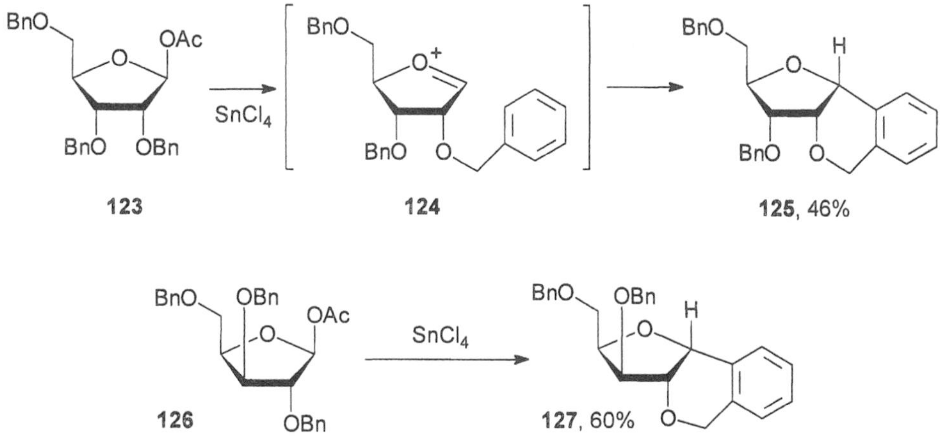

Scheme 25

It was subsequently found that the yields for this reaction could be raised up to the 80-90% range by using boron trifluoride etherate in dichloromethane.[43]

Araki[44] also found that the same product was formed from treatment of the anomeric fluoride **128** with boron trifluoride etherate (eq. 14).

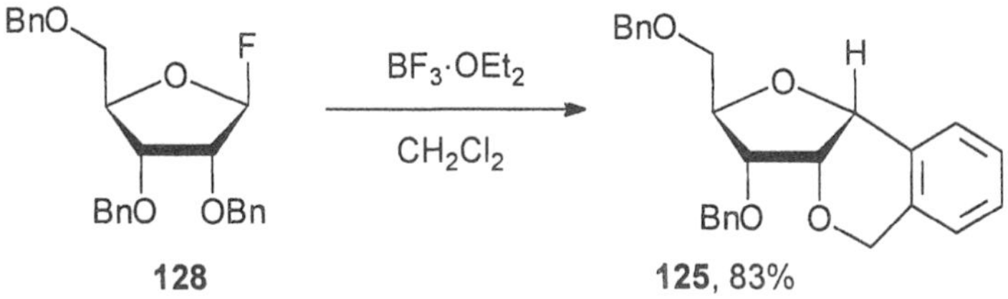

Further work by Martin and co-workers[45] has shown that 3-methoxybenzyl groups facilitate the reaction. When **129** was treated with tin tetrachloride **130** was formed in 80% yield. Oxidation with ruthenium tetraoxide gave lactone **131** which could be saponified to the *C*-glycoside **132**.

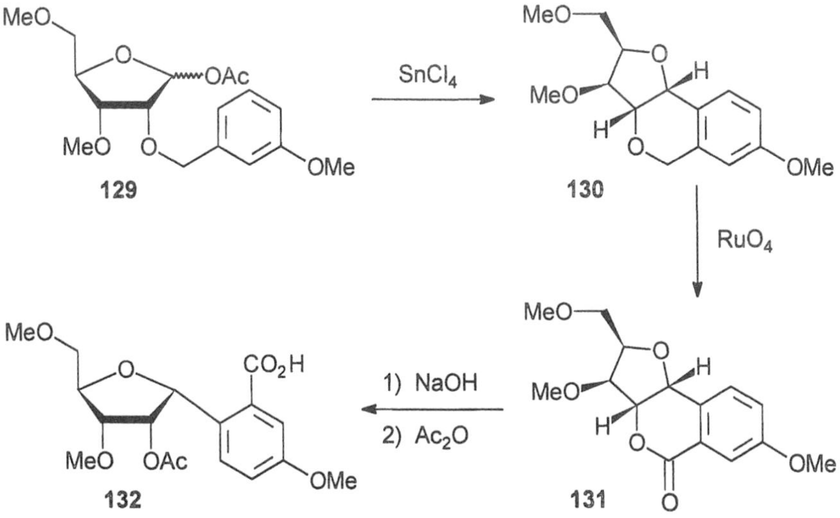

Scheme 26

The same lactone could be accessed by starting with acetate **133** and treating with tin tetrachloride to obtain **131** in 46% yield, eq. 15.

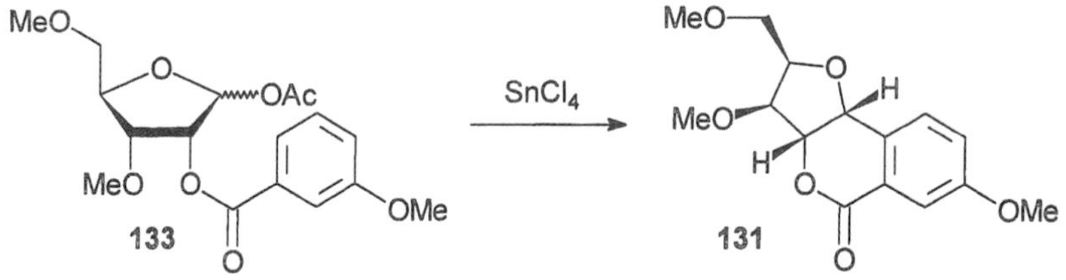

eq. 15

In the pyranose series, the workers chose to examine the manno compound **134**. Treatment with 3M sulfuric acid and acetic acid (conditions for glycosidic hydrolysis) gave the product *C*-glycoside **135** in 50% isolated yield, eq. 16.

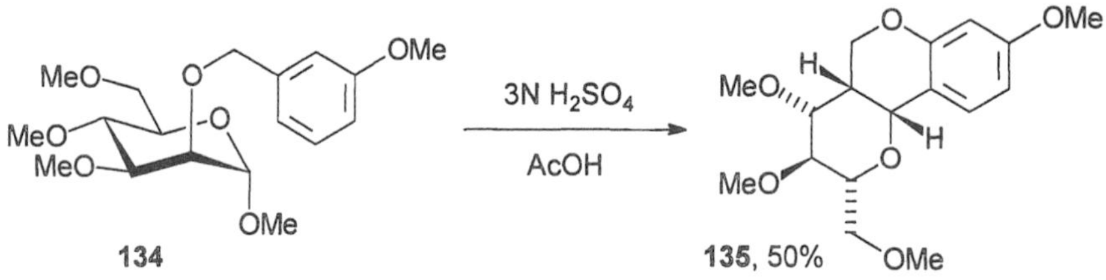

eq. 16

Studies dealing with multiple and long range participation of benzyl groups have also been published.[46]

Further work by Martin and collaborators[47] has focused on the synthesis of *C*-glycosides stereospecifically through the use of internal nucleophile delivery. Scheme 27 shows that when the *C*-2 hydroxyl is free, as in **136**, the product formed is the *C*-glycoside **138** with the aryl unit and the hydroxyl *syn*.

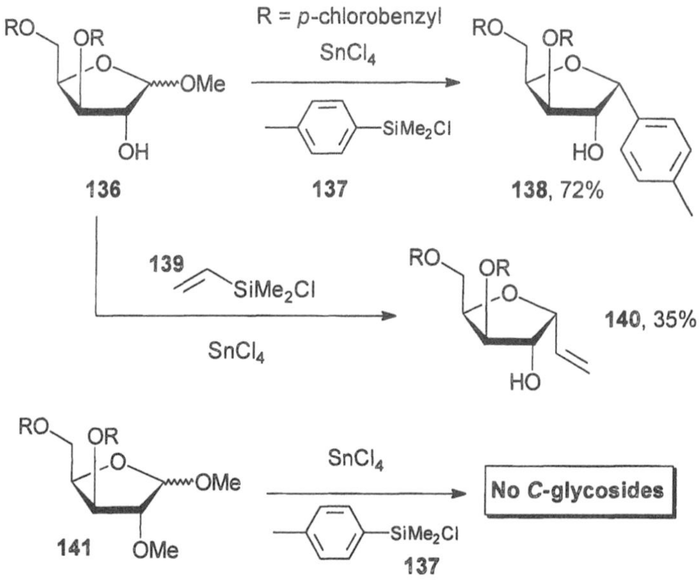

Scheme 27

However, when the hydroxyl is protected, as in **141**, no *C*-glycoside is formed. Conversion of **136** to **140** provides another example that supports the effect of a free hydroxyl.

VI. CONCLUSION

These last two chapters have illustrated and tabulated the methods employed to synthesize *C*-glycosides from electrophilic sugar partners with appropriate nucleophiles. Lactols, acetates, alkyl glycosides, thioglycosides, and glycosyl halides are all suitable precursors to *C*-glycosides depending on the nature of the Lewis acid and nucleophile used. For pyranoses the α-isomer is often the major product, dictated by the anomeric effect, but not in all the cases. Precursors to sugar lactones provide a convenient starting material for the selective production of β-*C*-glycosides via nucleophile addition and stereoselective reduction. Arylation with electron rich aromatics in the pyranose series leads to the thermodynamically more stable β-*C*-glycosides by a ring opening-closure process. Reaction of nucleophiles with glycals give an allylic transposition product; with allylsilanes, the product *C*-glycosides have been used as synthons in natural product synthesis. Anhydro sugars and enitols have also found some synthetic use in this domain as well. Further refinements in this area will probably involve milder conditions used to generate the requisite carbonium ions and the development of conditions that favor the formation of only the desired isomer.

VII. REFERENCES

1. Dawe, R.D.; Fraser-Reid, B. *J. Chem. Soc., Chem. Commun.* **1981**, 1180.
2. Tsang, R.; Fraser-Reid, B. *J. Org. Chem.* **1985**, *50*, 4659.
3. Grynkiewicz, G.; BeMiller, J.N. *J. Carbohydr. Chem.* **1982**, *1*, 121.
4. Yougai, S.; Miwa, T. *J. Chem. Soc., Chem. Commun.* **1983**, 68.
5. Danishefsky, S. J.; Kerwin, J.F., Jr. *J. Org. Chem.* **1982**, *47*, 3805.
6. Ichikawa, Y.; Isobe, M.; Konobe, M.; Goto, T. *Carbohydr. Res.* **1987**, *171*, 193.
7. Tanaka, S.; Tsukiyama, T.; Isobe, M. *Tetrahedron Lett.* **1993**, *34*, 5757.
8. Tsukiyama, T.; Peters, S.C.; Isobe, M. *Synlett* **1993**, 413.
9. Ichikawa, Y.; Isobe, M.; Goto, T. *Tetrahedron* **1987**, *43*, 4749.
10. Wincott, F.E.; Danishefsky, S.J.; Schulte, G. *Tetrahedron Lett.* **1987**, *28*, 4951.
11. Danishefsky, S.J.; DeNinno, S.; Lartey, P. *J. Am. Chem. Soc.* **1987**, *109*, 2082.
12. Nicolaou, K.C.; Duggan, M.E.; Hwang, C.-K.; Somers, P.K. *J. Chem. Soc., Chem. Commun.* **1985**, 1359.
13. Toshima, K.; Ishizuka, T.; Matsuo, G.; Nakata, M. *Chemistry Lett.* **1993**, 2013.
14. Paquette, L.A.; Dullweber, U.; Cowgill, L.D. *Tetrahedon Lett.* **1993**, *34*, 8019.
15. Herscovici, J.; Muleka, K.; Boumaîza, L.; Antonakis, K. *J. Chem. Soc., Perkin Trans. I* **1990**, 1995.
16. Herscovici, J.; Boumaiza, L.; Antonakis, K. *J. Org. Chem.* **1992**, *57*, 2476 and Herscovici, J.; Delatre, S.; Antonakis, K. *J. Org. Chem.* **1987**, *52*, 5691.
17. Casiraghi, G.; Cornia, M.; Colombo, L.; Rassu, G.; Fava, G.G.; Belicchi, M.F.; Zetta, L. *Tetrahedron Lett.* **1988**, *29*, 5549.
18. Czernecki, S.; Dechavanne, V. *Can. J. Chem.* **1983**, *61*, 533.
19. Casiraghi, G.; Cornia, M.; Rassu, G.; Zetta, L.; Fava, G.G.; Belicchi, F. *Carbohydr. Res.* **1989**, *191*, 243.
20. Grynkiewicz, G.; BeMiller, J.N. *Carbohydr. Res.* **1982**, *108*, C1.
21. Grynkiewicz, G.; BeMiller, J.N. *Carbohydr. Res.* **1982**, *108*, 229.

22. De Las Heras, F.G.; San Felix, A.; Fernández-Resa, P. *Tetrahedron* **1983**, *39*, 1617.

23. Grierson, D.S.; Bonin, M.; Husson, H.-P.; Monneret, C.; Florent, J.-C. *Tetrahedron Lett.* **1984**, *25*, 4645.

24. Tulshian, D.B.; Fraser-Reid, B. *J. Org. Chem.* **1984**, *49*, 518.

25. Ogihara, T.; Mitsunobu, O. *Tetrahedron Lett.* **1983**, *24*, 3505 and Mitsunobo, O.; Yoshida, M.; Takiya, M.; Kubo, K.; Maruyama, S.; Satoh, I.; Iwami, H. *Chemistry Lett.* **1989**, 809.

26. Maruoka, K.; Nonoshita, K.; Itoh, T.; Yamamoto, H. *Chem. Lett.* **1987**, 2215.

27. Orsini, F.; Pelizzoni, F. *Carbohydr. Res.* **1993**, *243*, 183.

28. Tatsuta, K.; Hayakawa, J.; Tatsuzawa, Y. *Bull. Chem. Soc. Jpn.* **1989**, *62*, 490.

29. Kunz, H.; Weißmüller, J.; Müller, B. *Tetrahedron Lett.* **1984**, *25*, 3571.

30. Bellosta, V.; Czernecki, S. *Carbohydr. Res.* **1987**, *171*, 279.

31. Yunker, M.B.; Plaumann, D.E.; Fraser-Reid, B. *Can. J. Chem.* **1977**, *55*, 4002.

32. Goodwin, T.E.; Crowder, M.C.; White, B.R.; Swanson, J.S.; Evans, F.E.; Meyer, W.L. *J. Org. Chem.* **1983**, *48*, 376.

33. Booma, C.; Balasubramanian, K.K. *Tetrahedron Lett.* **1992**, *33*, 3049.

34. Klein, L.L; McWhorter, W.W., Jr.; Ko, S.S.; Pfaff, K.-P.; Kishi, Y.; Uemura, D.; Hirata, Y. *J. Am. Chem. Soc.* **1982**, *104*, 7362.

35. Bellosta, V.; Czernecki, S. *J. Chem. Soc., Chem. Commun.* **1989**, 199 and *ibid.* *Carbohydr. Res.* **1993**, *244*, 275.

36. Lewis, M.D.; Cha, J.K.; Kishi, Y. *J. Am. Chem. Soc.* **1982**, *104*, 4976.

37. Lancelin, J.-M.; Zollo, P.H.A.; Sinaÿ, P. *Tetrahedron Lett.* **1983**, *24*, 4833.

38. Kraus, G.A.; Molina, M.T. *J. Org. Chem.* **1988**, *53*, 752.

39. Czernecki, S.; Ville, G. *J. Org. Chem.* **1989**, *54*, 610.

40. Dondoni, A.; Marra, A.; Scherrmann, M.-C. *Tetrahedron Lett.* **1993**, *34*, 7323.

41. Dondoni, A.; Scherrmann, M.-C. *Tetrahedron Lett.* **1993**, *34*, 7319.

42. Martin, O.R. *Tetrahedron Lett.* **1985**, *26*, 2055.

43. Anastasia, M.; Allevi, P.; Ciuffreda, P.; Fiecchi, A.; Scala, A. *Carbohydr. Res.* **1990**, *208*, 264.

44. Araki, Y.; Mokubo, E.; Kobayashi, N.; Nagasawa, J.; Ishido, Y. *Tetrahedron Lett.* **1989**, *30*, 1115.

45. Martin, O.R.; Hendricks, C.A.V.; Deshpande, P.P.; Cutler, A.B.; Kane, S.A.; Rao, S.P. *Carbohydr. Res.* **1990**, *196*, 41.

46. Martin, O.R.; Rao, S.P.; Hendricks, C.A.V.; Mahnken, R.E. *Carbohydr. Res.* **1990**, *202*, 49.

47. Martin, O.R.; Rao, S.P.; Kurz, K.G.; El-Shenawy, H.A. *J. Am. Chem. Soc.* **1988**, *110*, 8698.

Chapter 3

C-1 NUCLEOPHILES IN C-GLYCOSIDE PREPARATION

I. INTRODUCTION

Several workers have reversed the natural character of the anomeric carbon atom from electrophilic to nucleophilic and trapped these nucleophiles with appropriate carbon electrophiles to produce C-glycosides. Often the presence of a stabilizing group at C-1 is used such as a nitro, ester or sulfur based functional group. The remaining anomeric hydrogen is now sufficiently acidic to be deprotonated by a strong base and reaction with an electrophile gives a substituted C-glycoside. Some workers have developed methodology to remove the facilitating group and in some cases an existing functional group may be metallated, much like metal-halogen exchange, to furnish a reactive anomeric anion. Until recently, the use of C-1 anions was limited to 2-deoxy sugars since elimination to the glycal was a competitive process. Vinyl anions of glycals have also seen utility in the preparation of C-glycoside compounds. This chapter will first address nucleophilic approaches to C-glycosides through the use of stabilized anomeric anions, 2-deoxy anomeric anions, and then discuss C-1 vinyl anions both stabilized and non-stabilized.

II.   C-1 STABILIZED ANIONS

A.   ANOMERIC NITROSUGARS

Much of the work directed at C-glycoside preparation via C-1 nitrosugars can be attributed to the work of Vasella. The subject of nitroaldol reactions of nitrosugars has recently been outlined.[1] Anomeric C-1 nitrosugars undergo Henry reaction rather easily and the reaction tolerates a heteroatom substituent at C-2. The ribo derivative 1 undergoes a fairly stereoselective condensation to give a 74% yield of 2.[2]

This methodology has been applied to the synthesis of a C-disaccharide (see Chapter 8) by condensation of a C-1 nitro compound with a C-6-aldehydo sugar.[2] The versatility of this tertiary nitro group has been exploited for further carbon-carbon bond forming processes or reduction (nitro group replaced by hydrogen) via, for example, radical methods (see

Chapter 7). If the reaction conditions for the condensation are altered slightly it becomes possible to hydrolytically remove the nitro group to obtain keto-sugars (eq. 2).[2] The reaction is also applicable to furanoses.

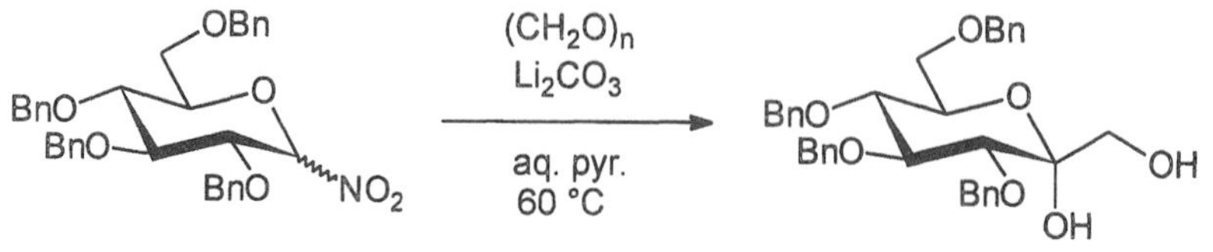

eq. 2

Once the desired condensation has been carried out it is also possible to replace the nitro group by a nitromethyl group which can eventually be turned into an aldehyde function. For example,[3] exposure of both isomers of **7** to basic photolysis conditions, to induce radical anion formation, gave the anomeric (bis)-*C*-glycosides **8** and **9**. Transformation into the aldehydes was affected by ozonolysis of the respective anions to the aldehydo sugars **10** and **12** which were then converted in several steps into the phosphonates **11** and **13**.

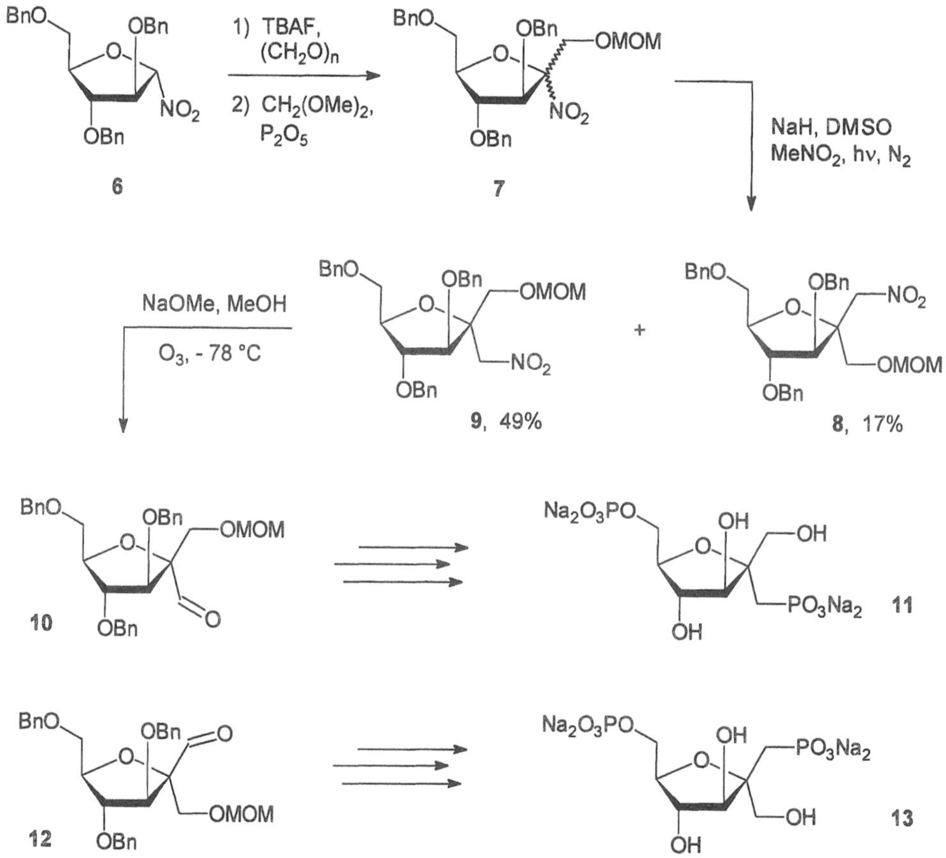

Scheme 1

C-glycosidic lactones of the type **15** are also available by this methodology.[2] Condensation of the weakly basic anion derived from **14** under the conditions shown in eq. 3 gives rise to the lactonic C-glycoside **15**.

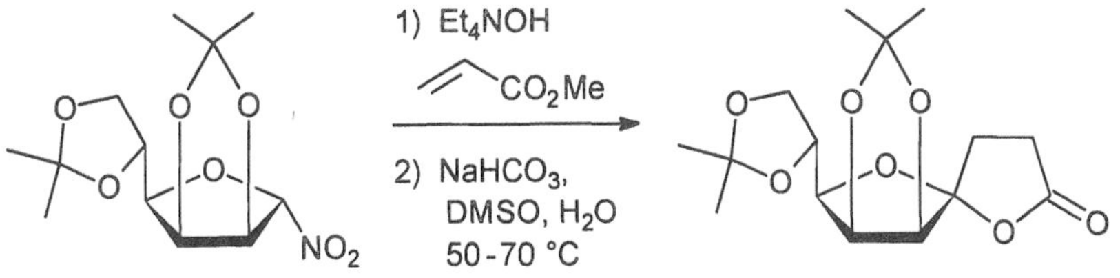

eq. 3

Vasella and Mirza[4] used nitro anion chemistry to synthesize methyl shikimate (**21**) from the 1-deoxy-nitro aldose **1**. Base catalyzed addition of **1** to **22** gave a mixture of anomeric phosphates **16** (87%). Detritylation and one carbon periodate cleavage was followed by silylation, separation, and separate treatment with n-butyllithium followed finally by exposure to methylchloroformate to deliver **19**. Anomeric deprotection and cyclization then gave **20** in 86% yield which was then deprotected to provide the target (**21**).

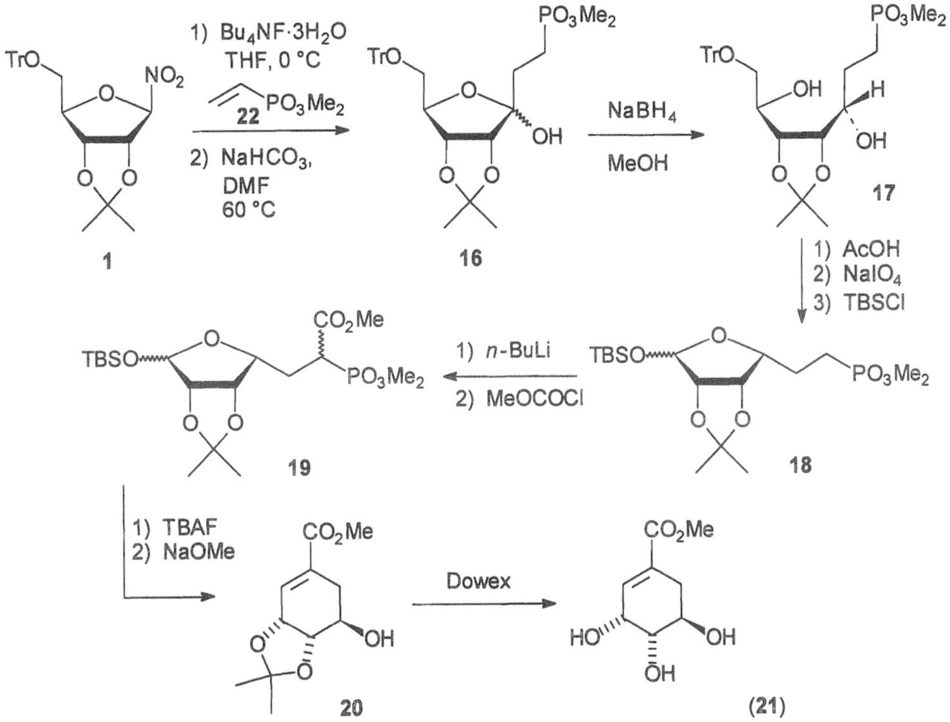

Scheme 2

B.   ANOMERIC ESTERS

It was seen above that with 1-deoxy nitrosugars the acidity of the nitro group was sufficient that heteroatom substituents at the 2 position were tolerated. This is not the case with the corresponding ester compounds, which are themselves C-glycosides. Further work by Vasella[5] involved the preparation of a C-glycoside analog of N-acetylneuraminic acid. Deprotonation of **23** with LCIA and trapping with formaldehyde gave a 3:1 ratio of **24** and

25 in 41% combined yield. Both recovered and isomerized starting material were also isolated from this reaction. Deprotection of **24** and **25** then provided **26** and **27** which proved to be rather weak inhibitors of *Vibrio cholerae* sialidase. Compound **28** was converted to the amine **29** by mesylation, displacement of the mesylate with azide, hydrogenation, and deprotection. While the hydroxy analogs proved to be very weak inhibitors, compound **29** seemed to stimulate the enzyme.

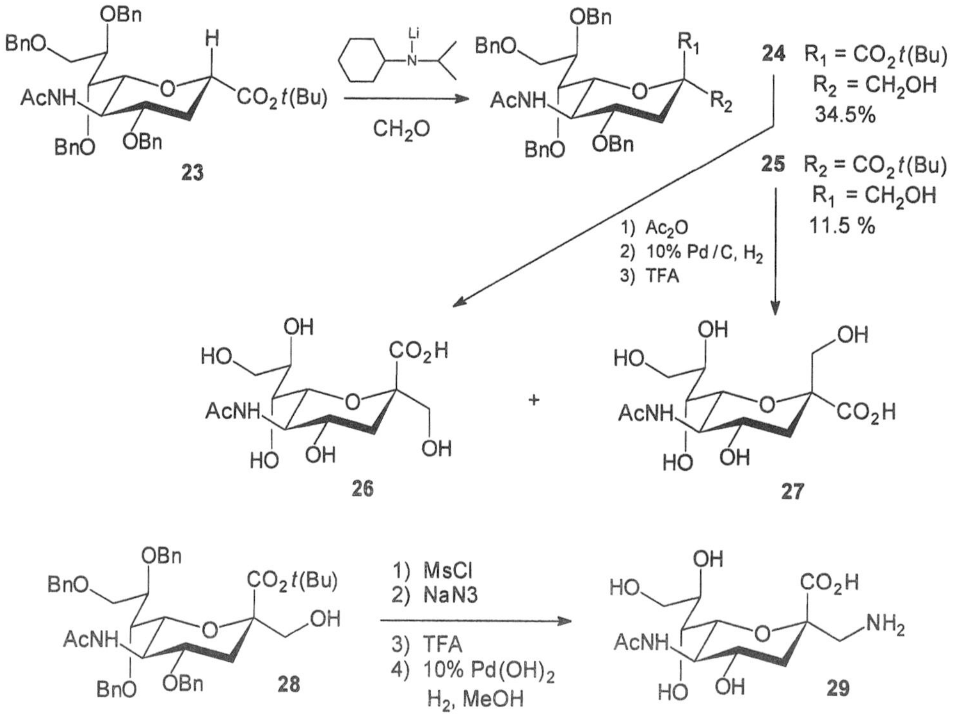

Scheme 3

Claesson and co-workers[6] also examined the behavior of a KDO analog (**30**) in alkylation reactions. The general reaction is shown in eq. 4 and the results of alkylation with various electrophiles in Table 1. The β-isomer (**32**) was found to be the major product in all the cases with the ratios being as high as 95:5.

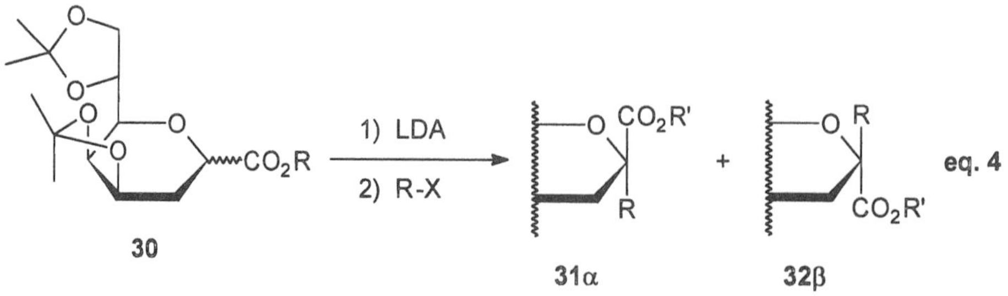

Table 1: Alkylation of 30.

Entry	R	R'	Yield	α:β Ratio
1	-CN	Et	55	90:10
2	CH_2OH	Et	47	90:10
3	$COCH_3$	Et	60	70:30
4	CH_3	Et	50	>95:5
5	CH_2CCH	Et	50	90:10
6	CH_2CO_2-t-Bu	Et	30	>95:10
7	CH_2Ph	Me	67	95:5
8	$CH_2CH_2CO_2Me$	Et	27	75:25

Crich and Lim's[7] approach began with either the sulfone ester 33 or the anomeric sulfone 34. Both compounds could be transformed to the desired anomeric enolate 35. Compound 33 was desulfonylated directly with lithium naphthalenide (LN) to give 35, while 34 had to first be carbonylated and then desulfonylated. Enolate 35 then underwent alkylation with methyl iodide from the axial direction to deliver the (bis)-C-glycoside 36. Other electrophiles were examined and proved to work quite well. The thrust of this methodology was not only the alkylation step, but rather radical decarboxylation of the corresponding thiohydroxamic acid to give β-C-glycosides exclusively (see Chapter 7).

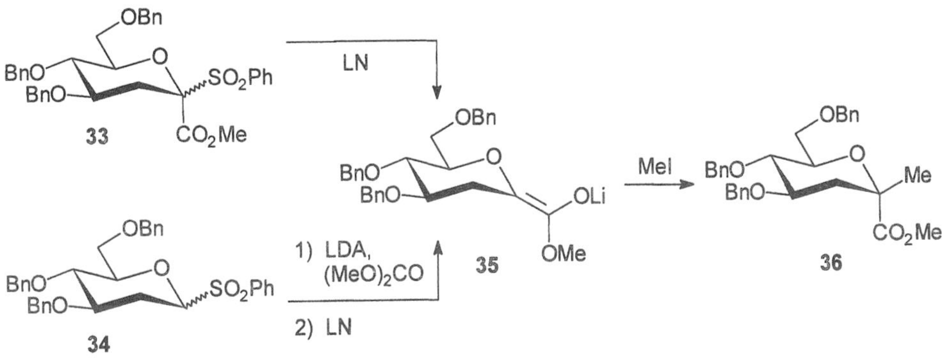

Scheme 4

C. STABILIZATION BY PHOSPHOROUS

Ley and Lygo have developed methodology to construct spiroketals through the use of α-oxygen anions stabilized by a diphenylphosphine oxide function.[8] Enol ether 37 was converted to 38 by exposure to triphenyl phosphine in the presence of hydrogen bromide gas to give an intermediate phosphonium salt that was transformed to 39 by the action of sodium hydroxide. Deprotonation of 39 with LDA was followed by exposure to aldehyde 40 to furnish a mixture of compounds 41 that cyclized to 42 (after concomitant removal of the THP group) when exposed to a trace of CSA in methanol (Scheme 5).

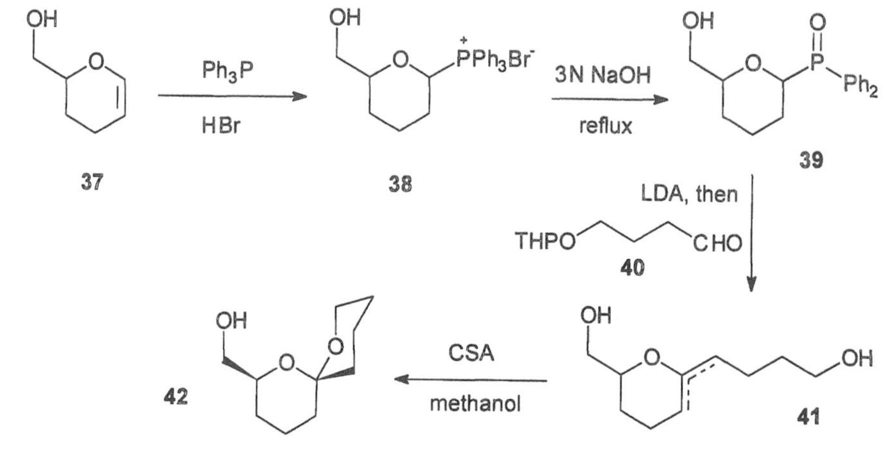

Scheme 5

Falck, Mioskowski, and co-workers[9] used a conceptually similar approach in their preparation of a 2-deoxy C-glycoside. Glycal **43** was converted to the phosphonium bromide **44** by exposure to triphenylphosphine hydrobromide in dichloromethane. Ylide formation and exposure to n-octanal then gave the exocyclic enol ether **45**.

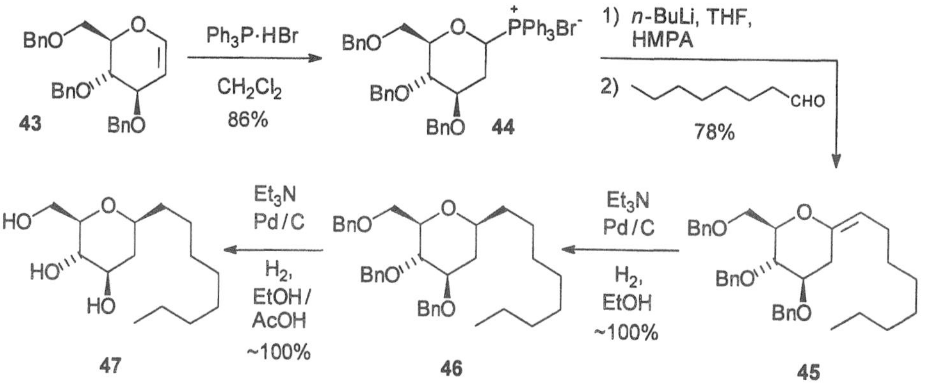

Scheme 6
Reproduced from ref. 9, Copyright 1984, pg. 5903-5904 with kind permission
from Elsevier Science Ltd, The Boulevard, Langford Lane, Kidlington, OX5 1GB, UK.

The double was stereoselectively hydrogenated over Pd/C in the presence of 1% triethylamine to **46**. Removal of the benzyl groups then provided the eight carbon C-glycoside **47** that shows interesting detergent properties. Compounds of this nature may prove effective for the activation and/or solubilization of membrane proteins.

D.   USE OF SULFUR

Fraser-Reid and collaborators[10] alkylated the phenylthio hex-2-enopyranoside **48** by deprotonation with n-butyllithium and exposure to the cyclic sulfate **49** to give an intermediate adduct, presumably **50**, which was not isolated but transformed into **51** by hydrolysis at pH >3.5. Compound **51** arises from allylic rearrangement of the phenylthio group. Treatment of **51** with TBAF then gave the spiro compound **52** as a mixture of isomers.

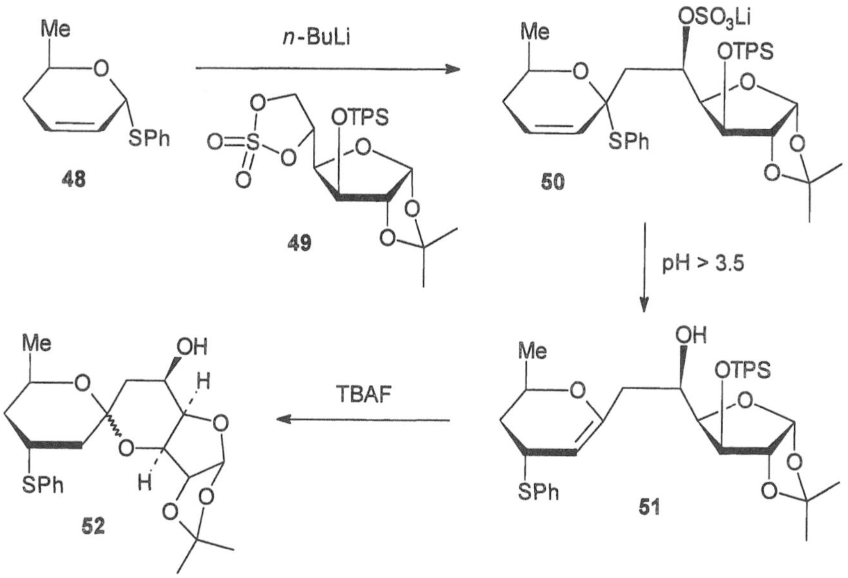

Scheme 7

Reproduced with permission from ref. 10, Copyright 1991, pg. 1207 courtesy of the Royal Society of Chemistry, Thomas Graham House, Science Park, Milton Road, Cambridge, CB4 4WF, UK.

Use of similar chemistry with the sulfate **53** led to adduct **54** which cyclized to the spiroketals **56** (65%) by the action of NIS.

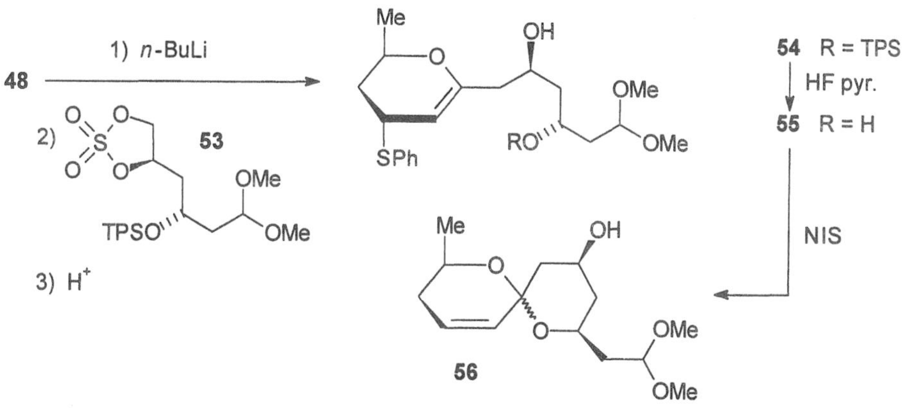

Scheme 8

Work carried out by Valverde *et al.*[11] has demonstrated that alkylation of **57** with *O*-4 unprotected leads to the alkylated product **58** exclusively. The reaction is thought to proceed via a chelated boat transition state with an α-lithio intermediate.

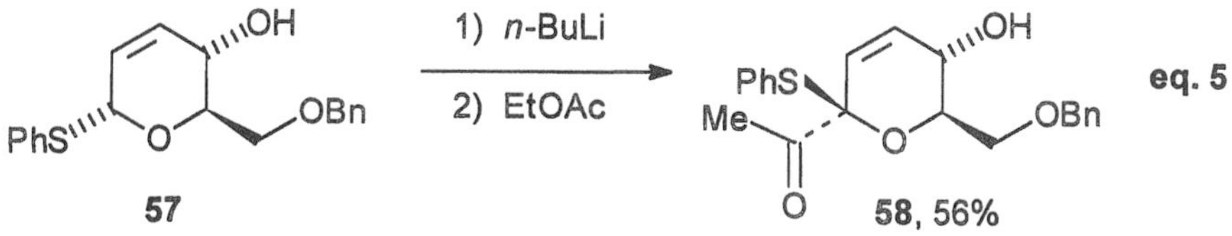

Ley[12] has found that sulfones of the type **60** undergo deprotonation at the 2-position and condense with electrophiles to give adducts possessing an internal olefin. The crystalline sulfone **60** is available by reaction of benzenesulfinic acid with **59**. Deprotonation of **60** with *n*-butyllithium and condensation of the formed anion with aldehydes gives the enol ethers **62** or **63**. When the anion is condensed with methylchloroformate sulfone **61** is isolated (Scheme 9). Sulfone **60** serves as an equivalent to 2-lithiodihydropyran.

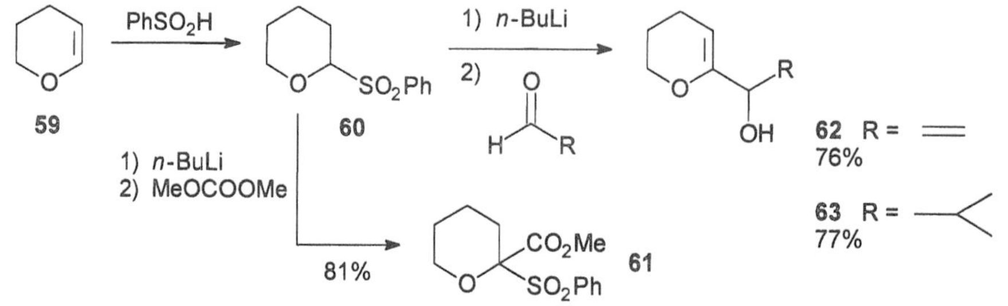

Scheme 9

Work by Ley[13] directed at milbemycin$_{\alpha 1}$ involved an intermediate *C*-glycosidic compound that was made in an anionic manner. Sulfone **64** was deprotonated with *n*-butyllithium at -78°C and the formed anion allowed to condense with epoxide **65** to presumably give intermediate **66** which eliminated benzenesulfinic acid at room temperature to give **67**. Treatment of enol ether **67** with CSA then furnished spiroketal **68**.

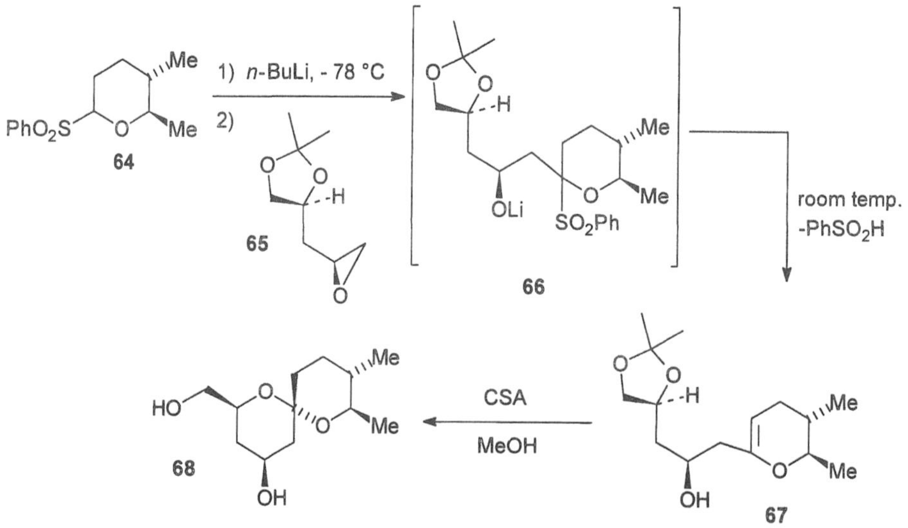

Scheme 10

Reproduced from ref. 13, Copyright 1986, pg. 5277 with kind permission from Elsevier Science Ltd, The Boulevard, Langford Lane, Kidlington, OX5 1GB, UK.

III. ANOMERIC ANIONS

A. REDUCTIVE METALLATION

1. Sulfides

Cohen and Lin[14] have utilized α-phenylsulfides cyclic ethers as precursors to the corresponding lithio derivatives. The sulfides **69** were reductively lithiated with lithium 1-

(dimethylamino)naphthalenide (LDMAN) to yield the α-lithio species **70** which then added in a 1,2-fashion to methacrolein to give **71**, which was not isolated, but trapped with carbon disulfide to give, after methylation, **72**. Compound **72** arises from a [3,3] sigmatropic rearrangement of the intermediate adduct. The synthesis was completed by reductive desulfurization to give racemic *trans*-rosoxide **73** which is a substance used in the perfume industry.

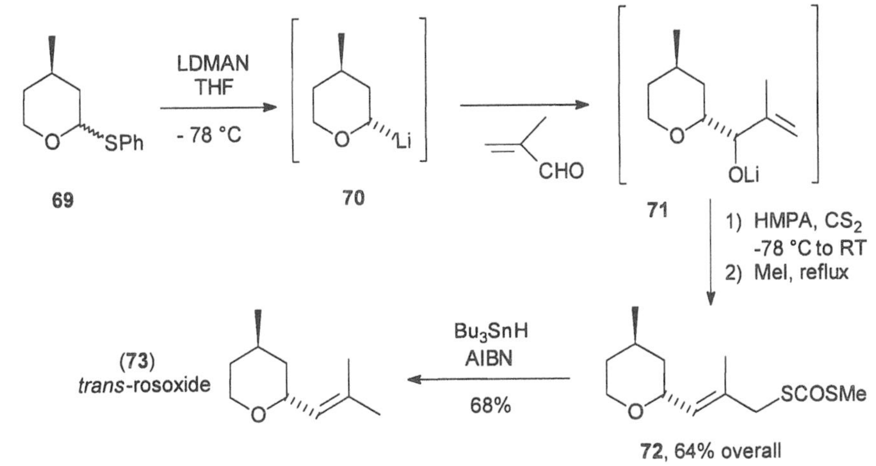

Scheme 11

Reprinted with permission from ref. 14, Copyright 1984 American Chemical Society.

Rychnovsky and Mickus[15] have extended this methodology to the stereoselective preparation of *C*-glycoside compounds by taking advantage of the fact that the α-lithio intermediate can be converted to the β-lithio intermediate simply by warming. Exposure of the anomeric sulfide **74** to reductive lithiation conditions (lithium di-*tert*-butylbiphenylide, LiDBB) at -78°C gave anion **75** which could be trapped with acetone to furnish the α-*C*-glycoside **76**. If the α-lithio species was allowed to warm to -20°C for 45 minutes to **77** and then trapped with acetone the major compound formed was now the β-*C*-glycoside **78**.

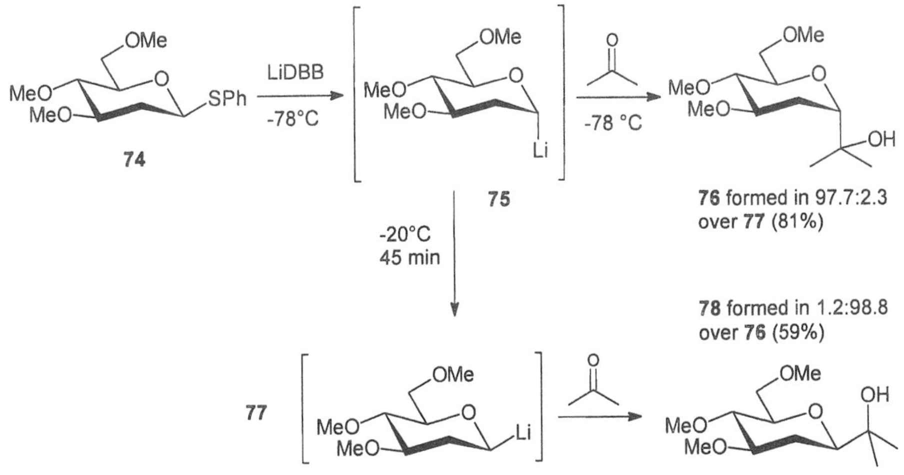

Scheme 12

Reproduced from ref. 15, Copyright 1989, pg. 3013 with kind permission
from Elsevier Science Ltd, The Boulevard, Langford Lane, Kidlington, OX5 1GB, UK.

2.   Sulfones

Sinaÿ and collaborators published three consecutive papers dealing with the reductive lithiation of anomeric sulfones to yield anomeric anions that were then trapped with suitable electrophiles to give a variety of *C*-glycosides.

When sulfones **79** were deprotonated with LDA and quenched with D_2O a mixture of anomeric sulfones **80** (α:β 4) were obtained that when treated with LN (2 equivalents) and quenched gave the β-D-deuterated compound **83** as the exclusive product.   The initial homolytic cleavage of the carbon-sulfur bond gives an intermediate radical that adopts an axial configuration and the second electron transfer gives a stable anion (at least at -78°C in THF) that does not isomerize.   This methodology is successful because of the radical anomeric effect and also the stability of the formed anion under the reaction conditions.[16]

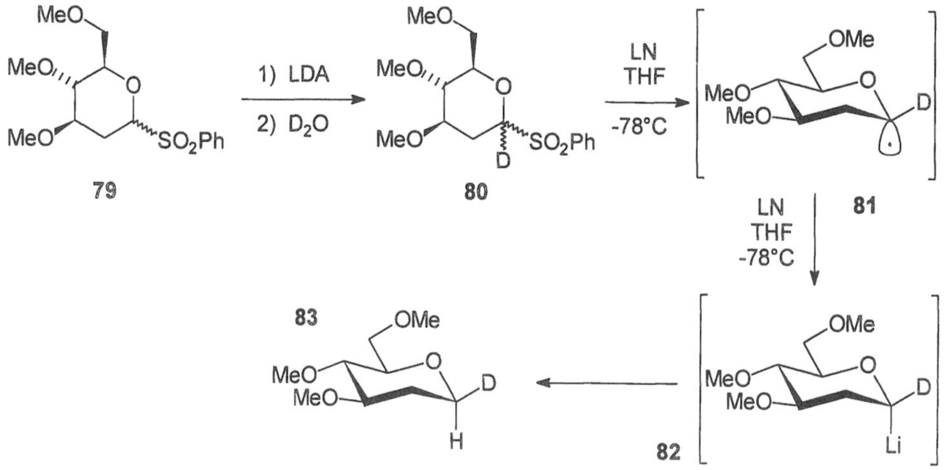

Scheme 13

The ability to produce axial anions selectively lends itself to the preparation of α-*C*-glycosides.   Scheme 14 shows some of the examples prepared by this methodology.[17]

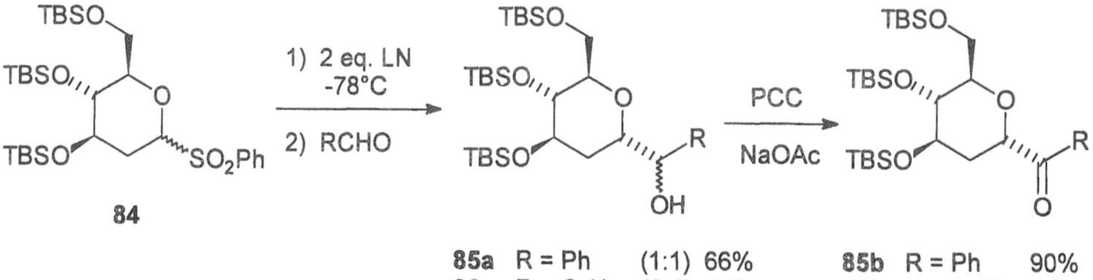

85a R = Ph (1:1) 66%	85b R = Ph 90%
86a R = C5H11 (3:1) 45%	86b R = C5H11 84%
87a R = i-Pr (2:1) 59%	87b R = i-Pr 83%

Scheme 14

If the order of events in the above methodology is altered slightly then access to β-*C*-glycosides becomes possible.   If the sulfones are first deprotonated with LDA, permitted to react with an electrophile, and then reductively desulfonylated an intermediate axial radical is formed and this is then reduced to a stable axial anion that is quenched from the axial direction to provide a β-*C*-glycoside.   The reaction tolerates reactive electrophiles such as methyl iodide and several aldehydes.   In the case of the aldehydes the product alcohols were oxidized to the corresponding ketones. Scheme 15 shows some of the C-glycosides made by this methodology.[18]

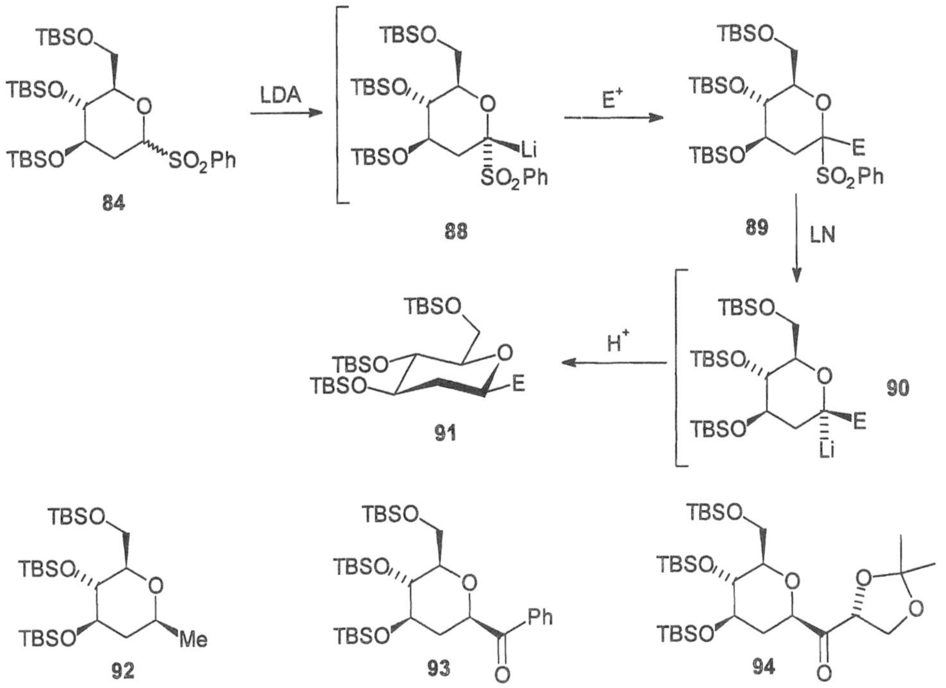

Scheme 15

If the electrophile in the above sequence is changed from an aldehyde to an ester then the corresponding α-C-glycoside is produced selectively. Reductive desulfonylation of 95 gives an enolate that, at low temperature, is quenched from the *exo* side to give the kinetically preferred α-product 97.

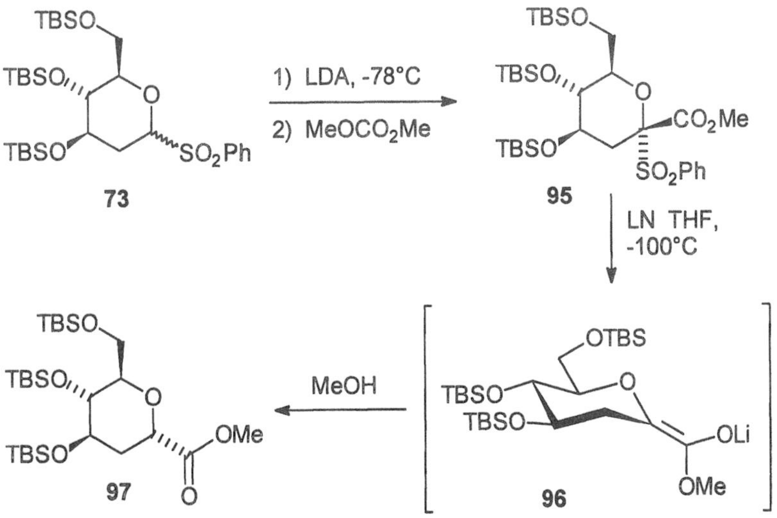

Scheme 16

3.   Anomeric Lithio Derivatives

a.   *From Chlorides*

Sinaÿ and co-workers[19] have utilized derivative **98** to form carbon-carbon bonds at the anomeric center by reversing the natural polarity of this center from electrophilic to nucleophilic. The gluco derivative **98** when treated with lithium naphthalenide gave rise to the glycal **100**. This result seems to suggest that an initial fast reductive lithiation at the anomeric center was followed by elimination to **100**. This is exactly the reason why many–but not all, *vide infra*–utilize 2-deoxy sugars as starting materials since a protected oxygen at *C*-2 undergoes facile elimination. Compound **100** was then hydrochlorinated to give the α-pyranosyl chloride **101**. Reductive lithiation at -78°C presumably gave the axial anion that was allowed to condense with appropriate electrophiles to give the α-*C*-glycosides **103** and **104**.

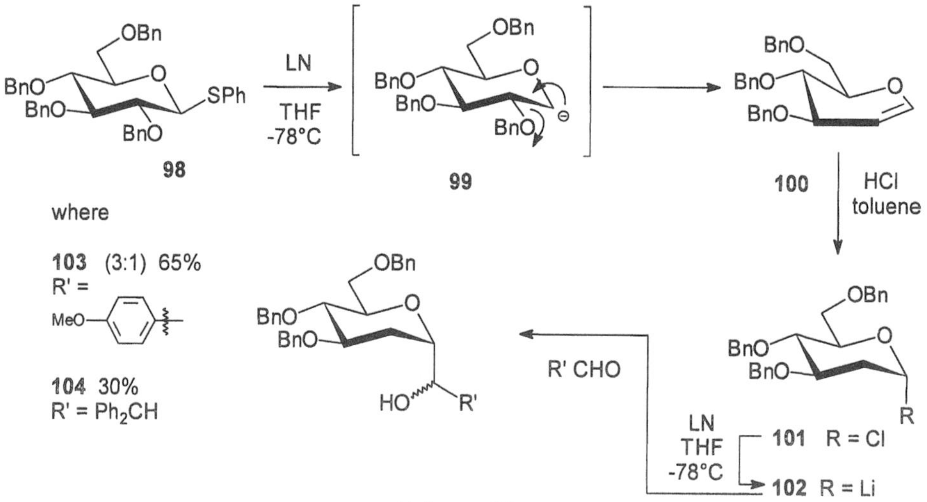

Scheme 17

b.   *Anomeric Stannanes*

In further work directed at stereoselective *C*-glycoside formation, Sinaÿ, Beau, and Lesimple[20] started with the anomeric chloride **101** that was selectively transformed into the corresponding α or β-lithio derivative via the corresponding α- or β-anomeric stannane.

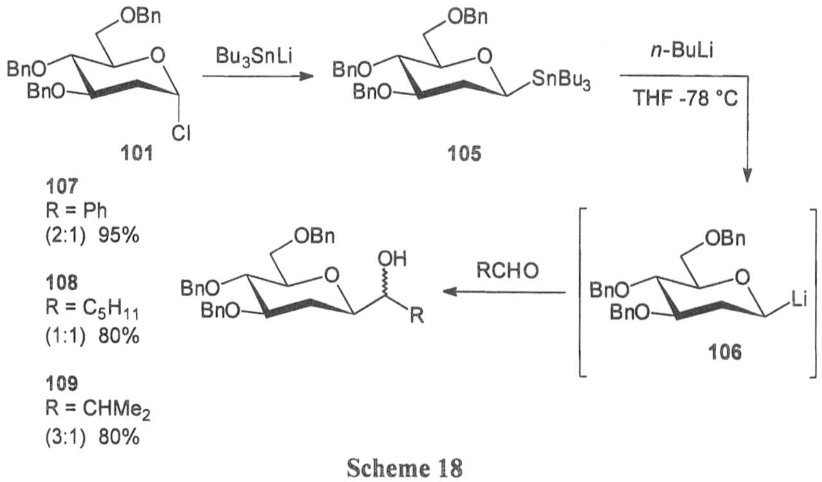

Scheme 18

The α-chloride **101** was treated with tri-*n*-butylstannyl-lithium to give the β-D-tri-*n*-butylstannyl derivative **105** (85% from the corresponding glycal **100**) along with 1% of the α-isomer **110**. When **105** was treated with *n*-butyllithium and the resulting anion trapped with benzaldehyde, hexanal or isobutyraldehyde the β-*C*-glycosides **107**, **108**, and **109** (respectively) were the exclusive products, Scheme 18. No α-*C*-glycosidic products were isolated from the reaction mixture.

The corresponding α-*C*-glycosides were available by trapping of the axial anion derived from lithium naphthalenide reduction of chloride **101** with tributyltin chloride to furnish the α-organostannane **110**. Lithiation and trapping then led to the desired α-*C*-glycosides **118** to **120**, Scheme 19. The diastereoselectivities were much higher for trapping reactions of **111** than **106** and the authors offer no explanation for this observation.

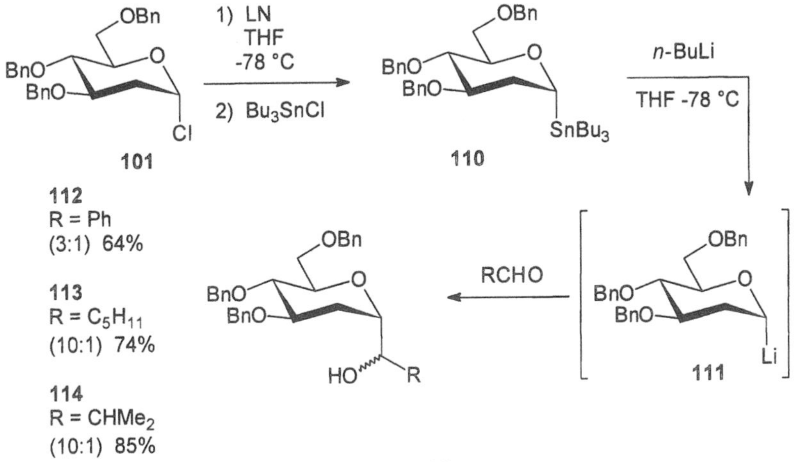

Scheme 19

Until recently, it was thought that the above approach to *C*-glycoside synthesis was limited to 2-deoxy sugars. In seminal work, Kessler and Wittman[21] have shown that similar reactions can be carried out on sugars carrying an oxygen atom at *C*-2, the only requirement being that it must be a free hydroxyl. Chloride **115** was treated with 1.1 equivalents of *n*-butyllithium at -100°C to generate the alcoholate that was then converted to the dianion **117** by quick addition of lithium naphthalenide. The dianion was then trapped with several electrophiles to generate the α-*C*-glycosides shown below in Scheme 20.

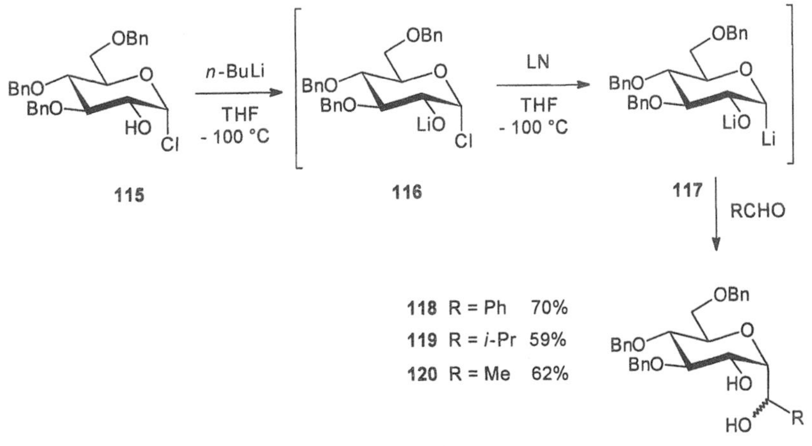

Scheme 20

4.   Anomeric Copper Derivatives

It is not surprising to find that several workers have transmetallated the anomeric lithio species to the corresponding copper compounds and used the special reactivity of these intermediates in *C*-glycoside synthesis.   Hutchinson and Fuchs[22] have used anomeric cuprates in 1,4-additions to enones to produce *C*-glycosides. The known α and β anomeric lithio compounds were transmetallated with copper bromide dimethyl sulfide complex in 1:1 diisopropylsulfide-THF at -78°C to give anomeric cuprates that added in a 1,4-manner to the enones shown in Scheme 21.   No observable diastereoselectivity was found in these additions as the adducts were formed as a 1:1 mixture of isomers.   The yields of *C*-glycosides are generally good, but the reaction does not tolerate heavily substituted acceptors since reaction with mesityl oxide gave only a low yield of 1,4-adduct.

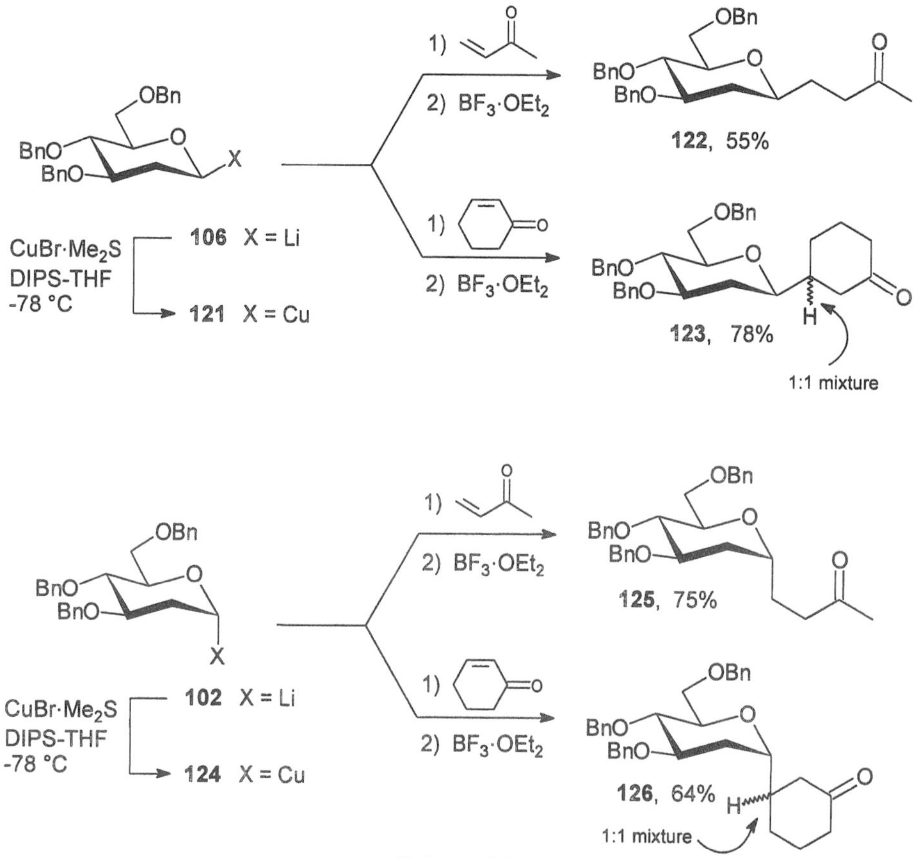

Scheme 21

The anionic approaches described thus far result in the preparation of *C*-glycosides either with oxygenation at the first carbon atom on the carbon aglycone by reaction with a carbonyl compound or by Michael type additions to provide *C*-glycosides with oxygenation on *C*-3. In studies directed at assigning the stereostructure of nystatin A$_1$, Beau and Prandi[23] employed the coupling of an anomeric cuprate with an epoxide to give *C*-glycosides with an oxygen atom on *C*-2 of the carbon aglycone chain The lithio derivative **128** was converted to the mixed cuprate **129** and treatment with racemic epoxide **130** gave a mixture of isomers **131** (epimeric at the hydroxyl bearing carbon) in 36% yield. The α-*C*-glycosidic products were the only ones isolated and this was proven by oxidation of the **131** to a single ketone.

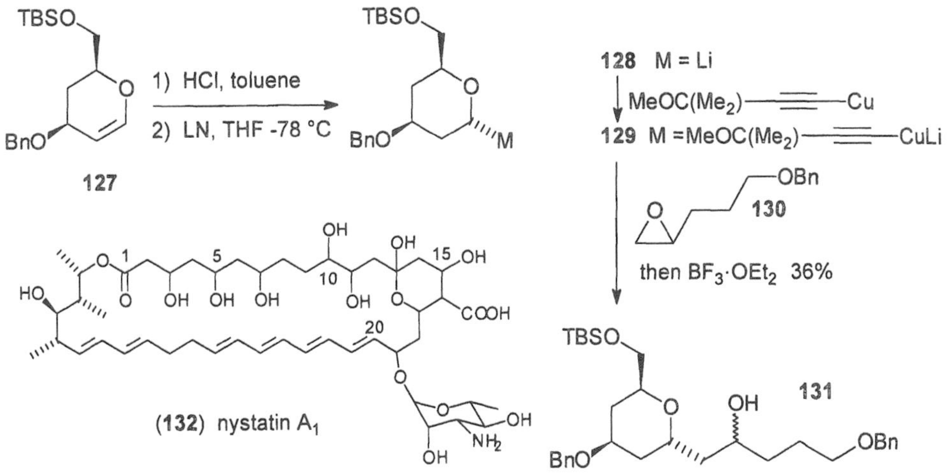

Scheme 22

The yield in the coupling step was improved by changing to a higher order cuprate.[24]

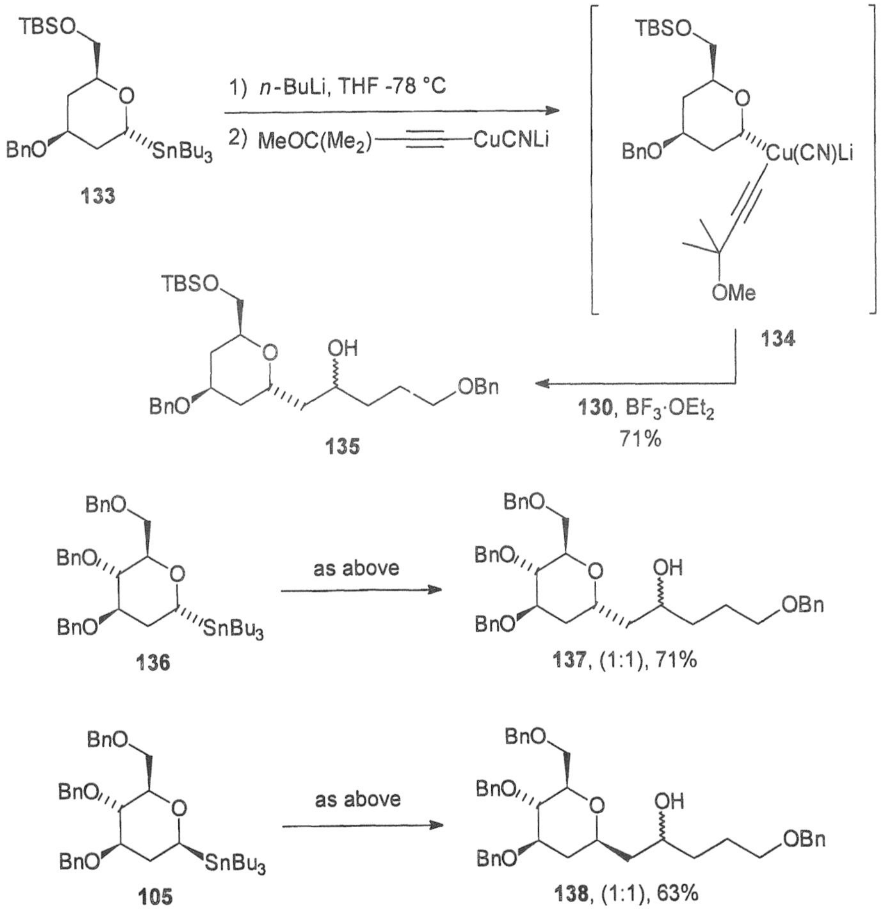

Scheme 23

Accordingly, treatment of **133** with lithium 2-thienylcyanocuprate gave **134** which reacted with epoxide **130** in the presence of boron trifluoride etherate to give a 71% yield of alcohols **135** in 2:3 ratio. Both the α- and 2-deoxy-β-C-glycosides **137** and **138** were readily accessible by application of this technology, Scheme 23.

5.   Anomeric Organosamarium Derivatives

Sinaÿ[25] has also studied the reaction of the transient organosamarium species **139** which is generated by reaction of samarium iodide and chloride **101** in THF usually with some HMPA present. The workers found that addition of a mixture of cyclopentanone and **101** to a THF-HMPA solution of samarium iodide gave a 70% yield of the α-C-glycoside **142**, 17% of the β-C-glycoside **141**, 1% of glycal **144**, and 4% of reduced product **143**. Generally, the reaction did not proceed well with aldehydes, but use of pivaldehyde as the electrophilic component gave 53% yield of **140** as a single undetermined diastereomer. It is interesting to note that reaction with an aldehyde gave only the α-C-glycoside.

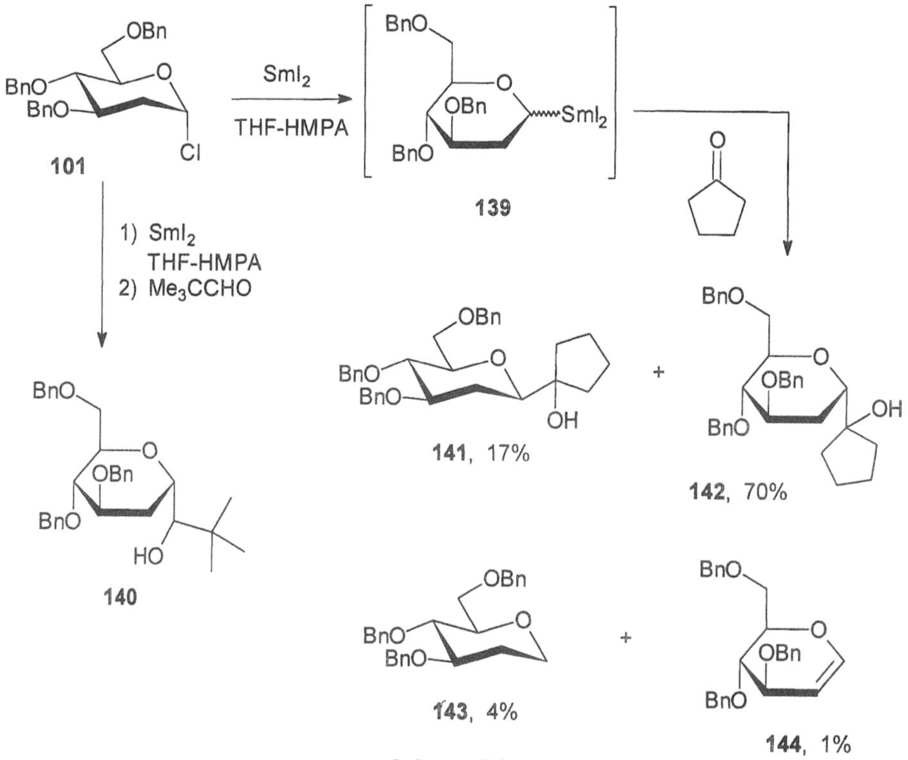

Scheme 24

This is probably due to the fact that the formed α-anomeric radical gives a α-samarium species which reacts quickly with the aldehyde to give the product of axial addition. For ketones the addition is slower and at room temperature the anomeric organosamarium species probably has time to convert to the seemingly more stable β-configuration.

In a crucial experiment, the workers observed a very interesting result. Sulfone **145** could be converted to the glycal by treatment with samarium iodide in THF-HMPA at ambient temperature in 98% yield. When the same reaction was conducted at -40°C the glycal was formed in 56% yield, but the product of reduction **146**, without concomitant elimination, was also formed in 33% yield. This result indicated that this methodology had the potential to be applied to the preparation of C-2 oxygenated C-glycosides.

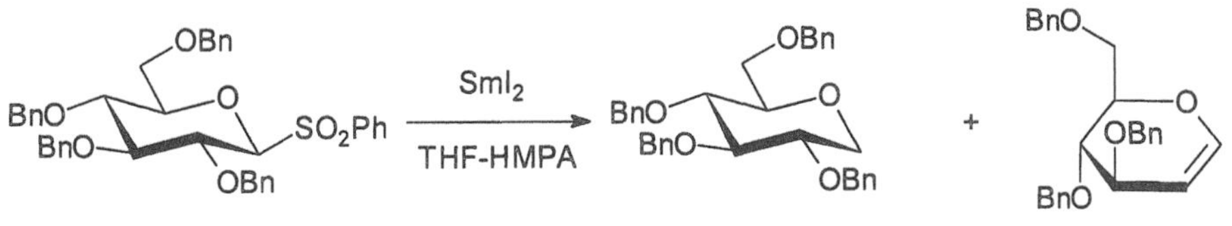

Scheme 25

Barbier type reactions of the sulfone **145** with pivaldehyde gave glycal **147** (39%) and β-C-glycoside **149** (24%) as a single diastereomer. Similar reaction with cyclopentanone gave **147** (57%), reduction product **146** (10%) and the β-C-glycoside **149** in 28% yield. These results are quite interesting and imply that the nature of the metal at C-1 may affect the reaction pathway either by favoring addition/reduction or by favoring elimination. Another interesting fact is that the C-glycosides formed were only those that had the β-configuration. This may be due to the fact that a Grignard type mechanism is operational through an equatorial organosamarium species, but there is still much debate over the exact nature of the mechanism in these types of additions.

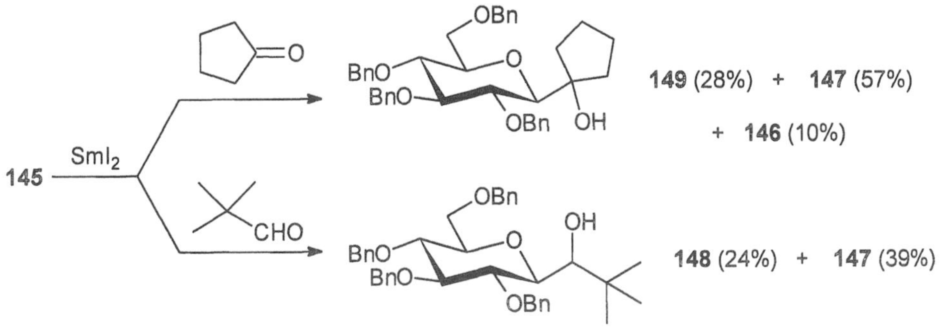

Scheme 26

IV. TRANSITION METAL ANOMERIC COMPLEXES

A. ANOMERIC MANGANESE COMPLEXES

DeShong and co-workers have developed the chemistry of anomeric manganese complexes in the preparation of C-glycosides.[26]

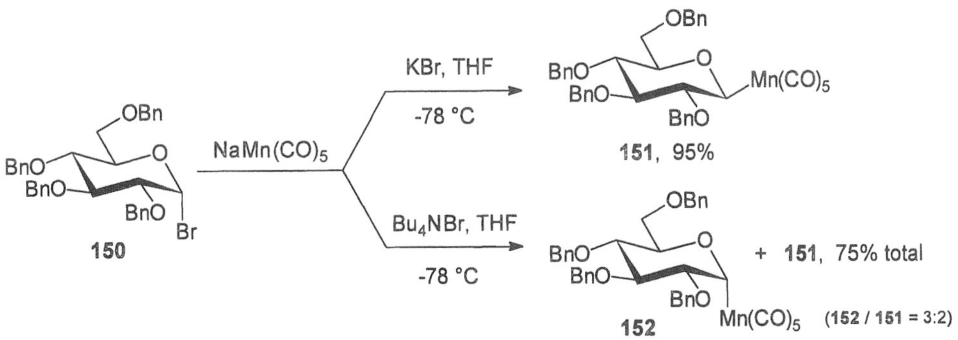

Scheme 27
Reprinted with permission from ref. 26, Copyright 1985, American Chemical Society.

This subject has been reviewed[27] by the workers in a recent monograph and therefore only a brief overview will be presented here. In the gluco series, the β-manganese anomer 151 is available by reaction of sodium pentacarbonyl manganate [NaMn(CO)$_5$] with the α-bromide 150. The α-anomer 152 is available by reaction of 150 with NaMn(CO)$_5$ in the presence of tetra-n-butylammonium bromide to give a mixture of anomers, Scheme 27. Once formed, these complexes have the ability to undergo insertion reaction that lead to C-glycosides. For example, carbon monoxide insertion of 151 occurs to give intermediate 153 which can then be converted to either ester 154 or amide 155 by Reppe reaction, Scheme 28. Compound 154 and 155 are useful precursors for C-nucleoside synthesis.

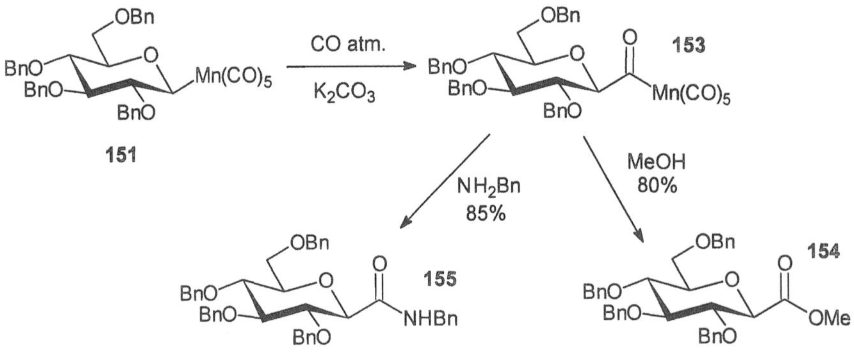

Scheme 28
Reprinted with permission from ref. 26, Copyright 1985, American Chemical Society.

The versatility of the complex 151 is illustrated by the insertions depicted in Scheme 29.

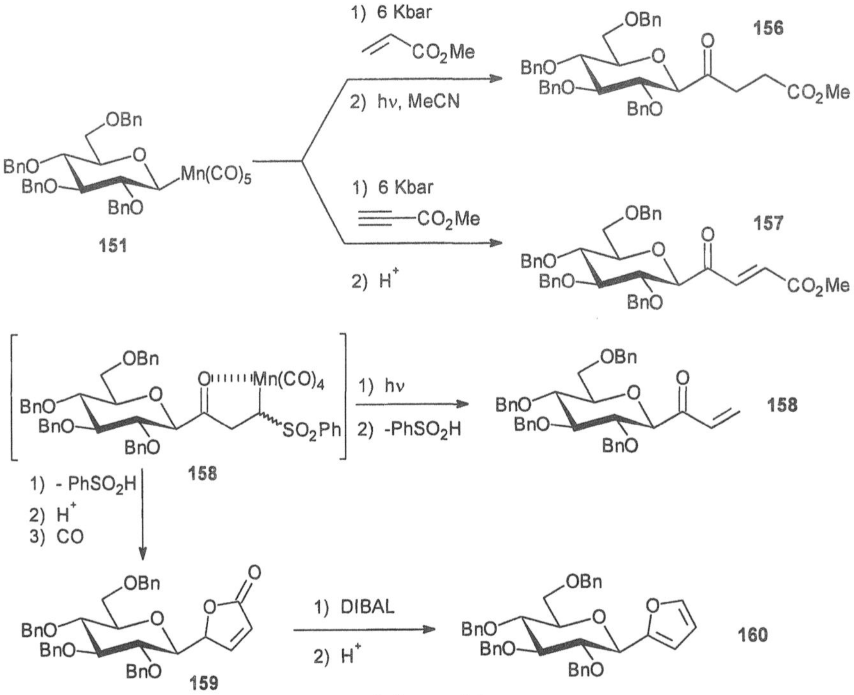

Scheme 29
Reprinted with permission from ref. 26, Copyright 1985, American Chemical Society.

The insertions retain the starting stereochemistry of the glycosylmanganese pentacarbonyl complexes and generally give good yields of C-glycosides. The reaction has also been applied to the furanosylmanganese pentacarbonyl complex **162**, (eq. 6).

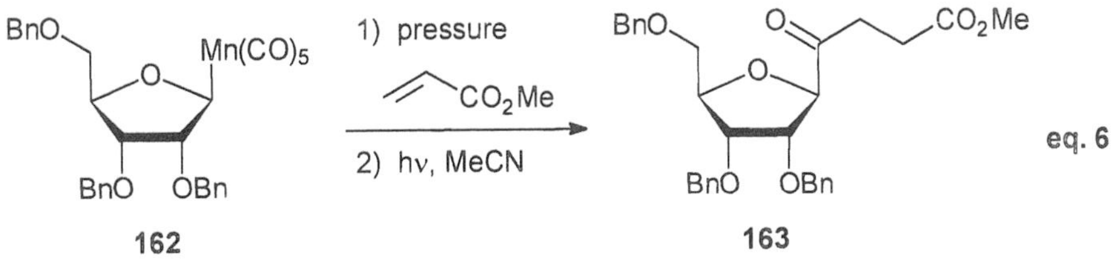

eq. 6

162   **163**

B.   ANOMERIC COBALT COMPLEXES

Cobalt chemistry has also been applied to C-glycoside preparation.[28]   Reaction of 1,2,3,4,6-penta-O-acetyl-β-D-glucose (**164**) with $Co(CO)_8$ in the presence of triethylsilane and carbon monoxide gave the β-C-glycoside **165**.

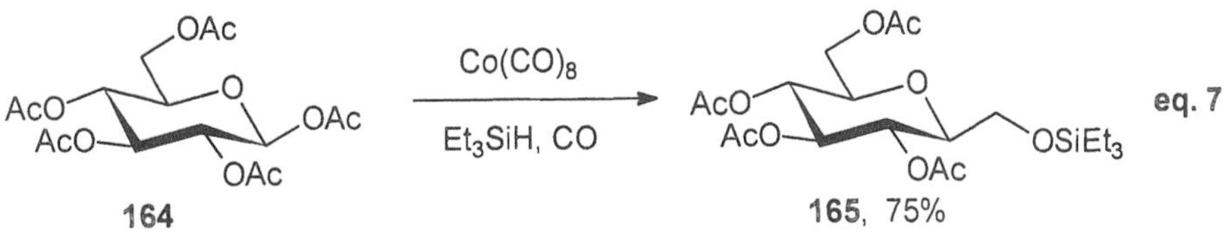

eq. 7

164   **165**, 75%

V. C-1 VINYL ANIONS

A.   NON-STABILIZED C-1 ANIONS

1.   From Glycals

Nicolaou, Hwang, and Duggan[29] utilized a C-vinyl anion derived from **166** in synthetic work directed at Brevetoxin B.   Glycal **166** was deprotonated with t-butyllithium, transmetallated, and then reacted with allyl bromide to give **169**, after protecting group manipulations.

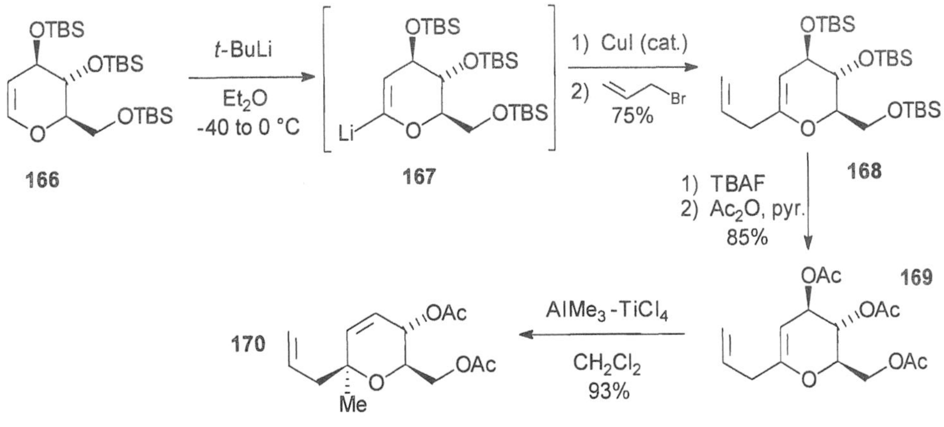

166   **167**   **168**

Scheme 30

A second carbon group was introduced via a Lewis catalyzed approach, as shown above in Scheme 30, to furnish the 1,1-dialkyl-*C*-glycoside **170**.

Hanessian and co-workers[30] have also deprotonated a glycal and used the concept of polarity inversion at the anomeric center to synthesize a *C*-glycoside. They found that optimum conditions for *C*-1 deprotonation entailed the use of *n*-butyllithium and potassium *t*-butoxide to give a vinyl anion that was then trapped with tributyltin chloride to give vinyl stannane **171**. After protecting group modification, treatment of the product with *n*-butyllithium cleanly gave a vinyl anion that was trapped with methyl iodide to provide **172**, 88%. Hydroboration followed by oxidation then furnished the β-*C*-glycoside **173**. The formed anion also reacts with aldehydes and silicon based electrophiles (**175** and **174**), Scheme 31. The merits of the sequence include high yielding steps and the versatility to synthesize various *C*-glycosides by simply changing the identity of the electrophile in the alkylation step.

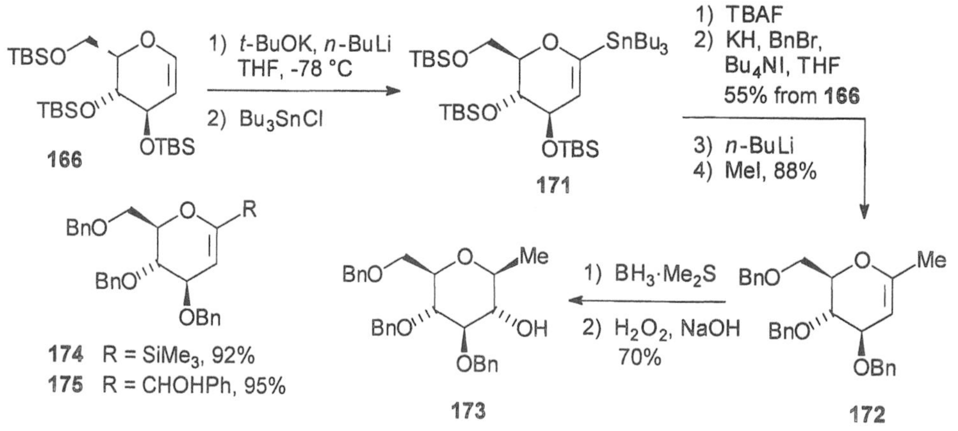

Scheme 31

Beau[31] has also observed that it is more convenient to use the stannane **178** as a *C*-1 vinyl anion precursor. Lithiation of silyl protected glycals proceeds well (*vide supra*), but when benzyl protecting groups are used, low yield of alkylated products are obtained.

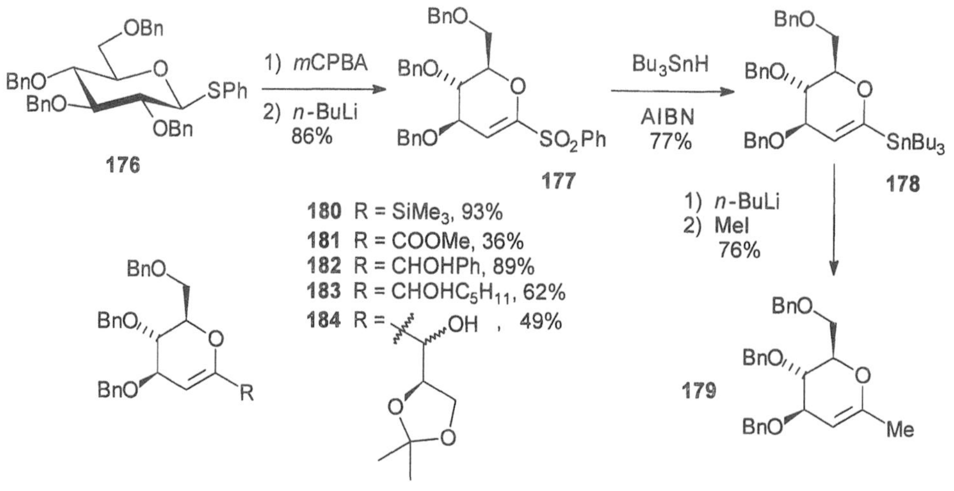

Scheme 32

This is probably due to competing deprotonation of the benzylic protons. The workers therefore developed an efficient synthesis of stannane **178** from sulfide **176**. Oxidation of **176** gave an anomeric sulfone that was eliminated to the vinyl sulfone **177**. Exposure of **177** to stannyl radicals then gave the requisite vinyl stannane **178** accompanied by some recovered **177**. Treatment of **178** with *n*-butyllithium gave a *C*-1 lithio species that was condensed with several electrophiles to give the *C*-glycosides shown in Scheme 32.

Workers at Merck-Frosst[32] used the umpolung approach in their synthesis of a 2-amino-*C*-glycoside. Vinyl stannane **185** underwent tin lithium exchange and the resulting anion condensed with MeI to give **186**, after the TIPS groups were exchanged for TBS groups and glucal **186** then underwent cycloaddition with bis(trichloroethyl)azido-dicarboxylate to give adduct **187** in 80% yield. Exposure of **187** to NaBH$_3$CN/ZnI$_2$ then gave **188** which was subsequently transformed into the acetylated 2-amino-*C*-glycoside **189**. The methodology was versatile in the sense that *C*-glycoside **190** was also made as were 2-amino-*C*-glucoside spiroketals (not shown).

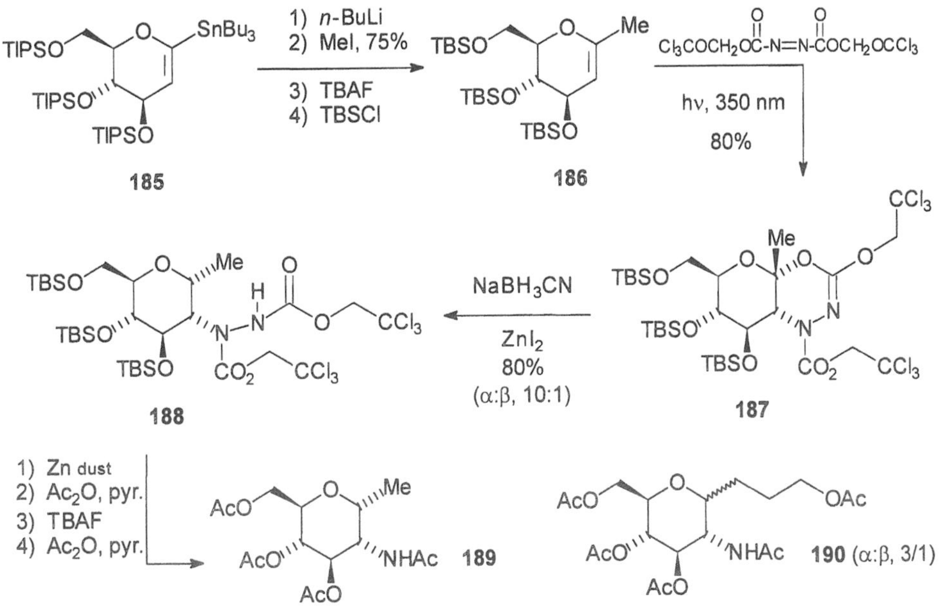

Scheme 33

Parker[33] has devised a very ingenious method for synthesizing *C*-aryl glycosides by the use of *C*-1 vinyl glycal anions. The lithio species **191** was coupled with the quinone derivative **192** to furnish **193** in good yield (eq. 8).

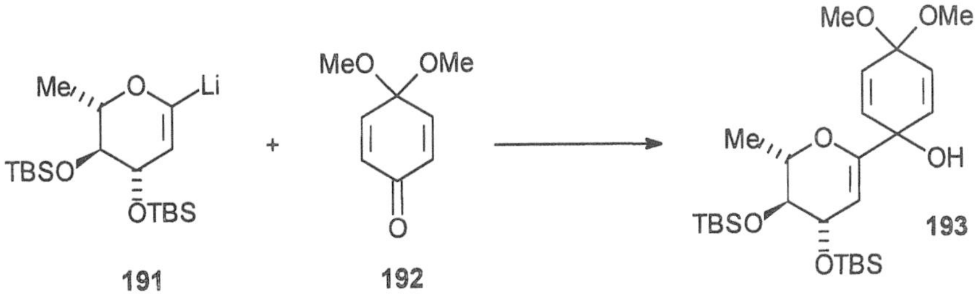

Transformation into the requisite *C*-aryl glycoside was achieved by DIBAL reduction of **193** to give a mixture of **194** and **195**. This mixture was then treated with POCl₃ in pyridine to give **195** in 94% overall yield. Hydroboration and oxidation then secured the target **196**, in which silyl migration had occurred, Scheme 34.

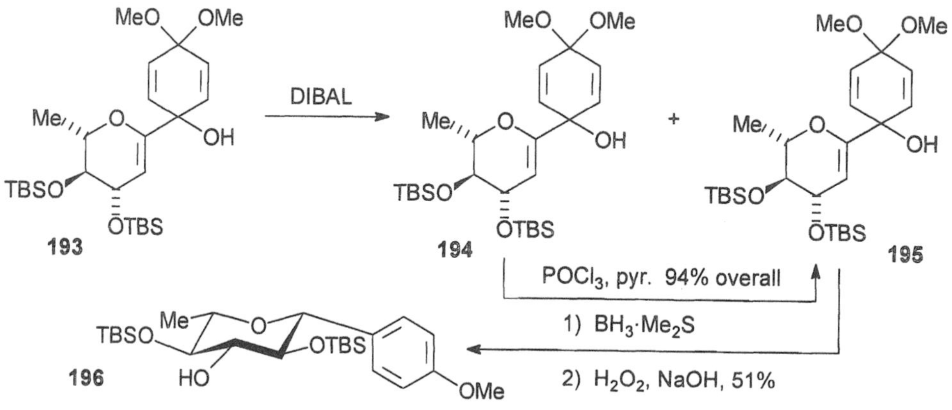

Scheme 34

This methodology has been extended to the preparation of phenolic *C*-aryl glycosides.[34] Adduct **199** (Scheme 35) could be transformed into the desired product **200** by simple treatment with sodium dithionite.

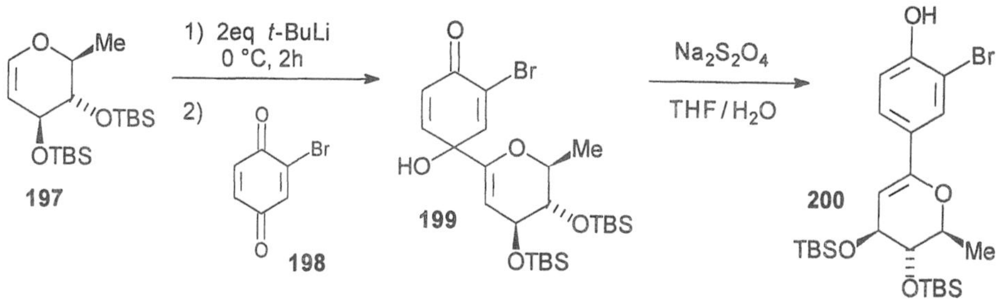

Scheme 35

In the naphthol series it was necessary to reduce adduct with aluminum amalgam to carry out the desired transformation (**201→202**) (eq. 9).

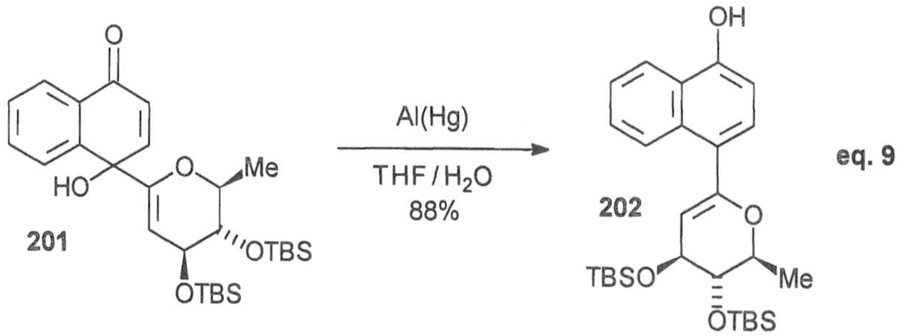

B. *C*-2 STABILIZED *C*-1 ANIONS

Schmidt and co-workers[35] have used the stabilizing effect of *C*-2 sulfur based groups to direct lithiation at the anomeric carbon. The required precursor is available from D-glucal by treatment with phenylsulfuryl chloride followed by elimination. Deprotonation with LDA gave a *C*-1 anion that was then trapped with several aldehydes as shown below in Scheme 36. The methodology is flexible in the sense that the phenylthio group can be removed (208→209) and hydroboration/oxidation of the remaining double bond then provides β-*C*-glycoside **210** with the gluco configuration.

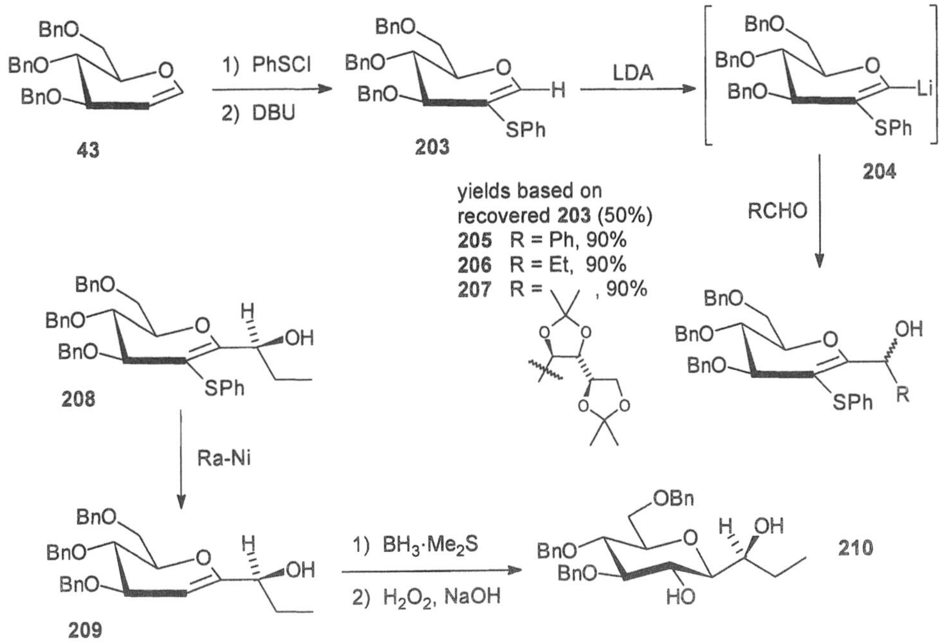

yields based on
recovered **203** (50%)
205 R = Ph, 90%
206 R = Et, 90%
207 R = , 90%

Scheme 36

Similar methodology,[36] using a sulfoxide as the *C*-2 directing group, was applied to the synthesis of a *C*-disaccharide, Scheme 37 (See Chapter 8 for additional discussion).

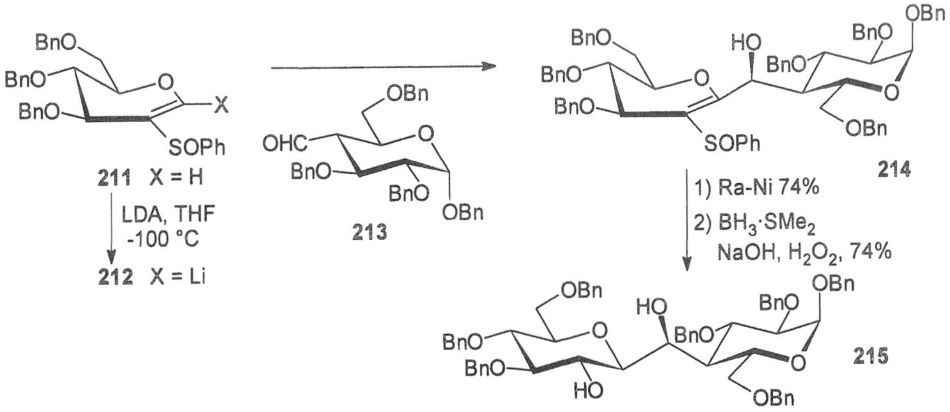

Scheme 37

Schmidt, Preuss, and Maier[37] have applied the same methodology to the synthesis of the basic structure of Ezomycin A (**223**). Reaction of the lithio species with aldehyde **224** gave a 2:1 ratio of isomers. Both were carried on, but the *R*-isomer was eventually used in the preparation of **222**. The free hydroxyl was benzylated and the sulfoxide removed (Raney nickel) to give **218**. Hydroboration and oxidation of the intermediate borane was followed by acetal hydrolysis to provide **219**. Treatment of **219** with pyridine and acetic anhydride was followed by conversion to pentaacetate **220**. Coupling with **221** then afforded the target structure **222**.

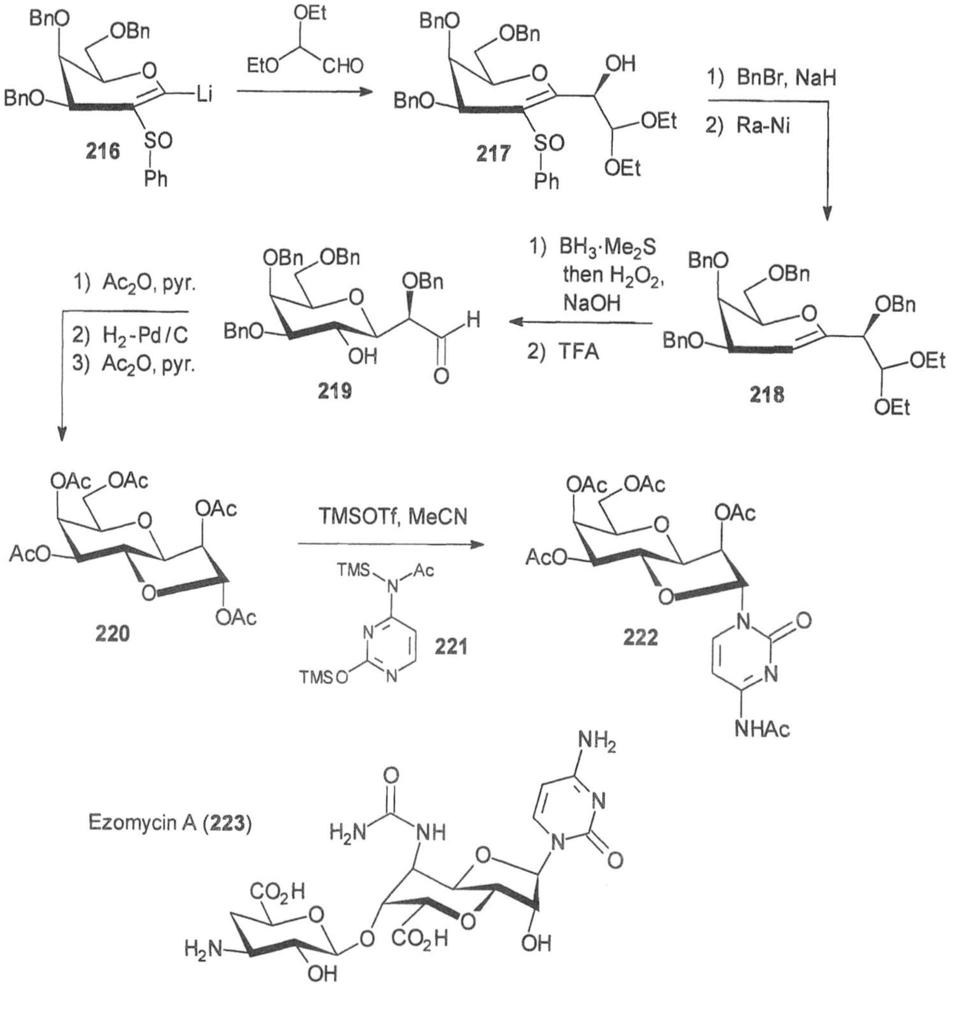

Scheme 38

VI. CONCLUSION

This chapter has summarized the various anomeric nucleophiles that have been prepared. Both the α- and β-2-deoxy-anomeric metallo species can be prepared selectively. The technology has been developed to such a level that it will tolerate oxygenation at *C*-2 without elimination to the glycal. Transition metal anomeric complexes have also begun to emerge as useful precursors for *C*-glycoside preparation. It is certain in the next few years more complexes with new and different reactivity will also be discovered. Glycals also serve as precursors to *C*-glycosides, since after alkylation of the anion, hydroboration selectively provides a *C*-glycoside possessing the β-gluco configuration.

VII. REFERENCES

1. Wade, P.A.; Giuliano, R.M. *The Role of the Nitro Group in Carbohydrate Chemistry, in Nitro Compounds: Recent Advances in Synthesis and Chemistry*, Feuer, H. and Nielsen, A.T. Eds., VCH, New York, 1990, Chapter 2.
2. Aebischer, B.; Bieri, J.; Prewo, R.; Vasella, A. *Helv. Chim. Acta* **1982**, *65*, 2251.
3. Meuwly, R.; Vasella, A. *Helv. Chim. Acta* **1986**, *69*, 751.
4. Mirza, S.; Vasella, A. *Helv. Chim. Acta* **1984**, *67*, 1562.
5. Wallimann, K.; Vasella, A. *Helv. Chim. Acta* **1991**, *74*, 1520.
6. Luthman, K.; Orbe, M.; Waglund, T.; Claesson, A. *J. Org. Chem.* **1987**, *52*, 3777.
7. Crich, D.; Lim, L.B.L. *Tetrahedron Lett.* **1990**, *31*, 1897. This approach is based on earlier work published by Sinaÿ, see later.
8. Ley, S.V.; Lygo, B. *Tetrahedron Lett.* **1984**, *25*, 113.
9. Ousset, J.B.; Mioskowski, C.; Yang, Y.-L.; Falck, J.R. *Tetrahedron Lett.* **1984**, *25*, 5903.
10. Gomez, A.M.; Valverde, S.; Fraser-Reid, B. *J. Chem. Soc., Chem. Commun.* **1991**, 1207.
11. Valverde, S.; García-Ochoa, S.; Martín-Lomas, M. *J. Chem. Soc., Chem. Commun.* **1987**, 383.
12. Ley, S.V.; Lygo, B.; Wonnacott, A. *Tetrahedron Lett.* **1985**, *26*, 535.
13. Greck, C.; Grice, P.; Ley, S.V.; Wonnacott, A. *Tetrahedron Lett.* **1986**, *27*, 5277.
14. Cohen T.; Lin, M.-T. *J. Am. Chem. Soc.* **1984**, *106*, 1130.
15. Rychnovsky, S.D.; Mickus, D.E. *Tetrahedron Lett.* **1989**, *30*, 3011.
16. Beau, J.-M.; Sinaÿ, P. *Tetrahedron Lett.* **1985**, *26*, 6185.
17. Beau, J.-M.; Sinaÿ, P. *Tetrahedron Lett.* **1985**, *26*, 6189.
18. Beau, J.-M.; Sinaÿ, P. *Tetrahedron Lett.* **1985**, *26*, 6193.
19. Lancelin, J.-M.; Morin-Allory, L.; Sinaÿ, P. *J. Chem. Soc., Chem. Commun.* **1984**, 355.
20. Lesimple, P.; Beau, J.-M.; Sinaÿ, P. *J. Chem. Soc., Chem. Commun.* **1985**, 894.
21. Wittman, V.; Kessler, H. *Angew. Chem. Intl. Ed. Engl.* **1993**, *32*, 1091.
22. Hutchinson, D.K.; Fuchs, P.L. *J. Am. Chem. Soc.* **1987**, *109*, 4930.
23. Prandi, J.; Beau, J.-M. *Tetrahedron Lett.* **1989**, *30*, 4517.
24. Prandi, J.; Audin, C.; Beau, J.-M. *Tetrahedron Lett.* **1991**, *32*, 769.
25. de Pouilly, P; Chénedé, A. Mallet, J.-M.; Sinaÿ, P *Bull. Soc. Chim. Fr.* **1993**, *130*, 256.
26. DeShong, P.; Slough, G.A.; Elango, V.; Traivor, G.L. *J. Am. Chem. Soc.* **1985**, *107*, 7788.
27. DeShong, P.; Slough, G.A.; Sidler, D.R.; Elango, V.; Rybczynski, P.J.; Smith, L.J.; Lessen, T.A.; Le, T.X.; Anderson, G.B. *Glycosylmanganese Pentacarbonyl Complexes. An Organomanganese Approach to the Synthesis of C-glycosyl Derivatives, in Cycloaddition Reactions in Carbohydrate Chemistry*, ACS Press, 1992, 97.
28. Chatani, N.; Ikeda, T.; Sano, T.; Sonoda, N.; Kurosawa, H.; Kawasaki, Y.; Murai, S. *J. Org. Chem.* **1988**, *53*, 3387.

29. Nicolaou, K.C.; Hwang, C.-K.; Duggan, M.E. *J. Chem. Soc., Chem. Commun.* **1986**, 925.
30. Hanessian, S.; Martin, M.;Desai, R.C. *J. Chem. Soc., Chem. Commun.* **1986**, 926.
31. LeSimple, P.; Beau, J.-M.; Jaurand, G. and Sinaÿ, P. *Tetrahedron Lett.* **1986**, *27*, 6201.
32. Grondin, R.; Leblanc, Y.; Hoogsteen, K. *Tetrahedron Lett.* **1991**, *32*, 5021.
33. Parker, K.A.; Coburn, C. A. *J. Am. Chem. Soc.* **1991**, *113*, 8516.
34. Parker, K.A.; Coburn, C. A.; Johnson, P.D.; Aristoff, P. *J. Org. Chem.* **1992**, *57*, 5547.
35. Schmidt, R.R.; Preuss, R.; Betz, R. *Tetrahedron Lett.* **1987**, *28*, 6591.
36. Schmidt, R.R.; Preuss, R. *Tetrahedron Lett.* **1989**, *30*, 3409.
37. Maier, S.; Preuss, R.; Schmidt, R.R. *Liebigs Ann. Chem.* **1990**, 483.

Chapter 4

WITTIG APPROACHES TO *C*-GLYCOSIDE FORMATION

I. INTRODUCTION

The idea of carrying out Wittig type chemistry on lactols is fairly mature and this field has seen considerable application in the literature. Early work was carried out by Zhdanov.[1] Many of the approaches to *C*-glycoside formation by this strategy involve the use of ylides to give open chain sugars that, under judicious choice of conditions, cyclize in a Michael fashion onto the formed double bond system, the requirement being that an electron withdrawing group be located on the end of the olefin. If the olefin is unsubstituted, then other types of cyclizations can be carried out to affect ring closure, such as halocyclization or mercuriocyclization. Both of these methods have been utilized in *C*-glycoside synthesis. If sugar lactones are chosen as the starting material then olefin formation furnishes an exomethylenic *C*-glycoside that is poised for further chemistry. This chapter will cover the aforementioned topics as well as some of the applications of the formed *C*-glycosides.

II. USE OF STABILIZED PHOSPHORANES

The early work of Moffatt and Hanessian has served to open the field of Wittig olefinations of reducing sugars. The examples discussed in this section will deal mainly with ester containing ylides that react with lactols to give open chain α,β-unsaturated ester that cyclize to the corresponding *C*-glycosides.

A. *C*-FURANOSIDES

Reaction of the ylide **3** with the ribofuranose **1** gave a 65% yield of β-*C*-glycoside **2**. Similar reaction with the benzoylated derivative **4** gave only a 38% yield of the β-*C*-glycoside **5**.[2]

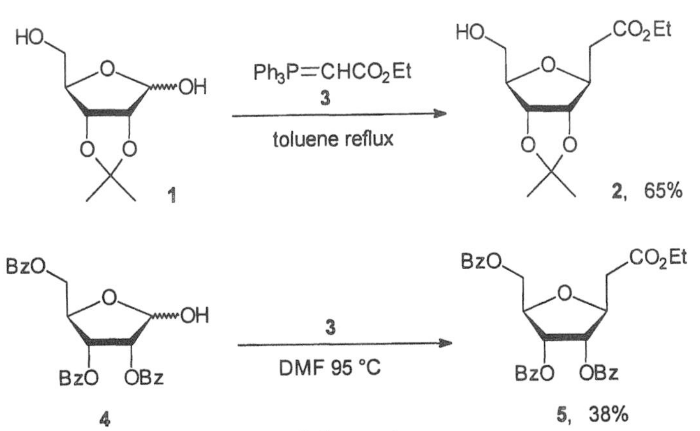

Scheme 1

The reaction is highly stereoselective and these compounds are useful precursors for *C*-nucleoside synthesis. If the ylide is changed to **6** then the open chain compound **8** is obtained, eq. 1.

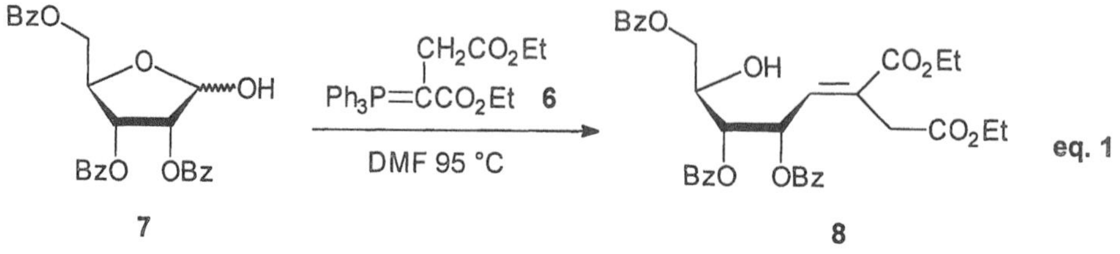
BzO
OH
BzO OBz
7
CH2CO2Et
Ph3P=CCO2Et 6
DMF 95 °C
BzO
OH
CO2Et
CO2Et
BzO OBz
8
eq. 1

Work by Moffatt[3] has shown that the product distribution can be altered. For example, reaction of **10a** with ylide **9** in acetonitrile gave 3:1 ratio of **11a** to **12a**. When the 5-hydroxyl was free as in **10b** a ratio of 22:1 of **11b** to **12b** was obtained, Scheme 2.

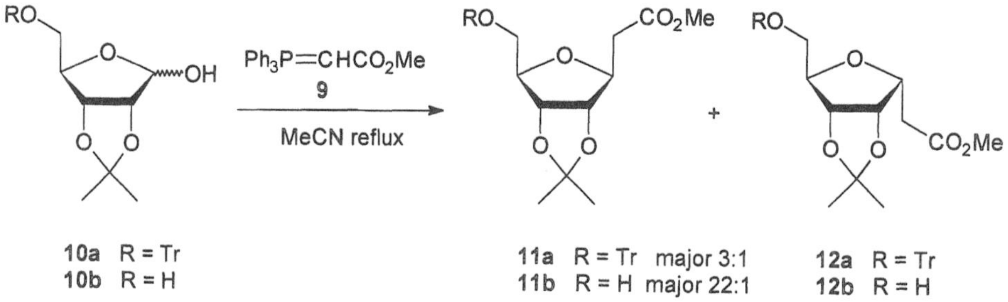
RO
OH
10a R = Tr
10b R = H
Ph3P=CHCO2Me
9
MeCN reflux
RO
CO2Me
11a R = Tr major 3:1
11b R = H major 22:1
+
RO
CO2Me
12a R = Tr
12b R = H

Scheme 2

When the reaction is carried out in chloroform then the corresponding open chain sugar was isolated. It is clear from these results that the *C*-glycosides are formed by base induced Michael cyclization of *O*-4 onto the double bond system.

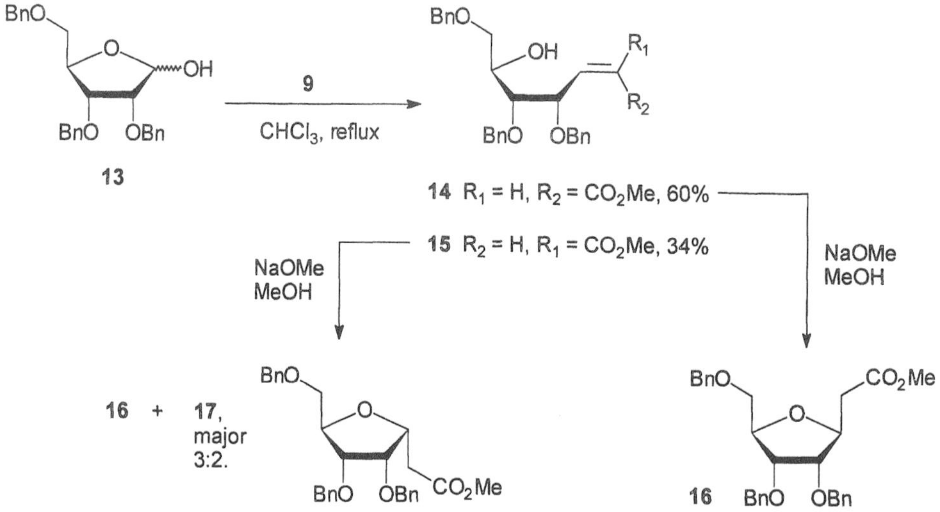
BnO
OH
BnO OBn
13
9
CHCl3, reflux
BnO
OH
R1
R2
BnO OBn
14 R1 = H, R2 = CO2Me, 60%
15 R2 = H, R1 = CO2Me, 34%
NaOMe
MeOH
NaOMe
MeOH
16 + 17, major 3:2.
BnO
BnO OBn
CO2Me
BnO
CO2Me
16 BnO OBn

Scheme 3

The isoproylidene function also plays an important role in the cyclization reaction because when **13** was exposed to similar conditions as outlined above only open chain products **14** and **15** were isolated. These could be cyclized to the corresponding *C*-glycosides by exposure to a catalytic amount of sodium methoxide in methanol. The *cis* olefin gave pure β-*C*-glycoside whereas cyclization of the *trans* olefin gave a mixture of *C*-glycosides, (α:β of 3:2). The product ratio obtained above (Scheme 3) can be altered by base catalyzed process but it does not significantly affect the product ratio. When the same base catalyzed anomerization is carried out with the isopropylidene compounds the ratio of β:α changes from 3:1 to 2:5, respectively. The kinetic product of the reaction is the β-*C*-glycoside **16** while the thermodynamic product is the sterically encumbered α-*C*-glycoside **17**.

An application of this reaction in the synthesis of both (+) and (-) methyl nonactate is shown below in Scheme 4.[4] Ketone **18** was hydrogenated to a mixture of alcohols and the major (**19**) was hydrolyzed to the triol which was then reprotected as **21**. Wittig reaction with ylide **25** then provided a 1:3 mixture of **22** and **23**. Equilibration shifted the ratio to 3:2. The synthesis was completed by dexoygenation and deprotection to provide (-) methyl nonactate (**24**).

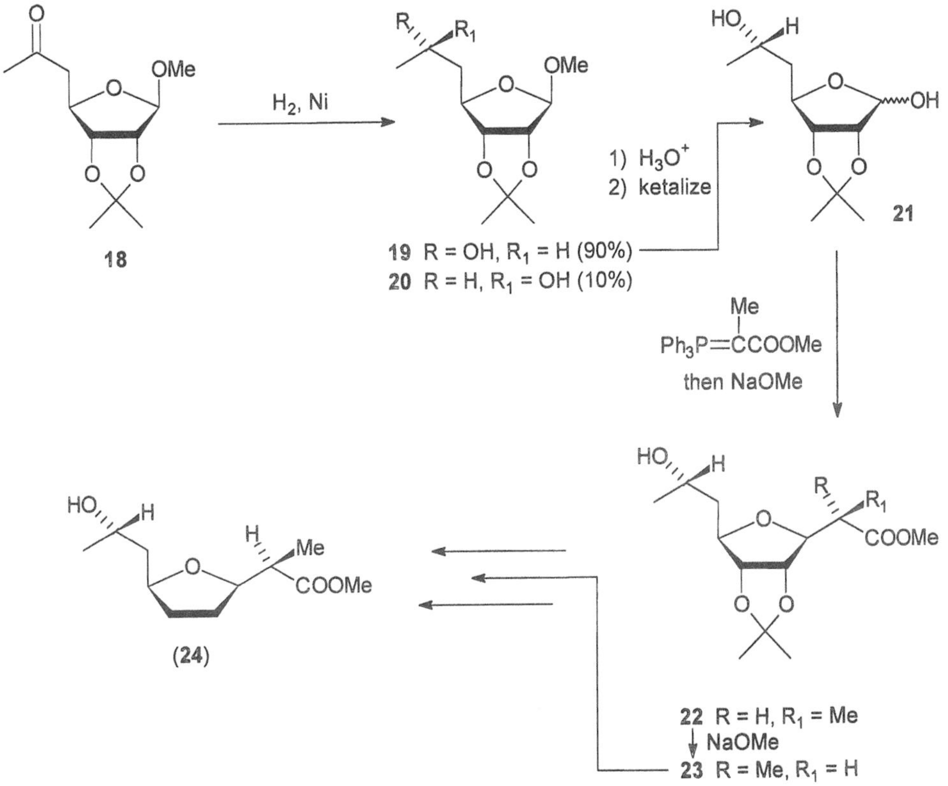

Scheme 4

In some cases only one isomer is obtained directly from the Wittig reaction. Lactol **26** was exposed to ylide **9** to give a good yield of the β-*C*-glycoside **27** exclusively, eq. 2.[5]

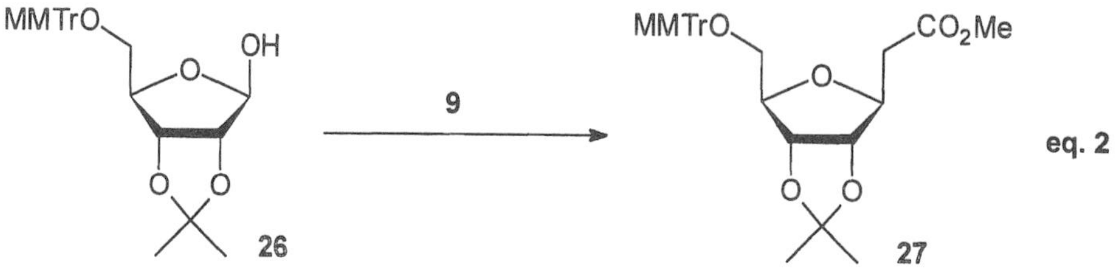

eq. 2

Workers at Merck-Frosst[6] have used the Wittig reaction on lactols in their synthesis of a leukotriene. Lactol **28** when treated with **3** in refluxing THF gave an 80% yield of the open chain compound **29**, whereas using two equivalents and a reflux time of 2 days gave a 80% yield of the *C*-glycosides **30**. Compound **29** was convertible to **30** by simple exposure to sodium ethoxide in ethanol. The primary alcohol was tosylated to **31** and treatment with LDA gave an open chain alcoholate that cyclized to epoxide **32**. Hydrogenation and base catalyzed epoxide formation then gave the LTA$_4$ precursor **33**, Scheme 5.

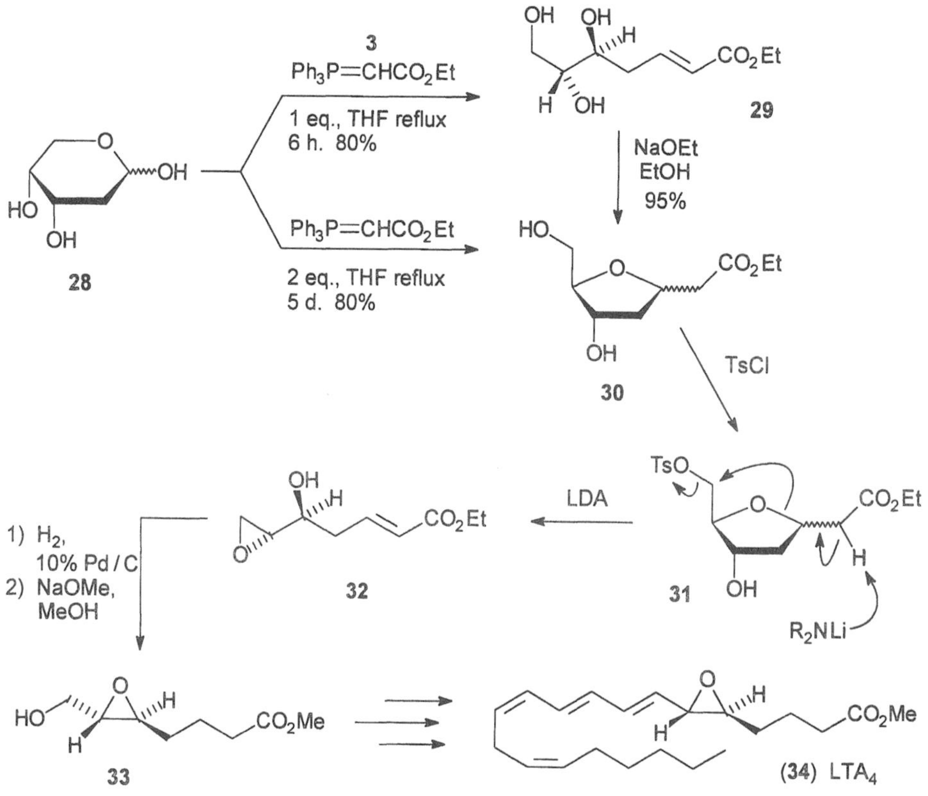

Scheme 5

In an effort to prepare functionalized *C*-glycosides Dondoni and Marra[7] had to resort to the use of a thiazole-armed phosphorane in their Wittig approach. Reaction of the furanose **35** with the ylide **41** in various solvents failed to give reaction, but exposure to the 2 equivalents of ylide **36** in refluxing toluene with molecular sieves provided a 93% yield of separable α- and β-*C*-glycosides **37** in a 35:65 ratio. As expected the ratio could be altered by stirring these compounds in a potassium hydroxide solution of methanol for 16 hours to 15:85 in favor of **37**β. Five other sugars were also successfully converted to their corresponding *C*-glycosides by use of this methodology. The formed *C*-glycoside **37** could be converted to the protected hydroxy ester **40** as shown in Scheme 6.

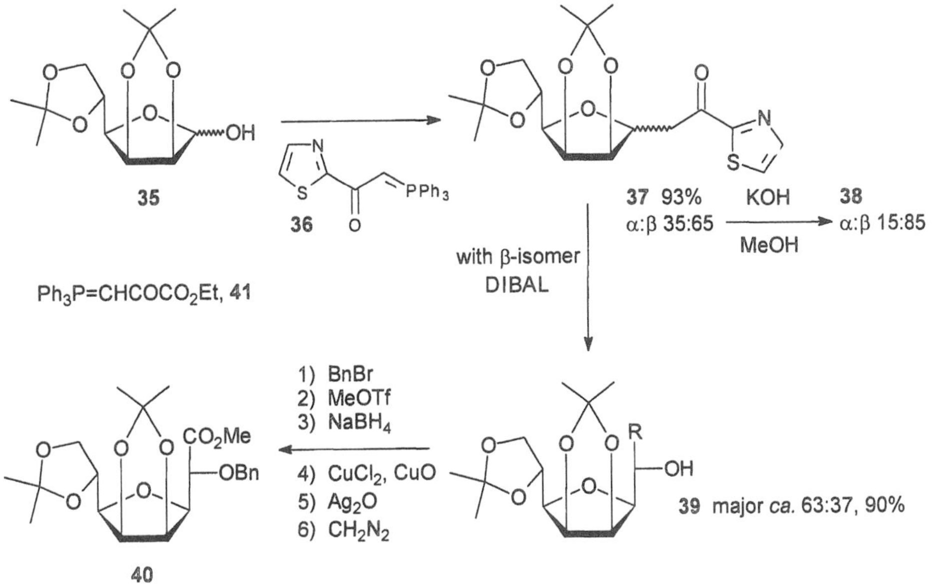

Scheme 6

Work by Nicolaou[8] applied to synthetic studies towards macrolide antibiotics demonstrates how in some cases open chain products can be isolated. Lactol **42** (available from D-glucose in several steps) was allowed to react with **3** in toluene at room temperature to give an 86% yield of the open chain compound **43**.

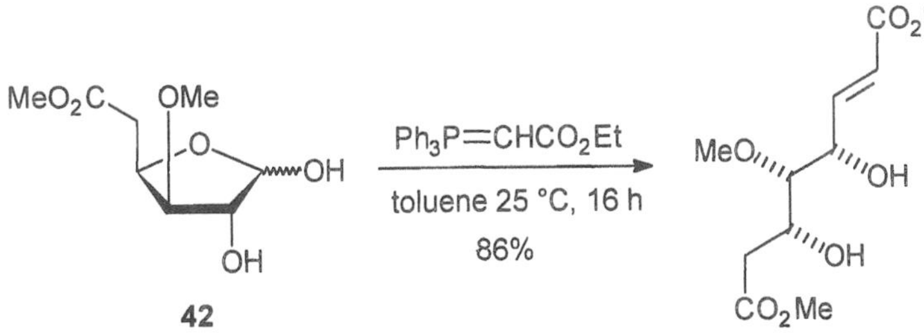

eq. 3

B.   *C*-PYRANOSIDES

Wittig approaches to *C*-pyranosides have also been extensively studied. The concepts involved are similar to those for furanose sugars. Wittig reaction gives an α,β-unsaturated ester and in some cases the open chain or mixtures of *C*-glycosides are isolated. Equilibration and other methods to improve stereoselection have also been reported.

Lactol **44** was exposed to acid-washed **9** (to avoid premature cyclization) in acetonitrile and refluxed for 60 h to provide an 87% yield of ester **45**. Base induced cyclization gave, after 1 hour, a 1:1 mixture of α and β anomers, but if the process was allowed to continue for approximately six hours then a quantitative yield of the thermodynamically more stable β-anomer **46** could be isolated. The α-anomer could be isolated by trapping as lactone **48** which arose from treatment of **45** with imidazole in water to give **48** in 24% yield.[9]

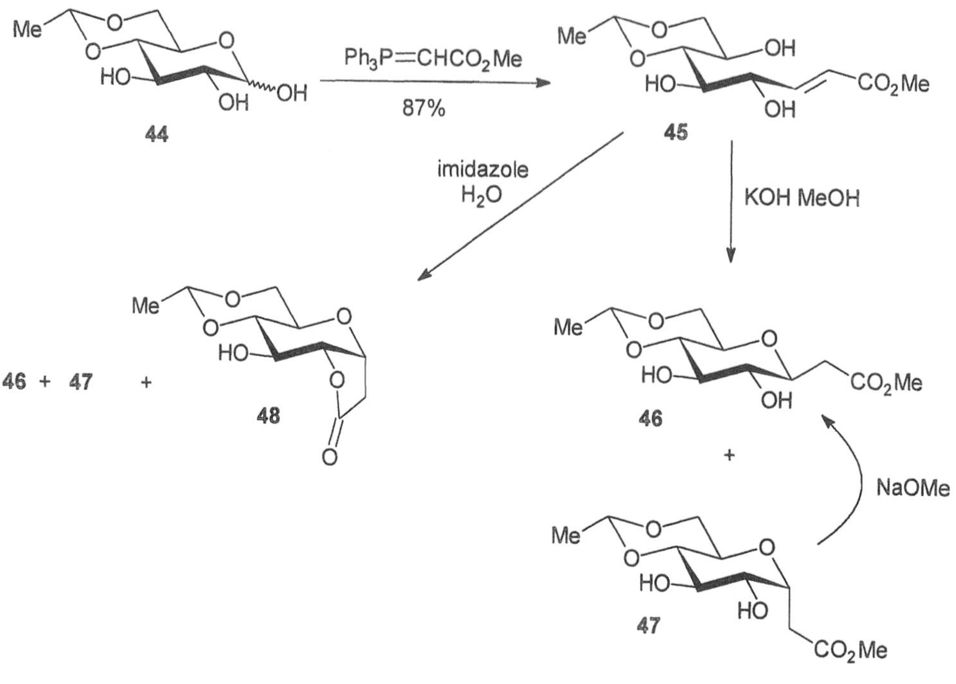

Scheme 7
Reprinted with permission from ref. 9, Copyright 1984 American Chemical Society.

There have also been a few examples of Wittig approaches to 2-deoxy-2-amino sugars. Russo, Nicotra and co-workers[10] reacted the lactol **49** with ylide **3** in boiling acetonitrile for 30 hours and obtained a 50% yield of the α-isomer **50**, eq. 4.

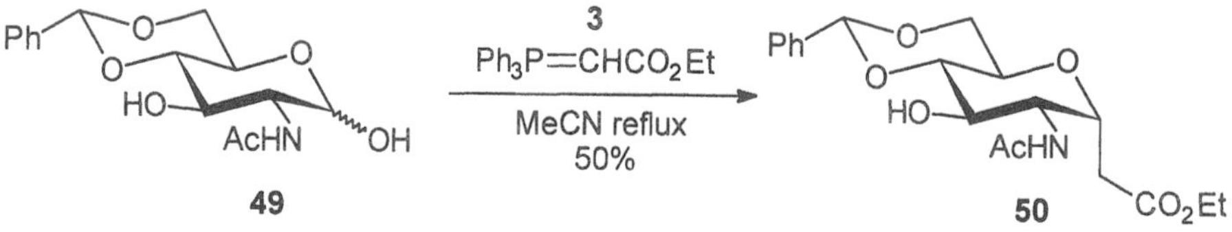

Giannis and Sandhoff[11] found that they could isolate the open chain form as a 3:1 mixture of *E/Z* isomers. It would seem that shorter reaction time and lower temperature sometimes favors the open chain form. Base treatment gave a mixture of anomeric *C*-glycosides that could be equilibrated to the more stable β-anomer. Treatment of **51** with 0.01M sodium ethoxide for 15 min gave a α:β ratio of 9:1. If the reaction time was increased to 36 hours then the ratio was altered to 3:10. The workers also converted the mixture of anomers to the triazenes **56** that could serve as a potential enzyme inhibitor, Scheme 8.

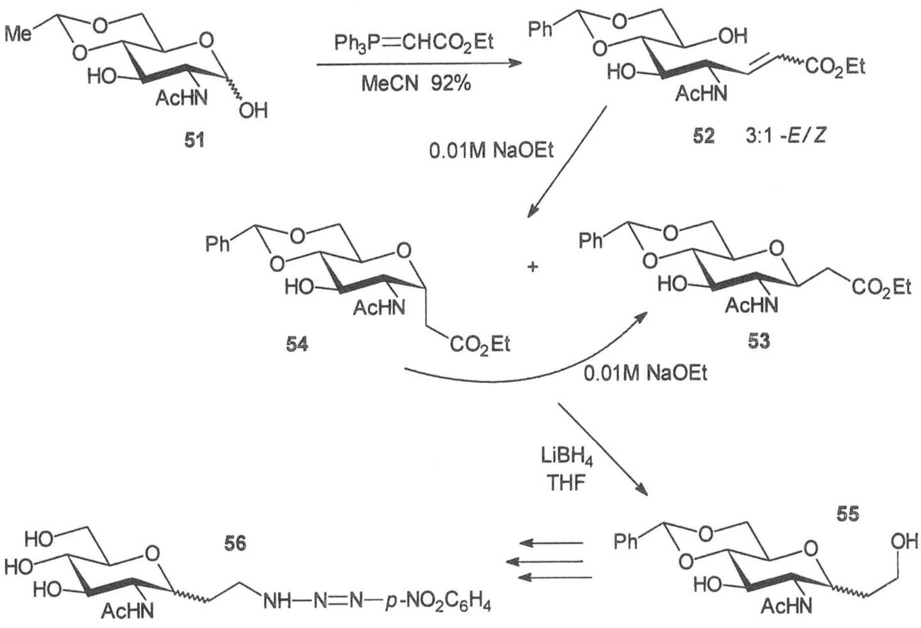

Scheme 8

Reprinted from ref. 11, Copyright 1987, pg. 202, with kind permission from
Elsevier Science Ltd, The Boulevard, Langford Lane, Kidlington, OX5 1GB, UK.

It has been observed[12] that subtle changes in the structure of the sugar drastically affects the outcome of the Wittig reaction. Attempted *C*-glycosidation of 2,3,4,6-tetra-*O*-benzyl-D-glucopyranose (**57**) with **3** under a variety of conditions gave only the product of elimination **58**, eq. 5.

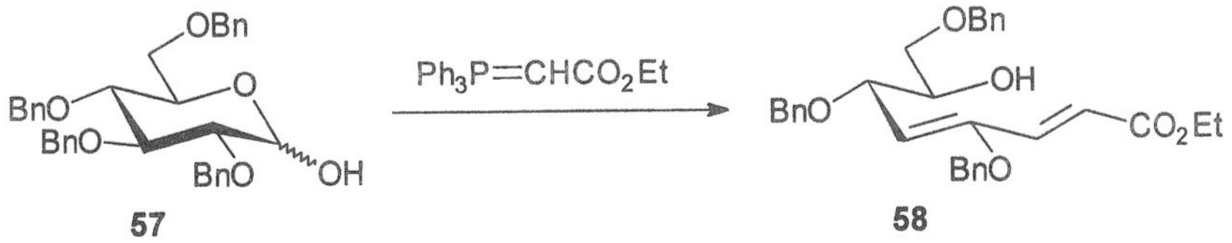

Use of acetate protecting groups or a combination of acetate and benzyl protecting groups also led to elimination. The corresponding manno, altro, and allo derivatives all afforded Wittig products **60a-b** under standard conditions, Scheme 9.

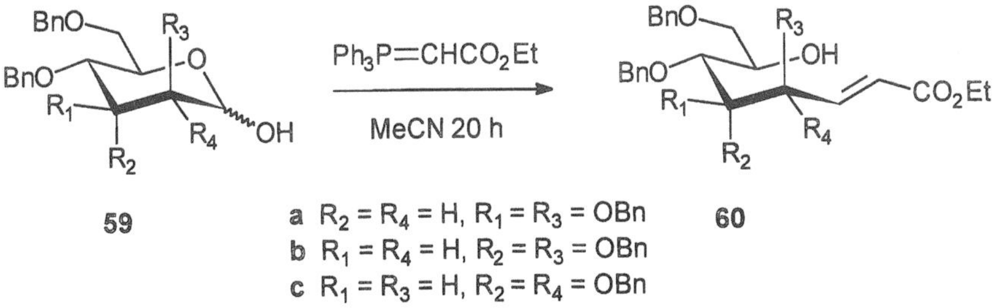

Scheme 9

These results stand in stark contrast to the behavior of the gluco derivative **57** which gives only elimination product (*vide supra*). It would seem that changes at *C*-2 or *C*-3 affect the course of the reaction. Changes at *C*-4 also affect the product distribution. For the case of the galacto sugar **61**, a mixture of *C*-glycosides was isolated plus a trace of the open chain Wittig product.

eq. 6

61

62 plus trace of open chain compound.

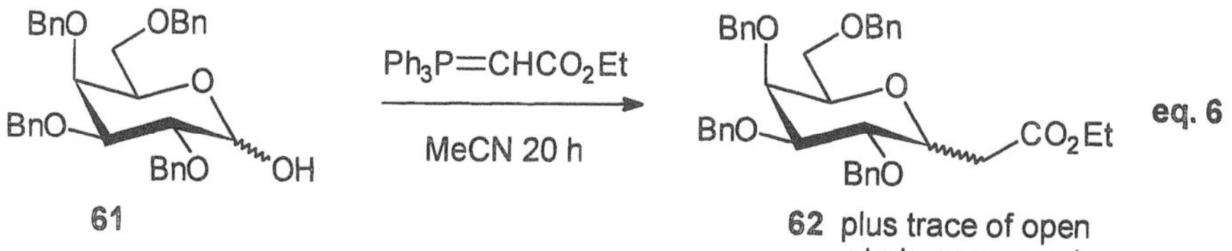

Success in the gluco series was realized with the 4,6-di-*O*-benzylidene protecting group with either benzyl or acetate protecting groups on *O*-2 and *O*-3, Scheme 10.

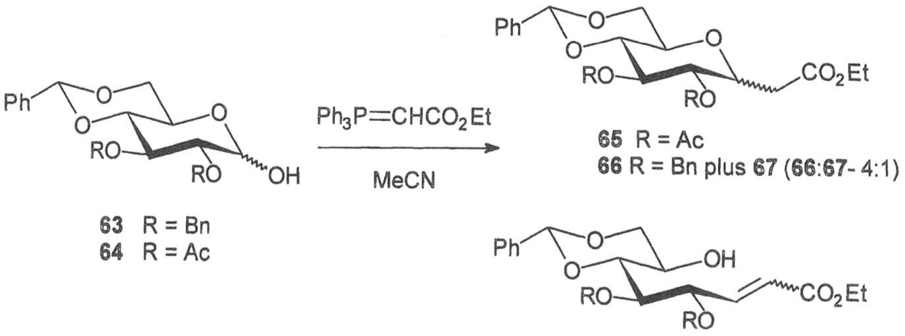

63 R = Bn
64 R = Ac

65 R = Ac
66 R = Bn plus **67** (**66:67**– 4:1)

67 R = Bn

Scheme 10

Masamune, Sharpless, and co-workers[13] have utilized titanium catalyzed asymmetric epoxidation to gain access to both the α- and β-*C*-glycoside from the same open chain Wittig product.

68

69

1) Et$_3$SiOTf
2) DIBAL

Ti(O*i*Pr)$_4$
(+)-DET

70

Ti(O*i*Pr)$_4$
(-)-DET

71

72

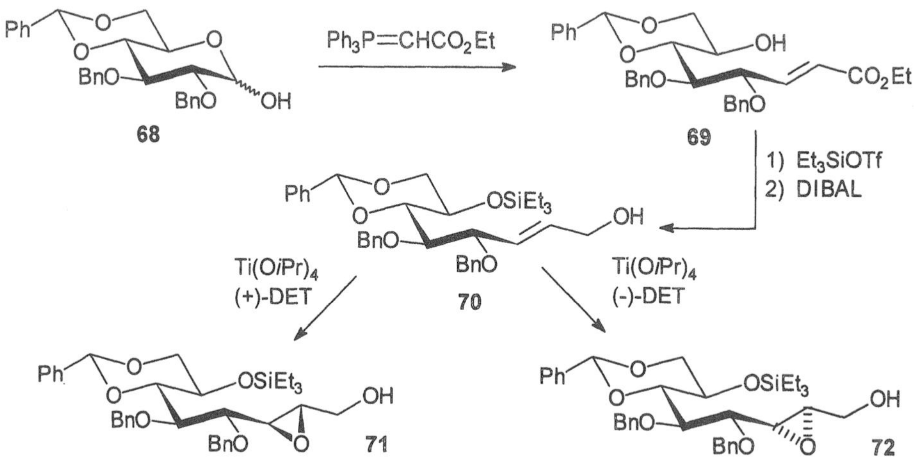

Scheme 11

Reprinted with permission from ref. 13, Copyright 1982 American Chemical Society.

Lactol was converted to the open chain derivative **69** under standard conditions and protection of the free hydroxyl followed by ester reduction provided **70**. Epoxide **71** or **72** could be obtained with excellent stereoselectivity by simply changing the tartrate (Scheme 11). Cyclization to the *C*-glycosides was effected by desilylation and treatment with sodium hydride in DMF to give the *C*-glycosidic diols **75** and **76**. Periodate cleavage and reduction of the formed aldehyde then furnished the α- and β-glycosides **78** and **77** as shown in Scheme 12.

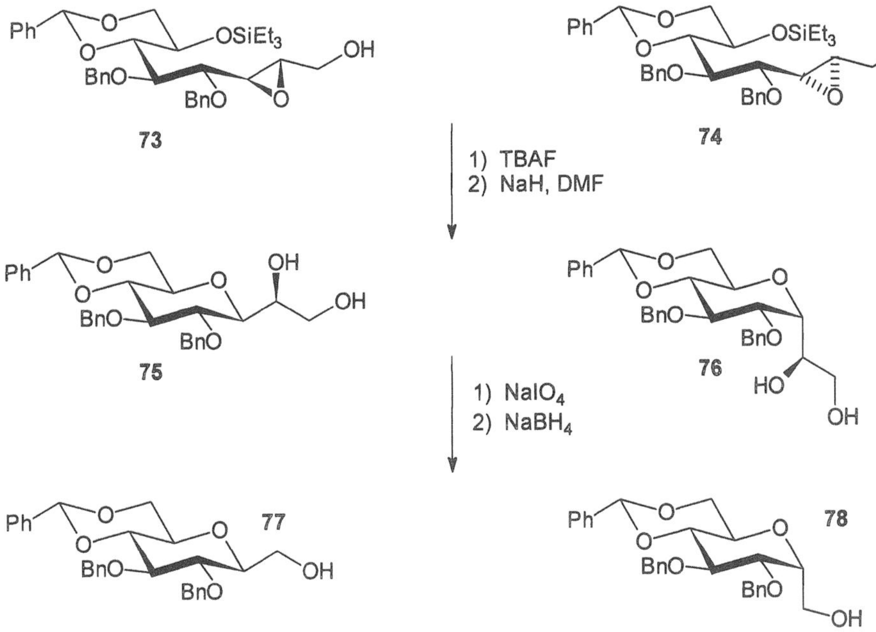

Scheme 12
Reprinted with permission from ref. 13, Copyright 1982 American Chemical Society.

Demailly *et al.*[14] have used the more reactive arsenate ylide **87** in their synthesis of *C*-glycosides from lactols. The gluco and manno derivatives **79a** and **79b** gave modest yields of open chain compounds with the *Z*-isomer predominating with some *C*-glycoside formation, while the galacto derivative **79c** gave an excellent yield of the *Z*-olefin exclusively along with some (9%) of a mixture of *C*-glycosides.

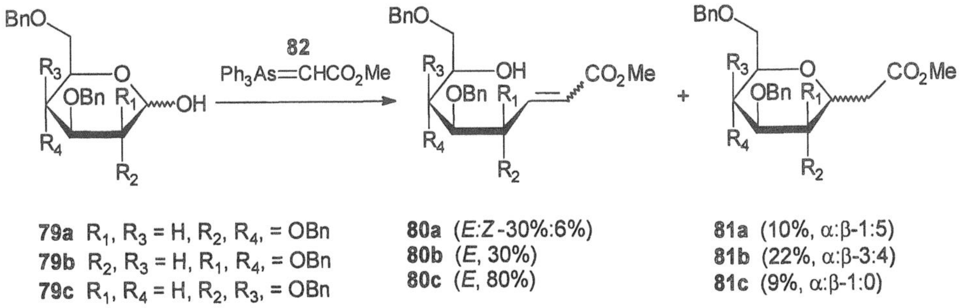

79a $R_1, R_3 = H, R_2, R_4, = OBn$ **80a** (*E:Z*-30%:6%)   **81a** (10%, α:β-1:5)
79b $R_2, R_3 = H, R_1, R_4, = OBn$ **80b** (*E*, 30%)   **81b** (22%, α:β-3:4)
79c $R_1, R_4 = H, R_2, R_3, = OBn$ **80c** (*E*, 80%)   **81c** (9%, α:β-1:0)

Scheme 13

The gluco **82a** and galacto **82c** *E*-olefins could be selectively cyclized to the corresponding β-*C*-glycosides by treatment with zinc bromide in refluxing benzene to give **83a** and **83c**. The authors explain the stereoselectivity by proposing the chelation model (**84**) shown below where the carbonyl oxygen and ring oxygen are held in a stable chair conformation by the zinc bromide.

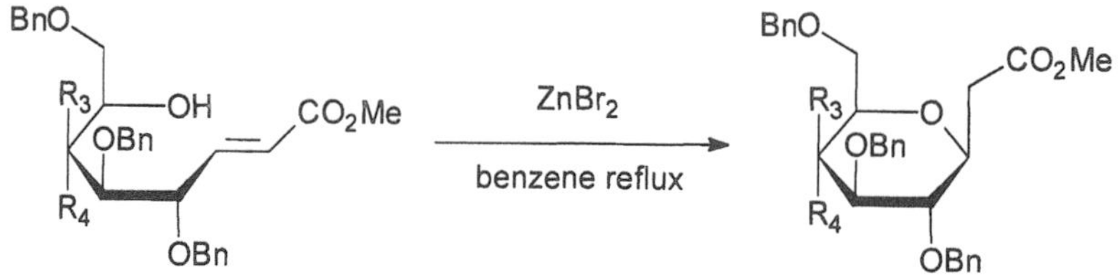

82 **83**

a R_3 = H, R_4 = OBn
c R_4 = H, R_3 = OBn

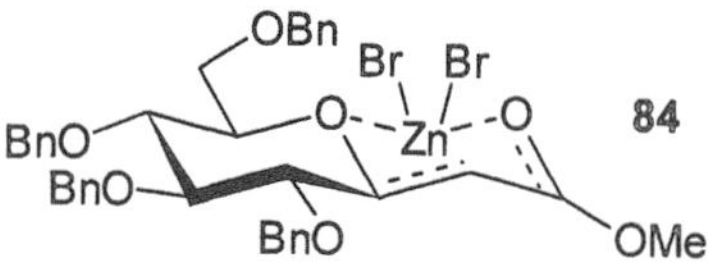

Scheme 14

Fréchou and co-workers[15] have employed a zinc promoted Wittig type reaction in their work towards *C*-glycosides to provide β-*C*-pyranosides stereoselectively. Heating of a mixture of lactol **85**, zinc dust, triphenylphosphine and methyl bromoacetate gave a 64% yield of the β-*C*-glycoside **86** as the exclusive product, Scheme 15.

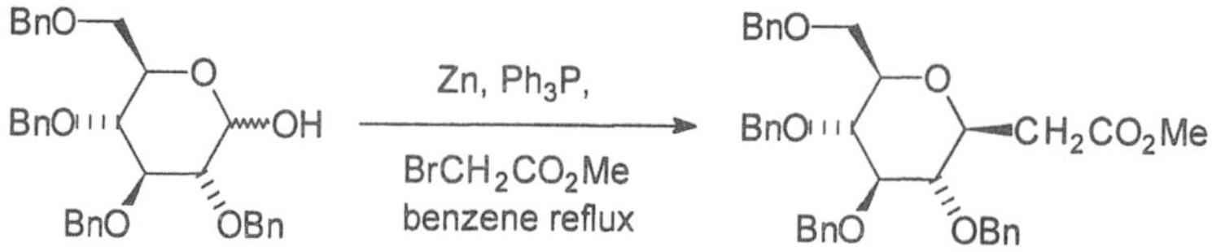

85 **86**, 64%

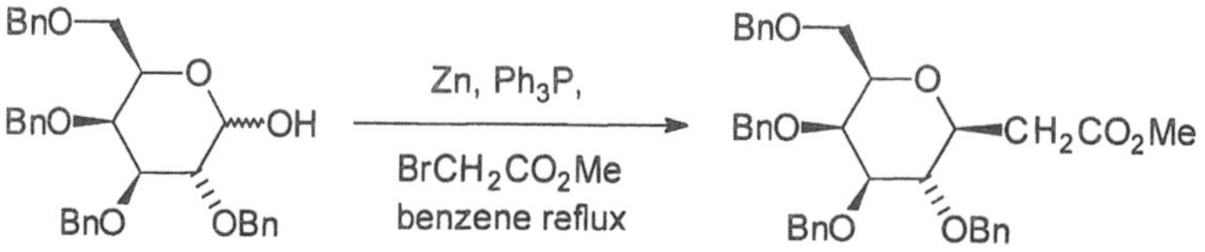

87 **88**, 75%

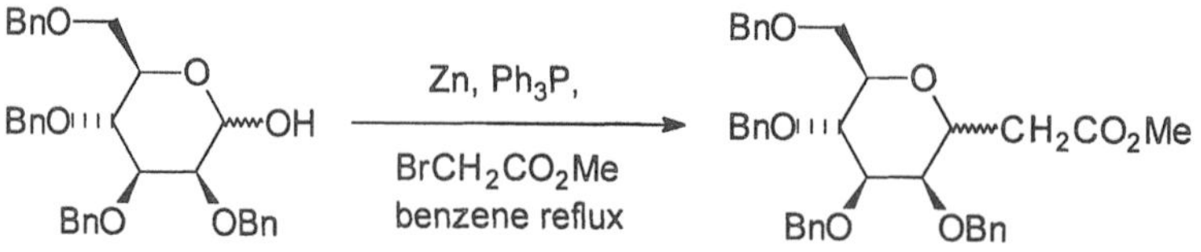

89 **90**, 49%, α:β -1:1

Scheme 15

The reaction is believed to proceed via the corresponding olefin and the mildness of the conditions would seem to indicate that no anomerization is taking place to account for the selectivity. The galacto derivative **87** also provided the β-*C*-glycoside **88** exclusively, but the manno compound **89** gave a 1:1 mixture of α- and β-*C*-glycosides **90**.

III. HORNER-EMMONS-WADSWORTH APPROACHES

Just as stabilized phosphoranes have found their place in *C*-glycoside synthesis so have phosphonates.

Reaction of **89** with triethyl phosphonoacetate sodium salt in THF at 50°C afforded four compounds of which the major ones **93a** and **94a** were obtained in 77% combined yield along with 8.6% of **91a** and **92a**. When the corresponding gluco compound was exposed to similar conditions, a mixture of four compounds **91b**, **92b**, **94b**, and **93b** (total yield 86%, ratio 13.2:5.6:2.3:1) was obtained. It seems that epimerization at *C*-2 is a problem and caution is indicated when conducting these reactions.[16]

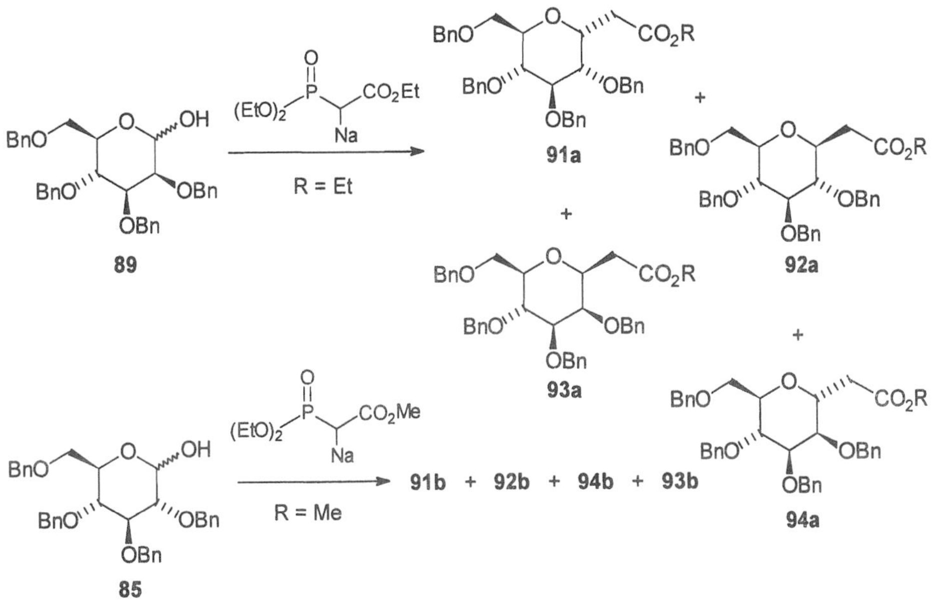

Scheme 16

Some stabilized phosphonates allow the use of unprotected sugars in *C*-glycoside formation.[17] Treatment of D-glucose with phosphonate salt **95** and acetylation gave a mixture of *C*-furanosides **97** and **98**. If however the reaction is treated with sodium methoxide prior to acetylation then a 50% overall yield of **99** can be obtained. When 4,6-ethylidene glucose (**100**) was exposed to the above conditions a 3:1 mixture of α- and β-anomers was obtained. The α-anomer could be converted to the β-anomer by treatment with sodium methoxide in methanol. Deprotection and acetylation of **102** then gave the acetylated β-anomer **99**, Scheme 17.

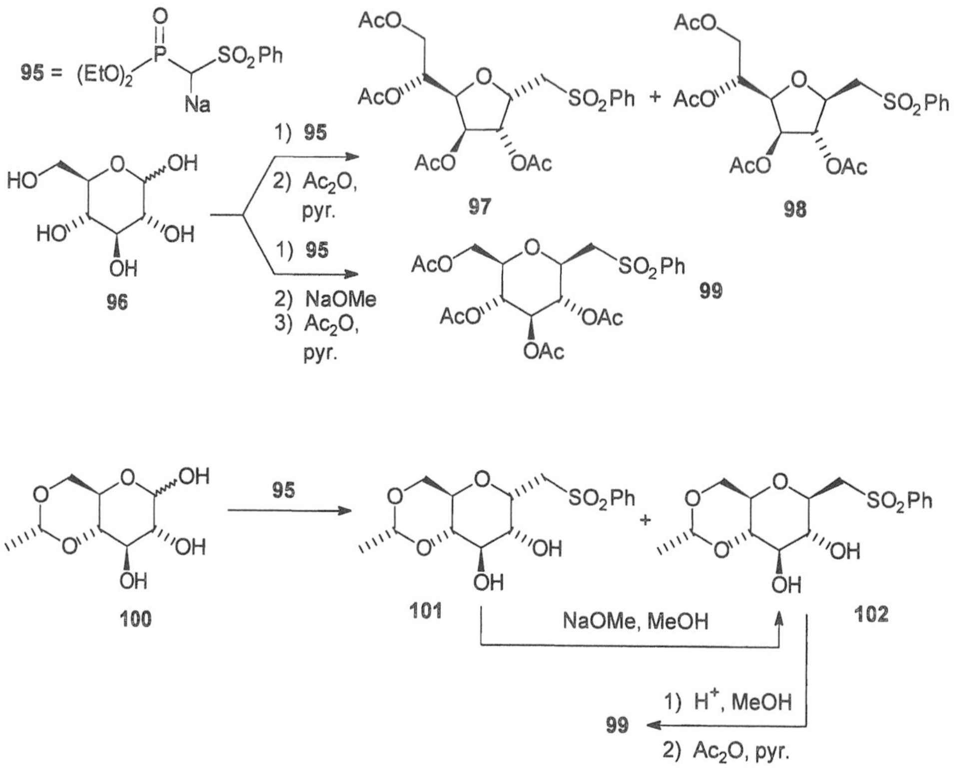

Scheme 17

Fréchou and co-workers[18] have also utilized the phosphonate salt **95** in their preparation of 2-deoxy-2-amino-*C*-glycosides, eqs. 7 and 8.

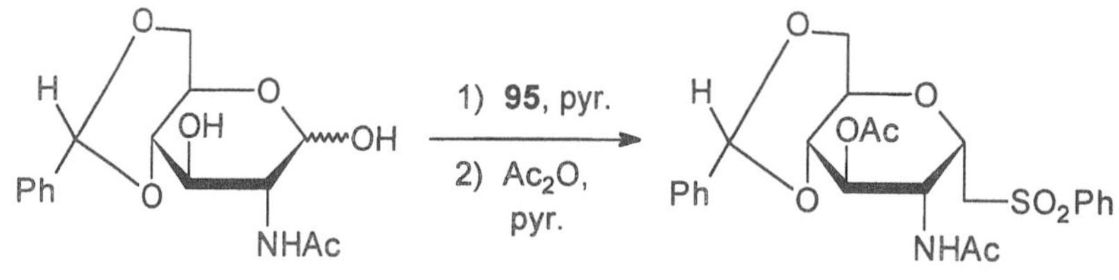

103 **104**, 50% eq. 7

In the furanose series, **105** was reacted with **95** to give a mixture of **106** and **107** in 35% and 10% yield, respectively.

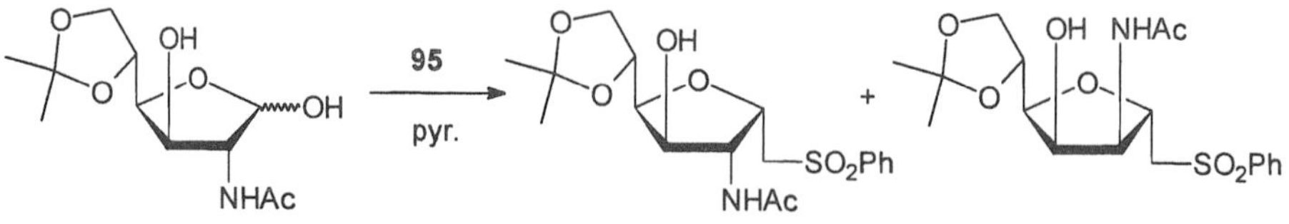

105 **106**, 35% **107**, 10% eq. 8

IV. OLEFINATION FOLLOWED BY ELECTROPHILIC CYCLIZATION

The two above sections dealt with *C*-glycoside synthesis by base induced cyclization of a suitably located oxygen atom on an Michael acceptor system, usually an α,β-unsaturated ester. If the initial olefination provides a monosubstituted olefin then base induced cyclization cannot be used. In order to affect cyclization an electrophilic process is required such as mercurio- or halocyclization both of which have seen use in the preparation of *C*-glycosides.

A. MERCURIOCYCLIZATION

The pioneering work of Sinaÿ[19] paved the way for future work in this area. Wittig reaction of **85** with the ylide **108** (Ph$_3$PMe$^+$Br$^-$, *n*-BuLi) gave an 80% yield of olefin **109**. It is interesting to note that use of Ph$_3$PMe$^+$Br$^-$ with sodium hydride in dimethylsulfoxide furnished diene **113** in 93% yield. Mercury-mediated cyclization with mercuric acetate followed by chlorination gave the α-anomer **110** exclusively. The mercury group could be reductively cleaved to **111** or oxidatively replaced with hydroxyl by exposure to oxygen in the presence of sodium borohydride in DMF to **112**. The cyclization is highly selective and this method constitutes a useful entry into α-*C*-glycosides via the use of a Wittig/cyclization strategy. The authors postulate that the adjacent *O*-benzyl (or other *O*-benzyl groups) coordinate the incoming mercury species, Scheme 18.

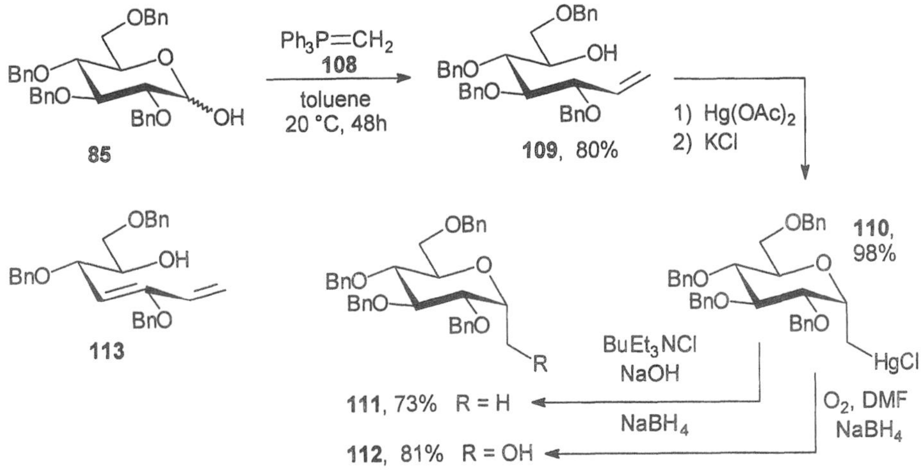

Scheme 18
Reproduced with permission from ref. 19, Copyright 1981, courtesy the Royal Society of Chemistry, Thomas Graham House, Science Park, Milton Road, Cambridge, CB4 4WF, UK.

Russo *et al.*[20] have studied the effect of changing the substituent adjacent to the olefin to see if the aforementioned hypothesis is indeed valid. When *O*-2 is free, as in **114**, then only the α-isomer **115** is obtained (eq. 9).

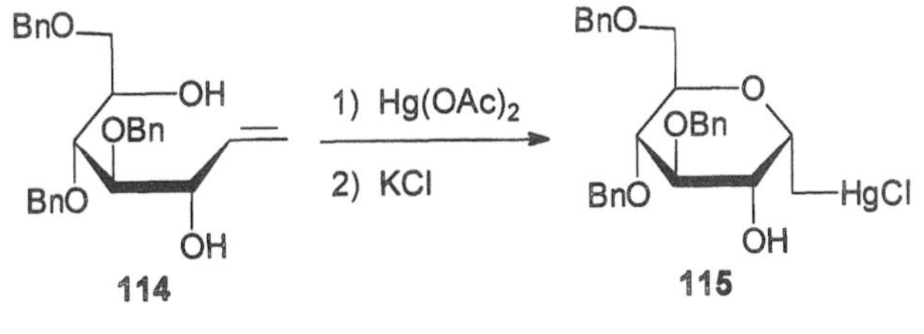

The manno derivative **116** gave a 3:7 ratio of α- and β-*C*-D-mannopyranosyl compounds **117** and **118**. It seems that reaction is sensitive to the disposition of the substituent at *C*-2. In the furanose series ribo derivative **119** gave an inseparable mixture of α- and β-*C*-glycosides in a 85:15 ratio, respectively. The 2-deoxy derivative **122** however showed no selectivity and afforded a 1:1 mixture of anomers upon treatment with mercuric acetate and potassium chloride.

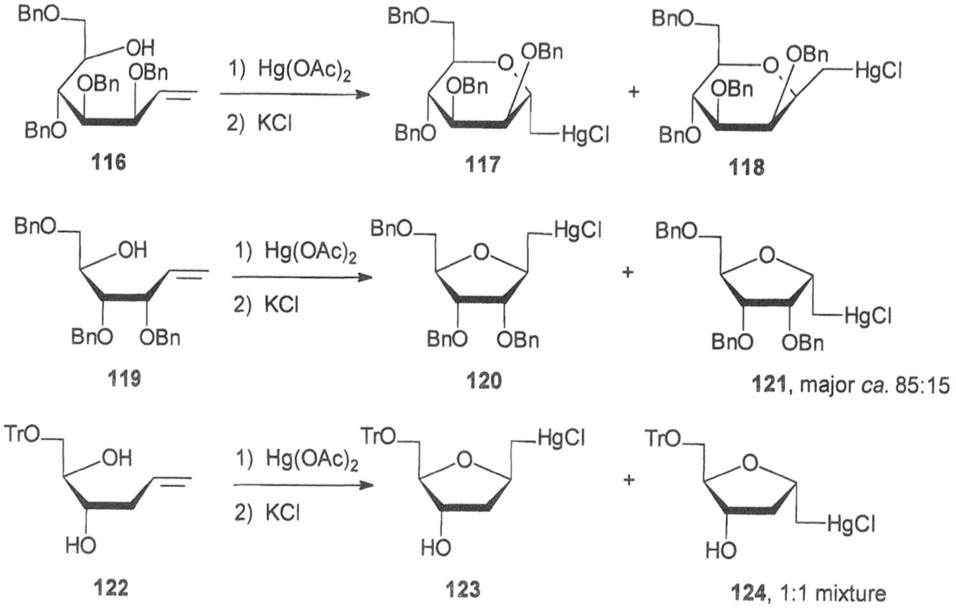

Scheme 19

Russo and collaborators[21] also applied this methodology to the preparation of a *C*-glycoside analog of α-D-glucose-1-phosphate. Treatment of **110** with bromine gave halide **125** which when exposed to refluxing triethyl phosphite gave the phosphonate **126** in 60% yield. Conversion to **127** was realized by treating **126** with trimethylsilyl iodide.

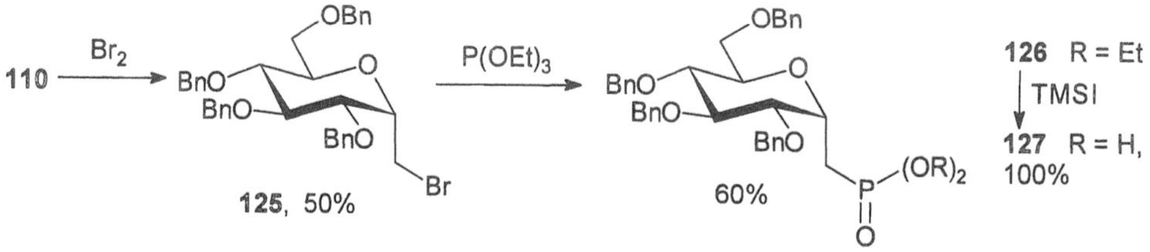

Scheme 20

The interest in phosphono analogs of naturally occurring phosphates as potential inhibitors or regulators of biological processes has resulted in the preparation of *C*-glycoside analogs.[22] Nicotra and collaborators[23] observed that mercuriocyclization of the ribo sugar **129** gave a 82% yield of mainly (α/β 95:5) the α-anomer **130**. Treatment with iodine followed by Arbuzov reaction with triethyl phosphite then gave the protected phosphono analog **131**.

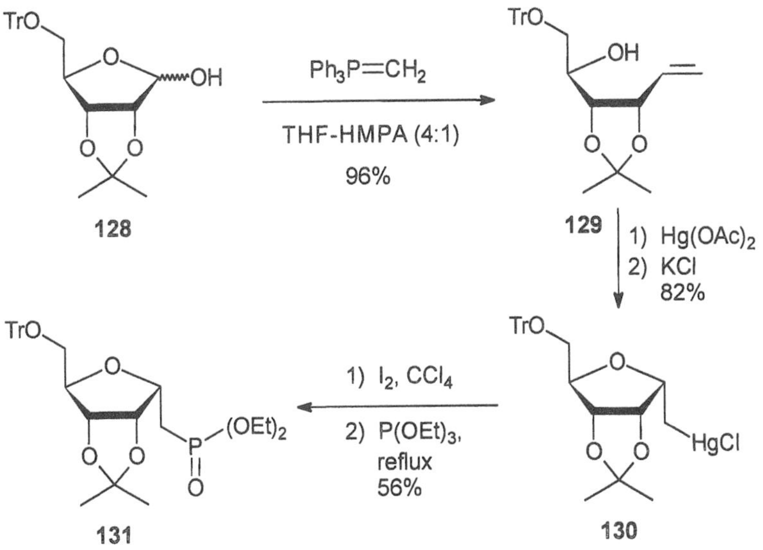

Scheme 21

Strictly speaking the following is not a Wittig approach to *C*-glycosides but the use of mercuriocyclization warrants its placement in this section. Nicotra[24] has cleverly applied the mercuriocyclization strategy to the synthesis of a 2-deoxy-2-amino-*C*-glycoside from the arabinose derivative **132**. Treatment with benzylamine gave a mixture of anomers which were transformed into **134** in 71% yield (88% diastereomeric excess) by exposure to vinylmagnesium bromide. Cyclization with $Hg(CF_3CO_2)_2$ followed by exposure to KCl then provided **135** which was reduced and deprotected to furnish the target structure **136**.

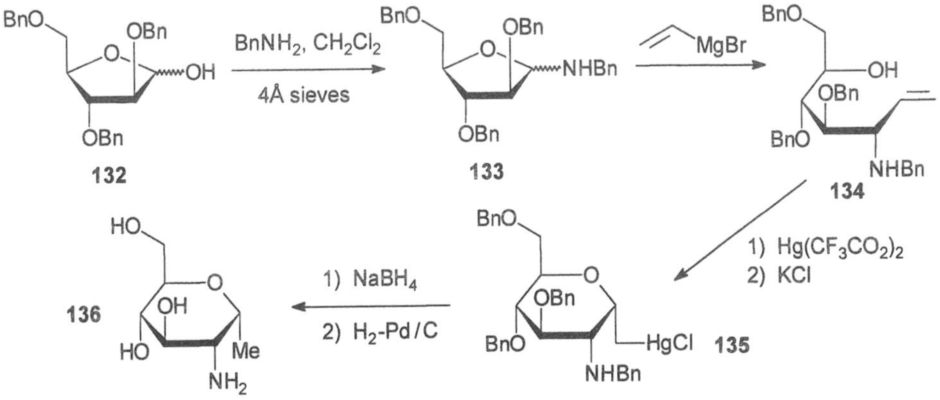

Scheme 22

B.   HALOCYCLIZATION

Russo and co-workers[25] have shown that control over the ring size can be achieved by judicious choice of reagent. They have found that iodocyclization favors *C*-furanosyl formation while mercuriocyclization favors *C*-pyranosyl generation. The olefins were prepared in a straightforward manner by Wittig olefination of a suitable lactol using ylide $Ph_3P=CH_2$ in the solvent system of THF-HMPA or just THF. The added HMPA sometimes served to improve the yields of the products. Exposure of **137** to iodocyclization under slightly acidic conditions caused cyclization to the *C*-furanoside **138** with concomitant debenzylation of *O*-5.   No *C*-pyranoside was isolated!   Further evidence for this debenzylation is found in Scheme 23. Both the benzylated (**139**) and free (**140**) hydroxyl compounds cyclize to **141**. The results also show that five membered ring formation is highly favored and in most cases the 1,2-*cis* adduct is either the exclusive or major product of the reaction. If the 2-position is unsubstituted (as in **144**) then no or little anomeric selectivity is observed in the cyclization step.

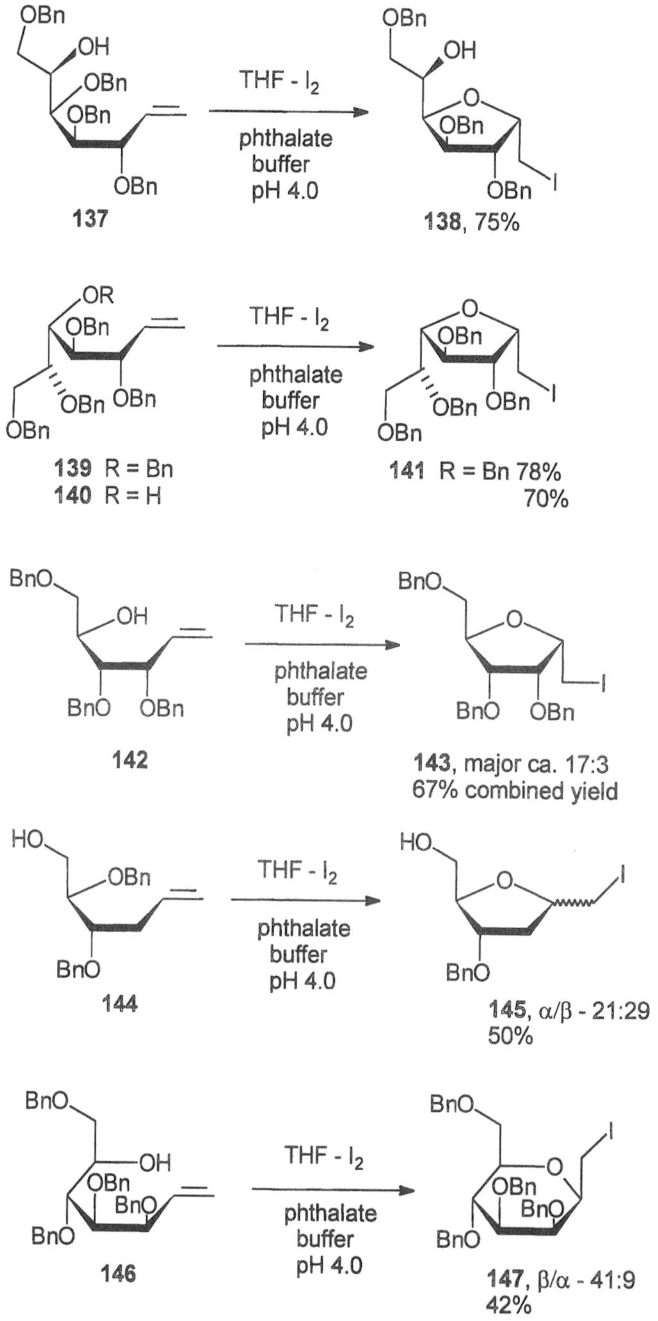

Scheme 23

The only exception to the five-membered ring formation occurs with the manno derivative 146 that cyclizes to a mixture of α- and β-*C*-pyranosides 147. The authors attribute this to the fact that the transition state leading to the formation of the five-membered ring compound would suffer from three 1,2-*cis* interactions. The above results clearly demonstrate that mercuriocyclization and iodocyclization are complementary methods for ring formation.

Freeman and Robarge[26] have examined the effect of various electrophilic reagents on the anomeric ratios of cyclization for the eneitol 148. Iodocylization, bromocyclization and mercuriocyclization all led predominantly to the allo or 1,2-*cis* product 149, eq. 10 and Table 1. Buffered phenylselenocyclization gave similar results. The results are rationalized via an A-strain model.

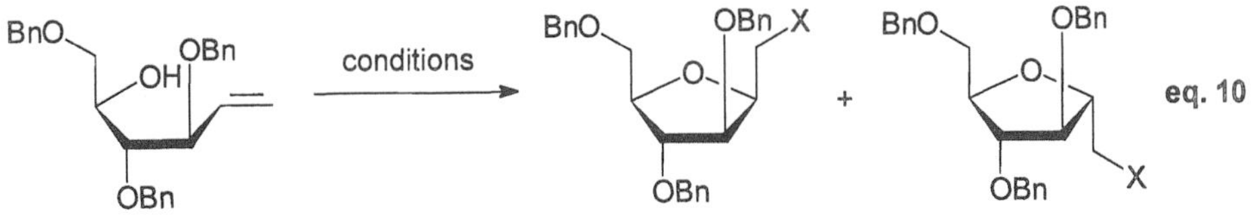

eq. 10

Table 1: Effect of Reagent on Cyclization Stereochemistry of 148.

CONDITIONS	YIELD AND RATIO OF 149:150
I_2, NaHCO$_3$, THF	79%, 82:18 X = I
NBS, DMF	75%, 84:16 X = Br
Hg(OAc)$_2$; KCl	87%, 89:11 X = HgCl
PhSeCl, K$_2$CO$_3$	48%, 85:15 X = SePh

Reitz and co-workers[27] have also examined the difference between halocyclization and mercuriocyclization as well an epoxidation–cyclization route. Lactol 151 was converted to olefin in the usual way and exposure of 152 to NBS gave a 90% yield of 153a and 154a in a ratio of 1:7.8, whereas exposure of 152 to *m*-CPBA gave a mixture of alcohols that were converted to the corresponding bromides. As expected, mercuriocyclization of 152 also gave a preponderance of the β-anomer 154.

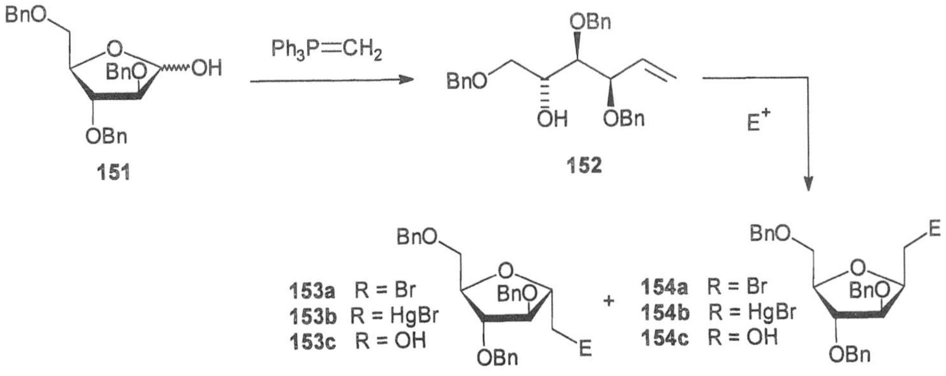

Scheme 24

The anomeric mixture **153a** and **154a** (7.8:1) was selectively acetolyzed to acetate **155**, hydrolyzed and phosphorylated to furnish **156**. Arbuzov reaction was followed by three step deprotection to furnish the potential enzyme inhibitor **158**.

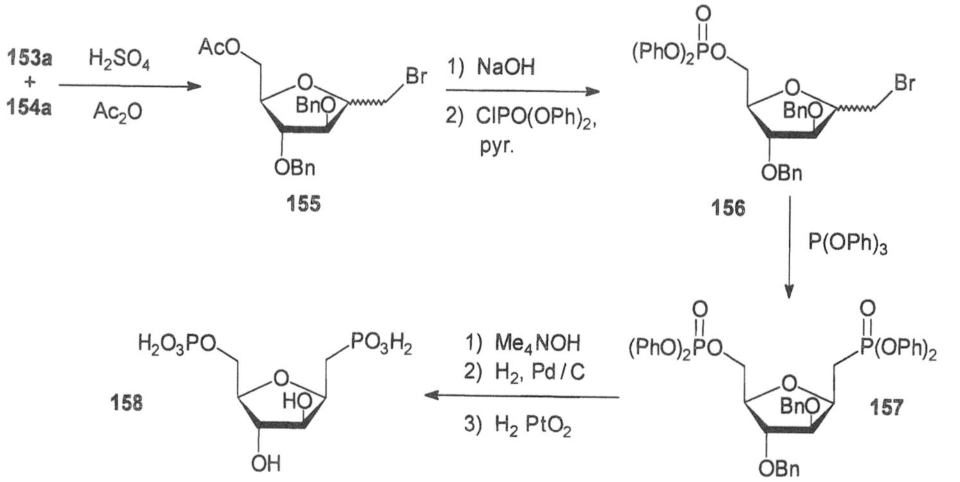

Scheme 25

Freeman and Robarge[28] have shown that the stereochemistry of iodocyclization is directly related to olefin geometry. Reaction of **159** with **3** in dichloromethane gave the *Z*-isomer **160** exclusively. This was proven by conversion of **160** into the *E*-isomer **161** by THP protection, isomerization, and hydrolysis. Iodocyclization of the *Z*-isomer gave the β-product **162** exclusively while cyclization of the *E*-isomer gave the α-isomer **163**. The authors explain that the stereoselectivity can be attributed to interactions between the isopropylidene group and the ester function.

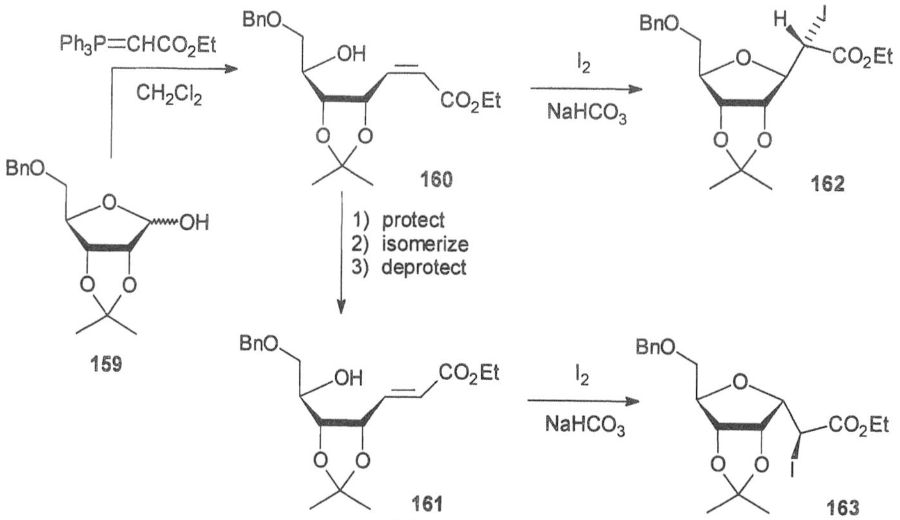

Scheme 26

Cyclization of the corresponding *E*-alcohol gave a mixture of α- and β-isomers, while the *Z*-alcohol gave a pyranosidic *C*-glycoside (not shown).

C. SELENOCYCLIZATION

Sinaÿ[29] has employed a Wittig-selenocyclization strategy to quickly assemble the exomethylenic *C*-glycoside **166**. Previously it was shown that Wittig reaction on the perbenzylated lactol **85** was problematic and led mainly to the elimination product (see eq. 5). Sinaÿ *et al.* have found that if the lactol **85** is first converted to its lithium salt followed by reaction with ylide **108** then a 74% yield of enitol **109** can be obtained. Exposure to *N*-phenylselenophthalimide affected cyclization to a mixture of α-*C*-pyranoside **165** (60%) and the *C*-furanoside **164** (30%). Tributyltin hydride mediated reduction provided the methyl *C*-glycoside **167** while oxidation and selenoxide fragmentation gave the exomethylenic sugar **166**.

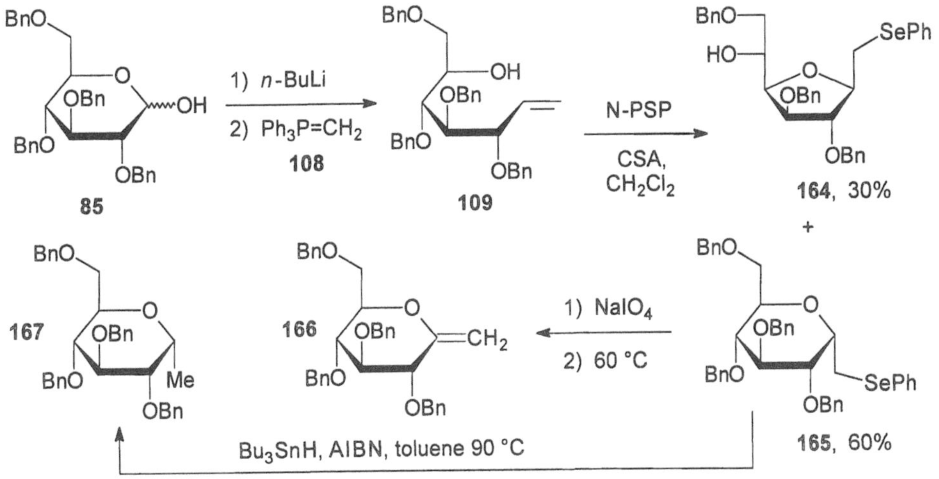

Scheme 27

In model studies directed at the chrysomicins Hart and co-workers[30] used selenocyclization to construct a model of the *C*-aryl glycoside core. Alcohol **168** was exposed to phenylselenyl chloride in dichloromethane and a 6-*endo* cyclization occurred to provide a 74% yield of pyran **169**. Oxidation and fragmentation of the resulting selenoxide gave olefin **170** (80%) which hydroxylated to a single isomer in 88% yield. Deprotection then furnished the target **172**.

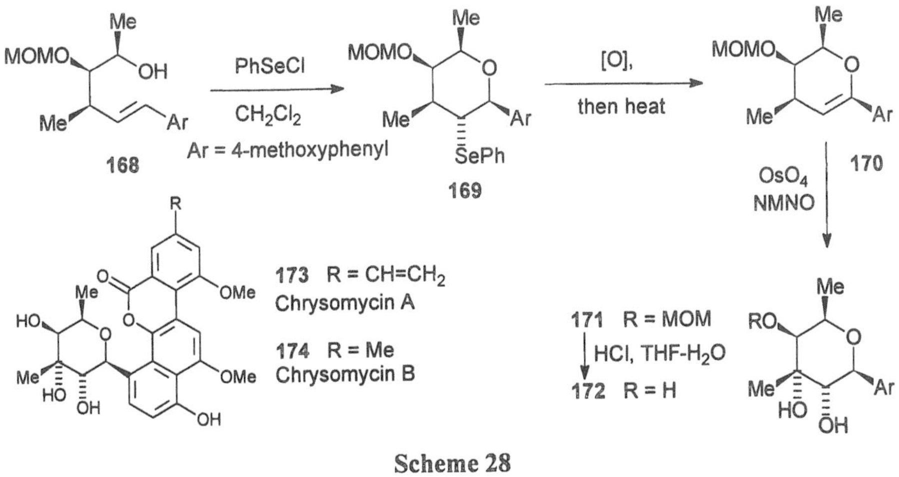

Scheme 28

Work by Kane and Mann[31] further illustrates the use of selenocyclization in *C*-glycoside preparation. Exposure of the ribose derivative to buffered phenylselenyl chloride gave a 40% yield of the β-isomer **176** exclusively, eq. 11.

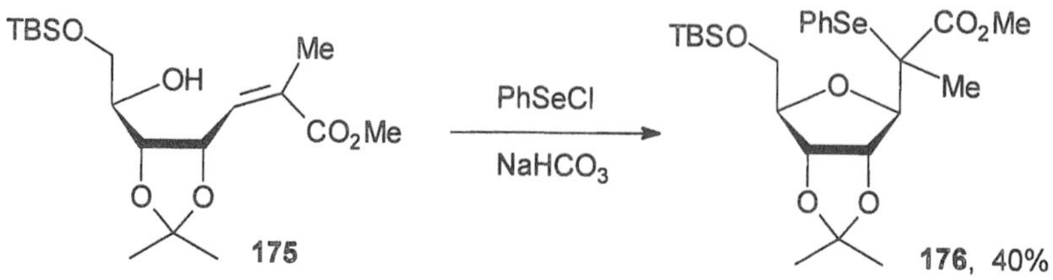

V. USE OF SULFUR YLIDES

A few workers have used the reaction of sulfur ylides with lactols (carbonyl compounds) to yield hydroxy-epoxides that cyclize to the corresponding hydroxy *C*-glycosides.

Valpuesta, Durante and López-Herrera[32] have found that reaction of the ylide **178** with ribo sugar **177** gave an 85% yield of the α-*C*-glycoside**180**. Small amounts of the β-*C*-glycoside **179** were also formed, but it was found difficult to purify. It is interesting to note that no intermediate epoxides were isolated in this reaction indicating that in this case cyclization was a facile process.

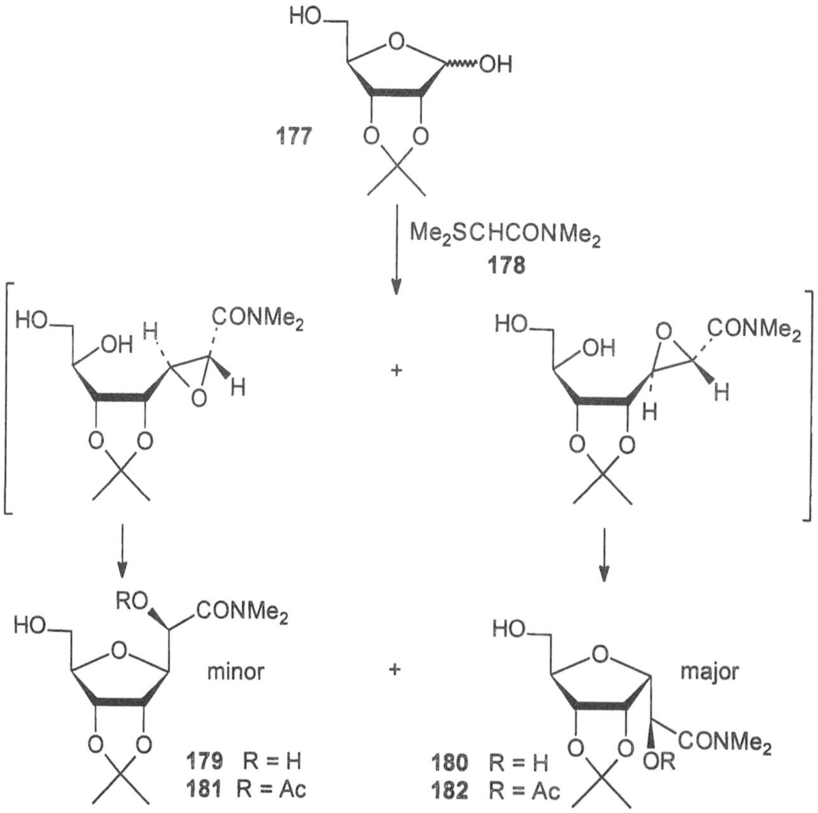

Scheme 29

The manno derivative **35** upon reaction with **178** furnished a mixture of two *C*-glycosides **183** and **184** in a 3.1:1.0 ratio and the *trans*-epoxide **185** in practically quantitative yield. The ratio of 3.1:1 could be increased to 6.5:1 by increasing the reaction time.

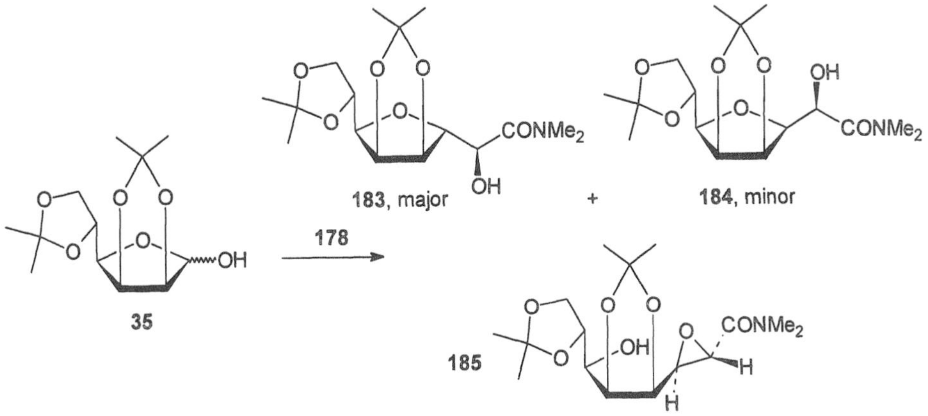

Scheme 30

Fréchou and co-workers[33] have found that similar reactions could be carried out with the related ylide **186** in DMSO at ambient temperature. The gluco sugar **85** gave the epoxide olefin **187** that arises from epoxidation and elimination of benzyl alcohol. Similar products were obtained with the manno and galacto sugars (not shown). Two furanose sugars were examined. The manno derivative **35** gave a mixture of *C*-glycosides while reaction with the ribo sugar **128** furnished the α-*C*-glycoside **189** in 58% yield.

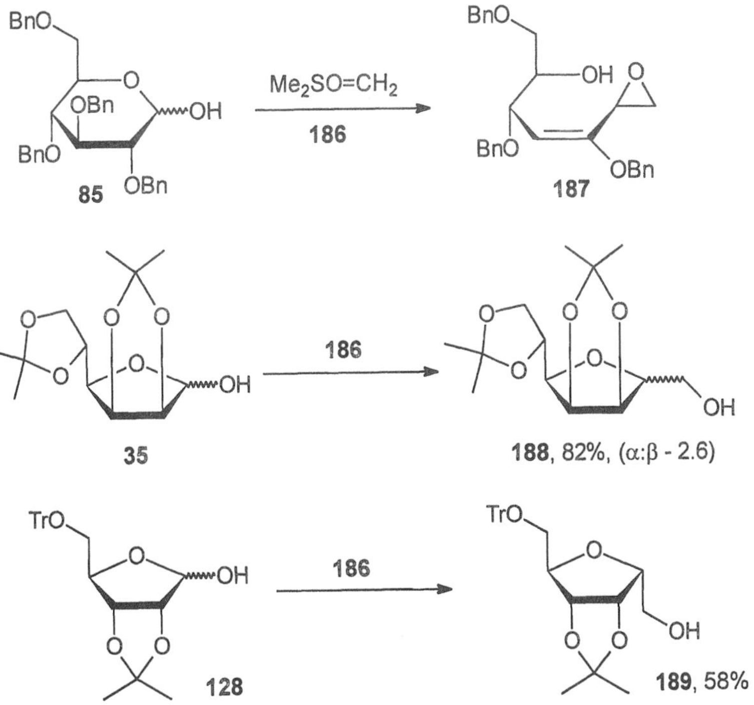

Scheme 31

VI. OLEFINATION OF SUGAR LACTONES

All the previous examples discussed involve olefination reactions on lactols to give hydroxy olefins that cyclize *in situ* or by the aid of another reagent to *C*-glycosides. There have been several approaches to *C*-1 exomethylenic sugars that are in their own right *C*-glycosides; many of the approaches use olefination technology on a pre-existing sugar lactone.

Wilcox, Long, and Suh[34] used Tebbe's reagent (191)[35] to construct an exomethylenic sugar that served as an intermediate in synthetic work towards citreoviridin (190). Olefination of 192 with 191 gave an 85% yield of 193. Acetolysis followed by Lewis acid catalyzed allylation provided the bis-*C*,*C*-furanoside 194. Oxidative cleavage of the double bond followed by desilylation then furnished the lactol 196.

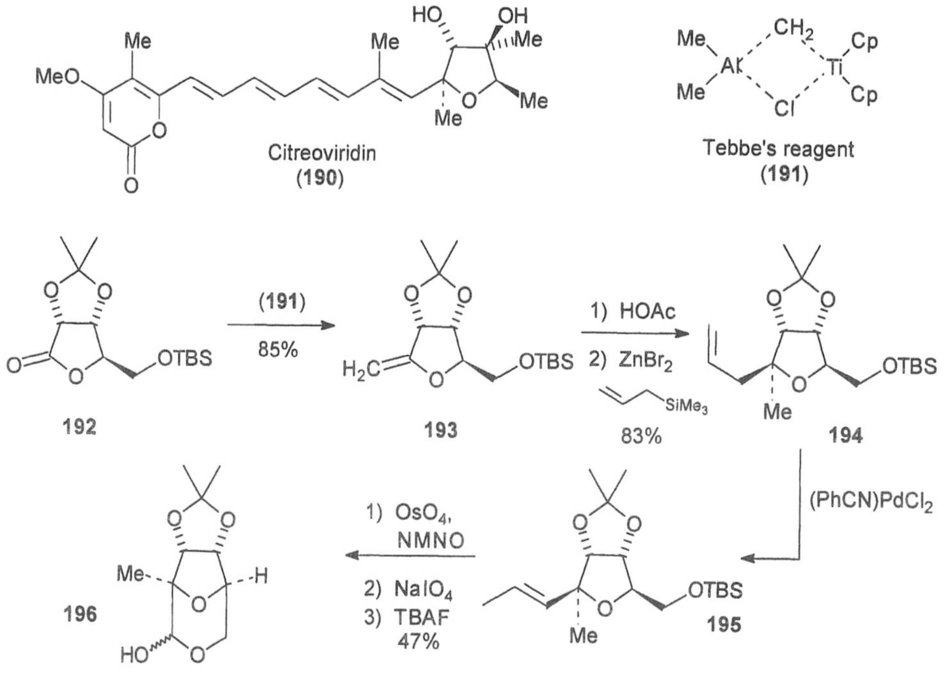

Scheme 32

RajanBabu and Reddy[36] have also utilized Tebbe's reagent in *C*-glycoside synthesis, Scheme 33. The gluconolactones (197 or 198) reacted readily with (191) to give the corresponding exomethylenic sugars 199 and 200. Compounds 203 and 204 were also prepared in a similar fashion. The exomethylenic sugars proved to be useful precursors to *C*-glycosides. In one example, 200 could be selectively hydroborated to the β-*C*-glucopyranoside 202 with 9-BBN. When the same reaction was carried out with borane-THF complex a 1:1 mixture of α- and β-*C*-glycosides was produced.

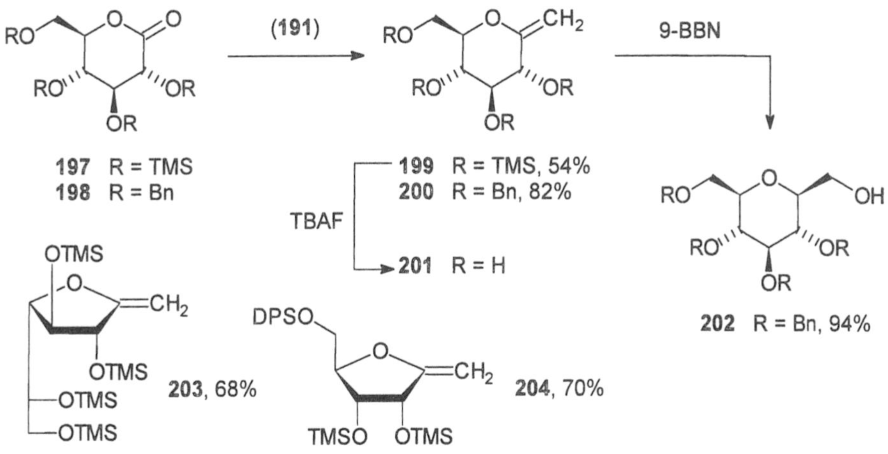

Scheme 33

Csuk and Glänzer[37] have applied 211 as a convenient methylenation reagent for sugar lactones. The reactions proceed in good yield (Scheme 34) and the advantage is that 211 is stable in the dark at -20°C and has a considerable shelf life.

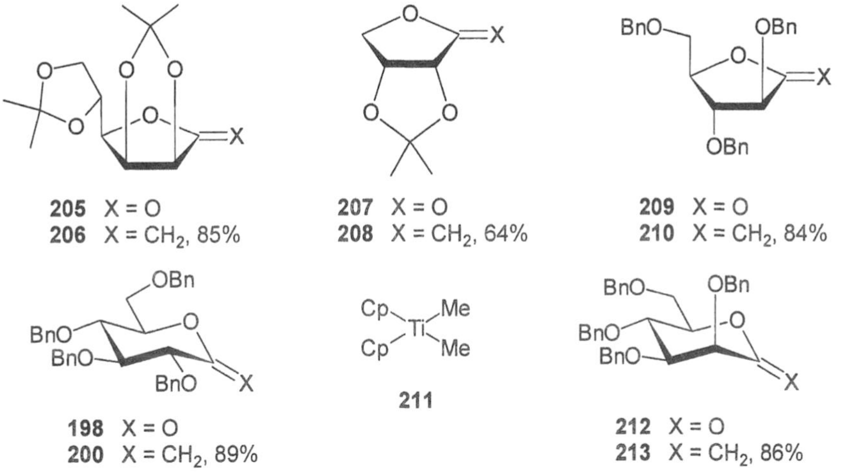

Scheme 34

Chapleur and Bandzouzi[38] have used the hexamethylphosphorous triamide-carbon tetrachloride system to synthesize dichloroolefinic C-glycosides in one step from sugar lactones. Addition of the phosphine to CCl$_4$ followed by addition of the lactone provided good yields of the dichloroolefins 214, 217, and 220. It was found that treatment of these compounds with Raney nickel caused reduction as shown in Scheme 35 with the addition of hydrogen occurring from the less hindered face. Exposure of the dichloroolefin 214 to LDA caused elimination to the corresponding ketone 222 which when treated with Raney nickel gave a mixture of diastereomeric products 223 and 224. The corresponding dibromoolefination reaction is a well known olefination process for ketones and aldehydes, but it failed to give any isolable product when applied to lactones.

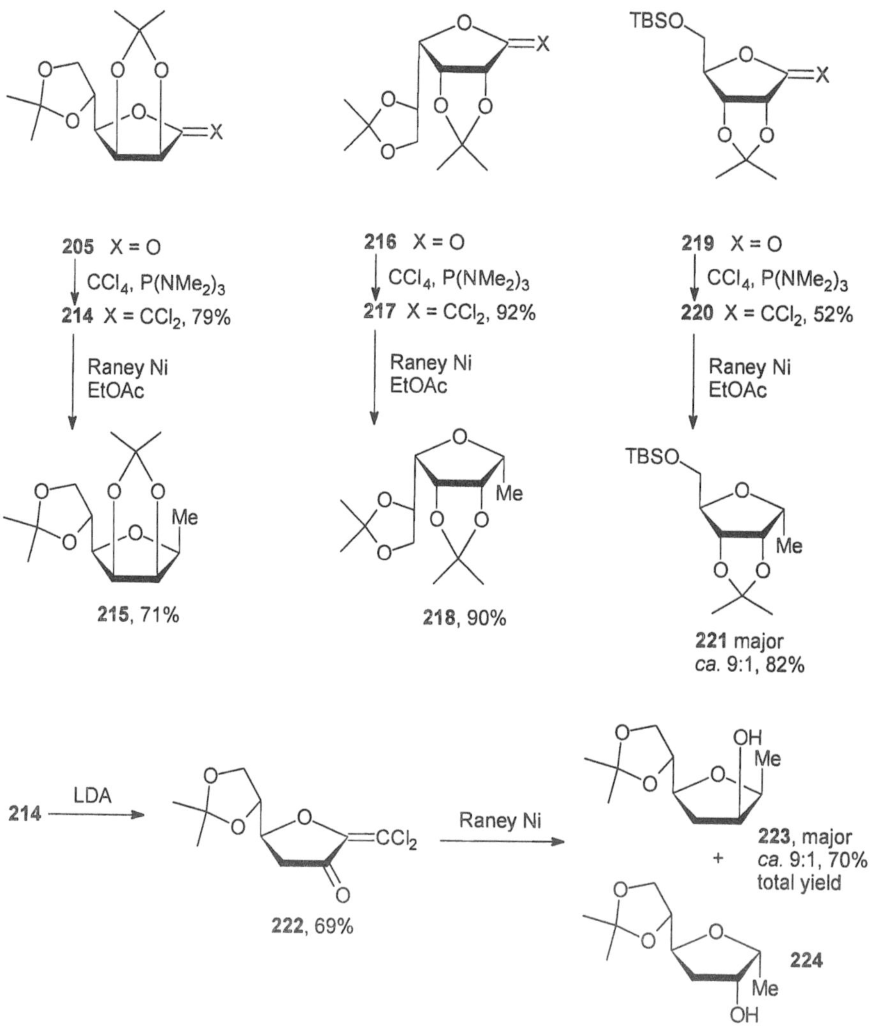

205 X = O
| CCl₄, P(NMe₂)₃
214 X = CCl₂, 79%

Raney Ni
EtOAc

215, 71%

216 X = O
| CCl₄, P(NMe₂)₃
217 X = CCl₂, 92%

Raney Ni
EtOAc

218, 90%

219 X = O
| CCl₄, P(NMe₂)₃
220 X = CCl₂, 52%

Raney Ni
EtOAc

221 major
ca. 9:1, 82%

214 —LDA→

222, 69%

Raney Ni →

223, major
ca. 9:1, 70%
total yield

+

224

Scheme 35

Other examples support the viability of this method to synthesize dichloroolefinic *C*-glycosides in improved yields from sugar lactones (Scheme 36).[39]

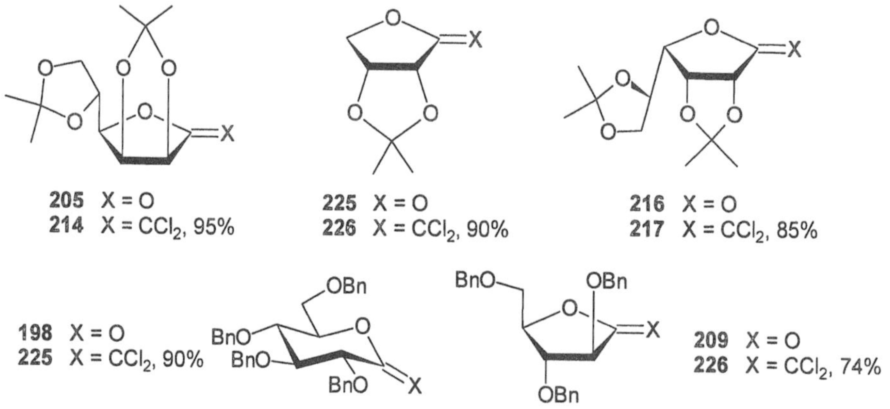

205 X = O
214 X = CCl₂, 95%

225 X = O
226 X = CCl₂, 90%

216 X = O
217 X = CCl₂, 85%

198 X = O
225 X = CCl₂, 90%

209 X = O
226 X = CCl₂, 74%

Scheme 36

Motherwell[40] has used a conceptually similar approach to produce the corresponding difluoroolefins from sugar lactones. This is illustrated with the lactone 219. Treatment of the reagent derived from reaction of tris(dimethylamino)phosphine with dibromofluoroethane and the substrate in THF followed by addition of zinc and 10% phosphine provided the difluoromethylene compound 227 in 68% yield. Hydrogenation then furnished the α-isomer 228 (95%) as the sole product of reaction. Four other furanose lactones were also studied and led to the difluoroenol ethers 229-232. The gluconolactone 233 was transformed into 235 as shown in Scheme 37.

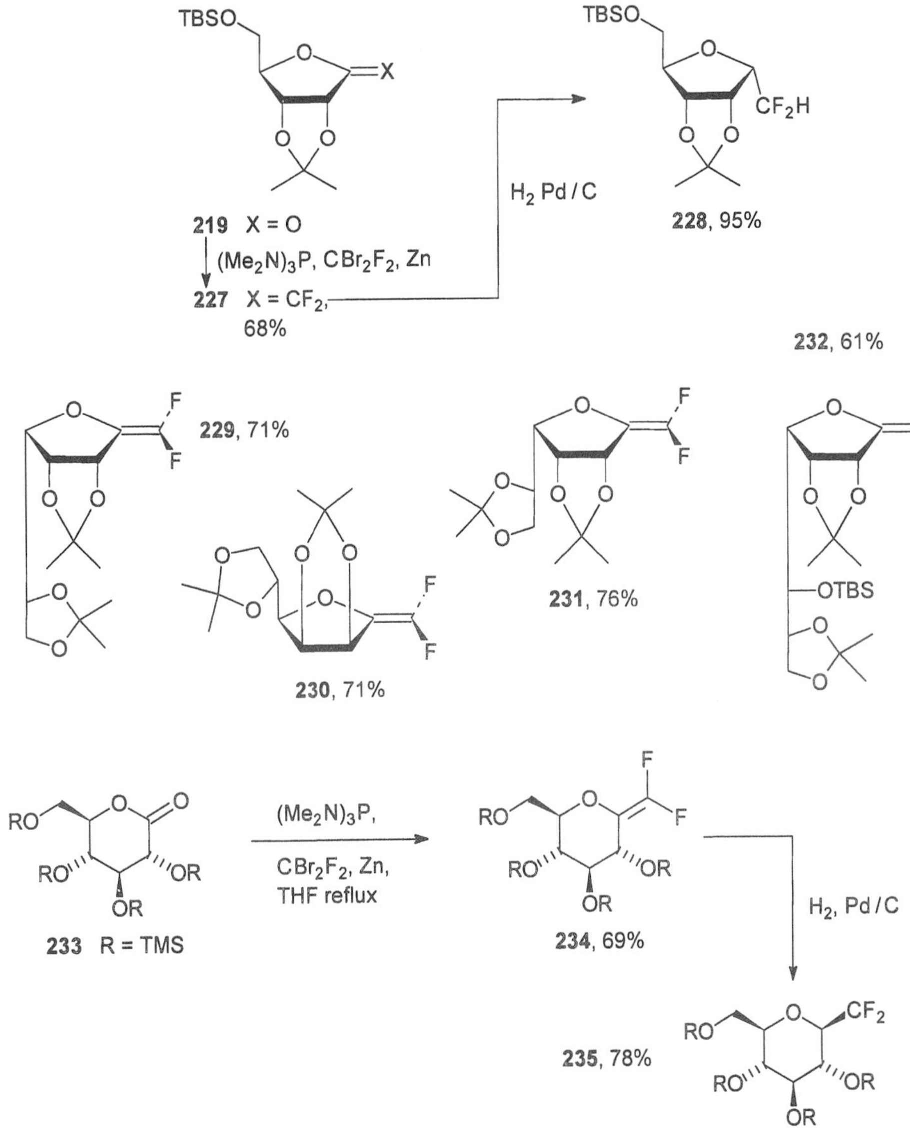

Scheme 37

VII. CONCLUSION

This chapter has illustrated the use of Wittig olefination on sugar lactols to synthesize *C*-glycosides. Many different types of ylides have been utilized with the ester phosphoranes being the most popular. Horner-Emmons-Wadsworth approaches have also seen some utility. If the ylide is unstabilized (*i.e.* $Ph_2P=CH_2$), then the open chain *C*-glycoside is isolated.[41] Various reagents have been employed to cyclize these compounds to the corresponding *C*-glycosides. The methods for stereoselective construction have been highlighted and there exists a fair number of ways to gain access to either the α- or β-anomer. Generally, for *C*-pyranosides the equatorial or β-anomer is the thermodynamically more stable and with ester substituted *C*-glycosides they can be favored by equilibration.[42] Access to the α-anomer can usually be achieved through the use of mercuriocyclization, whereas other electrophilic reagents have the tendency to provide mixtures of anomers. Epoxidations on lactols, via sulfur ylides, have also been recently utilized for *C*-glycoside formation. Finally, the utility of exomethylenic sugars, both substituted and unsubstituted, has been reviewed. The field of Wittig olefination on sugars is a mature one, but work is still being carried out that holds the promise of milder and more stereoselective methods for *C*-glycoside assembly.

VIII. REFERENCES

1. Zhdanov, Y. A.; Polenov, V.A. *J. Gen. Chem. USSR* **1969**, *39*, 107. These workers were the first to use Wittig olefination to synthesize *C*-glycosides. For example:

See also: Zhdanov, Y.A.; Alexeev, Y.E.; Alexeeva, V.G. *Adv. Carbohydr. Chem. Biochem.* **1972**, *27*, 227.
2. Hanessian, S.; Ogawa, T.; Guindon, Y. *Carbohydr. Res.* **1974**, *38*, C12.
3. Ohrui, H.; Jones, G.H.; Moffatt, J.G.; Maddox, M.L.; Christensen, A.T.; Bryam, S.K. *J. Am. Chem. Soc.* **1975**, *97*, 4602. See also: Jones, G.H. and Moffatt, J.G. *J. Carbohydr., Nucleosides and Nucleotides* **1979**, *6*, 127.
4. Sun, K.M.; Fraser-Reid. B. *Can. J. Chem.* **1980**, *58*, 2732.
5. Mandal, S.B.; Achari, B. *Synthetic Commun.* **1993**, *23*, 1239.
6. Rokach, J.; Lau, C.-K.; Zamboni, R.; Guindon, Y. *Tetrahedron Lett.* **1981**, *22*, 2763.
7. Dondoni, A.; Marra, A. *Tetrahedron Lett.* **1993**, *34*, 7327.
8. Nicolaou, K.C.; Pavia, M.R.; Seitz, S.P. *Tetrahedron Lett.* **1979**, *25*, 2327.
9. Dawe, R.D.; Fraser-Reid, B. *J. Org. Chem.* **1984**, *49*, 522.
10. Nicotra, F.; Russo, G.; Ronchetti, F.; Toma, L. *Carbohydr. Res.* **1983**, *124*, C5.
11. Giannis, A.; Sandhoff, K. *Carbohydr. Res.* **1987**, *171*, 201.

12. Nicotra, F.; Ronchetti, F.; Russo, G.; Toma, L. *Tetrahedron Lett.* **1984**, *25*, 5697. For a synthetic application on the use of a related diene product see: Reitz, A.B.; Jordan, Jr., A.D.; Maryanoff, B.E. *J. Org. Chem.* **1987**, *52*, 4800.

13. Reed, III, L.A.; Ito, Y.; Masamune, S.; Sharpless, B.K. *J. Am. Chem. Soc.* **1982**, *104*, 6468.
14. Dheilly, L.; Lièvre, C.; Fréchou, C.; Demailly, G. *Tetrahedron Lett.* **1993**, *34*, 5895.
15. Dheilly, L.; Fréchou, C.; Beaupère, D.; Uzan, R.; Demailly, G. *Carbohydr. Res.* **1992**, *224*, 301.
16. Allevi, P.; Ciuffreda, P.; Colombo, D.; Monti, D.; Speranza, G. Manitto, P. *J. Chem. Soc., Perkin Trans. I* **1989**, 1281. See also: Monti, D.; Gramatica, P.; Speranza, G.; Manitto, P. *Tetrahedron Lett.* **1987**, *28*, 5047.
17. Davidson, A.H.; Hughes, L.R.; Qureshi, S.S.; Wright, B. *Tetrahedron Lett.* **1988**, *29*, 693.
18. Mbongo, A.; Fréchou, C.; Beaupère, D.; Uzan, R.; Demailly, G. *Carbohydr. Res.* **1993**, *246*, 361.
19. Pougny, J.-R.; Nassr, M.A.M.; Sinaÿ, P. *J. Chem. Soc., Chem. Commun.* **1981**, 375.
20. Nicotra, F.; Perego, R.; Ronchetti, F.; Russo, G. ; Toma, L. *Gazz. Chem. Ital.* **1984**, *114*, 193.
21. Nicotra, F.; Ronchetti, F.; Russo, G.; *J. Org. Chem.* **1982**, *47*, 4459. See also Nicotra, F.; Perego, R.; Ronchetti, F.; Russo, G.; Toma, L. *Carbohydr. Res.* **1984**, *131*, 180 for similar chemistry in the manno series.
22. Engel, R. *Chem. Rev.* **1977**, *77*, 349.
23. Nicotra, F.; Panza, L.; Ronchetti, F.; Toma, L. *Tetrahedron Lett.* **1984**, *25*, 5937.
24. Carcano, M.; Nicotra, F.; Panza, L.; Russo, G. *J. Chem. Soc., Chem. Commun.* **1989**, 297.
25. Nictora, F.; Panza, L.; Ronchetti, F.; Russo, G.; Toma, L. *Carbohydr. Res.* **1987**, *171*, 49. See also: López Herrera, F.J.; Pino Gonzalez, M.S.; Nieto Sampedro, M. ; Dominguez Aciego, R.M. *Tetrahedron* **1989**, *45*, 269.
26. Freeman, F.; Robarge, K.D. *Carbohydr. Res.* **1987**, *171*, 1.
27. Maryanoff, B.E.; Nortey, S.O.; Inners, R.R.; Campbell, S.A.; Reitz, A.B.; Liotta, D. *Carbohydr. Res.* **1987**, *171*, 259.
28. Freeman, F.; Robarge, K.D. *Tetrahedron Lett.* **1985**, *26*, 1943.
29. Lancelin, J.-M.; Pougny, J.-R.; Sinaÿ, P. *Carbohydr. Res.* **1985**, *136*, 369.
30. Hart, D.J.; Leroy, V.; Merriman, G.H.; Young, D.G.J. *J. Org. Chem.* **1992**, *57*, 5670.

31. Drew, M.G.B.; Kane, P.D.; Mann, J.; Naili, M. *J. Chem. Soc., Perkin Trans. I* **1988**, 433.

32. Valpuesta, M.; Durante, P.; López-Herrera, F.J. *Tetrahedron* **1993**, *49*, 9547.

33. Fréchou, C.; Dheilly, L.; Beaupère, D.; Uzan, R.; Demailly, G. *Tetrahedron Lett.* **1992**, *33*, 5067.

34. Wilcox, C.S.; Long, G.W.; Suh, H. *Tetrahedron Lett.* **1984**, *25*, 395.

35. Tebbe, F.N.; Parshall, G.W.; Reddy, G.S. *J. Am. Chem. Soc.* **1978**, *100*, 3611.

36. RajanBabu, T.V.; Reddy, G.S. *J. Org. Chem.* **1986**, *51*, 5458.

37. Csuk, R.; Glänzer, B.I. *Tetrahedron* **1991**, *47*, 1655.

38. Bandzouzi, A.; Chapleur, Y. *Carbohydr. Res.* **1987**, *171*, 13.

39. Lakhrissi, M.; Chapleur, Y. *Synlett* **1991**, 583.

40. Houlton, S.J.; Motherwell, W.B.; Ross, B.C.; Tozer, M.J.; Williams, D.J.; Slawin, A.M.Z. *Tetrahedron* **1993**, *49*, 8087.

41. For a recent article dealing with the preparation of open chain sugars via Wittig olefination from unprotected sugars see: Henk, T.; Giannis, A.; Sandhoff, K. *Liebigs Ann. Chem.* **1992**, 167.

42. For a study on epimerization of α- to β-*C*-glucopyranosides see: Alllevi, P.; Anastasia, M.; Ciuffreda, P.; Fiechi, A.; Scala, A. *J. Chem. Soc., Perkin Trans. I* **1989**, 1275.

Chapter 5

PALLADIUM MEDIATED APPROACHES TO *C*-GLYCOSIDES

I. INTRODUCTION

The extensive development of palladium chemistry over the last two decades has impacted on the field of *C*-glycoside synthesis. A few groups have made significant contributions to the understanding and application of this mild and powerful method for forming carbon-carbon bonds. This chapter will be divided into two main sections. The first part will cover the chemistry of π-allyl complexes and how this type of chemistry has been successfully applied to *C*-glycoside formation. The second section will address the use of Heck type couplings in anomeric carbon-carbon bond formation. Simple glycals, stannanes, and halides will all be examined and the generality of the methods compared. Some mechanistic discussion will be presented, but only as far as to rationalize the observed results and allow the reader to make reasonable stereochemical predictions.

II. π-ALLYL COMPLEXES

Dunkerton was the first to apply the chemistry of π-allylic systems to the preparation of simple *C*-glycosides.[1] Initial work focused on a simple system, dihydrofuran (**1**). Treatment of **1** with Pd(II) followed by addition of the stabilized anion gave the *C*-glycosides **2a** to **2c**.

eq. 1

Table 1: Nucleophiles and Products for Equation 1.

RM	R	Compound, Yield
$NaC(NHCHO)(CO_2Et)_2$	NHCHO	**2a**, 76%
$NaC(NHAc)(CO_2Et)_2$	NHAc	**2b**, 73%
$NaC(Me)(CO_2Et)_2$	Me	**2c**, 50%

If an unsubstituted malonate was used then the exocyclic olefin **3** was isolated in good yield. This is due to the acidity of the α-proton that is eliminated during the β-elimination of palladium hydride instead of one of the ring protons, eq. 2. This constitutes a fairly quick method for making exomethylenic sugars since normally they are prepared from the corresponding lactones by olefination.

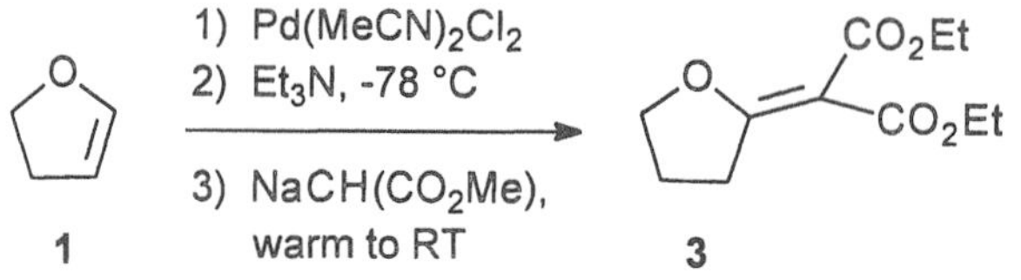

1) Pd(MeCN)$_2$Cl$_2$
2) Et$_3$N, -78 °C

3) NaCH(CO$_2$Me),
   warm to RT

eq. 2

Further work by the same group[2] extended this methodology to the dihydropyran **4**. In this case, the starting material contains a stereogenic center that allows for some insight into the mechanism of the reaction. When **4** is treated with Pd(Ph$_3$P)$_4$, Ph$_3$P in DMF followed by addition of a suitable anion the product of net retention **5** is obtained in excellent yield. If no Ph$_3$P is added and only the catalyst and starting material are allowed to stir in DMF for 2 h prior to addition of the anion, then the product of net inversion **6** is obtained. The stereoselectivity of the reaction can thereby be controlled simply by altering the timing of reagent addition. Presumably the palladium is isomerizing the starting material to a stable *trans* arrangement and then displacement occurs, with retention, to provide the product of net inversion. If a non-stabilized carbanion is used and the reaction is conducted in THF then the product of net inversion is obtained exclusively. The reaction was not limited to aromatic nucleophiles as shown by formation of the vinyl *C*-glycoside **8**, Scheme 1.

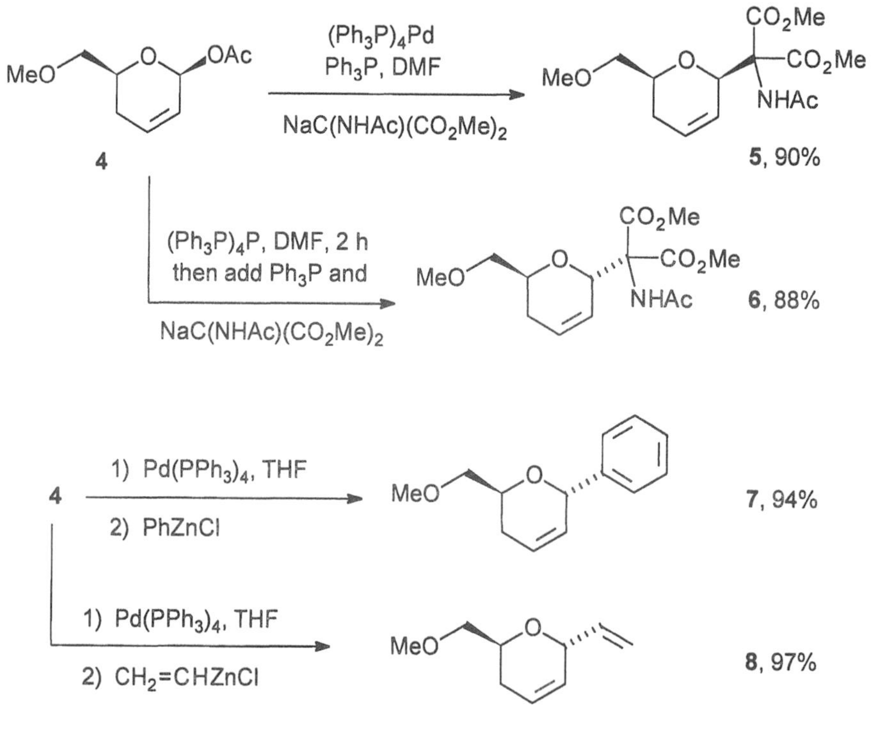

Scheme 1

The more complex substrate **9** (containing 14% of the other isomer) was treated under the arylzinc chloride protocol and gave as the major product **10** in a 86:14 ratio in 16% yield. Compound **11** when exposed to the same conditions furnished **12** in 23% isolated yield, Scheme 2.

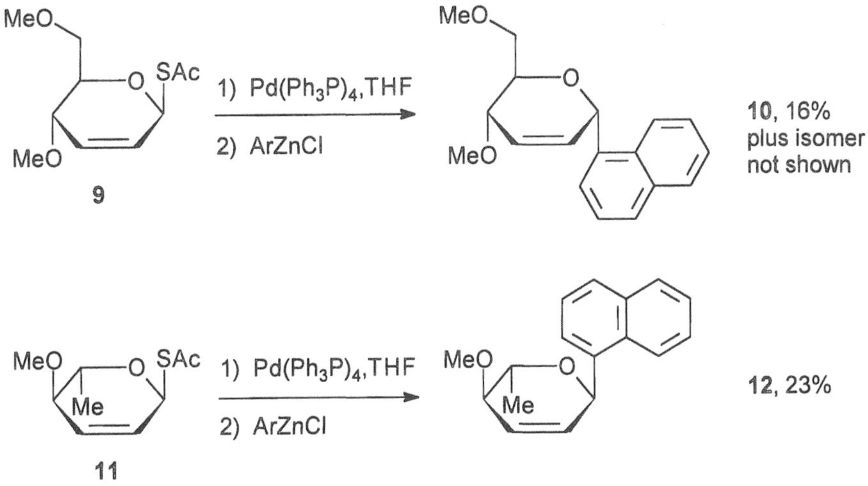

Scheme 2

Miwa and Yougai[3] examined the palladium catalyzed reaction of dicarbonyl compounds with various glycals and some of their results are summarized in Scheme 3. In almost all of the cases the α-anomer was the major or exclusive product formed.

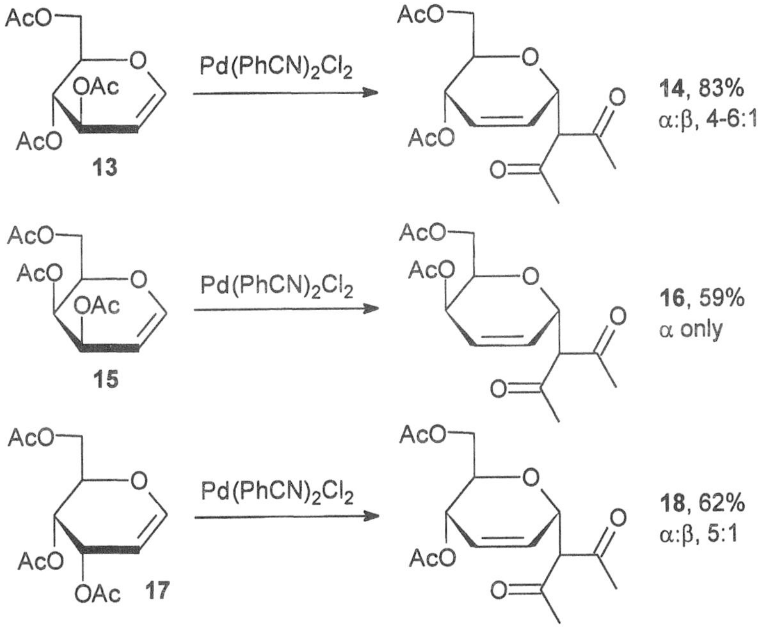

Scheme 3

The reaction of xylal and arabinal gave opposite stereochemical results to the other examples. In this case the β-anomer was formed as the major product in a 4:1 ratio over the α-anomer. This means that the reaction must be proceeding via the same intermediate (**21**) and that the addition of the nucleophile occurs from the top face, Scheme 4.

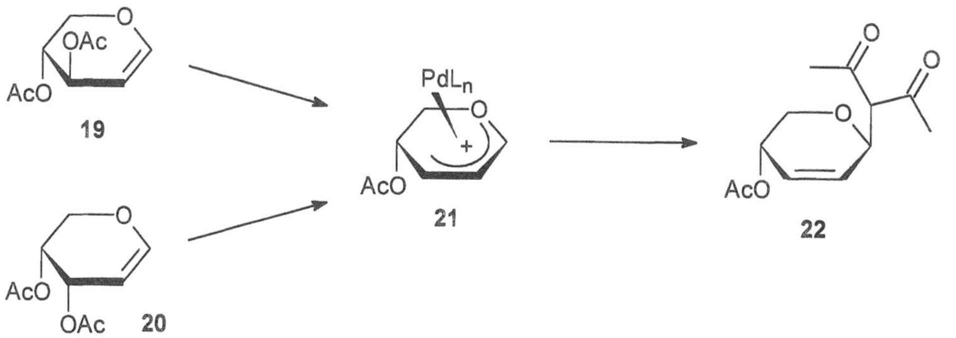

Scheme 4

RajanBabu[4] has found that if the *O*-3 protecting group is changed to trifluoroacetate, as in **23**, then only mild conditions are required for carbon-carbon formation to take place. Reaction of the corresponding acetate or phosphate failed to give any product under previously described conditions. It is interesting to note that no exomethylenic sugar is isolated from the reaction mixture. Also, compound **24** is stable enough to undergo basic benzylation to **25** in which there has been no anomerization. The key reaction proceeds by addition of the nucleophile from the same face as the leaving group. The reaction also works with β-diketones as nucleophiles, Scheme 5.

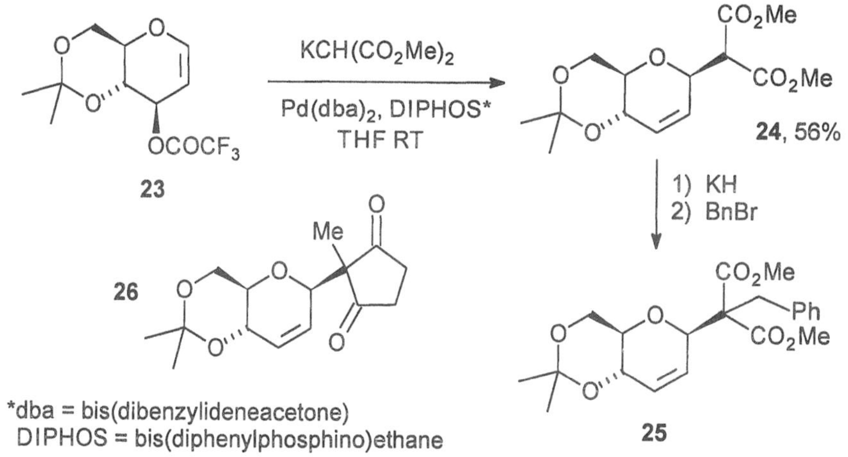

*dba = bis(dibenzylideneacetone)
DIPHOS = bis(diphenylphosphino)ethane

Scheme 5

Sinou and co-workers[5] have carried out a comprehensive study on a similar type of reaction. They found that reaction of **13** under various conditions gave none of the expected *C*-glycosidic products. They therefore chose to use the 2,3-unsaturated phenyl glycoside **29** as the starting material to make the π-allyl complex. Reaction of **13** under standard conditions gave an 85:15 ratio of α and β anomers **29** and **27**, respectively. Again reaction under standard conditions gave no *C*-glycoside. The acetate groups were substituted for benzyl groups and reaction of the β-anomer **28** with various nucleophiles gave the expected products of retention, Table 2.

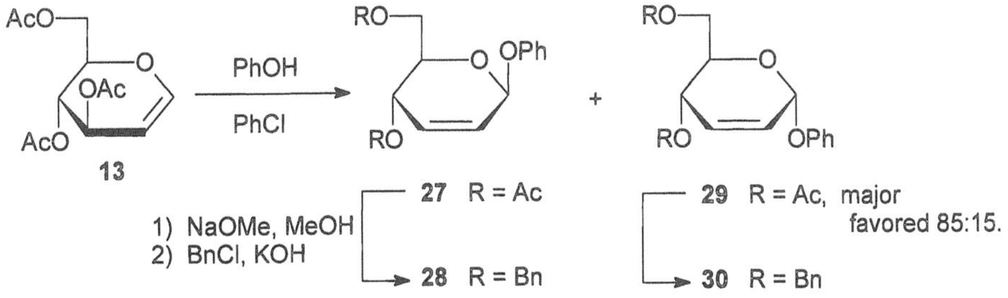

Scheme 6

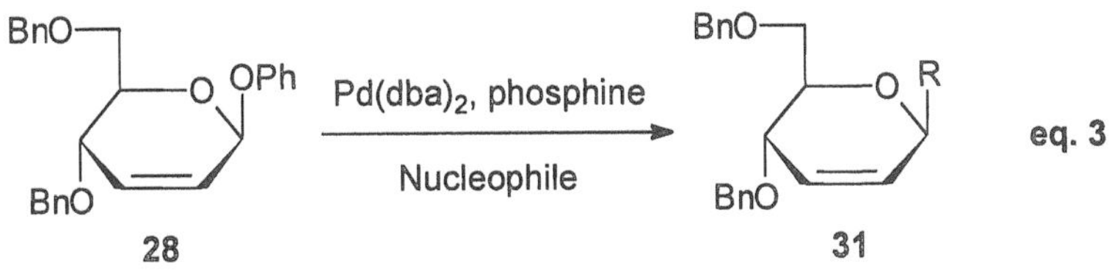

eq. 3

Table 2: Reactions with β-anomer 28.

Entry	R-H	Solvent T, °C / Phosphine	Product, Yield, α:β
1	$CH_2(CO_2Et)_2$	MeCN, 70 / dppb	**31a**, 82%, 0:100
2	$CH_2(COMe)_2$	MeCN, 70 / dppb	**31b**, 84%, 0:100
3	$NO_2CH_2CO_2Et$	THF, 60 / PPh$_3$	**31c**, 72%, 0:100
4	$NO_2CH(CO_2Et)_2$	MeCN, 70 / dppb	**31d**, 92%, 0:100

dppb = 1,4-bis(diphenylphosphinobutane)

The reaction of the α-anomer **30** was stereoselective only when methyl, ethyl nitroacetate or ethyl nitromalonate was used as the nucleophile.

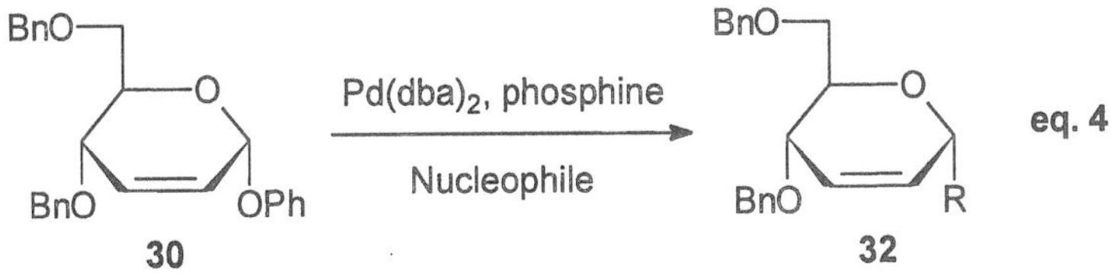

eq. 4

Table 3: Reactions with α-anomer 30.

Entry	R-H	Solvent T, °C / Phosphine	Product, Yield, α:β
1	$CH_2(CO_2Et)_2$	MeCN, 70 / dppb	**32a**, 60%, 75:25
2	$CH_2(COMe)_2$	MeCN, 70 / PPh$_3$	**32b**, 40%, 24:76
3	$NO_2CH_2CO_2Et$	THF, 60 / PPh$_3$	**32c**, 75%, 100:0
4	$NO_2CH(CO_2Et)_2$	THF, 60 / dppb	**32d**, 90%, 100:0
5	$NO_2CH(CO_2Et)_2$	MeCN, 70 / dppb	**32e**, 93%, 100:0

dppb = 1,4-bis(diphenylphosphinobutane)

Access to the α-anomer **33** was not a concern since reductive denitration of **32e** gave **33**.

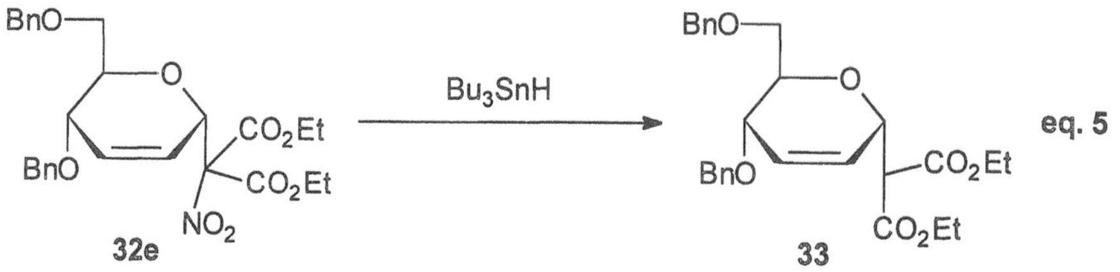

eq. 5

No explanation was put forward to account for the difference in selectivity between the formation of the α- and β-anomers. In further work[6] the authors postulated that the liberated phenoxide can act as a base and ring open the α-*C*-glycoside to an intermediate such as 34 which can recyclize to give some of the thermodynamically favored β-anomer.

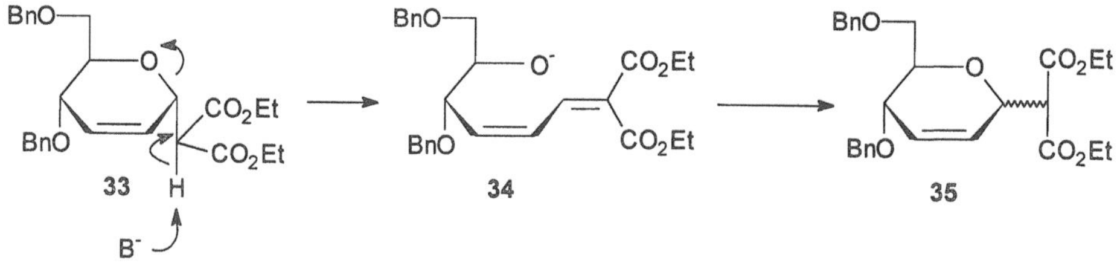

Scheme 7

The results in Table 4 support this hypothesis. Mixtures were obtained when an excess of phosphine was used with malonate or acetylacetone as the nucleophile (entries 1 to 4). Entries 5 and 6 show the effect of temperature. At lower temperature no anomerization is observed since ring opening is favored at higher temperatures. The final confirmation comes from entries 7, 8, and 9 where no acidic proton is available for abstraction to have ring opening occur. In these cases only the α-anomer is observed.

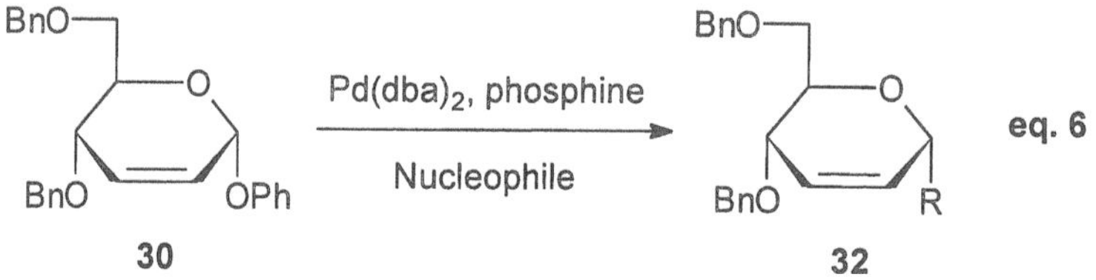

eq. 6

Table 4: Reactions with α-anomer 30, Results of Anomerization.

Entry	NuH	solvent T, °C / Phosphine	Product, Yield, α:β
1	$CH_2(CO_2Et)_2$	MeCN, 60 / dppb	32a, 60%, 75:25
2	$CH_2(CO_2Et)_2$	MeCN, 60 / dppb + 3PPh$_3$	32a, 67%, 75:25
3	$CH_2(COMe)_2$	MeCN, 60 / dppb	32b, 67%, 75:25
4	$CH_2(COMe)_2$	MeCN, 60 / PPh$_3$	32b, 40%, 24:76
5	$NO_2CH_2CO_2Et$	MeCN 60 / dppb	32c, 80%, 100:0
6	$NO_2CH_2CO_2Et$	MeCN, 80 / dppb	32c, 80%, 79:21
7	$CH(COMe)(CO_2Me)_2$	MeCN, 80 / dppb	32e, 64%, 100:0
8	$NaC(CO_2Et)_2Me$	THF, 60 / dppb	32f, 88%, 100:0
9	$NaC(CO_2Et)_2CH_2$-CH=CH$_2$	THF, 60 / dppb	32g, 92%, 100:0

dppb = 1,4-bis(diphenylphosphinobutane)

Reprinted from ref. 6, pg. 612 by courtesy Marcel Dekker Inc.

Engelbrecht and Holzapfel[7] improved this type of reaction by employing pseudoglycals that have a carbonate group at the anomeric position. Treatment of the **36** with isobutyl chloroformate gave a 75% yield of the α-anomer **37**. Exposure of **36** to a catalytic amount Pd(PPh₃)₄ and PPh₃ to 10 eq. of diethyl malonate furnished α-*C*-glycoside **38** in 81% yield. Similarly, **39** and **40** were prepared by reaction with the appropriate nucleophile, Scheme 8. The β-isomer **41** also reacted by giving the product of retention **42** and **43**. These reactions proceeded well, but the excess of nucleophile was impractical and caused purification problems. It was subsequently found that reaction with the anion (of the nucleophiles) required only 2 equivalents of nucleophile to ensure complete reaction.

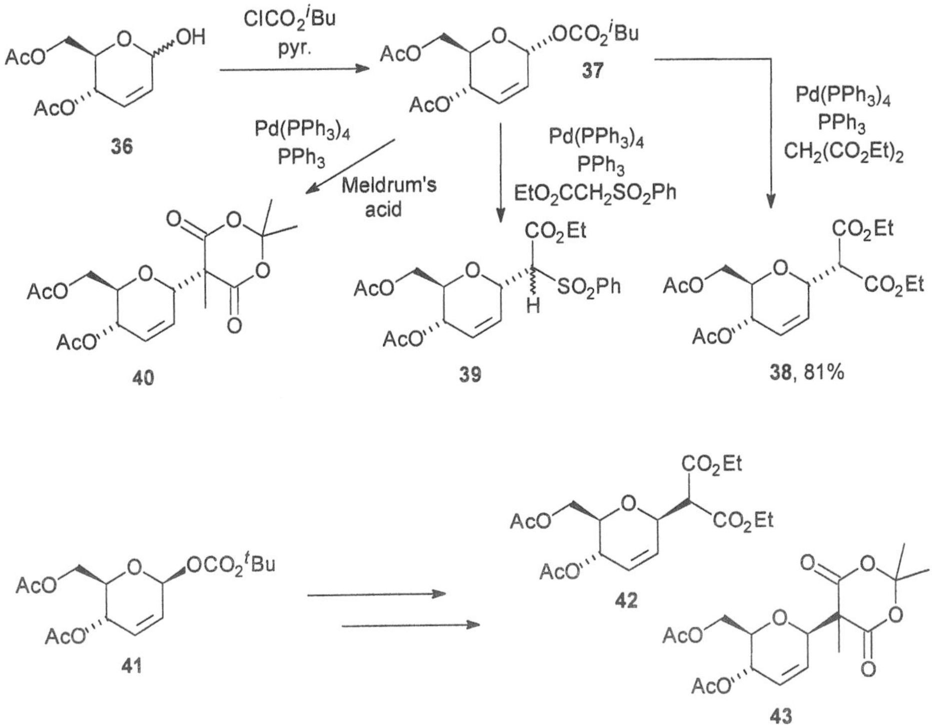

Scheme 8

Under these basic conditions some anomerization was observed probably via the open chain compound. Treatment of the *C*-glycoside **38** with triethylamine resulted in anomerization to a mixture of anomers (not shown). Tertiary nucleophiles reacted in the expected manner by giving only the products resulting from retention of configuration, eq. 7.

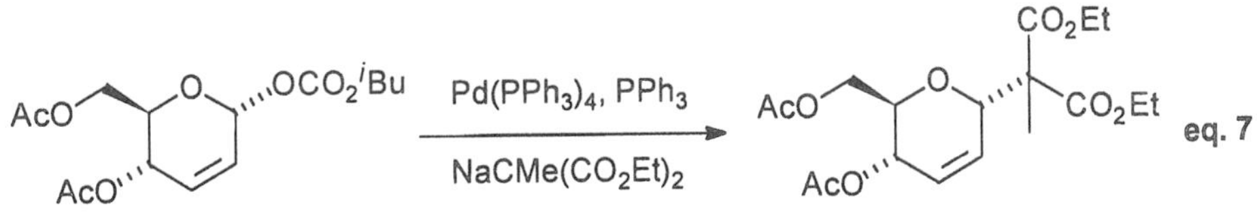

Although not *C*-glycosides in the strictest sense, compounds **47** and **49** possess some resemblance to the skeleton of a *C*-glycoside; they have a pyranoid ring with an adjacent carbon chain and a hydroxyl function. Hirama and co-workers[8] have assembled these compounds using palladium-mediated cyclization. Treatment of the silyl ether **45** with

TBAF followed by exposure to Pd(PPh$_3$)$_4$ in chloroform gave **47** favored over **49** in a 99:1 ratio in 90% yield. Similarly, **49** was available from **48** in 89% yield also in a ratio of 98:2 over **47**.

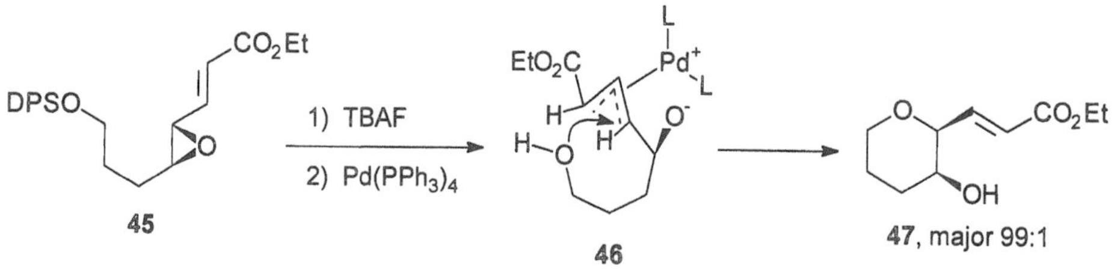

Scheme 9

III. HECK TYPE COUPLINGS

This section will be sub-divided into two main sections. The first will address the coupling of arylmercurials with glycals while the second section will focus on tin and zinc palladium mediated couplings with aryl halides. In the Heck reaction[9] of simple olefins regioisomers are often formed. In the coupling reaction with enol ethers the regiochemistry of the addition is often highly predictable. The first step in the accepted mechanism[10] is insertion of the aryl species into the palladium to give an active palladium species **51** that then complexes the olefin to give the π-complex **52**.

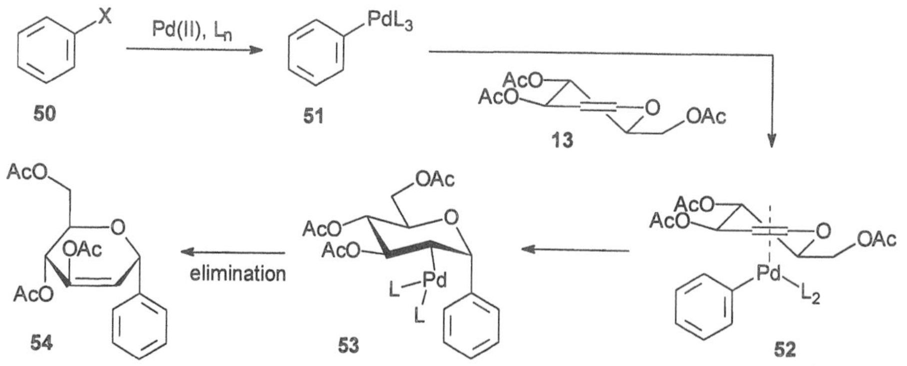

Scheme 10
Reprinted with permission from ref. 10b, Copyright 1990 American Chemical Society.

The adduct is then formed and elimination gives the product olefin, Scheme 10. The carbon-carbon bond that is formed occurs between *C*-1 of the enol ether and the incoming carbon group with *syn* addition of the palladium complex. The electropositive palladium atom is now bonded to the electron rich β-carbon atom and not the electron poor α-carbon atom.

A.   USE OF GLYCALS

In early work Daves and Arai[11] coupled the organomercurial acetate **56** with 3,4-dihydro-2*H*-pyran (**55**) and obtained two compounds **57** (66%) and **58** (24%). Compound **57** arises from elimination of palladium hydride from the intermediate adduct, while **58** arises by palladium-mediated isomerization of the olefin **57**.

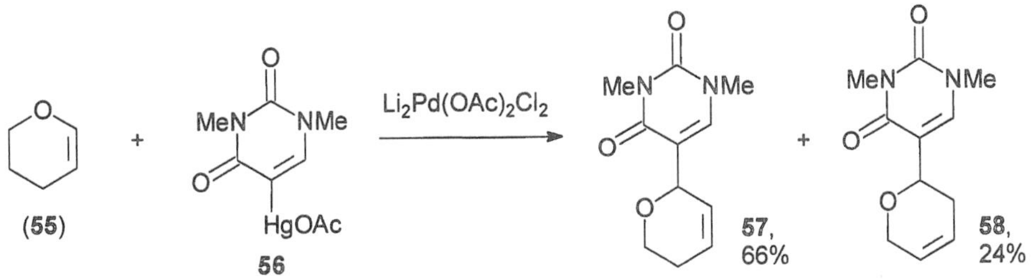

Scheme 11

Coupling with glycals[12] proceeded in moderate yield, but generally gave good facial selectivity. Often mixtures of products were obtained indicating that the elimination of palladium did not always go via the same mechanism, Scheme 12. In some cases *syn* elimination occurs (*e.g.* **59**) and in others *anti* elimination as in the formation of **60**. *Anti* elimination can also occur with elimination of alkoxide since in this case (**20**→**62**) the ring opened *C*-glycoside is isolated as the main constituent of the reaction mixture.

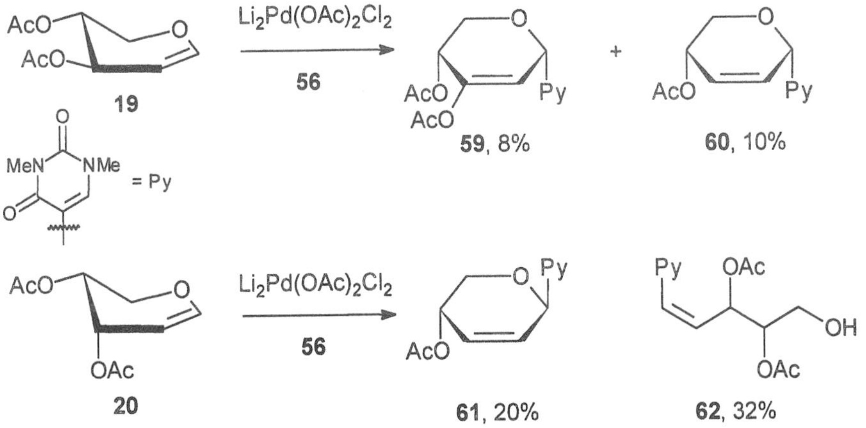

Scheme 12

The workers[13] then decided to probe the reason for the variation in the elimination step that leads to the different products since a predictable reaction is needed for synthetic usefulness. Coupling of **56** with the conformationally rigid benzylidene **63** gave only **64** which is formed by *anti* elimination palladium acetate from the intermediate adduct. If the conformationally mobile glycal was employed then two compounds **65** and **66** were isolated in 22% and 10% yield, respectively. When the leaving group ability of the *C*-3 substituent is reduced as in **67** then only the *syn* elimination product is isolated. Rigidity in the molecule does not permit the necessary conformational mobility required for a *syn* elimination. With a poor leaving group at *C*-3, *syn* elimination of palladium hydride is now observed.

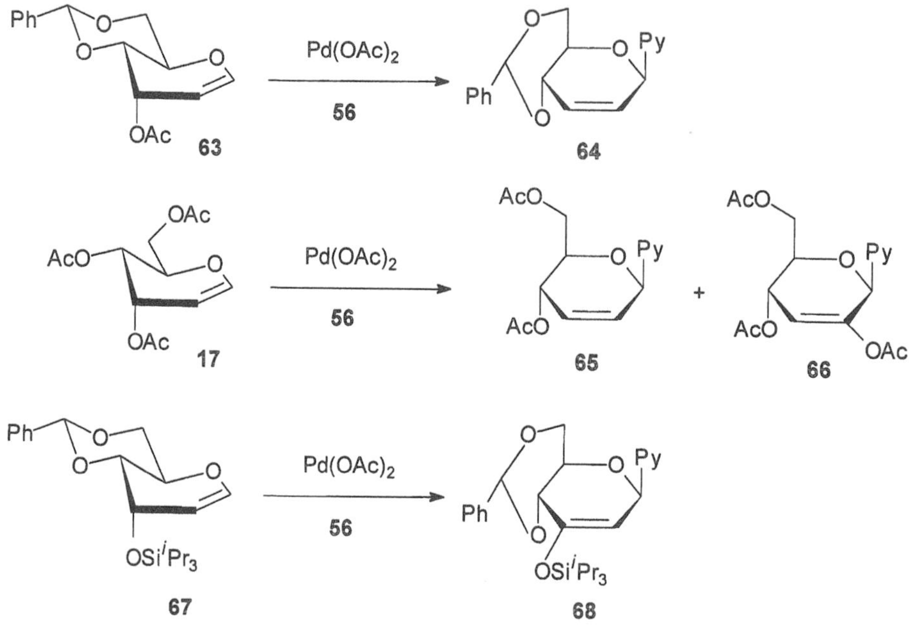

Scheme 13

Similar trends were observed in the furanoid series.[14] *Syn* elimination occurred to enol 70 with a poor leaving group at *C*-3 If an acetate group was used at *C*-3, then olefin 72 was isolated as the exclusive product. It is formed by *anti* elimination of palladium acetate, Scheme 14.

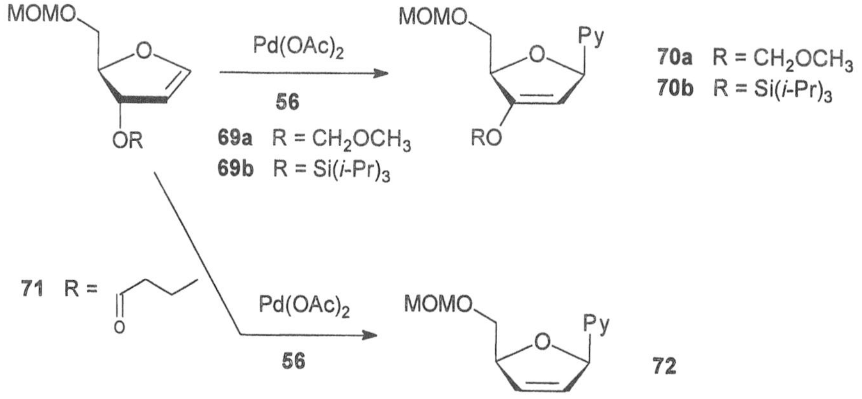

Scheme 14

If no protecting group is present on *O*-3 then a directed reaction takes place with delivery of the pyrimidine group from the bottom face. *Syn* elimination of palladium hydroxide then occurred to give the olefin 74, eq. 8. Other substrates were also studied.[15]

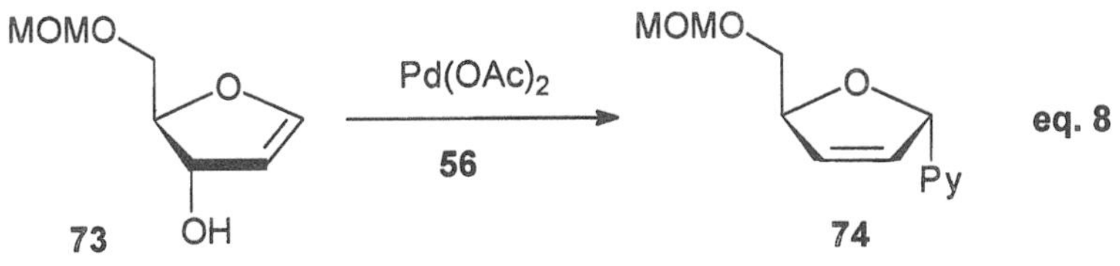

eq. 8

Work by Czernecki[16] has contributed to the understanding of these types of reactions. The palladium acetate catalyzed reactions of glycals **13** and **15** with benzene and the dimethoxy benzenoid compounds gave the *C*-glycosides as shown in Scheme 15. Other aromatic groups were also examined and gave comparable yields of products.

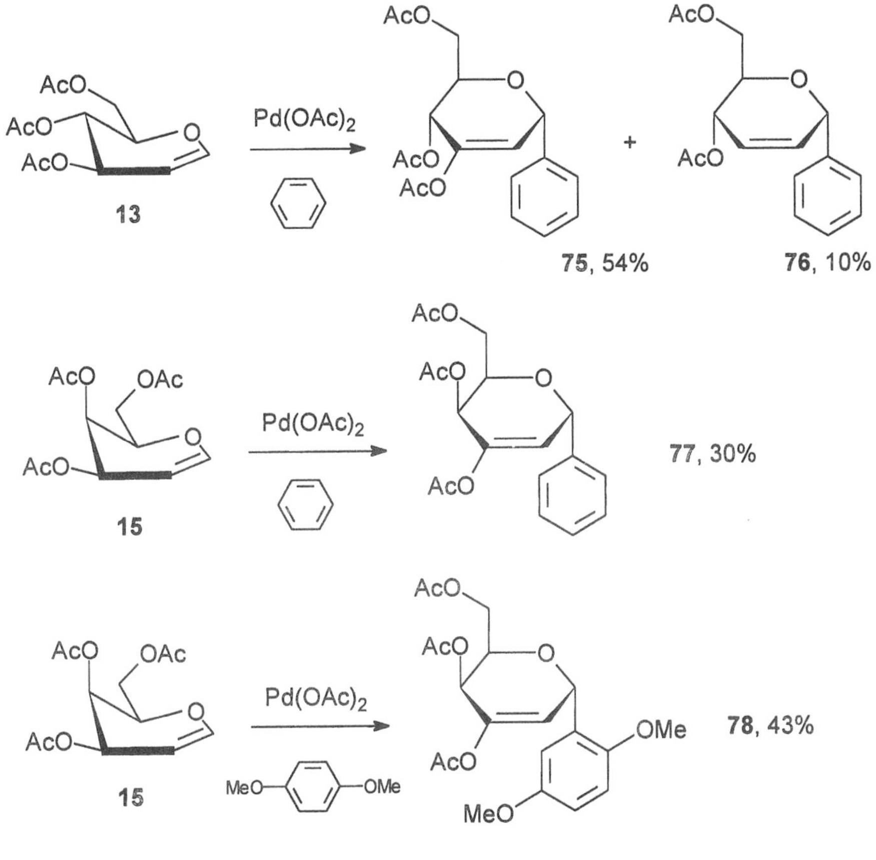

Scheme 15

In further work Czernecki[17] has examined the palladium mediated arylation of enones of the type **79**. The reactions were conducted in benzene in the presence of palladium acetate and acetic acid. In all the cases mixtures of products were formed, Scheme 16. The authors postulate that the intermediate adduct is epimerized to **89** that is now poised to undergo *syn* elimination of palladium hydride or protonolysis to either **80** or **81**. The enones (**81, 84,** and **87**) were then reduced by hydrogenation to give **90, 91,** and **92** in good yield, Scheme 17. The net result of these transformations is a good synthesis of aryl 2-deoxy-β-*C*-glycopyranosides from enones of the type **79**. This type of compound is important since it occurs in several naturally occurring compounds.

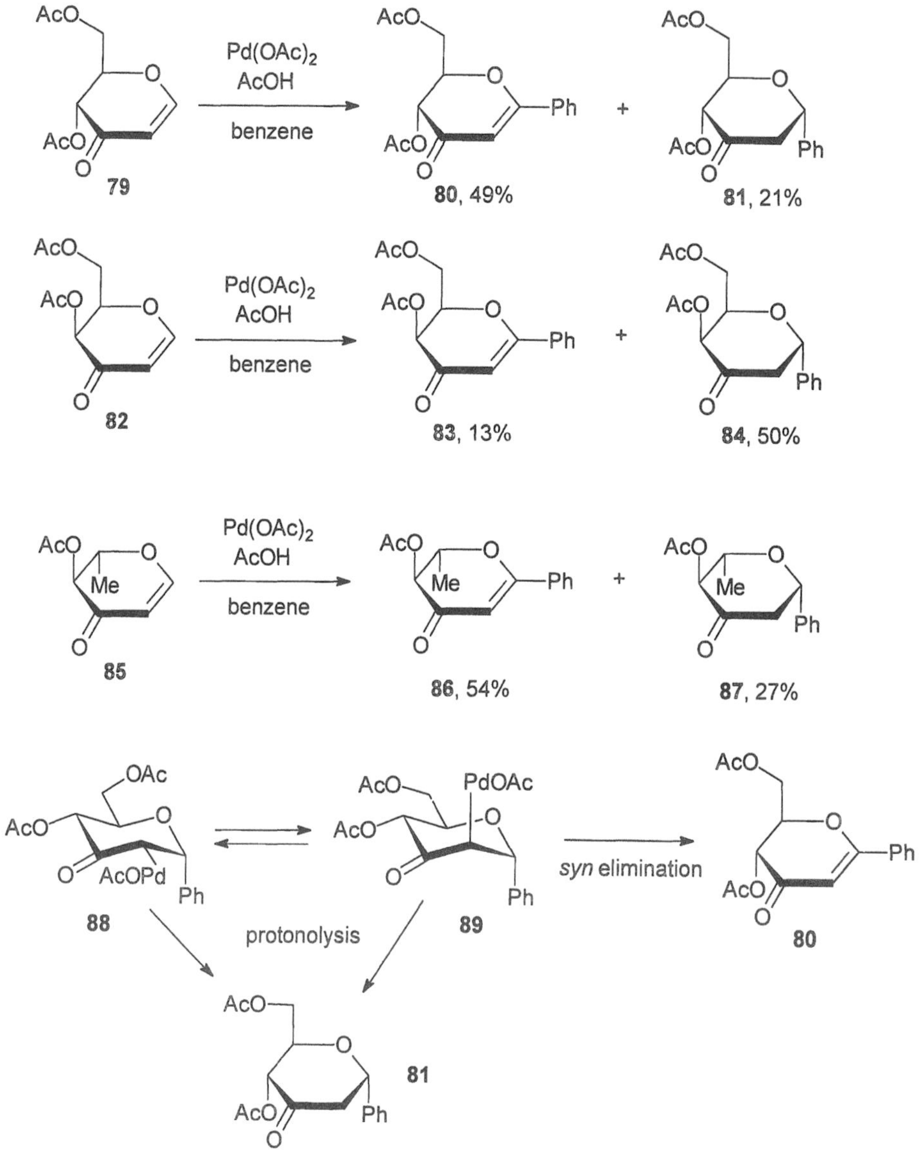

Scheme 16

Reprinted with permission from ref. 17, Copyright 1992 American Chemical Society.

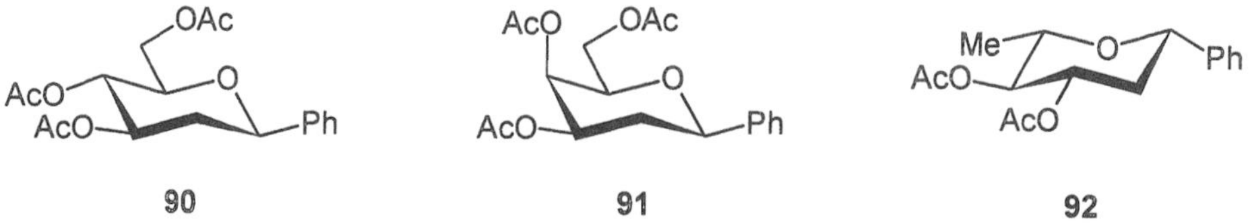

Scheme 17

All the examples of Heck type couplings have dealt with glycals or enones in which the double bond is endocyclic. RajanBabu and Reddy[18] have examined the Heck coupling of an aromatic ring and exomethylenic sugar 93. Two compounds were isolated from the reaction

94 and **95**. Acyclic compound **95** arises from elimination of palladium and the alkoxide while **94** comes from an elimination of palladium hydride. The open chain compound was cyclized via the mercury method to give, after reduction, the (bis)-*C,C*-glycoside **96**, Scheme 18.

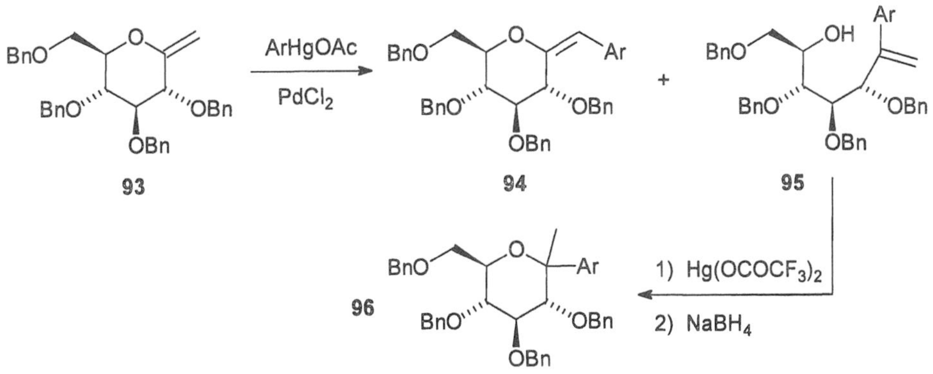

Scheme 18

Daves[19] has applied the technology developed in his laboratories in model studies towards the gilvocarcins. Coupling of iodide **98** with **97** under palladium acetate catalysis gave adduct **99** exclusively which was desilylated under the reaction conditions to **100**.

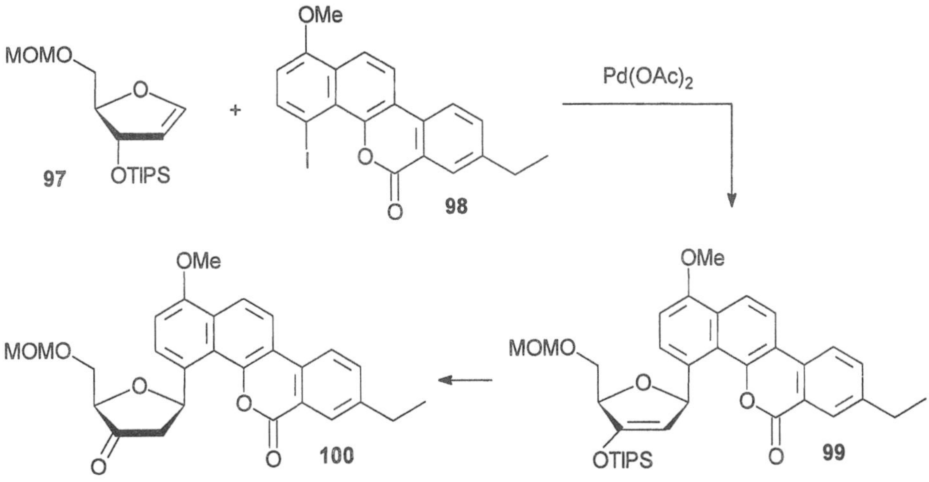

Scheme 19

A more impressive and efficient synthesis is shown in Scheme 20. Coupling of **98** with **101**, in which the 5-hydroxyl is free, gave **102** which was desilylated by the action of TBAF to give hydroxy-ketone **103** that was acetylated to **104** in 89% overall yield. Directed reduction of **103** with triacetoxyborohydride provided **106**, after acetylation. These two sequences show the power of the palladium mediated coupling for assembling *C*-glycosides in a very efficient and highly stereocontrolled manner.

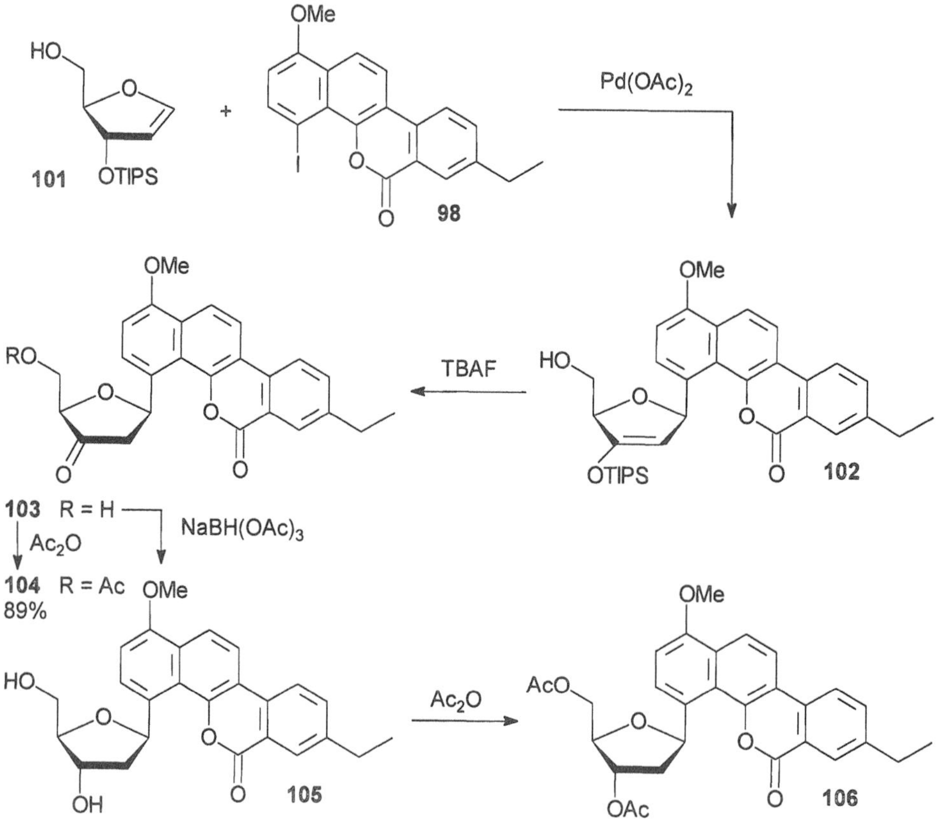

Scheme 20
Reprinted with permission from ref. 19, Copyright 1990 American Chemical Society.

In the pyranoid glycal series a 42% yield of the α-anomer **108** was isolated from reaction of **98** with glycal **107** under palladium catalysis, eq. 9.

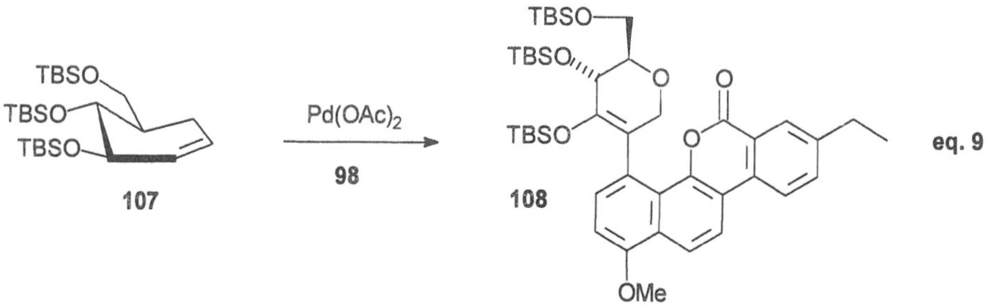

Work from the same group[20] has also focused on the use of the 3-deoxy-glycal **109**. In coupling with the pyrimidine mercurial **56** gave an 8:1 mixture of **110** and **111**. Coupling with iodide **112** gave a similar ratio of α and β-anomers **113** and **114**. The attack of the palladium reagent on the double bond when a *C*-3 oxygen substituent is present attack occurs from the opposite face. The absence of the *C*-3 substituent makes it clear that attack of the palladium species is favored from the bottom face, possibly due to an effect involving the lone pair of electrons on the ring oxygen. The formed double bond is well poised for further manipulation since it can be dihydroxylated or mono-oxygenated depending on the choice of reagent, Scheme 21.

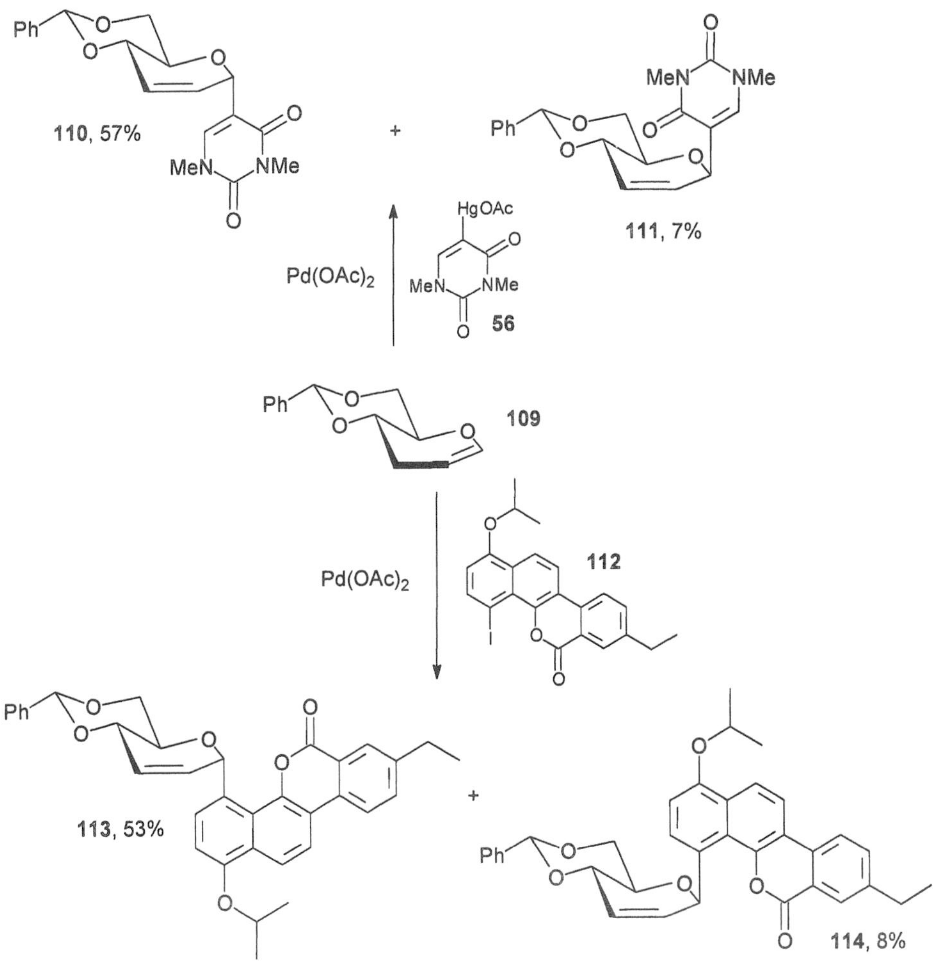

Scheme 21
Reprinted with permission from ref. 20, Copyright 1991 American Chemical Society.

Aryl stannanes also work well in the coupling reaction as shown in eq. 10. The aryl *C*-glycoside **115** is formed in 70% yield. The use of tin in the coupling step, as opposed to mercury, is slightly superior since the same reaction with the corresponding aryl mercurial gave only a 57% yield of adduct *C*-glycoside.[21]

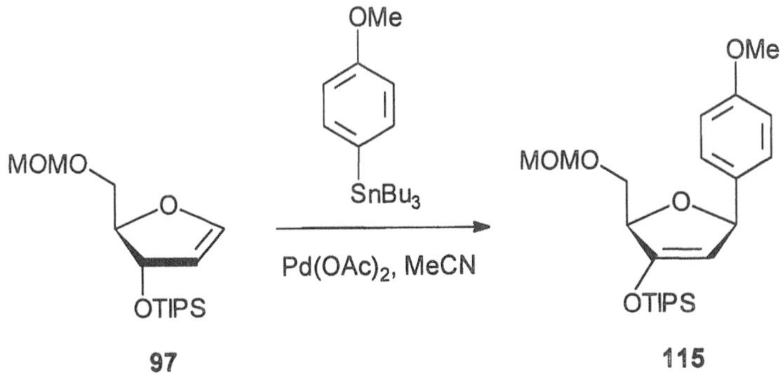

The workers also coupled the stannane **116** with the same glycal **97** under palladium catalysis to give the *C*-glycoside **117** in 66% yield, eq. 11.

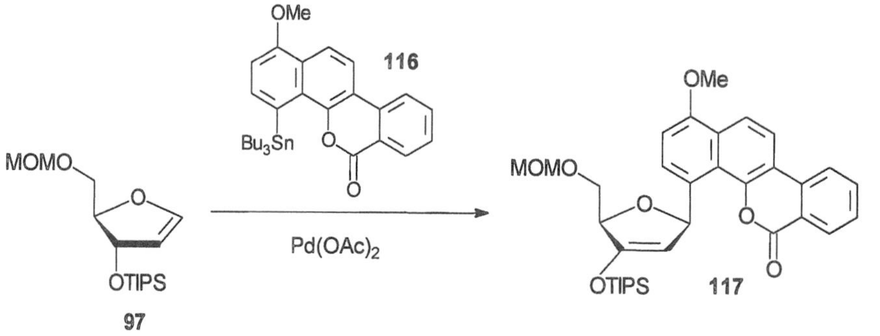

eq. 11

B. USE OF ANOMERIC STANNANES

Several workers have utilized *C*-1 stannyl glycals in the palladium-mediated reaction with suitable aromatic halides. This is a variant of the Stille[22] coupling reaction and generally leads to good yields of adducts. The product, unlike in the π-allyl palladium couplings and Heck type couplings, is a glycal with the olefin in the same position as the starting material. The thrust of this work has been directed at synthetic studies towards some of the naturally occurring aryl *C*-glycosides. A few of the targets include the papulacandins and chaetiacandin as well as vineomycinone B_2 methyl ester which itself is not a natural product, but a product of methanolysis of vineomycine B_2. A detailed survey of total syntheses and some model studies towards naturally occurring aryl *C*-glycosides can be found in Chapter 10 of this text.

Friesen and Sturino[23] used palladium catalysis to affect carbon-carbon formation between *C*-1 of a vinyl glycal and various aromatic substrates, Scheme 22.

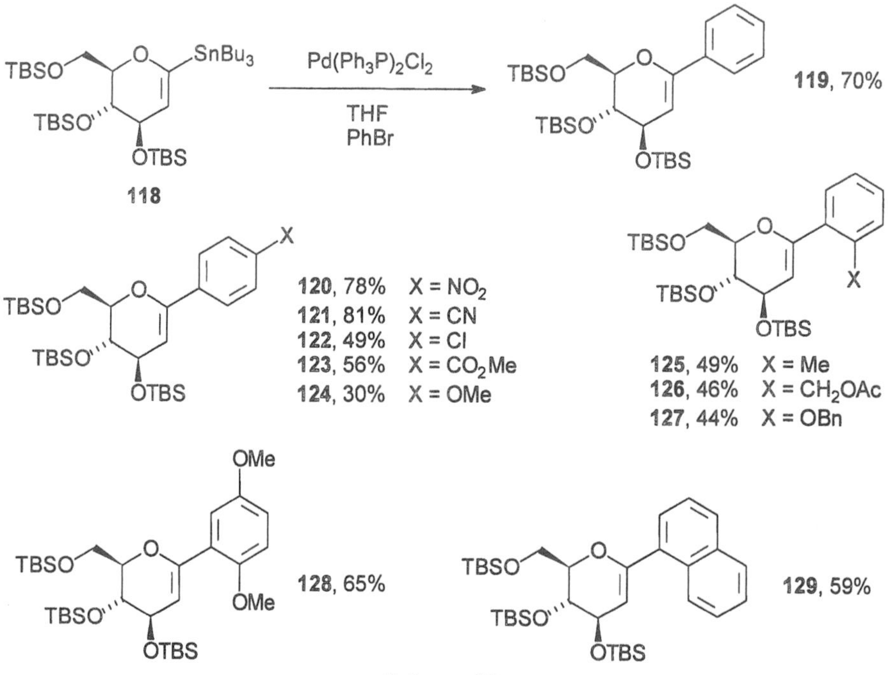

Scheme 22

The vinyl stannane was coupled with bromobenzene to give a 70% yield of adduct119. Various other aryl *C*-glycosides were synthesized using similar protocols as shown in Scheme 22. The isolated yields are a reflection of the electron withdrawing group *para* to the bromine atom. The highest yield was obtained with a nitro group at *C*-4 while a methoxy group at *C*-4 gave only a 30% yield of adduct **124**.

The workers[24] then applied this technology to a synthesis of the core of the papulacandins. Coupling of **118** with **130** gave **131** which was a key intermediate in the synthesis, eq. 12.

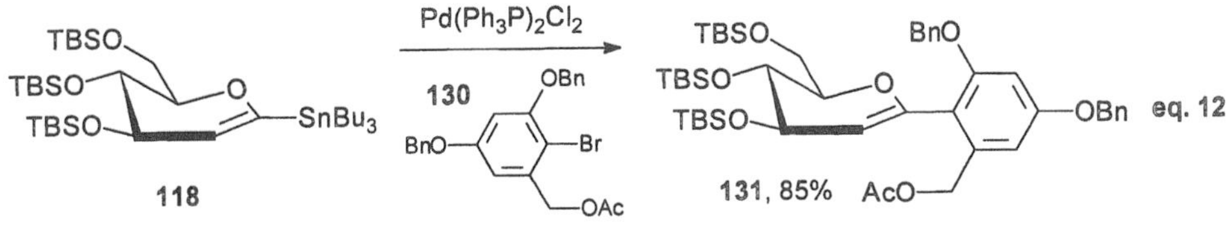

The workers[25] then studied the palladium-mediated coupling of various metallated aromatics with a vinyl glycal. This work came about due to the fact that glycal118 could not be prepared efficiently. So a change to the TIPS protecting group solved the problem of substrate preparation, but it was quickly discovered that the coupling reaction with 4-bromobenzonitrile proceeded in only 67% yield which is a drop from 81% with the TBS protected vinyl stannane. The workers then elected to reverse the polarity in the coupling step and use a vinyl iodide with a metallated aromatic. In the coupling of **132** with phenyl zinc chloride **133** was formed in 90% yield. The 4-methoxy compound **134** was also made in good yield, Scheme 23. The compounds shown were usually prepared from the organozinc or from the corresponding phenylboronic acid.

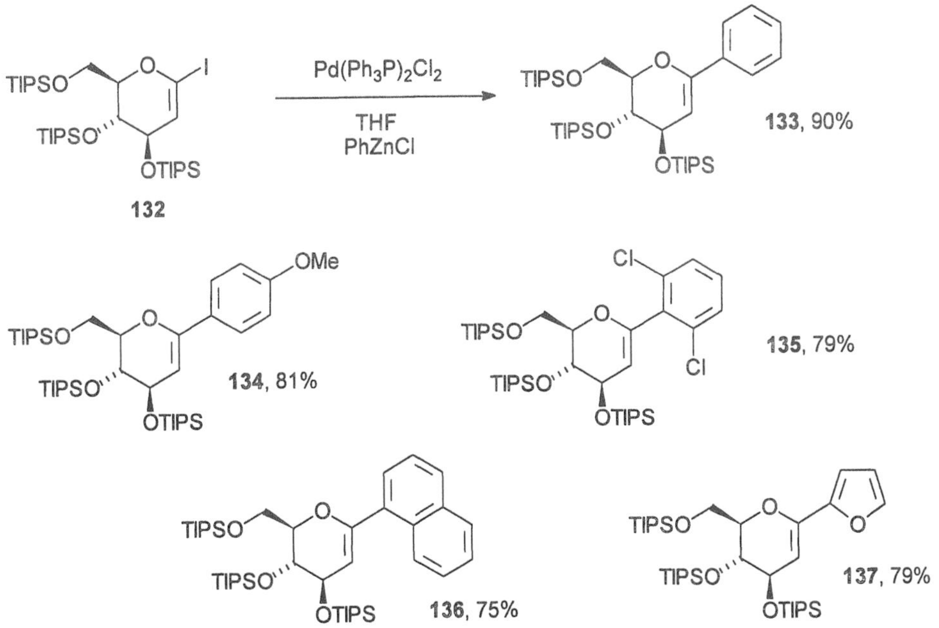

Scheme 23

Beau and Dubois[26] used similar chemistry to affect the key carbon-carbon bond formation. The nature of the product was dependent on the reaction conditions. Coupling of **138** with **139** under palladium catalysis gave a 72% yield of adduct **140**, which is probably formed by an acid catalyzed ketalization process. If sodium carbonate was added to buffer the reaction mixture then the vinyl aryl *C*-glycoside **141** was isolated in 78% yield, Scheme 24.

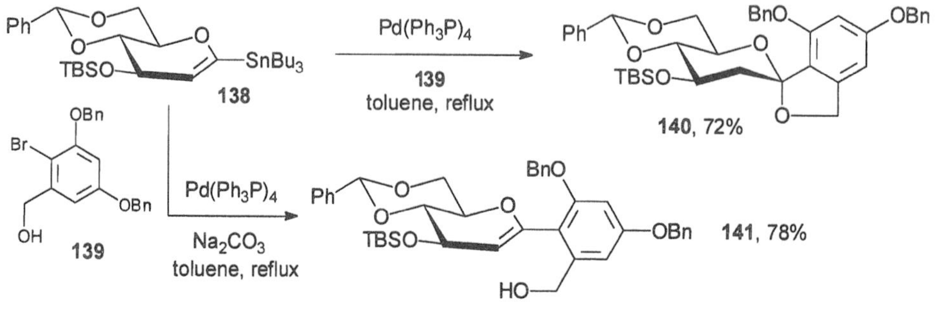

Scheme 24

The workers also studied this reaction in some detail.[27] It was found that either **142** or **148** could be used in the coupling reactions. Coupling of **142** with various halides gave **143** through **146**. In the absence of the halide dimerized product **147** was formed, Scheme 25.

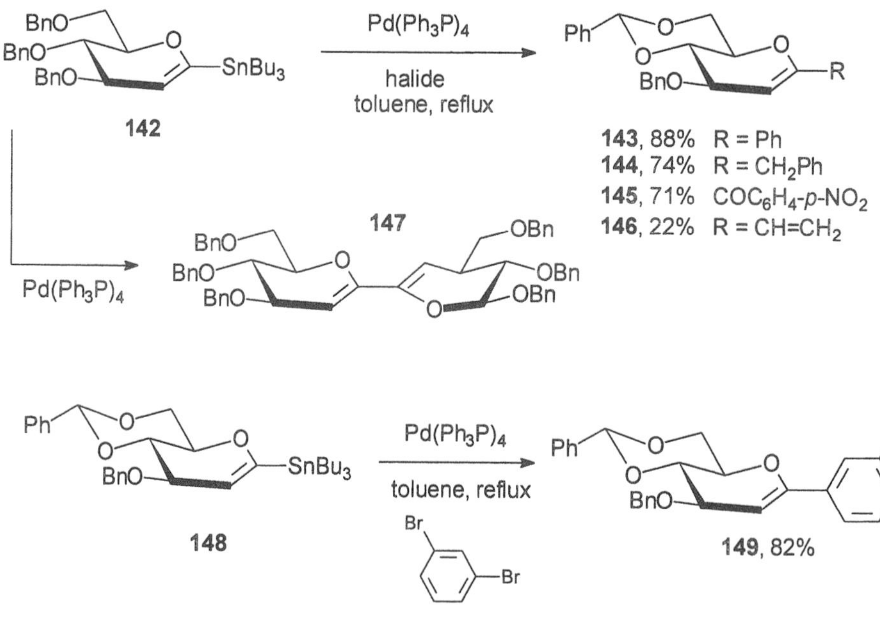

143, 88% R = Ph
144, 74% R = CH$_2$Ph
145, 71% COC$_6$H$_4$-*p*-NO$_2$
146, 22% R = CH=CH$_2$

149, 82%

Scheme 25

The workers then applied palladium coupling strategy to model studies of some di- and tri-*C*-glycosidic antitumor antibiotics such as hedamycin, kidamycin, and the pluramycins. The aryl *C*-glycoside **149** was coupled with the stannyl glycal **148** and gave a 79% yield of the 1,3-diglycosylbenzene **150**. A similar strategy was applied in the synthesis of **151** and **152**.

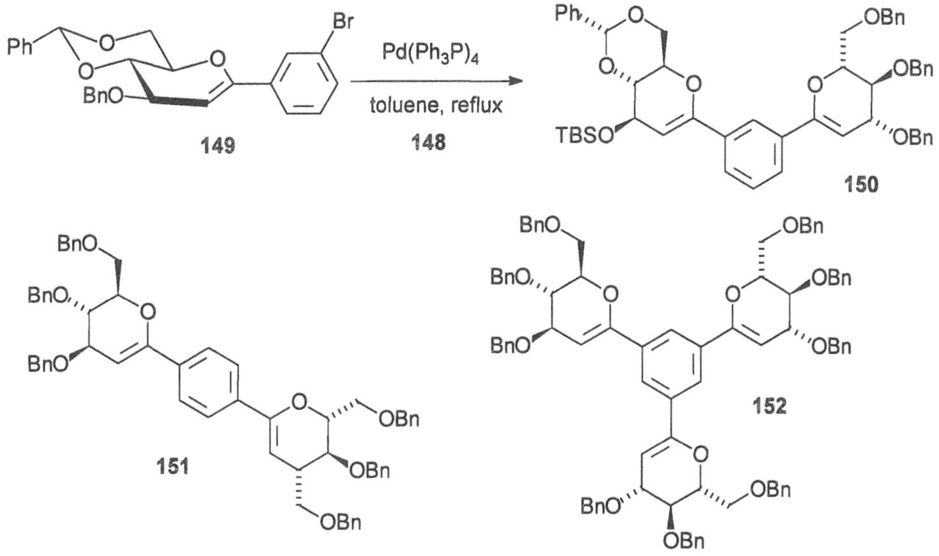

Scheme 26

Tius *et al.*[28] used the palladium-mediated coupling of aromatic iodide 154 with stannyl glycal **153** as the key step in their synthesis of vineomycinone B$_2$ methyl ester. After considerable experimentation it was found that the optimum conditions were as shown in eq. 27 through the use of activated palladium which was generated by *in situ* reduction of Pd(PPh$_3$)$_2$Cl$_2$ by DIBAL in THF.

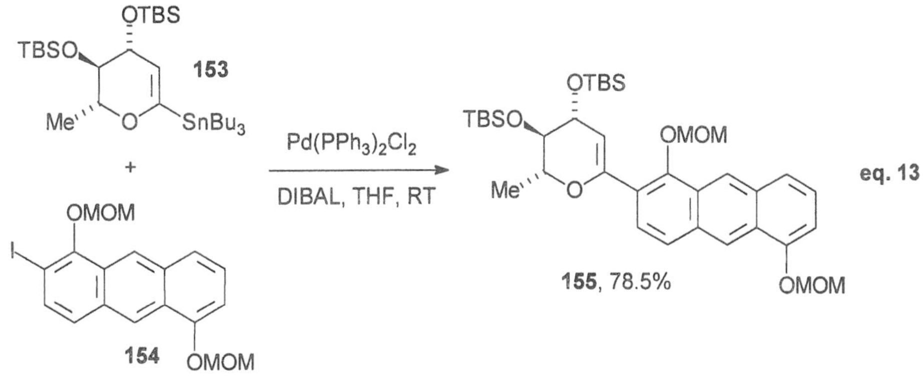

IV. CONCLUSION

This chapter has surveyed the three main methods for forming carbon-carbon bonds at the anomeric center by the use of palladium chemistry. The use of π-allyl complexes has proved useful for stereoselectively introducing malonate units at the anomeric center while the Heck coupling has been shown to be a versatile method for *C*-glycoside formation. Both of these methods result in a shift of the olefin in the product. The nature of the elimination step of palladium from the product can be controlled to afford enol acetates or olefinic *C*-glycosides. The final section concentrated on the use of the Stille coupling to make *C*-1 substituted glycals that are amenable to further transformation. The coupling reactions have been carried out with both simple and complex aromatic species. It seems reasonable to assume that palladium methodology will continue to be used since it is a mild and efficient method for forming carbon-carbon bonds at the anomeric center.

V. REFERENCES

1. Dunkerton, L.V. and Serino, A.J. *J. Org. Chem.* **1982**, *47*, 2814.
2. Dunkerton, L.V.; Euske, J.M. and Serino, A.J. *Carbohydr. Res.* **1987**, *171*, 89.
3. Yougai, S and Miwa, T. *J. Chem. Soc., Chem. Commun.* **1983**, 68.
4. RajanBabu, T.V. *J. Org. Chem.* **1985**, *50*, 3642.
5. Brakta, M.; Lhoste, P. and Sinou, D. *J. Org. Chem.* **1989**, *54*, 1890.
6. Chaguir, B.; Brakta, M.; Bolitt, V.; Lhoste, P. and Sinou, D. *J. Carbohydr. Chem.* **1992**, *11*, 609, Marcel Dekker Inc., N.Y.
7. Engelbrecht, G.J. and Holzapfel, C.W. *Heterocycles* **1991**, *32*, 1267.
8. Suzuki, T.; Sato, O. and Hirama, M. *Tetrahedron Lett.* **1990**, *31*, 4747.
9. Heck, R.F. *J. Am. Chem. Soc.* **1968**, *90*, 5518, 5526, 5531, 5535, 5538, 5542, 5546.
10. For recent reviews on this topic as applied to *C*-glycoside synthesis and relevant background references to palladium chemistry in general see: (a) Daves, G.D., Jr.; Hallberg, A. *Chem. Rev.* **1989**, *89*, 1433 and (b) Daves, G.D., Jr. *Acc. Chem. Res.* **1990**, *23*, 201.
11. Arai, I. and Daves, G.D., Jr. *J. Org Chem.* **1978**, *43*, 4110.
12. Arai, I. and Daves, G.D., Jr. *J. Am. Chem. Soc.* **1978**, *100*, 287.
13. Cheng, J. C.-Y. and Daves, G.D., Jr. *J. Org. Chem.* **1987**, *52*, 3083.
14. Cheng, J. C.-Y. and Daves, G.D., Jr. *Organometallics* **1986**, *5*, 1753.
15. Hacksell, U. and Daves, G.D., Jr. *J. Org. Chem.* **1983**, *48*, 2870.
16. Czernecki, S. and Dechavanne, V. *Can. J. Chem.* **1983**, *61*, 533.
17. Benhaddou, R.; Czernecki, S. and Ville, G. *J. Org. Chem.* **1992**, *57*, 4612.
18. RajanBabu, T.V. and Reddy, G.S. *J. Org. Chem.* **1986**, *51*, 5458.
19. Farr, R.N.; Outten, R.A.; Cheng, J.C.-Y. and Daves, G.D., Jr. *Organometallics* **1990**, *9*, 3151.
20. Kwok, D.-I.; Farr, R.N. and Daves, G.D., Jr. *J. Org. Chem.* **1991**, *56*, 3711.
21. Outten, R.A. and Daves, G.D., Jr. *J. Org. Chem.* **1989**, *54*, 29.
22. Milstein, D. and Stille, J.K. *J. Am. Chem. Soc.* **1979**, *101*, 4992.
23. Friesen, R.W. and Sturino, C.F. *J. Org. Chem.* **1990**, *55*, 2572.
24. Friesen, R.W. and Sturino, C.F. *J. Org. Chem.* **1990**, *55*, 5808. For a more detailed discussion of the synthesis see Chapter 10 of this text.
25. Friesen, R.W. and Loo, R.W. *J. Org. Chem.* **1991**, *56*, 4821.
26. Dubois, E. and Beau, J.-M. *Tetrahedron Lett.* **1990**, *31*, 5165.
27. Dubois, E. and Beau, J.-M. *Carbohydr. Res.* **1992**, *228*, 103.
28. Tius, M.A.; Gomez-Galeno, J.; Gu, X.-q. and Zaidi, J.H. *J. Am. Chem. Soc.* **1991**, *113*, 5775.

Chapter 6

CONCERTED REACTIONS APPLIED TO *C*-GLYCOSIDE PREPARATION

I. INTRODUCTION

This chapter will deal with a fairly recent number of approaches to the synthesis of *C*-glycosides, the use of concerted processes. It will be divided into two main sections. The first part will deal with the use of sigmatropic rearrangements, with heavy emphasis on [3,3] rearrangements. The merits of these methods and some of their applications to natural product synthesis will also be discussed. The second section will address the use of cycloaddition chemistry in the preparation of *C*-glycosyl compounds. Hetero Diels-Alder reactions will comprise a major portion of this section, but some work on photochemical cycloadditions and reactions that can be formally classified as cycloadditions (but may proceed in a stepwise fashion) will also be included. Finally, examples dealing with cycloaddition chemistry on the carbon portion of an existing *C*-glycoside will be examined.

II. SIGMATROPIC REARRANGEMENTS

The Claisen and Ireland rearrangements are powerful synthetic tools for the stereoselective construction of carbon-carbon bonds. The new carbon-carbon bond is formed *syn* to the pre-existing oxygen atom as illustrated below in eq. 1. The reaction also leaves behind a well placed double bond that, in the case of sugars, can be used as a handle to introduce hydroxyl groups.

A. CLAISEN REARRANGEMENT

Both Heyns and Fraser-Reid applied the Claisen rearrangement to *C*-glycoside synthesis. In Heyns' case, glycal **3** was treated with mercuric acetate and ethyl vinyl ether to give a modest yield of the vinyl ether **4** which underwent [3,3] sigmatropic rearrangement in refluxing nitromethane to provide the β-*C*-glycoside **5**. Once again, it is the stereochemistry of *O*-3 that controls the resulting stereochemistry in the formed *C*-glycoside, Scheme 1.[1]

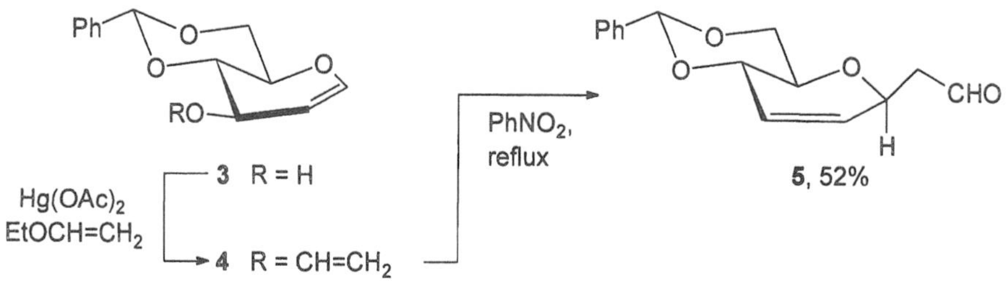

Scheme 1

Fraser-Reid and Tulshian[2] have employed a similar reaction for the synthesis of α-*C*-glycosides. They began with a compound that had the opposite configuration at *C*-3. Accordingly, transformation of 6 to the vinyl ether proceeded in low conversion (50%) and exposure to high temperature (refluxing nitromethane, 210°C), to affect rearrangement, resulted in the formation of dienic aldehyde 8. Lowering the temperature (185°C) gave a good yield of the expected α-*C*-glycoside 9. The Eschenmoser variant of the Claisen rearrangement was also examined. Exposure of 6 to the dimethyl acetal of *N,N*-dimethylacetamide in refluxing xylene led directly to the product of rearrangement 10 in excellent yield. When the amide group was reduced to an aldehyde function simple anomerization took place to give the β-*C*-glycoside 11 as the major product with a minor amount of the α-*C*-glycoside 9.

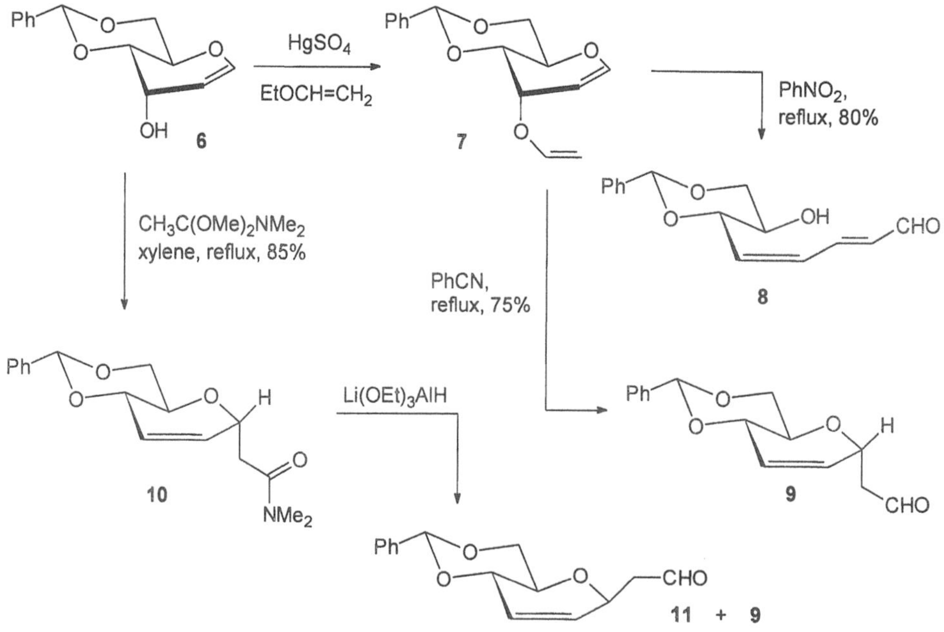

Scheme 2

B.   IRELAND-CLAISEN REARRANGEMENT

The ester enolate rearrangement is also a very popular carbon-carbon bond forming reaction. It has been applied to *C*-glycoside synthesis with high levels of success. Often the resulting *C*-glycoside is elaborated further for use in natural product synthesis. Not only does the relative stereochemistry of the oxygen control the stereochemistry of the resulting *C*-glycoside but, if for example a propanoate ester is used, the geometry of the formed enolate is directly related to the stereochemistry at C-2'. So theoretically both C-glycosides **15** and **17** are available from the same starting ester **13**, Scheme 3.[3]

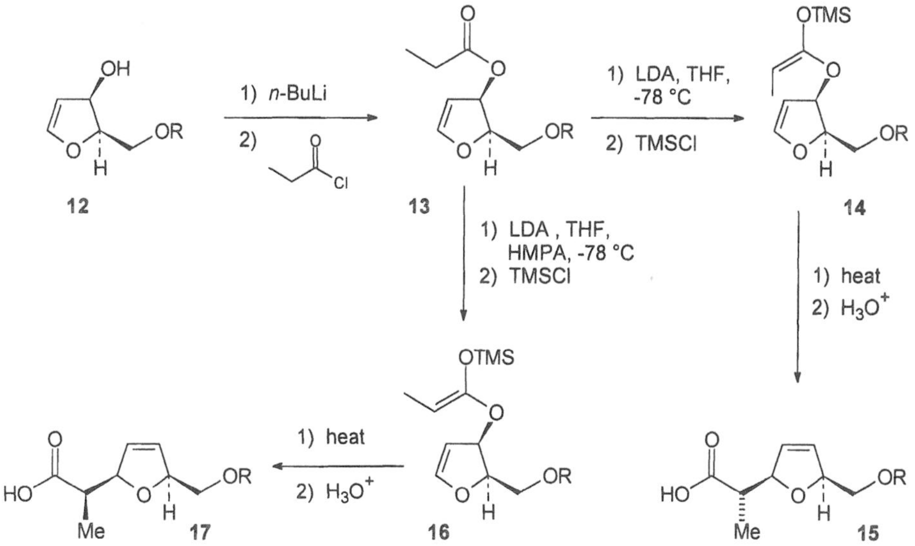

Scheme 3

Ireland has also applied the reaction to pyranoid glycals as shown below in Scheme 4. Mixtures were obtained in all the examples and this is a reflection of the isomeric purity of the corresponding ketene silyl acetals. For the sake of clarity only the major isomer is shown.[4]

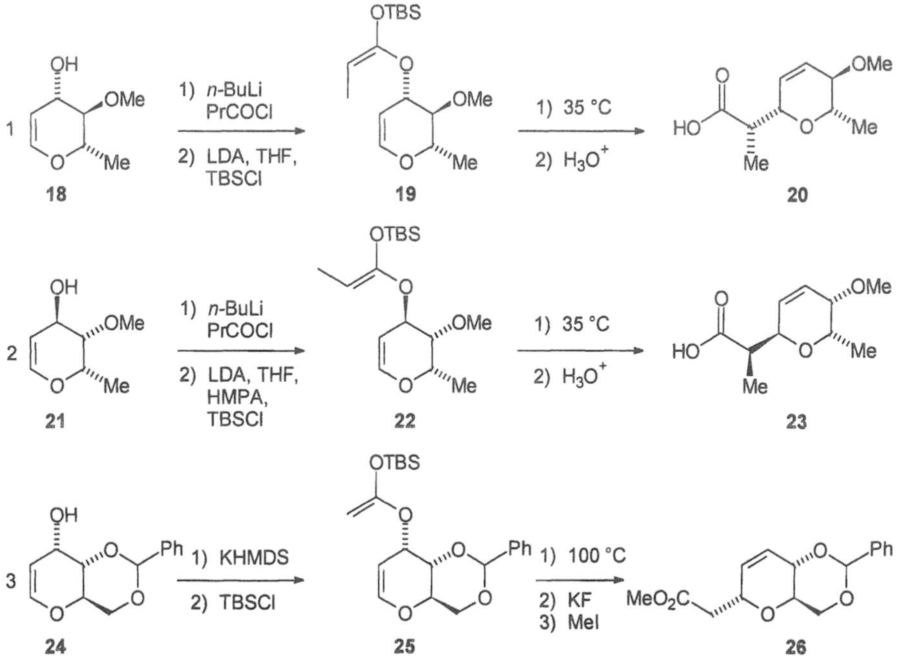

Scheme 4

This rearrangement was applied to the preparation of an early intermediate in Ireland's Tirandamycic acid (**30**) synthesis (Scheme 5).[5] Tirandamycic acid is an antibiotic that belongs to a small group of 3-acyltetramic acids. This compound is of particular importance due to its powerful inhibition of bacterial DNA-directed RNA polymerase. The sequence began with glycal **27** which was transformed into **28** in six steps. Enolate formation, silylation and rearrangement gave lactone **29** as the major product, after iodocyclization. Lactone **29** was then eventually converted to (**30**).

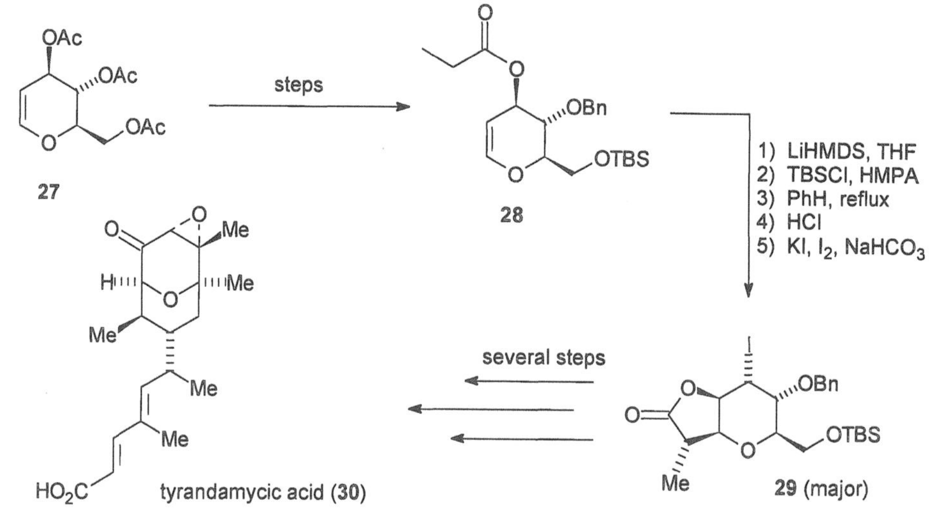

Scheme 5

Curran[6] has also applied a similar rearrangement in his formal total synthesis of pseudomonic acid C (for a detailed look at synthetic work directed towards this compound see Chapter 9) The bis(ketenesilyl)acetal has the potential to undergo two sequential Ireland-Claisen rearrangements to eventually provide **33**. It was found however that at 60° C a 60% yield of the mono-Claisen rearrangement product **32** was formed along with (<5%) of diester **33** (after hydrolysis and methylation). If access to the tandem product is desired, the reaction need only be conducted in refluxing toluene. In this case a 45% yield of diester **33** (after hydrolysis and methylation) is obtained.

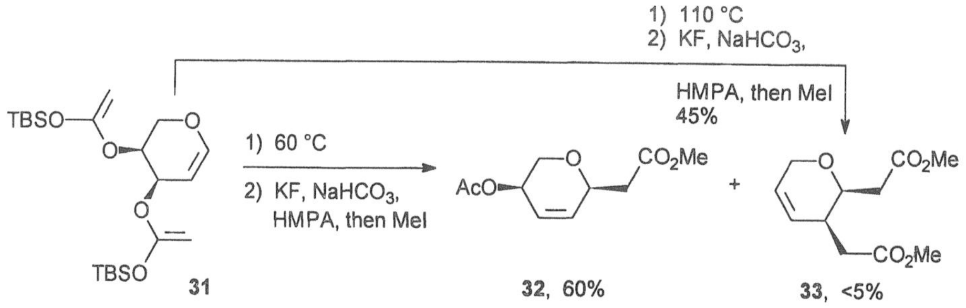

Scheme 6

The selective nature of this rearrangement is explained by invoking what the authors term as the vinylogous anomeric effect. The ring oxygen is thought to stabilize the partial positive charge that is formed during the initial bond breaking process. The workers have exploited the selectivity in the preparation of several *C*-glycosides (Scheme 7). The advantage of this method is that no protecting groups are required. This is illustrated by entry 3 as shown below in which a *C*-glycoside is produced stereospecifically in 55% yield from glycal **27**. Products from the tandem reaction could be isolated, albeit in low unoptimized yields, if the rearrangements were carried out for longer reaction times or at higher temperatures.

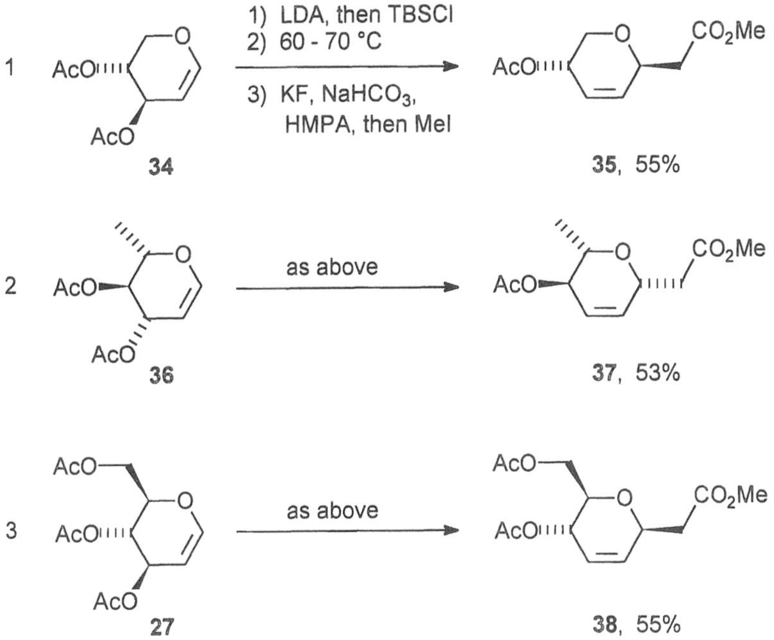

Scheme 7

Kishi and co-workers[7] also used the Ireland-Claisen rearrangement to synthesize an early intermediate in their synthetic work directed towards the halichondrins. They attempted to, like Curran, utilize the tri(ketenesilyl)acetal **39** to hopefully isolate the mono-Claisen rearrangement. In this case, the rates of the two rearrangements were too similar to have the reaction be synthetically useful. Therefore, the differentially protected glycal **40** was converted to its bis(ketenesilyl)acetal and rearrangement gave a 8:1 ratio of acids. The major product **41** underwent iodocyclization and reductive removal of iodine furnished lactone **42**, Scheme 8.

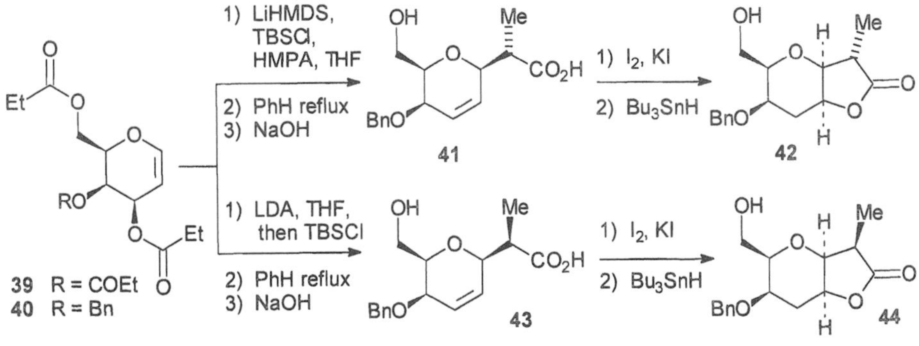

Scheme 8

Lactone **42** was carried on to the *C*-27–*C*-38 **45** segment of halichondrins, Scheme 9.

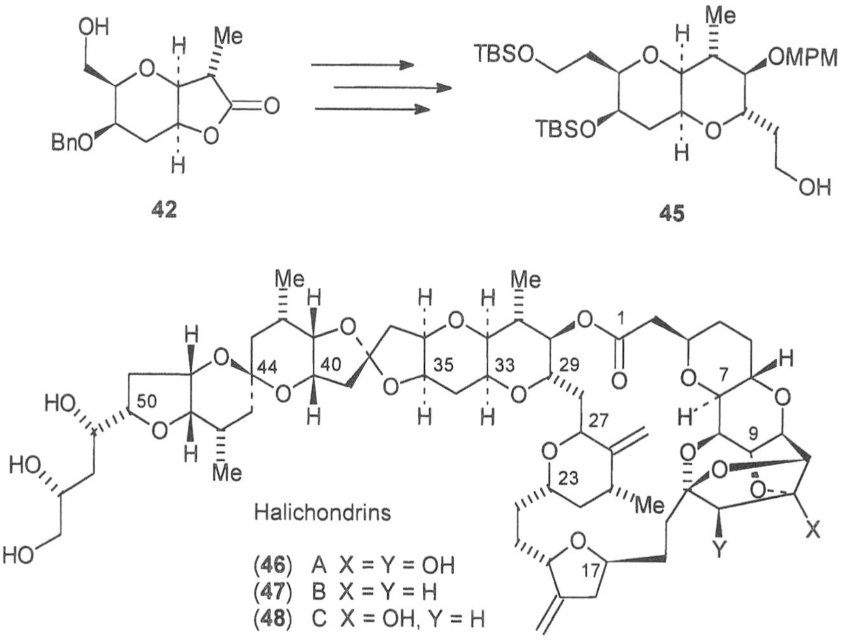

42   **45**

Halichondrins

(**46**) A X = Y = OH
(**47**) B X = Y = H
(**48**) C X = OH, Y = H

Scheme 9

Ireland has also converted a *C*-glycoside, that results from Ireland-Claisen rearrangement into the Prelog-Djerassi lactone (**51**). This rearrangement is well suited for setting up the relative stereochemistry of the methyl group on the ester bearing chain in the final product.[8]

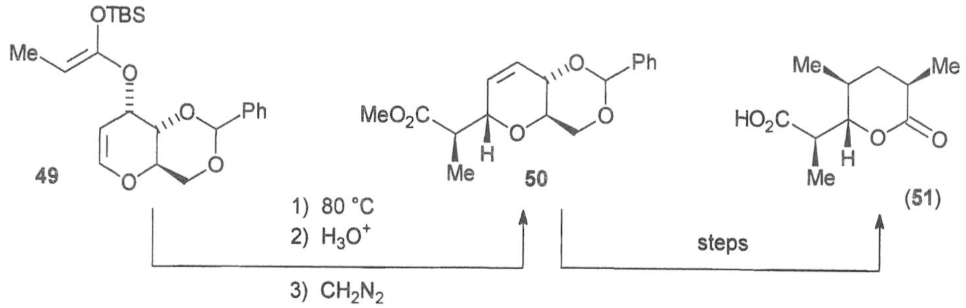

49   **50**   (**51**)

1) 80 °C
2) H$_3$O$^+$
                                   steps
3) CH$_2$N$_2$

Scheme 10

Ley and co-workers[9] have also employed the Ireland-Claisen rearrangement as the starting point for the preparation of the left hand portion of the antibiotic X-14547 A (**56**). Ester **52** (which is available from 1,6 anhydro-β-glucose in six steps) was converted to the *E*-ketenesilyl acetal by treatment with LDA in THF followed by trapping with trimethylsilyl chloride to give **53**. Heating to 50°C caused rearrangement and treatment with TBAF and methylation with diazomethane then gave ester **54** along with a small amount on the *C*-2 epimer (5:1) in 67% combined yield. Further steps provided **55** which was eventually coupled with an appropriate right hand side piece to furnish the target (**56**).

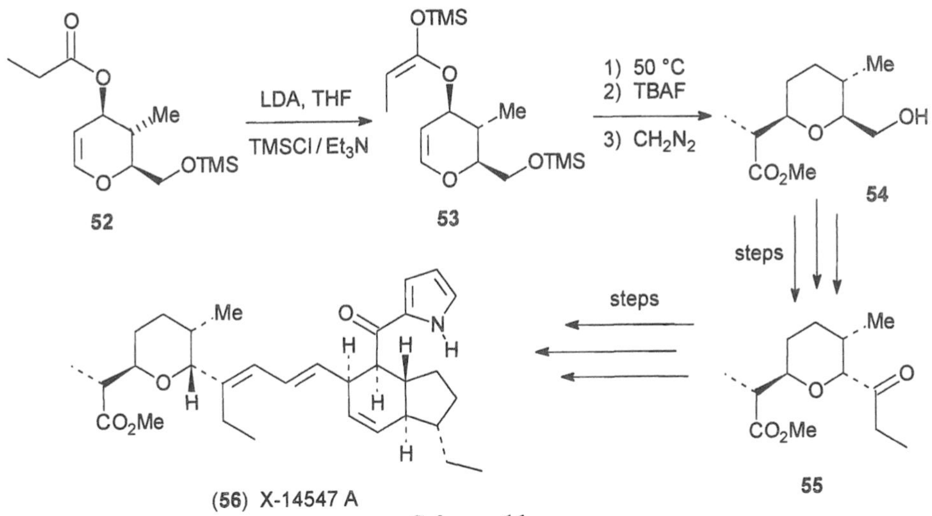

Scheme 11

The Ireland-Claisen rearrangement has also been applied to the synthesis of the polyether antibiotic lasalocid A (X537A). In this example the familiar propionate ester is replaced by a more complex cyclic group. The acid **57** and alcohol **58** were coupled (to give **59** which was not isolated) and enolate formation followed by trapping with trimethylsilyl chloride gave an intermediate ketenesilylacetal that rearranged to **60**, after saponification and esterification. Further steps then provided the target (**61**).[10]

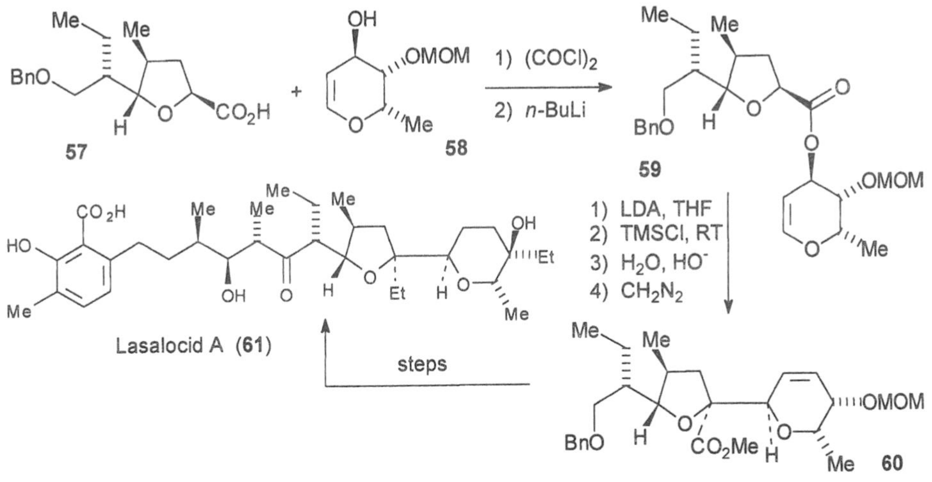

Scheme 12

The ketenesilyl acetal can also exist within a ring. Burke and co-workers[11] used this concept to synthesize the left hand side of the ionophore antibiotic X-14547 A which is also termed indanomycin. Lactone **62** was converted to **63** in the usual way and rearrangement proceeded to give acid **65**, after hydrolysis (Scheme 13). The reaction is believed to proceed through a boat-boat transition state (see **64**) and the geometric requirements for rearrangement are reflected in the relatively high temperatures required to affect rearrangement.

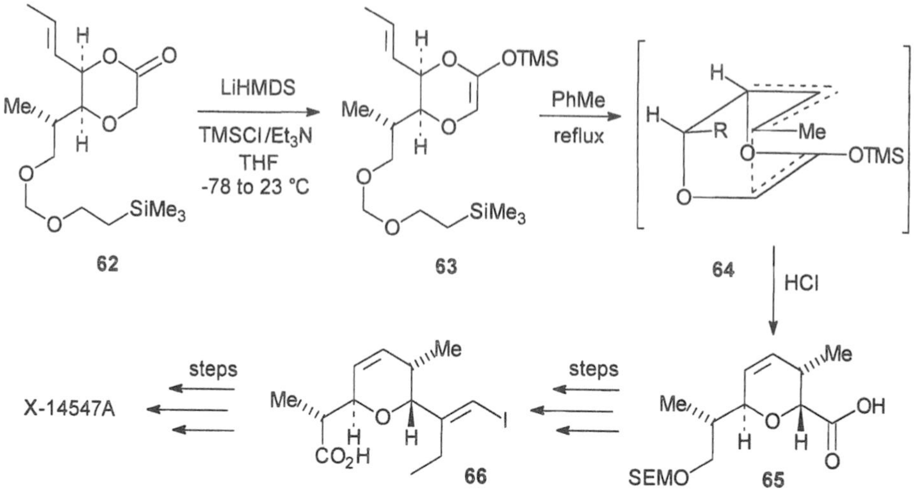

Scheme 13

The reaction tolerates substituents, but only to a certain extent (Scheme 14). Alkyl groups at the allylic position and on the distal end of the double are acceptable as long as the geometry of the olefin is *trans*. Entry 3 shows a compound with a *cis* double bond. In this case, the reaction does not proceed due to the geometric requirement for the transition state of the rearrangement (see **64**).[12]

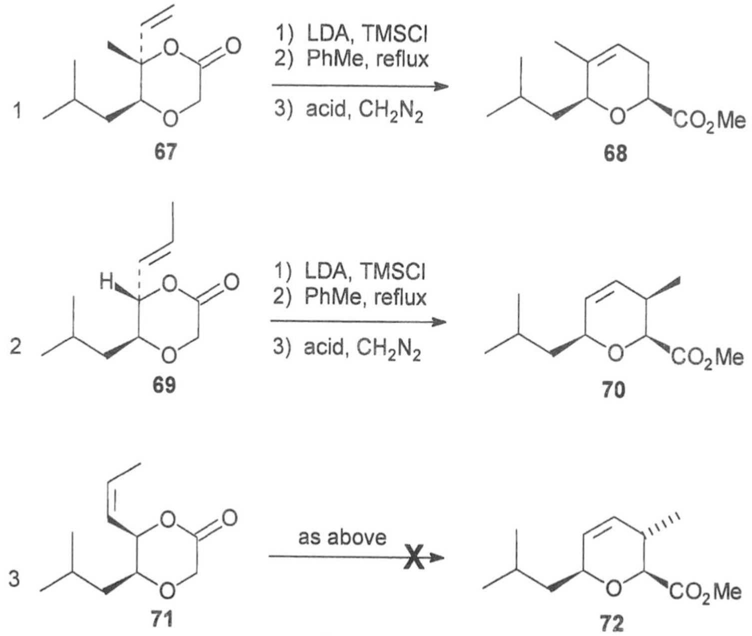

Scheme 14

In related work, Burke[13] has applied this genre of chemistry to construct a template that was elaborated into a C(1)-C(6) fragment of erythronolide B aglycone. Lactone **73** underwent Claisen rearrangement as expected to provide the ester **74**. Hydroboration of the olefin was followed by conversion of the ester to the corresponding iodide **75**. Exposure to zinc dust caused reductive fragmentation to provide the requisite segment 76. Although the sequence shown is quite short, lactone 73 was prepared from a derivative of ethyl (+) lactate in eight steps.

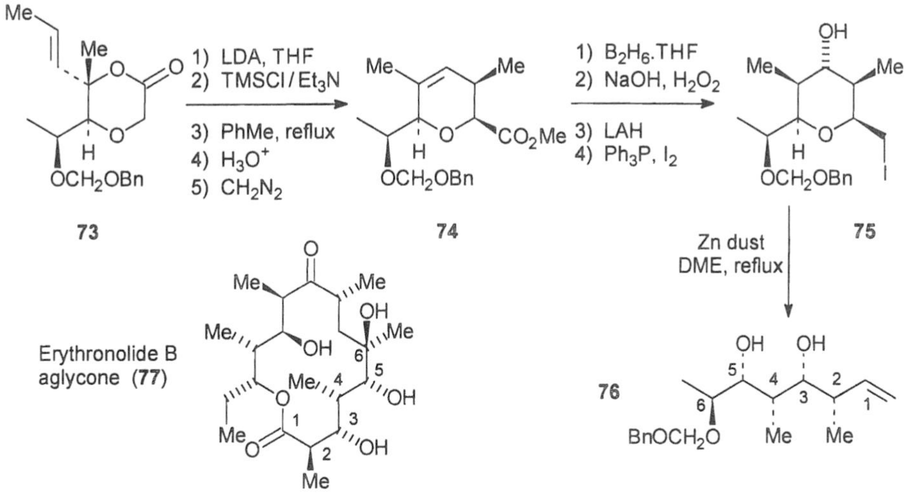

Scheme 15

C. WITTIG REARRANGEMENT

Two examples of [2,3] rearrangement have been applied to the preparation of *C*-glycosides. In the first instance, the rearrangement is conceptually similar to the Claisen example shown in eqs. 2 and 3. Tri-*n*-butylstannyl ether 78 was prepared by exposure of alcohol 6 to iodide 81 in the presence of sodium hydride in HMPA. Lithiation at low temperature followed by warming to room temperature gave a 3:1 mixture of methyl ether 79 and rearranged product, the α-*C*-glycoside 80.[2]

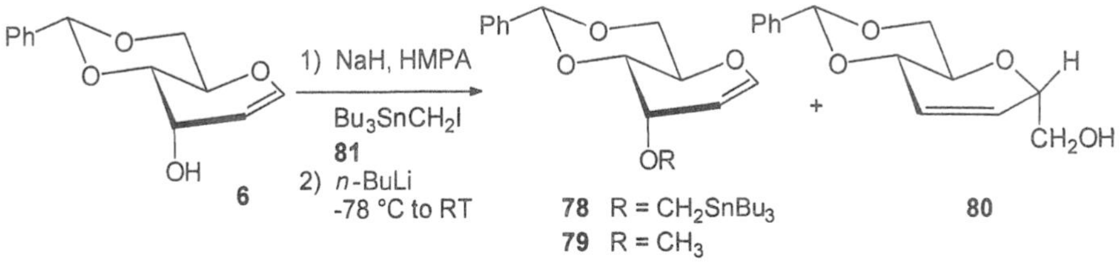

Equation 3 shows another application of the Wittig rearrangement to *C*-glycoside synthesis. Reaction of 82 with butyllithium gave 83 (11%) and 85 (8%) yield. Changing the base to LDA altered the product ratio to 83 (4%) and 85 (23%).

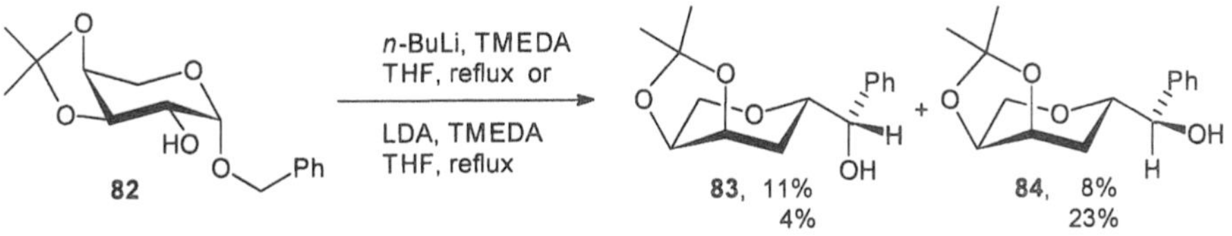

Other substrates were also examined. When diol 85 was metallated with an excess of *n*-butyllithium a 16% and 9% yield of 86 and 87 were isolated, respectively along with 5% of recovered starting material. Both the β and α anomers of the corresponding cyclohexylidene were also studied. In the case of the β-anomer **88**, a 15% yield of **89** was isolated while similar reaction with **90** afforded two compounds (**91** and **92**) in the ratios noted in Scheme 16.[14]

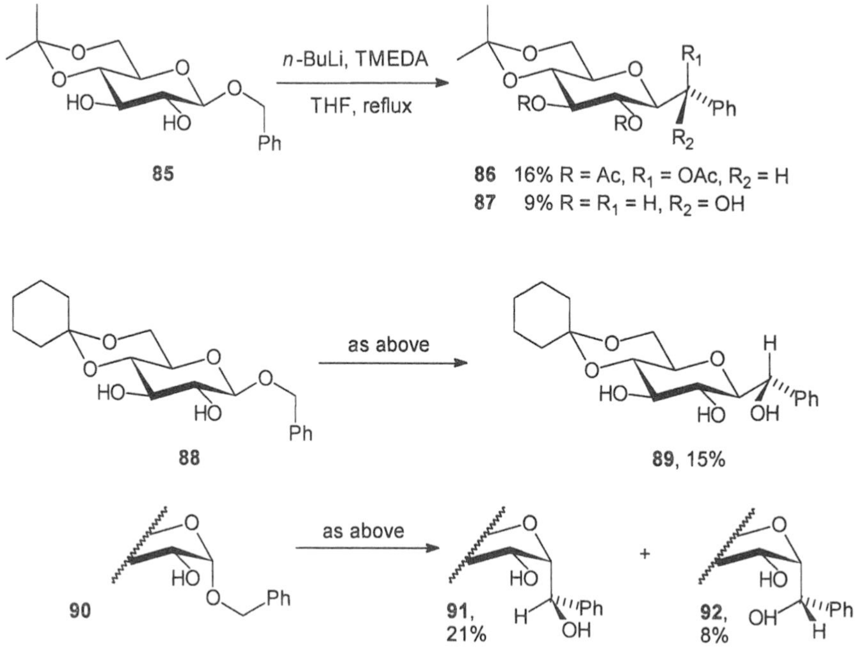

85

n-BuLi, TMEDA

THF, reflux

86 16% R = Ac, R$_1$ = OAc, R$_2$ = H
87   9% R = R$_1$ = H, R$_2$ = OH

88

as above

89, 15%

90

as above

91,
21%

+

92,
8%

Scheme 16

III. CYCLOADDITIONS

The Diels-Alder reaction, which is probably the most important reaction in organic chemistry, has also found application in *C*-glycoside synthesis. In general terms the reaction can be described as the cycloaddition of a suitable diene **93** with an appropriate dienophile **94** to form a six-membered ring.

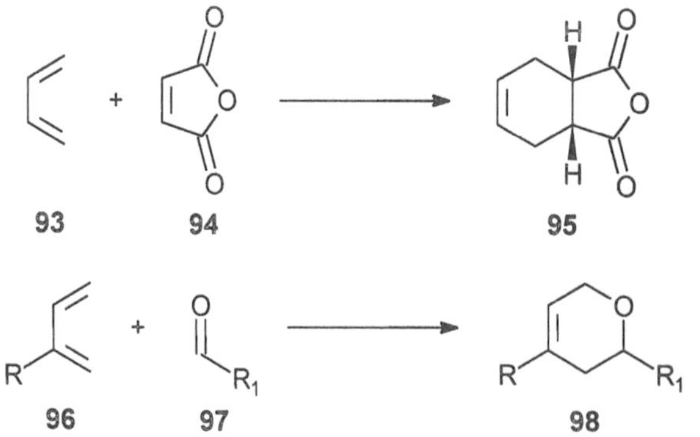

93   **94**   **95**

96   **97**   **98**

Scheme 17

If one of the carbon atoms participating in the cycloaddition is replaced by a heteroatom, such as oxygen, the reaction is now termed a hetero Diels-Alder reaction. This type of cycloaddition reaction has been applied to the construction of C-glycosides and C-glycoside precursors. Other types of cycloadditions that have been developed in this area include cycloadditions of suitable glycals and Diels-Alder reactions on the carbon portion of a pre-existing C-glycoside.

A. FORMATION OF THE PYRAN RING

The work of Schmidt and co-workers[15] illustrates the utility of hetero Diels-Alder for the preparation of aryl C-glycosides. Little reaction was observed when diene **99** and styrene **101** were heated to 70°C. However, high pressure Diels-Alder reaction (60°C) gave a 57% yield of the *endo* adduct **102a**. This dihydropyran was converted to the 2-deoxy aryl C-glycoside **104a** in the following manner: deacetylation, benzylation, desulfurization, and hydroboration followed by oxidative work-up.

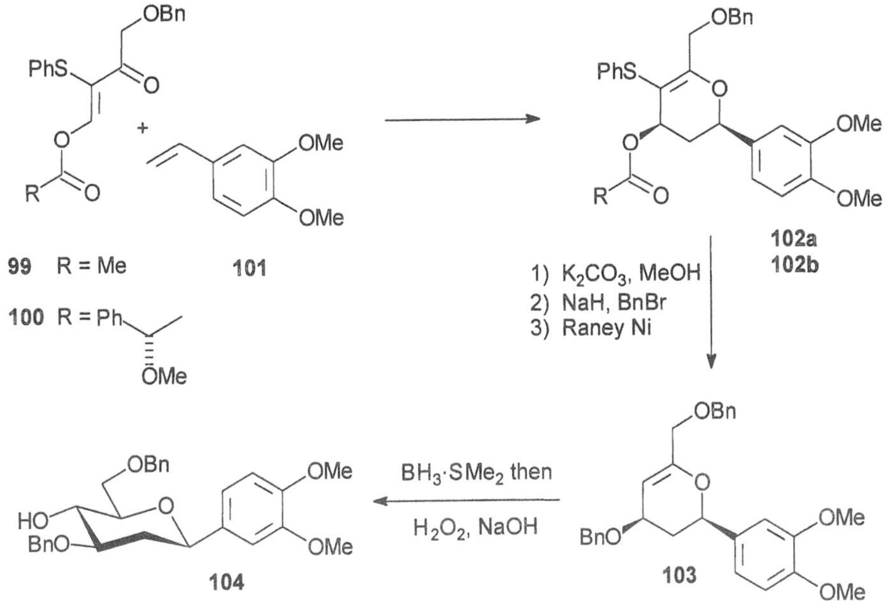

Scheme 18

Chiral induction was observed with diene **100** (R = *O*-methylmandeloyl) to give a 4:1 ratio of isomers. This cycloaddition strategy was flexible enough that it allowed for the production of fully oxygenated aryl C-glycosides by simply changing the identity of the dienophile. Accordingly, reaction of diene **100** with the styrene derivative **101** furnished **102b** and suitable steps then gave the aryl β-C-pyranoside **104** in optically pure form. When an *O*-alkyl protecting group was used on the styrene the incorrect regioisomer was obtained. Compound **104** was shown to be identical with a optically pure sample obtained by condensation of 1,3-dimethoxybenzene with the corresponding perbenzylated trichloroacetimidate. High pressure cycloaddition of diene **99** with methyl α-methoxyacrylate **109** gave a mixture of *endo* and *exo* isomers (3:1) in 60% combined yield. Separation was followed by transformation of the *endo* **110** isomer into the 3-deoxy-arabino-heptulosonate as shown in Scheme 19. This strategy is flexible and provides convenient access to the desired C-glycosides, although once the cycloaddition has been successfully carried out several steps are required to convert the adduct to a C-glycoside.

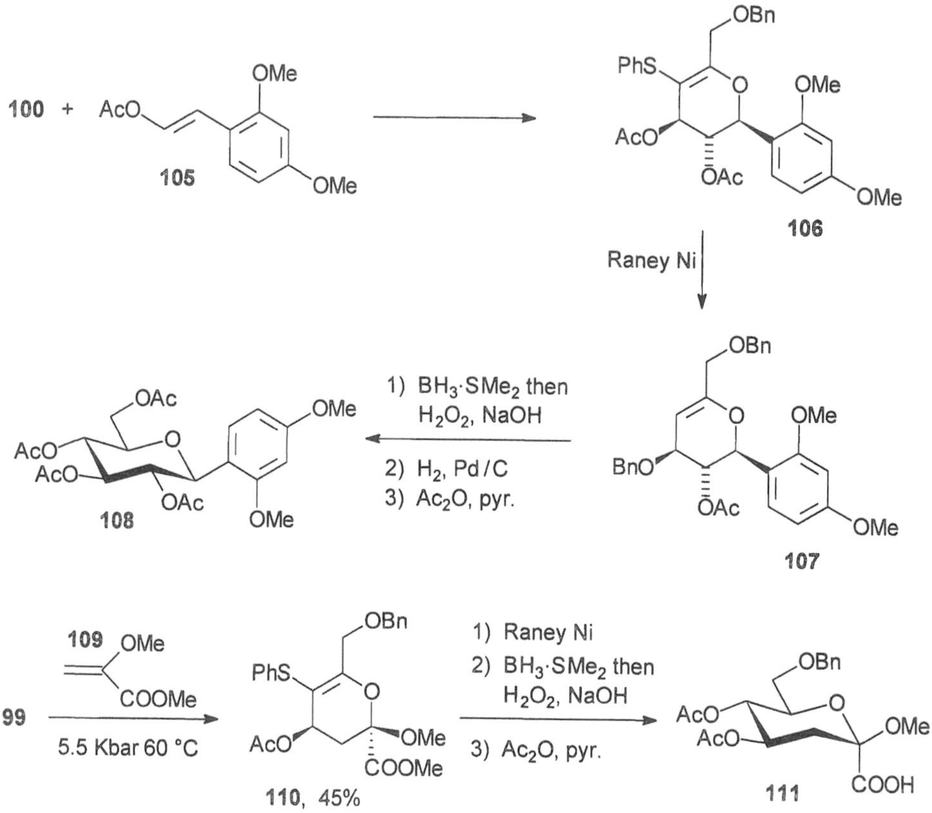

Scheme 19

Oxygenated butadienes have been used in cycloaddition reactions to give compounds that have been transformed into *C*-glycosides or derivatives thereof. Diels-Alder reaction of the diene **113** with the chiral aldehyde **112** occurs under chelation control to give a 47% yield of the adduct **114**.[16] Compound **114** is structurally reminiscent of a *C*-disaccharide (eq. 4).

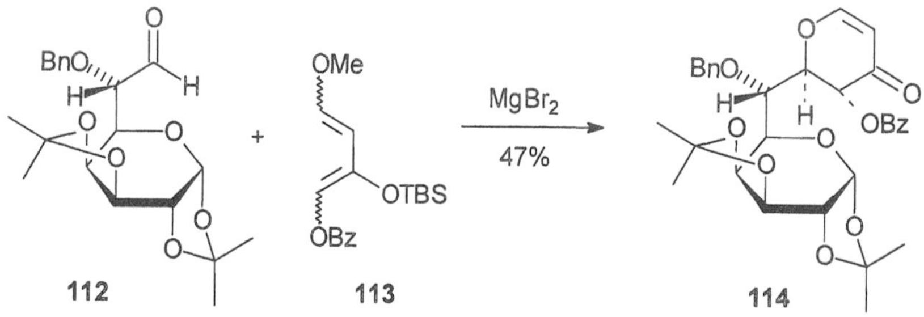

eq. 4

Work from the same laboratories further illustrates the usefulness of this methodology for *C*-glycoside preparation When diene **115** was condensed with aldehyde **116** under mild Lewis acid catalysis adduct **117** was obtained in 92% yield.[17] Although seemingly lacking the characteristics of a *C*-glycoside, this compound was appropriately oxygenated and eventually converted to **118** which is a protected model of the papulacandin core, Scheme 20.

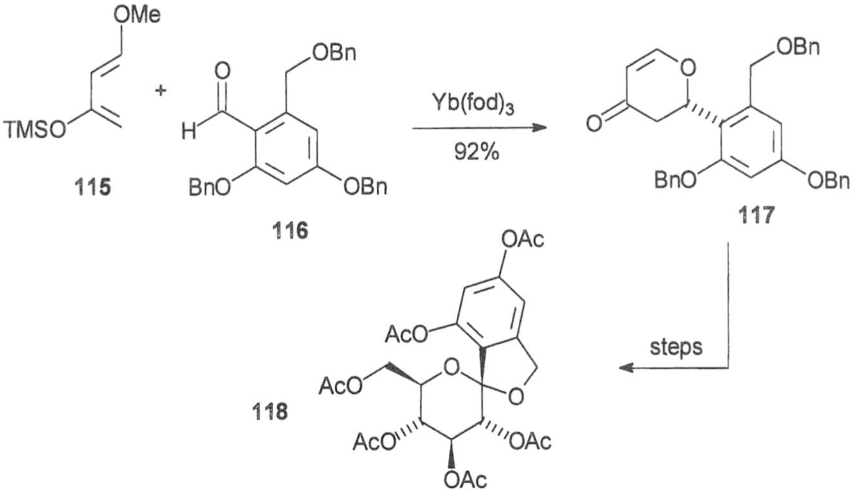

Scheme 20

Another example can be found in Danishefsky's[18] racemic synthesis of 3-deoxy-D-*manno*-2-octulopyranosate (KDO). KDO is an eight carbon sugar that is unique to Gram positive bacteria. One avenue of research that may lead to inhibitors that will block the production of lipopolysaccharides involves the preparation of KDO analogs. In order to gain access to these compounds a versatile high yielding synthesis of KDO is desirable. Diels-Alder reaction of diene **120** with aldehyde **119** gave a 58% yield of the *C*-glycoside **121** along with 24% of the corresponding *trans* isomer (not shown). Several steps then served to convert **121** into racemic KDO **122**.

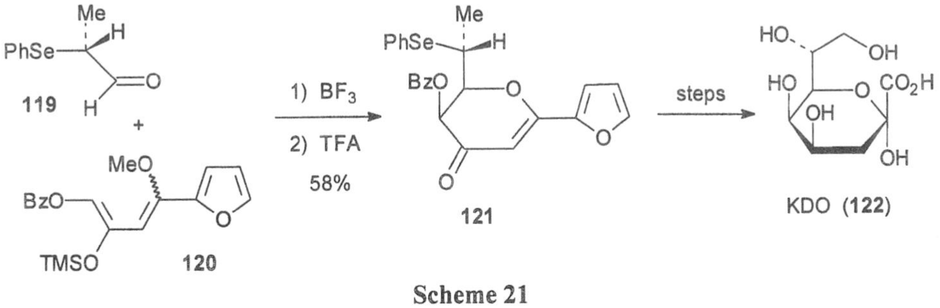

Scheme 21

Work headed by Yamamoto[19] has focused on chiral induction in these types of cyclocondensation reactions. The workers utilized the chiral organoaluminum catalyst **123** which is derived from the corresponding chiral binaphthol (fig. 1).

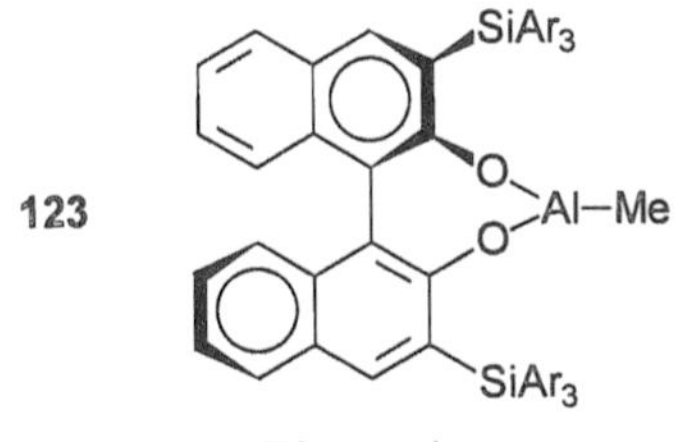

Figure 1

Reaction of siloxydiene **124** with benzaldehyde and catalyst **123** (where Ar = 3,5-xylyl) at -20°C, after treatment of the adduct with TFA, gave a 30:1 of **125** to **126**. The major *cis* isomer was obtained in 97% *ee*. Diene **127** reacted (Ar = Ph) with aldehyde **128** to give **129** as the major compound (76%:9%) in 93% *ee*. It is noteworthy that both enantiomers of the catalyst **123** are available thereby giving access to the antipodes of the compounds shown here.

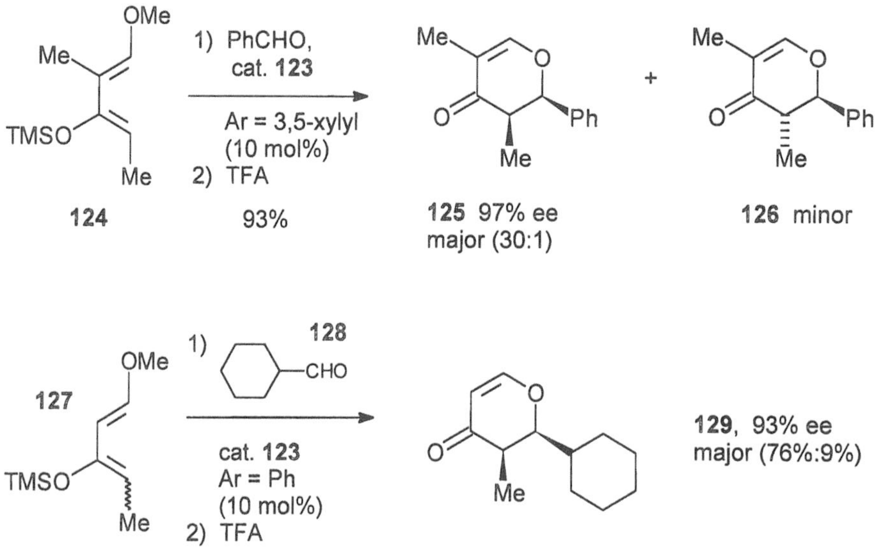

Scheme 22

Katagiri and collaborators[20] have used Diels-Alder cycloaddition to construct an appropriately functionalized [2.2.1] bicyclo unit that was further transformed into β-ribofuranosylmalonate derivative. Condensation of the furan **131** with the dienophile **130** at 40°C gave *exo* adduct **132** exclusively. Hydrogenation was selective and gave only **133** in quantitative yield. When **133** was exposed to potassium carbonate and sodium borohydride in methanol **134** was produced.

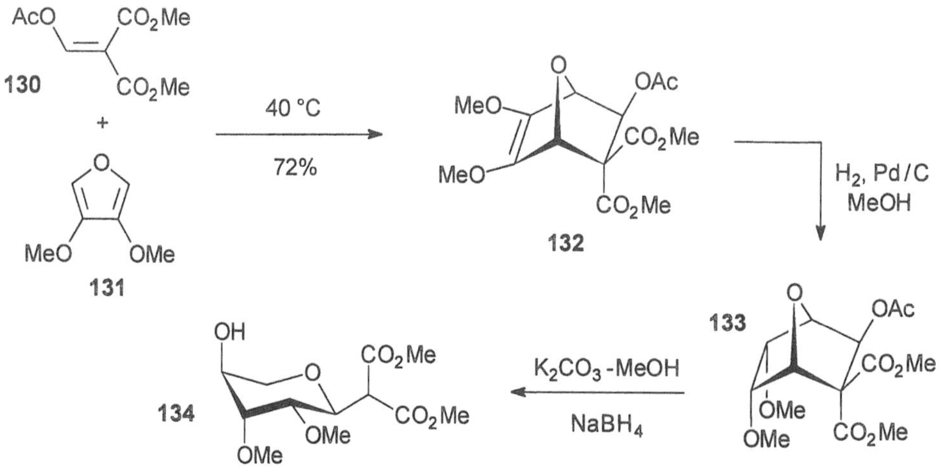

Scheme 23

The formation of **134** is outlined in Scheme 24. Cleavage of the acetate was followed by a retrograde aldol reaction to give the *C*-furanoside **136** which then underwent ring opening to **138**. The primary hydroxyl then cyclized, in a Michael fashion, to furnish the thermodynamically more stable β-lyxopyranoside **134**.

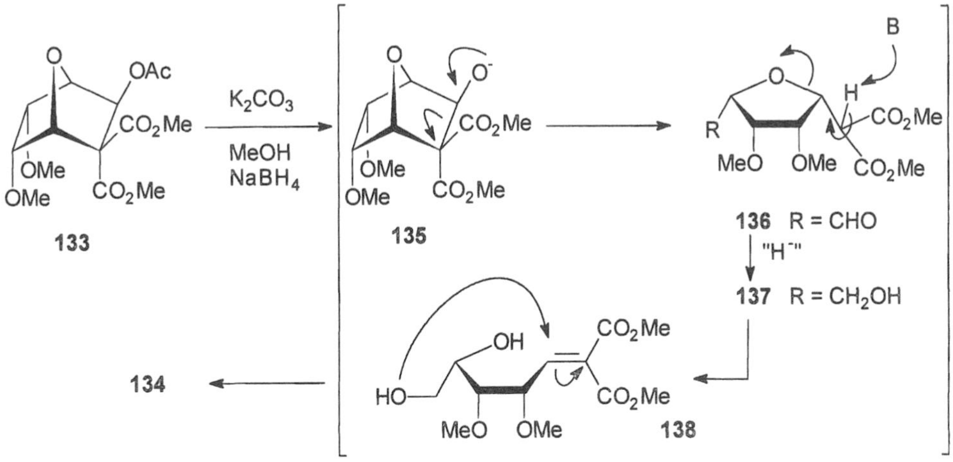

Scheme 24

The workers wished to gain access to β-ribofuranosylmalonates which are useful precursors in *C*-nucleoside synthesis. The dibenzyl lyxo derivative was synthesized as shown above, the benzyl groups were removed and the *cis* diol protected as an acetonide. The remaining hydroxyl was inverted to **140** and treatment of this compound with sodium methoxide in methanol gave a mixture of anomers **142** and **141** (3:2).

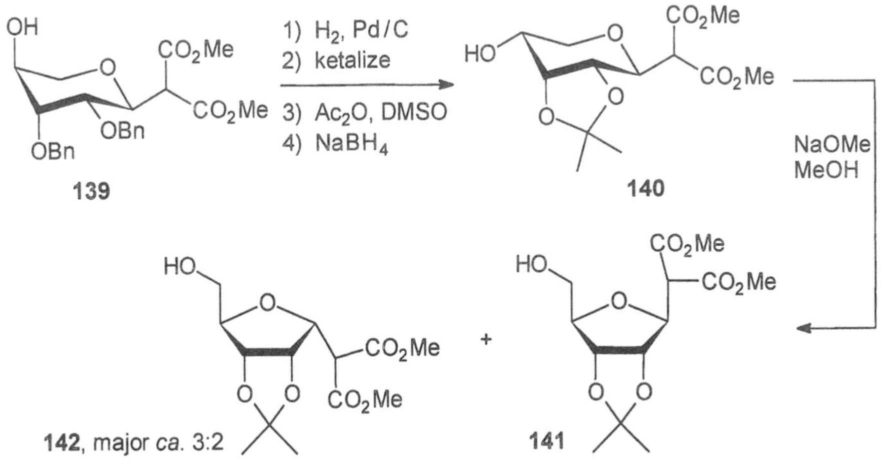

Scheme 25

A more direct approach to the requisite *C*-furanoside was realized by introducing the vicinal *syn* hydroxyl function after the cycloaddition reaction. Adduct **143** (available from high pressure Diels-Alder reaction between furan and (acetoxymethylene)malonate) was hydroxylated and protected to **144**. Exposure of this compound to the standard retrograde aldol conditions then provided **145**, a useful *C*-nucleoside precursor, in quantitative yield.

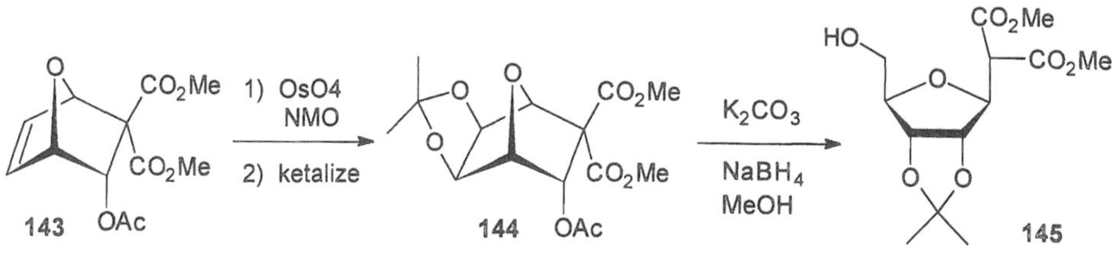

Scheme 26

Noyori and Hayakawa[21] utilized a polybromo ketone/iron carbonyl reaction as the key step for the formation of the *C*-ribose skeleton (Scheme 27). Cyclocoupling promoted by $Fe_2(CO)_9$ of **146** with furan gave, after reduction with Zn/Cu couple, **148**. *Syn* hydroxylation and acetonide formation then furnished **149**. Bayer-Villager oxidation served to introduce the requisite oxygen atom and hydrolysis gave *C*-glycoside **151**. Resolution via the cinchodine salt of acid **151** gave access to optically pure material, Scheme 27

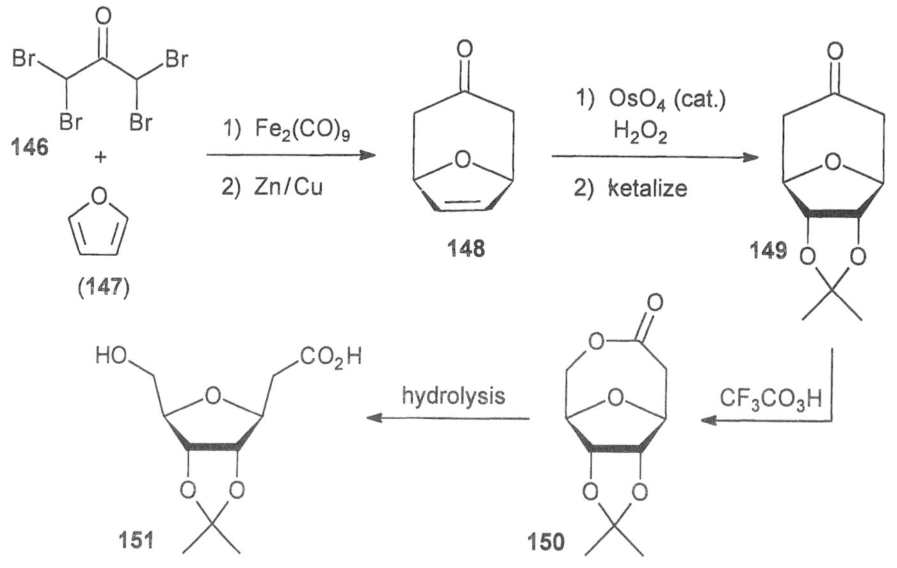

Scheme 27

B.   FORMATION OF THE *C*-GLYCOSIDE BOND

If a cycloaddition reaction occurs on a (substituted) glycal then this may lead to *C*-glycoside formation depending on the orientation of the addition and the nature of the partners. Two such approaches are illustrated below. The first employs a thermal Diels-Alder approach to carbon-carbon bond formation while the second is photochemical in nature.

1.   [4+2] Cycloaddition

In a novel approach, Lopez and co-workers[22] used the cycloaddition reaction of maleic anhydride with the dienic glycals **152** and **154** to efficiently assemble *C*-glycosides. The allylic methoxy group directs the facial selectivity of the cycloaddition reaction. Both cycloadditions proceed via an *endo* transition state with the addition occurring opposite to the allylic methoxy group. When the methoxy group is α, access to the annulated β-*C*-glycoside **153** becomes possible while a β-methoxy group provides access to the annulated α-*C*-glycoside **155**.

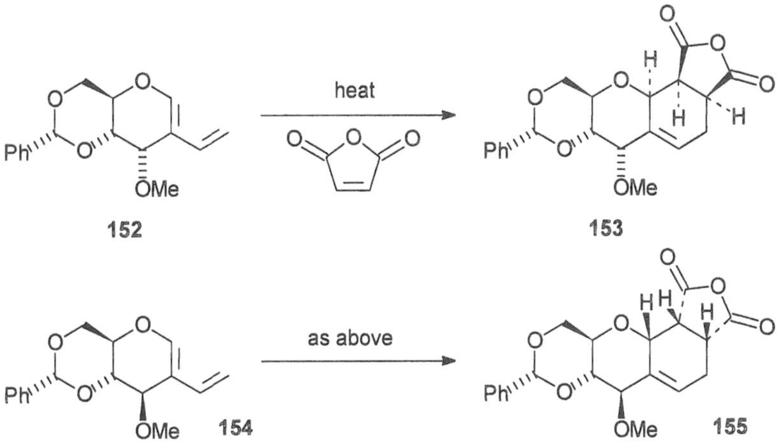

Scheme 28

These facts are supported by the fact that if there is no substituent on *C*-3 (OMe replaced by H) then a mixture of α and β isomers are formed, eq. 5.

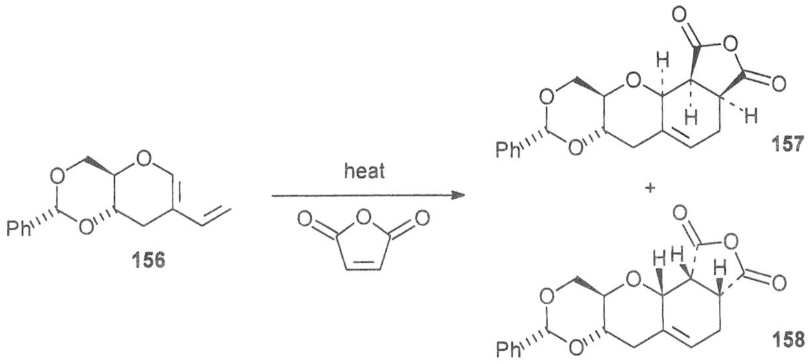

eq. 5

2. [2+2] Photocycloaddition

Work by Araki and collaborators[23] focused on solvent effects in the photochemical addition of acetone to 3,4,6-tri-*O*-acetyl-D-glucal.

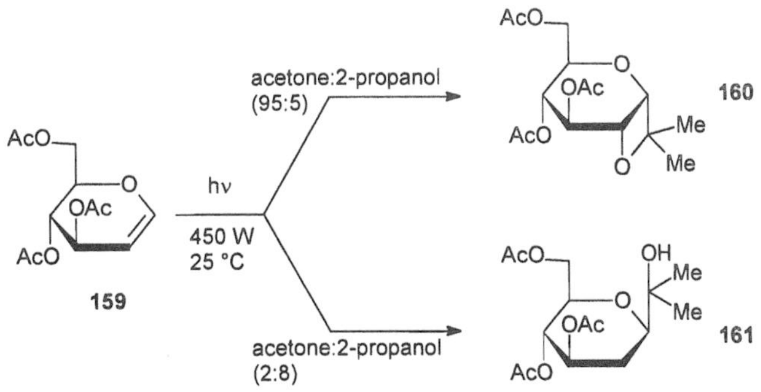

Scheme 29

When the reaction is carried out in 95:5 acetone:isopropanol oxetane **160** is formed in almost quantitative yield (99%). If isopropanol is in excess (8:2) now the 2-deoxy-β-*C*-glycoside **161** is formed quantitatively. The stereoselectivity in the formation of **160** was explained by assuming that the stability of biradical **162** should be greater than that of **163**. The argument used for explaining the formation of the β-*C*-glycoside **161** assumes that the adduct radical **165** is more stable than **164** because the anomeric group is equatorial.

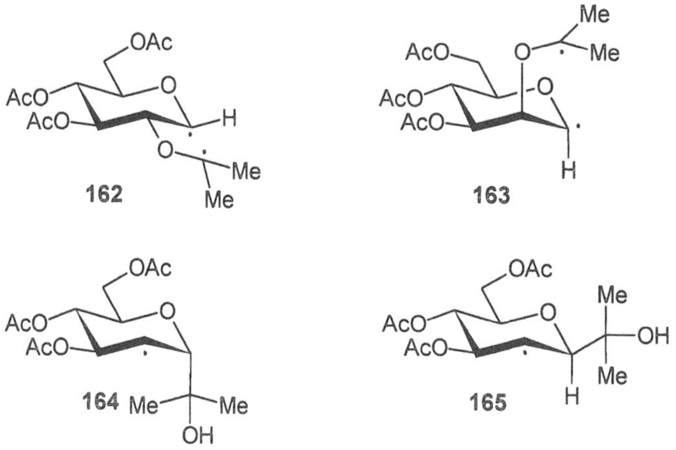

Scheme 30

Adduct **160** can be useful synthetically. Scheme 31 illustrates its conversion to the α-*C*-glucopyranoside derivative **168**.[24]

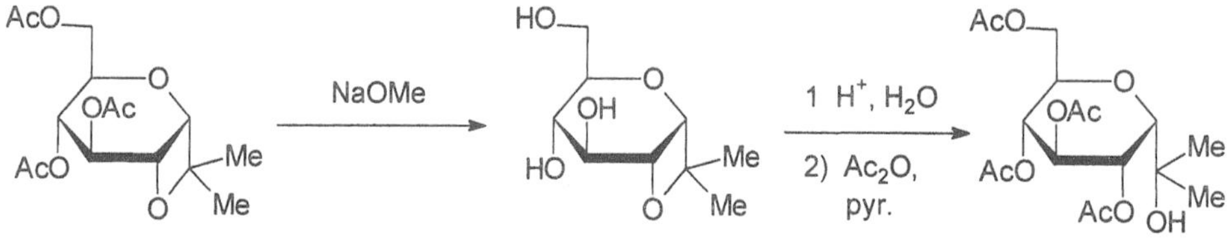

Scheme 31

C.   CYCLOADDITION REACTIONS OF *C*-GLYCOSIDES

There have been some reports involving cycloaddition reactions with the carbon portion of *C*-glycosides. This methodology serves to build up the pre-existing carbon portion of a *C*-glycoside. Three of the illustrated examples employ the Diels-Alder reaction while another example involves the use of nitrile oxide cycloaddition chemistry as a means of assembling *C*-aryl glycosides.

Work by Acton and co-workers[25] focused on the preparation of a *C*-glycoside analog of doxorubicin (**173**). They chose to use Diels-Alder chemistry to assemble this compound. The dienyl *C*-glycoside **169** was heated with diquinone epoxide **170** and gave a good yield of adduct **171**. A 1:1 mixture of diastereomers was obtained indicating, that at least in the thermal cycloaddition, no chiral induction from the sugar was observed. Treatment of **171** with sodium dithionite then gave **172** a *C*-glycoside analog of (**173**), Scheme 32.

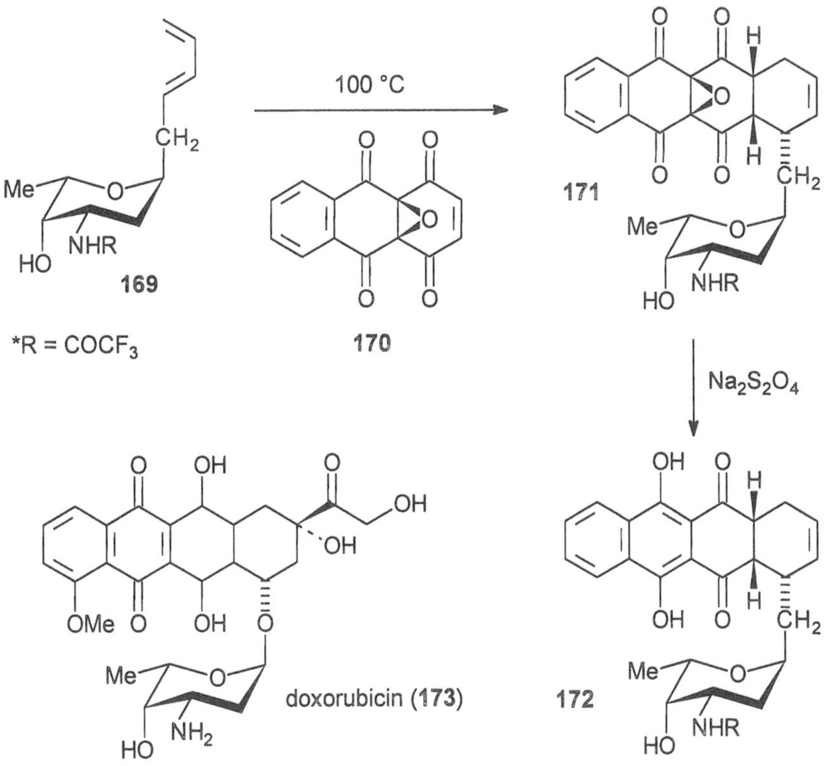

Scheme 32

Horton and Martin[26] also conducted Diels-Alder reaction on dienes attached to the anomeric position of sugars via a carbon atom. Diene **174** (prepared by allylation of the corresponding *p*-nitrobenzoate with *E*-penta-2,4-dienyltrimethylsilane under boron trifluoride etherate catalysis in 45% yield) underwent *endo* addition from the *re* face with maleic anhydride at 90°C to give a 78% yield of a single adduct, **175**!

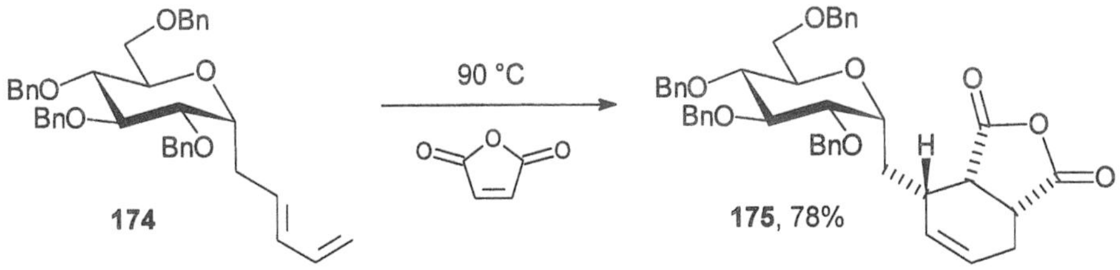

The workers also attempted to use the sugar portion as the dienophile in reactions with activated dienes. Thus *C*-glycoside **176** was converted to the α,β-unsaturated sulfone **177** and attempted cycloaddition reaction with Danishefsky's diene proved to be unsuccessful, Scheme 33.

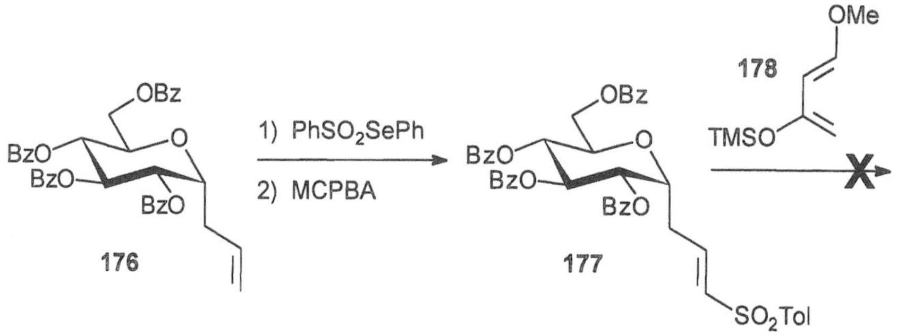

Scheme 33

Work by Danishefsky[27] involved the hetero Diels-Alder reaction between179 and **178** via what the authors term a chelation controlled reaction (see **179**) to provide **180** (major 5:1, 75% combined yield).

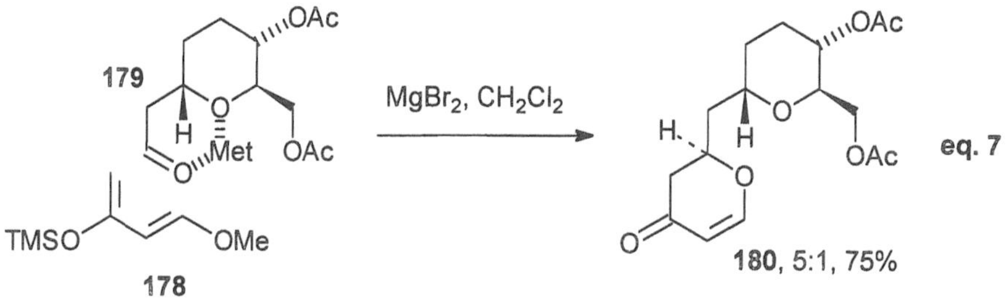

Work describing the [4+2] cycloaddition of singlet oxygen to the furan based *C*-glycoside has also been described (eq. 8).[28]

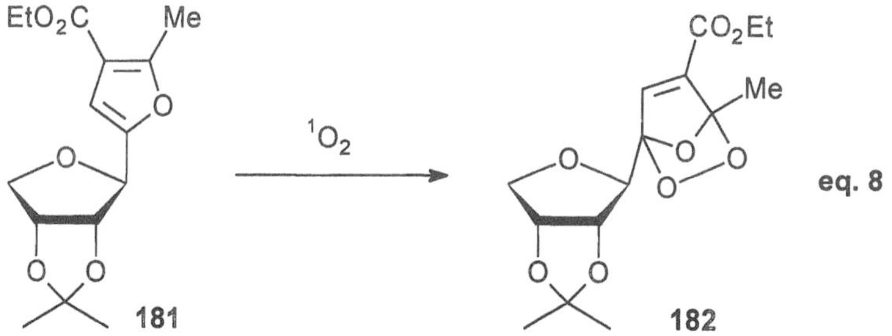

Kozikowski's nitrile oxide cycloaddition route is illustrated below in Scheme 34. In this methodology, the stereochemical disposition of the *C*-1 nitromethyl substituent is retained in the target aryl *C*-glycoside.[29]

The *C*-1 nitrosugar **179** was exposed to phenyl isocyanate and triethylamine in the presence of dienophile **181** to give the cycloadduct **182**. Treatment with Raney nickel served to cleave the oxygen-nitrogen bond to provide **183** which was cyclized and aromatized by the action of a trimethylsilyl triflate to aryl *C*-glycoside **184**. Scheme 35 shows some of the aryl *C*-glycosides synthesized by this technology.

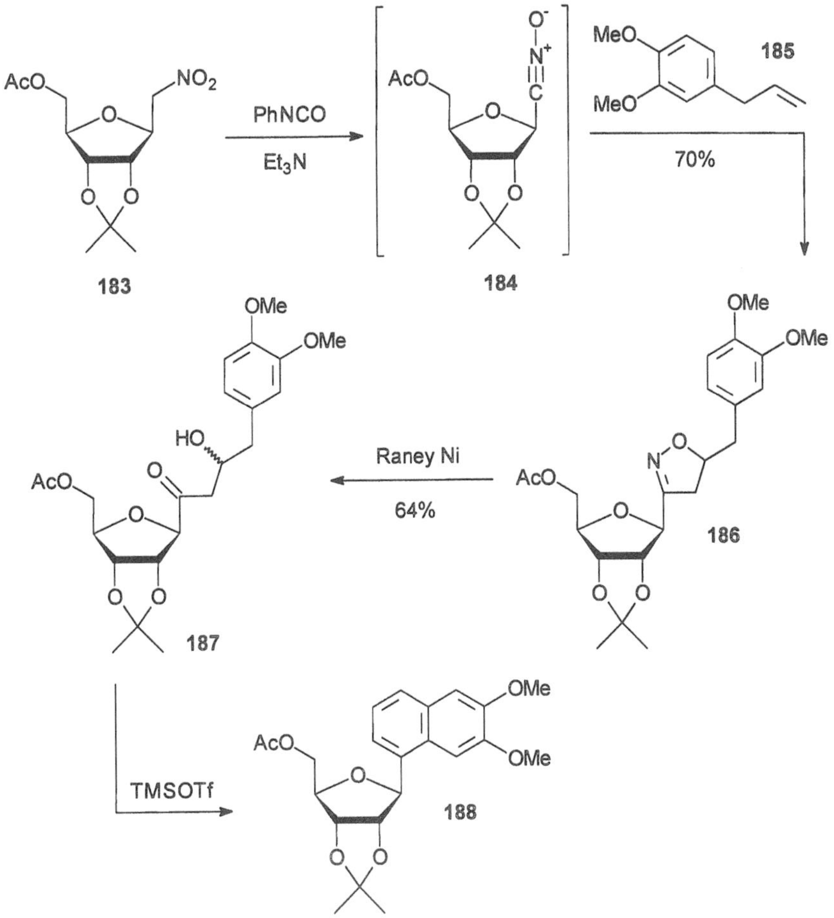

Scheme 34

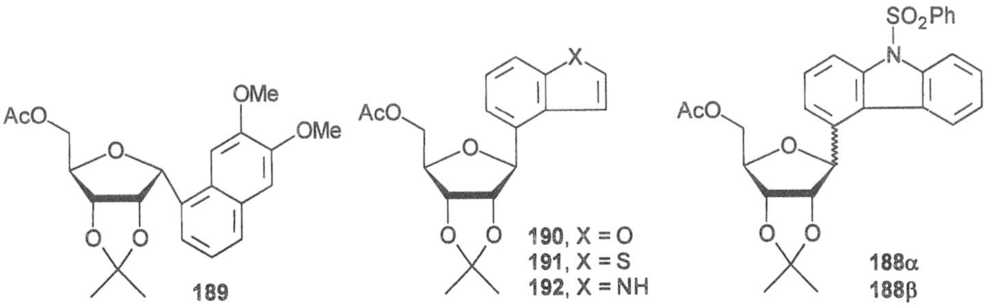

Scheme 35

Reproduced with permission from ref. 29, Copyright 1987, courtesy the Royal Society of Chemistry, Thomas Graham House, Science Park, Milton Road, Cambridge, CB4 4WF, UK.

IV. CONCLUSION

This chapter has illustrated the power of concerted processes in *C*-glycoside preparation. Both the Claisen and Ireland-Claisen [3,3] rearrangements provide stereospecific access to either α- or β-*C*-glycosides depending on the configuration of the involved oxygen atom. The [2,3] rearrangements have also seen limited use in this area.

The Ireland ester rearrangement has been used in several instances to quickly assemble *C*-glycosides to be used as synthetic intermediates in total synthesis. Cycloaddition chemistry has been used to prepare a *C*-substituted pyran ring that was further elaborated to a sugar like substance. Thermal and photochemical cycloadditions have as well been applied to the construction of *C*-glycosidic bonds. A few examples of cycloaddition reactions on the carbon portion of a pre-existing *C*-glycoside have also been presented.

V. REFERENCES

1.   Heyns, K.; Hohlweg, R. *Chem. Ber.* **1978**, *111*, 1632.
2.   Tulshian, D.B.; Fraser-Reid, B. *J. Org. Chem.* **1984**, *49*, 518.
3.   Boivin, T.L.B. *Tetrahedron* **1987**, *43*, 3309.
4.   Ireland, R.E.; Wilcox, C.S.; Thaisrivongs, S.; Vanier, N.R. *Can. J. Chem.* **1979**, *57*, 1743.
5.   Ireland, R.E.; Wuts, P.G.M.; Ernst, B. *J. Am. Chem. Soc.* **1981**, *103*, 3205.
6.   Curran, D.P.; Suh, Y.-G. *Carbohydr. Res.* **1987**, *171*, 161.
7.   Aicher, T.D.; Buszek, K.R.; Fang, F.G.; Forsyth, C.J.; Jung, S.H.; Kishi, Y.; Scola, P.M. *Tetrahedron Lett.* **1992**, *33*, 1549.
8.   Ireland, R.E.; Daub, J.P. *J. Org. Chem.* **1981**, *46*, 479.
9.   Edwards, M.P.; Ley, S.V.; Lister, S.G.; Palmer, B.D. *J. Chem. Soc., Chem. Commun.* **1983**, 630.
10. Ireland, R.E.; Anderson, R.C.; Badoud, R.; Fitzsimmons, B.J.; McGarvey, G.J.; Thaisrivongs, S.; Wilcox, C.S. *J. Am. Chem. Soc.* **1983**, *105*, 1988.
11. Burke, S.D.; Piscopio, A.D.; Kort, M.E.; Matulenko, M.A.; Parker, M.H.; Armistead, D.M.; Shankaran, K. *J. Org. Chem.* **1994**, *59*, 332.
12. Burke, S.D.; Armistead, D.M.; Schoenen, F.J. and Fevig, J.M. *Tetrahedron* **1986**, *42*, 2787.
13. Burke, S.D.; Schoenen, F.J.; Murtiashaw, C.W. *Tetrahedron Lett.* **1986**, *27*, 449.
14. Grindley, T.B.; Wickramage, C. *J. Carbohydr. Chem.* **1988**, *7*, 661.
15. Schmidt, R.R.; Frick, W.; Haag-Zeino, B.; Apparao, S. *Tetrahedron Lett.* **1987**, *28*, 4045.
16. Danishefsky, S.J.; Pearson, W.H.; Harvey, D.F.; Maring, C.J.; Springer, J.P. *J. Am. Chem. Soc.* **1985**, *107*, 1256.
17. Danishefsky, S.J.; Phillips, G.; Ciufolini, M. *Carbohydr. Res.* **1987**, *171*, 317.
18. Danishefsky, S.J.; Pearson, W.H.; Segmuller, B.E. *J. Am. Chem. Soc.* **1985**, *107*, 1280.
19. Maruoka, K.; Itoh, T.; Shirasaka, T. and Yamamoto, H. *J. Am. Chem. Soc.* **1988**, *110*, 310.
20. Katagiri, N.; Akatsuka, H.; Haneda, T.; Kaneko, C. and Sera, A. *J. Org. Chem.* **1988**, *53*, 5464 and references cited therein.
21. Noyori, R.; Sato, T.; Hayakawa, Y. *J. Am. Chem. Soc.* **1978**, *100*, 2561.
22. Burnouf, C.; Lopez, J.C.; Calvo-Flores, F.G.; Laborde, M.; Olesker, A.; Lukacs, G. *J. Chem. Soc., Chem. Commun.* **1990**, 823.
23. Matsuura, K.; Araki, Y.; Ishido, Y.; Murai, A. and Kushida, K. *Carbohydr. Res.* **1973**, *29*, 459.
24. Araki, Y.; Senna, K.; Matsuura, K. and Ishido, Y. *Carbohydr. Res.* **1978**, *60*, 389.
25. Acton, E.M.; Ryan, K.J. and Tracy, M. *Tetrahedron Lett.* **1984**, *25*, 5743.
26. Martin, M.G.C. and Horton, D. *Carbohydr. Res.* **1989**, *191*, 223.
27. Wincott, F.E.; Danishefsky, S.J.; Schulte, G. *Tetrahedron Lett.* **1987**, *28*, 4951.
28. Aparcio, F.J.L.; Sastre, L.A.J.; Garcia, J.I.; Diaz, R.R. *Carbohydr. Res.* **1984**, *132*, 19.
29. Kozikowski, A.P. and Cheng, X.-M. *J. Chem. Soc., Chem. Commun.* **1987**, 680.

Chapter 7

FREE RADICAL APPROACHES TO *C*-GLYCOSIDES

I. INTRODUCTION

Free radical chemistry, once only thought useful for polymerization reactions, has become an important synthetic tool over recent years. Carbon based free radicals have attracted the most attention and their reactivity and application in synthetic organic chemistry has been reviewed.[1] Free radical chemistry is a powerful method for forming carbon-carbon bonds. Thus, this methodology would seem to have applications in the generation of *C*-glycosides. The merits of free radical chemistry include mild reaction conditions for radical generation and tolerance of a wide range of functional groups. For instance, free radical cyclizations can be carried out in the presence of free hydroxyl and amino functionalities. The reactions can be carried out in refluxing benzene with a thermal initiator or at low temperature with the use of a photochemical initiator. Common radical precursors such as halides, selenides, acids, thiocarbonyl groups, and tertiary nitro compounds are fairly straightforward to prepare and are usually stable enough to be purified. It should also be mentioned that the most utilized radical acceptors are carbon-carbon multiple bonds.

The chain method is the most common and efficient way to carry out radical additions. A typical chain process using the 5-hexenyl radical is shown below in Schemes 1 and 2. The first step is generation of a free radical species by using an initiator. The most widely used is azobisisobutyronitrile (AIBN) (1) which extrudes nitrogen under thermal conditions to give rise to two carbon based radicals. Hexamethylditin has been used as a photochemical initiator. The advantage here is that the photolysis reaction can carried out at low temperature.

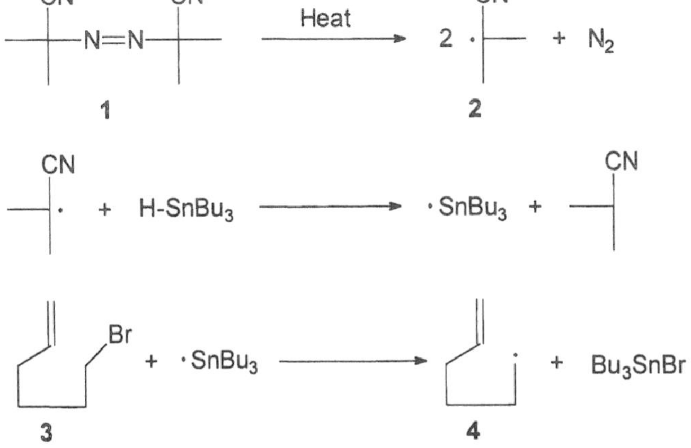

Scheme 1

161

These formed radicals react with a hydride source such as tributyltin hydride to give a stannyl radical which then attacks the homolyzable group such as bromide to give the desired free radical **4** which has several fates. It can abstract hydride from another molecule of tin hydride to give the product of reduction **5** or it can cyclize, in a 5-*exo* fashion, to give the product of closure **6** which is then reduced by tin hydride to the product **7**.

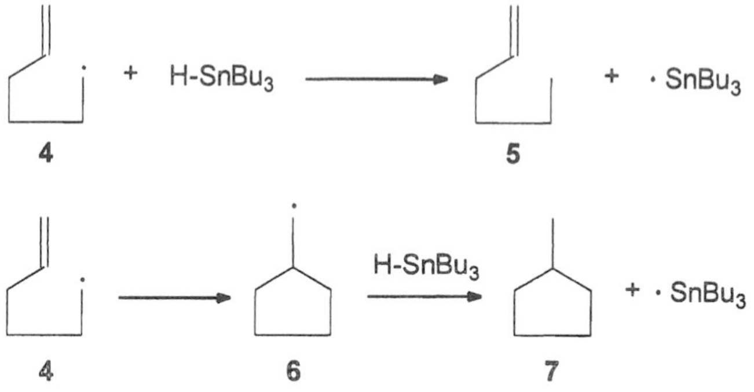

Scheme 2

Most radical reactions are done at low concentration or with syringe pump techniques to avoid radical coupling. The addition shown in Scheme 2 is an intramolecular process, but radical additions can also occur in an intermolecular sense as shown in eq. 1.[2]

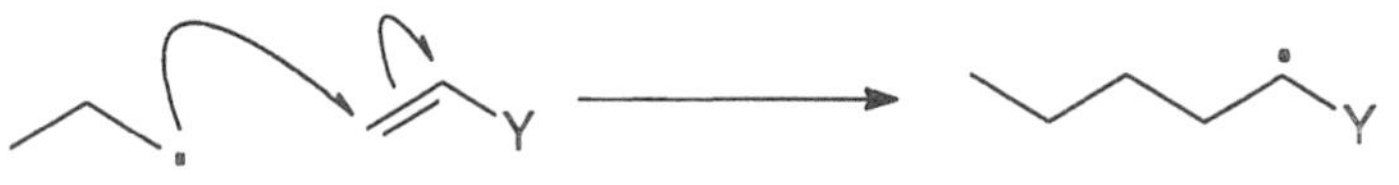

eq. 1

One requirement for intermolecular additions to be successful is that once addition occurs, the newly formed radical must be stable enough so that it will not undergo further additions but, will be reduced by tin hydride. An efficient radical acceptor (*i.e.* Y = CN) must also be employed so that the rate of addition will be faster than the rate of direct reduction of the initially formed radical. Often an excess of acceptor is utilized.

The one drawback of the tin hydride method is that the product formed has undergone reduction. This is in effect a loss of functionality. Keck has developed an alternative strategy to the tin hydride method based on allyltin chemistry.[3] The addition of a radical to tributylallyltin causes a fragmentation reaction to occur as shown below (eq. 2). The net result is the formation of a carbon-carbon bond along with the expulsion of a stannyl radical that goes on to carry the chain. The allyl group is fairly versatile in the sense that it can be modified into other functionalities. Also direct reduction of the initially formed radical is not of consequence since there is no hydride source available to facilitate reduction.

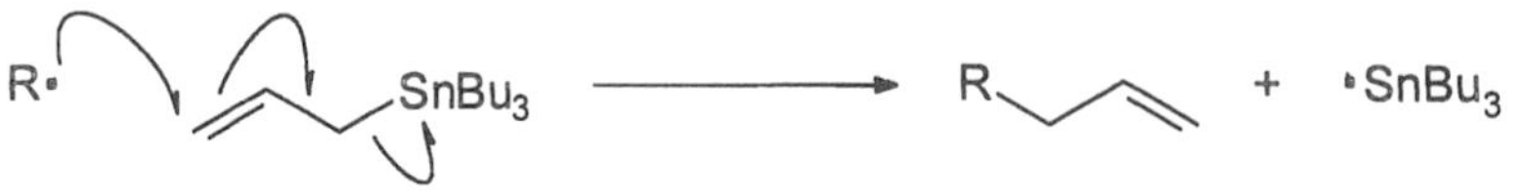

eq. 2

Both intermolecular additions and intramolecular additions have been used to assemble *C*-glycosides. The radical can be generated at the anomeric center and undergo addition to a suitable acceptor or a strategically located free radical can cyclize on an unsaturated anomeric carbon. The former method has been most widely used and this is probably due to the availability of glycosyl halides and selenides. Both pyranose and furanose sugars have been used in these types of additions. Glucosyl radicals have been found to react preferentially in the α or axial mode to give α-C-glycosides. This chapter will be divided into two main sections. The first part will deal with intermolecular additions while the second will address additions of an intramolecular nature. A small section at the end of the chapter will address a few examples dealing with anomeric carbenes.

II. INTERMOLECULAR ADDITIONS

A. MICHAEL ACCEPTORS

Giese was one of the first to react glycosyl radicals with suitable acceptors to produce C-glycosides in a highly stereoselective manner (Scheme 3).[4] The anomeric radical is generated from the α-glucopyranosyl bromide **8** by photolysis in boiling ether in the presence of tributyltin hydride and a ten to twenty fold excess of acrylonitrile to give the α-C-glucopyranoside **9** and the β-isomer **10** (97:3) in 75% combined yield. A small amount (5%) of reduced product **11** was also formed.

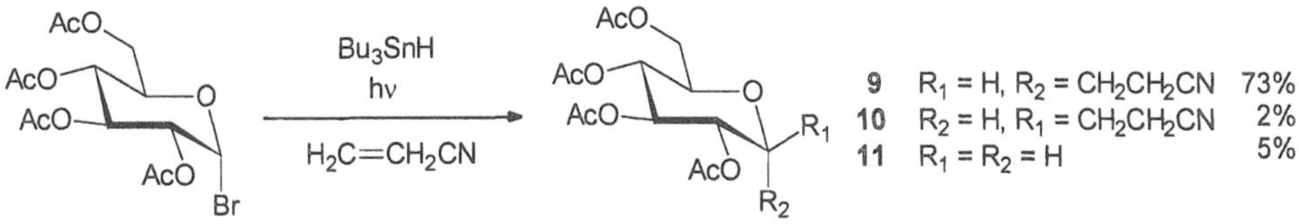

Scheme 3

The radical addition was also found to be highly diastereoselective with the corresponding manno and galactopyranosyl compounds.[5]

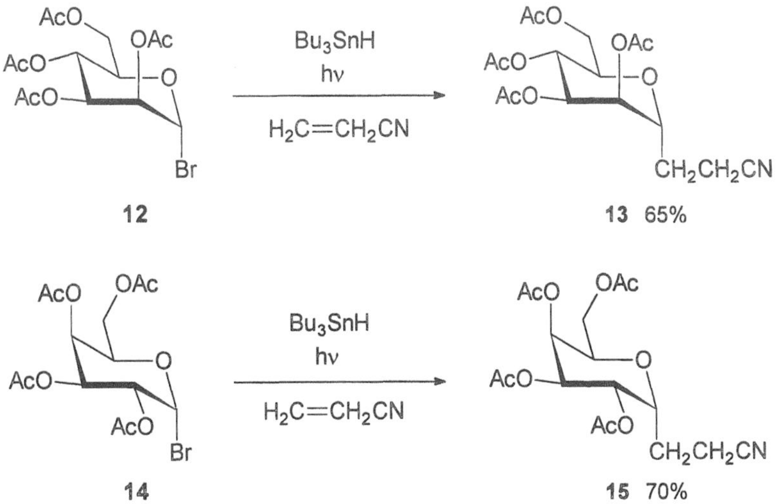

Scheme 4

The highly selective nature of the radical addition is somewhat surprising, since Giese has shown that radicals generated at other carbon centers on the glucose ring react from the equatorial side.[6] ESR spectroscopy has demonstrated that the most favorable conformation for the radical is to be a twist boat **16** as shown below (Scheme 5).[7] The twist boat is stabilized by the orbital interactions of the SOMO of the radical and the LUMO of the adjacent axial C-O bond. As a result of this conformation, attack occurs in an equatorial sense on the twist boat to give the product of net axial addition in the chair conformation. This effect is sometimes termed the radical anomeric effect.[8] This net result is similar to the stereoelectronic effect that is observed in the corresponding cationic glycosidation reactions. ESR measurements on the free radical derived from the manno derivative showed that the conformation of this species was a chair. This radical is stabilized by the orbital interactions of the adjacent axial C-O bond. In the gluco compound the stability from this interaction overcomes the steric repulsions inherent in the boat conformation.

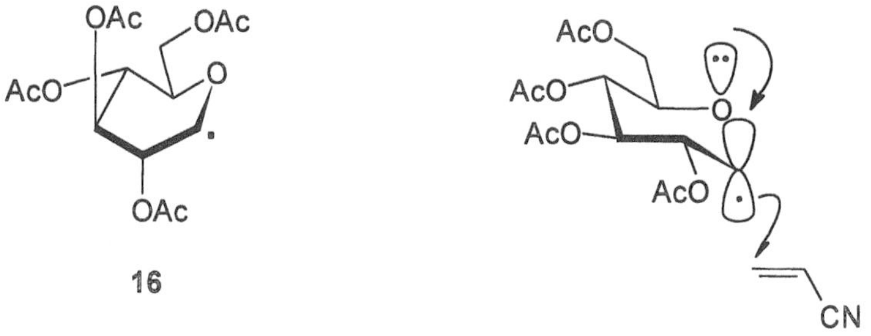

Scheme 5

When the 5-position is unsubstituted the ratio of axial to equatorial addition changes.[9] The xylosyl bromide **17** reacts in the opposite sense to that of the glucopyranosyl bromide to give the equatorial *C*-glycoside **18** as the major product. Some double addition product **19** is also formed (eq. 3).

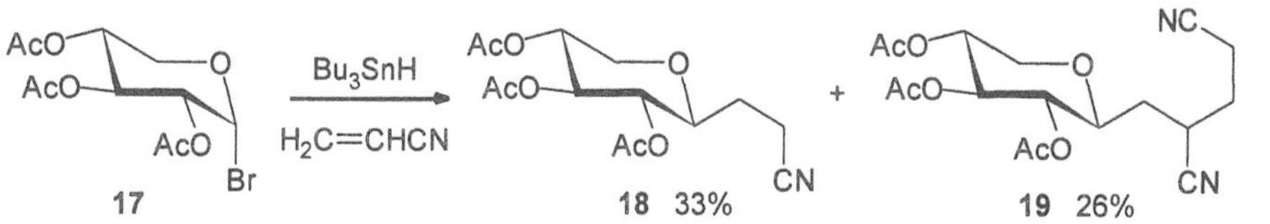

Araki has used a four carbon radical acceptor to synthesize a *C*-glycoside.[10] Reaction of 2,3,4,6-tetra-*O*-acetyl-α-D-glucopyranosyl bromide (**8**) with tributyltin hydride gives an anomeric radical that reacts with methyl vinyl ketone in axial fashion to give the *C*-glycoside **20**. The rest of the product is of the reduced form, compound **11** (eq. 4).

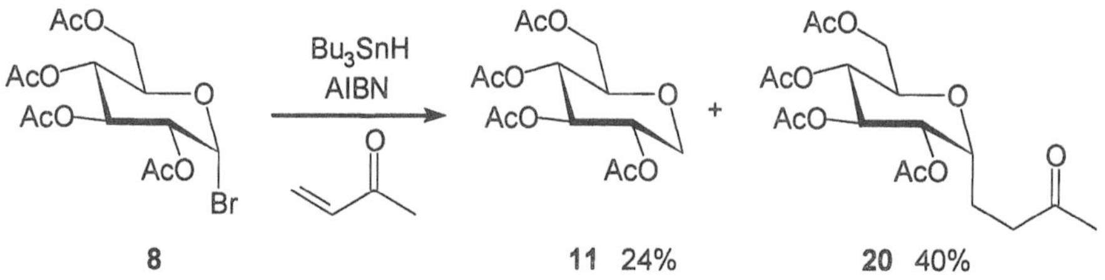

Baldwin has utilized glycosyl selenides as radical precursors, eq. 5.[11] They have the advantage of being more stable than anomeric bromides and iodides and can usually be purified by column chromatography. The yield of the addition product **22** is 40% and the addition is also highly stereoselective. The axial product **22** is accompanied by 10% of double addition product **23**.

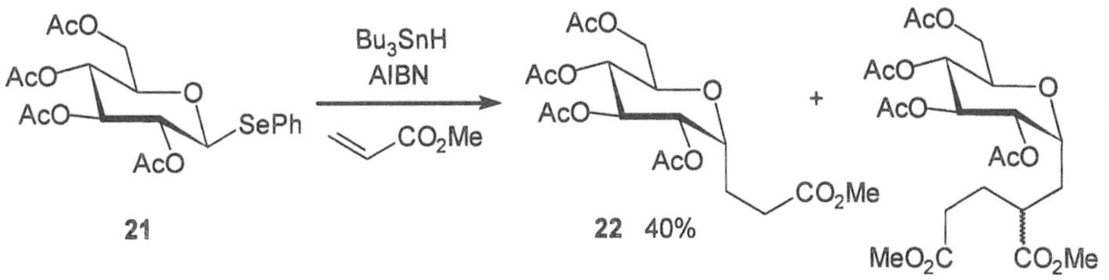

eq. 5

Hart has developed a free radical based one carbon homologation by radical addition to *O*-benzylformaldoxime.[12] The reaction involves the radical reaction with piconolate (**24**) in the presence of tin hydride to give, in the case shown in eq. 6, the α-*C*-glycoside **25**. It should be noted that the 2-*O* acetyl group is transferred to the oxime nitrogen under the reaction conditions.

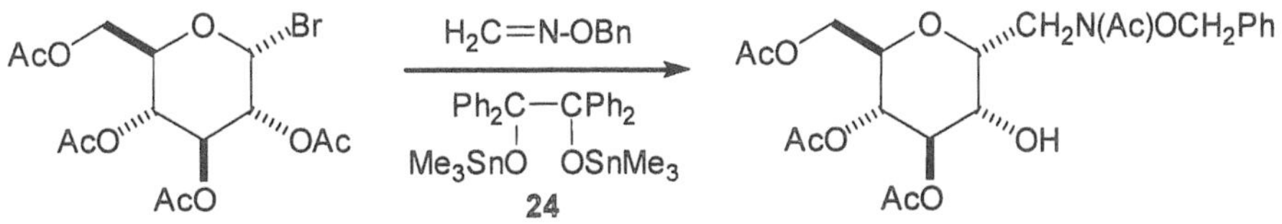

eq. 6

Araki has added glucosyl radicals to dimethyl maleate to give a mixture of products.[10] The equatorial *C*-glycoside **26** is formed in 9% yield, along with the expected products, the anomeric α-*C*-glycosides **27**, and some reduced product **11** in 26% yield.

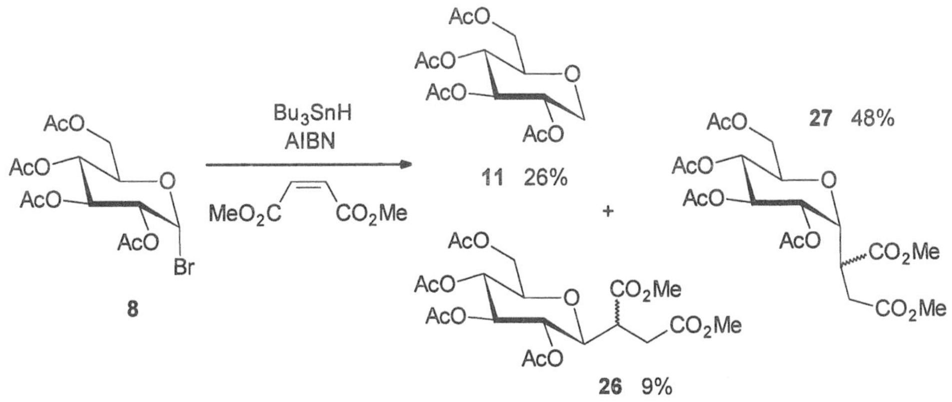

Scheme 6

It should be noted that if acetylenedicarboxylate is used as the radical acceptor then only starting material is recovered in 91% yield along with the addition product **28** that arises from the addition of tributyltin hydride to the triple bond. Addition of the anomeric radical to butyl vinyl ether results only in the formation of reduced product.[10]

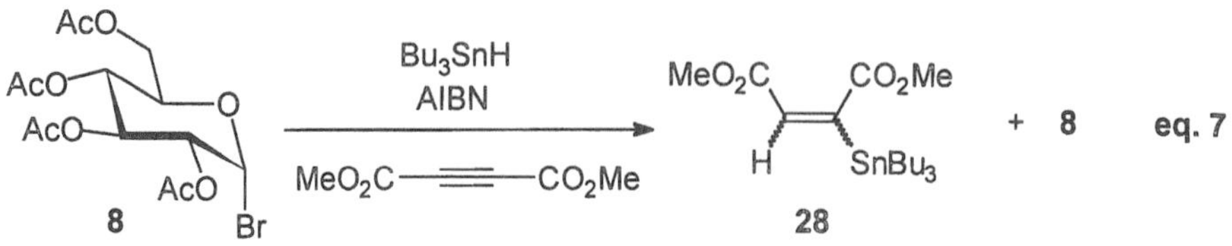

eq. 7

Nicolaou has employed a glycosyl fluoride to synthesize an axial *C*-glycoside (eq. 8).[13]   The reaction is carried out in the presence of magnesium bromide diethyl etherate complex which presumably converts the glycosyl fluoride to an anomeric bromide which then undergoes attack by stannyl radical, in the usual manner, to provide an anomeric radical which is then free to add to acrylonitrile to give the product **9**.

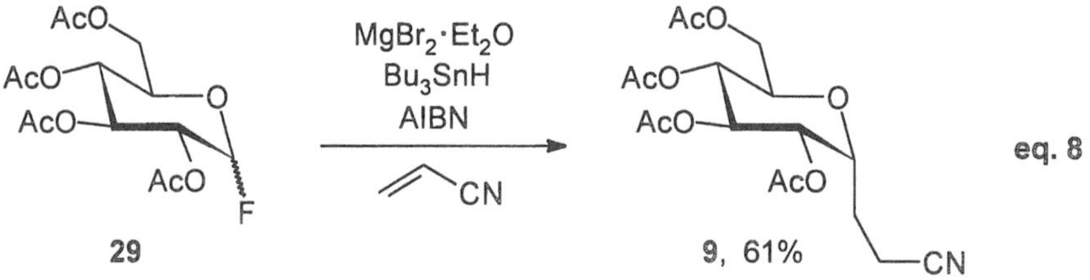

eq. 8

Ferrier has photolytically brominated tetra-*O*-acetyl-β-D-xylopyranose **17** to give as the major product compound **30** (Scheme 7).[14]   When this compound is treated with tributyltin hydride in the presence of acrylonitrile and in the presence of visible light two compounds are obtained.[15]   The L-ido adduct **31** is formed in 38% yield while the D-gluco isomer **32** is produced in 35% yield.   When methyl acrylate was used as the radical acceptor, similar ratios of addition products were formed.   The variation in reactivity of the pentos-5-yl radical, as compared to the stereoselective radical additions of the glucopyranosyl radical, has been ascribed to the lack of conformational stability of the former.

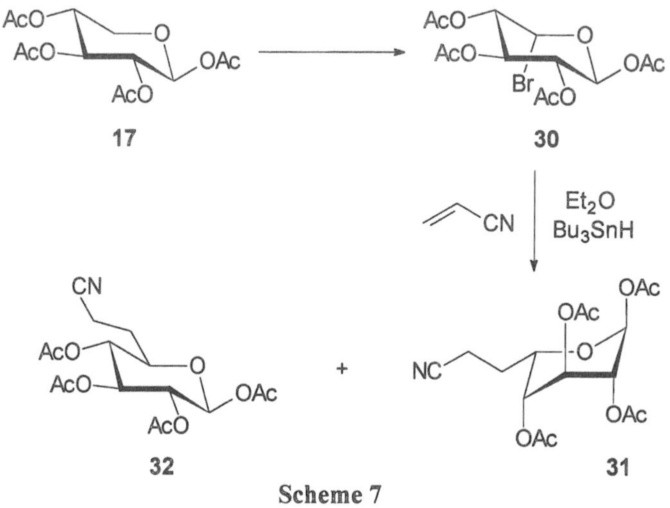

Scheme 7

Both the L-ido adduct **31** and the D-gluco compound **32** can be radically brominated at *C*-5 to give **33** which can then be reduced stereoselectively via the action of tributyltin hydride to give the D-gluco compound **32**, thereby increasing the overall efficiency in the conversion of **17** to **32**.[15]

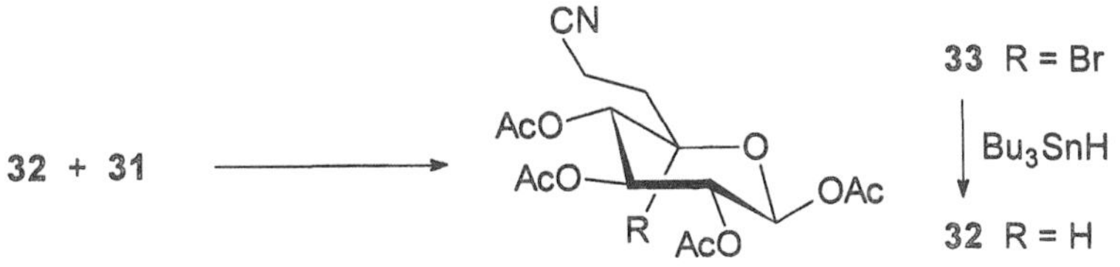

32 + 31 $\longrightarrow$

33 R = Br

$\downarrow$ Bu$_3$SnH

32 R = H

Scheme 8

If the anomeric acetate is replaced by bromine, one obtains compound **34** which can then add to a radical acceptor to give the dinitrile **35**. Using this methodology one has the ability to stereoselectively introduce two carbon chains at C-1 and C-5 of the original xylo sugar, Scheme 9.[15]

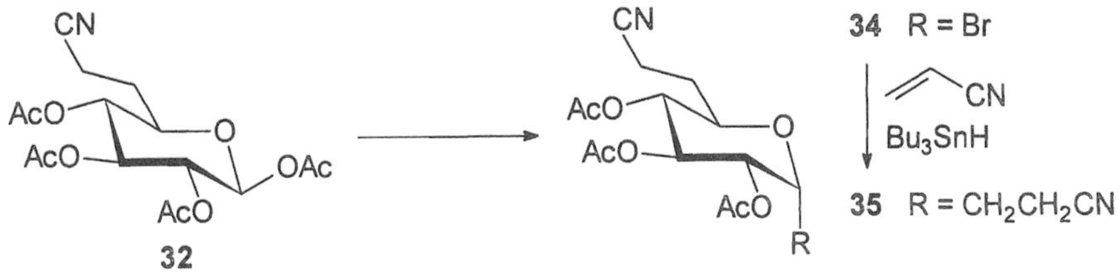

32

34 R = Br

$\downarrow$ Bu$_3$SnH

35 R = CH$_2$CH$_2$CN

Scheme 9

Barton has employed telluride as the radical precursor.[16] The anisyl telluride group is exceptionally radicophilic and its high nucleophilicity makes it facile to incorporate into organic molecules. The starting telluride can be accessed via displacement of an anomeric mesylate or tosylate with inversion of configuration to provide the sugar telluride **37** (eq. 9).

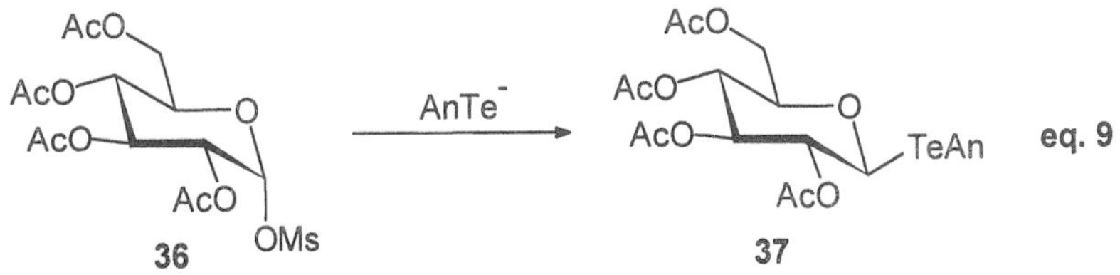

36 OMs $\xrightarrow{\text{AnTe}^-}$ **37** TeAn **eq. 9**

Since radical formation from tellurides is relatively facile, the photolabile acetyl compound **38** can be employed to generate the requisite carbon based radical. When exposed to tungsten light **38** generates methyl radicals as shown in eq. 10.

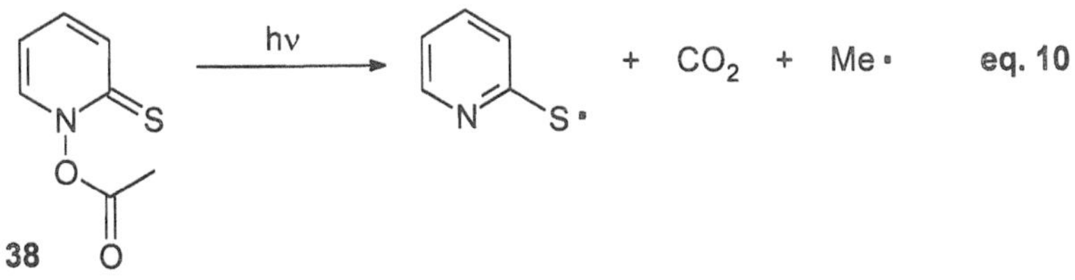

38 $\xrightarrow{h\nu}$ + CO$_2$ + Me• **eq. 10**

The methyl radical reacts with the telluride to provide an anomeric radical that adds to acrylonitrile. The secondary radical formed reacts with the thiocarbonyl compound **38** to give the product **39**. This group can be eliminated to provide the unsaturated *C*-glycoside **40**. The advantage of this method for radical generation is two-fold. Firstly, the product of addition has not undergone a net reduction as in the tin hydride method and secondly, the products of the reaction do not include toxic tin by-products. It is appropriate to mention that sometimes purification of the products from tin residues can be problematic with fairly non-polar products. However, in the synthesis of *C*-glycosides this should generally not be a problem.

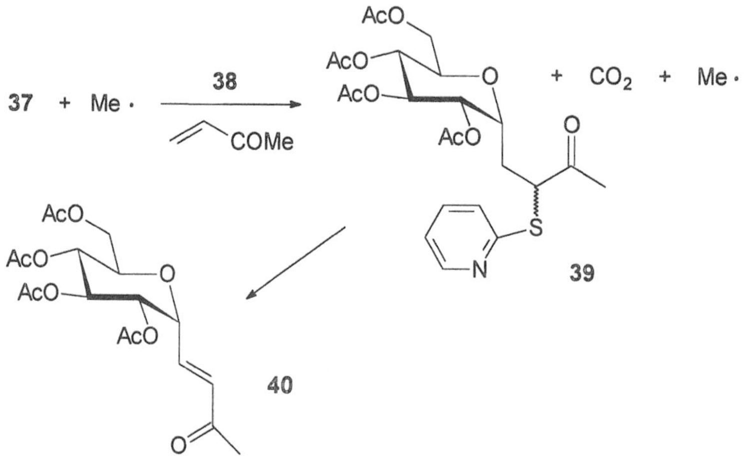

37 + Me· $\xrightarrow{\textbf{38}}$ [product] + CO_2 + Me·

Scheme 10

Giese has applied the use of an iron based initiator to generate iron based radicals which can abstract halide to form carbon radicals.[17] Irradiation of the dimeric iron complex **41** furnishes the monomeric radical species **42**.

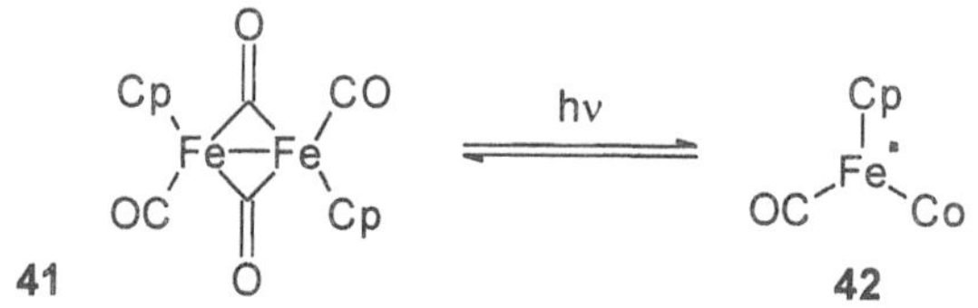

eq. 11

Radical **42** then reacts with a halide, in this case α-D-glucopyranosyl bromide **8**, to give an anomeric radical which then goes on to add to acrylonitrile to give the α-*C*-glycoside **9**, eq. 12.[17] Once addition is complete, the secondary adduct radical goes on to form an organometallic species which reacts with the solvent in an ionic manner to give the product of addition/reduction.

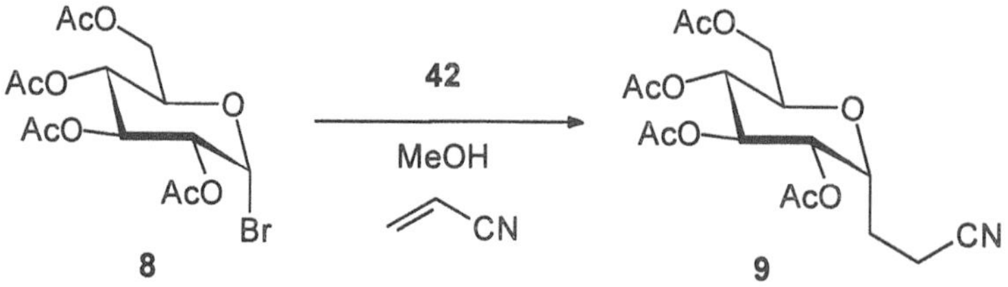

eq. 12

Glycosyl cobalt complexes, which can be prepared from glycosyl halides and Co(I) complexes (Scheme 11), react with olefins in a free radical fashion to give either the product of solvolysis **13** or the product of elimination **44**.[18] This method has the drawback that the radical precursor has to be prepared from a glycosyl halide. This is in contrast to the iron initiated reactions, *vide supra*, where glycosyl halides can be used directly in the carbon-carbon bond forming reaction.

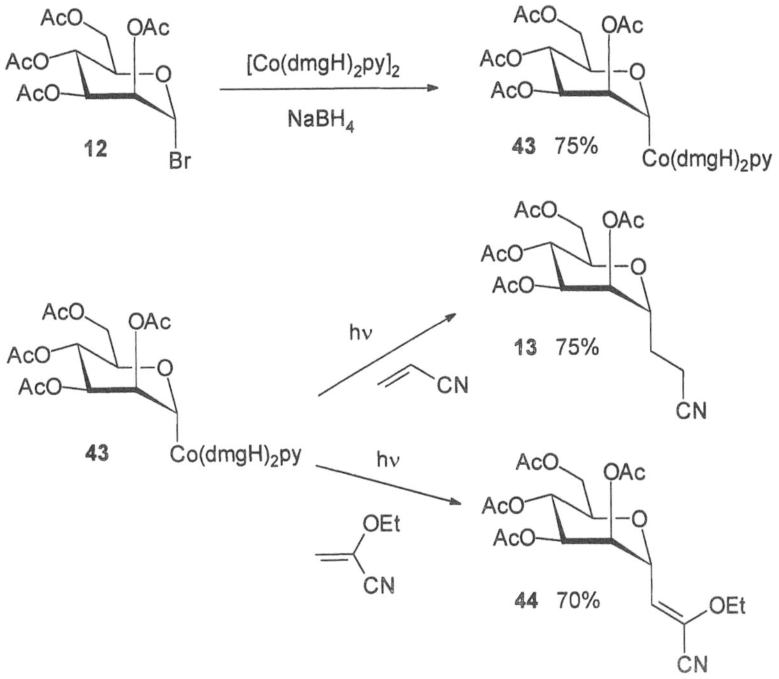

Scheme 11

Abrecht and Scheffold have used vitamin B$_{12}$ as an initiator to facilitate the addition of the pyranosyl radical derived from **8** to give the α-*C*-glycoside **20** in 41% yield (eq. 13).[19]

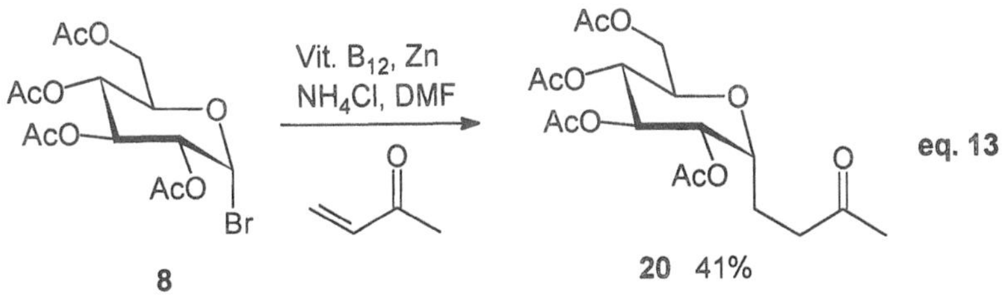

Up to now the use of glycosyl halides, selenides, and cobalt complexes as anomeric radical precursors has been illustrated. Tertiary nitro compounds have been shown to be reductively denitrated by the action of tin hydride via a free radical pathway.[20] The synthetic usefulness of the nitro group coupled with its ability as a radical precursor makes it an ideal group for use in *C*-glycoside synthesis. Both Vasella and Giese have used this group to advantage in the preparation of *C*-glycosides. Reaction of tributyltin hydride with compound **45** generates an anomeric free radical, which then adds to acrylonitrile to give the bis *C,C*-glycoside **46**, eq. 14.[20] The new carbon-carbon bond is formed from the bottom face.   This stereoselectivity has been proven by a trapping experiment with tributyltin deuteride.

eq. 14

Similar reactions can be carried out on the glucopyranose derivatives (eq. 15).[21]   The addition is stereoselective, but the stereochemistry of the new carbon-carbon bond could not be unambiguously assigned, although one would expect the addition to proceed in an axial fashion.

eq. 15

Thus far, the illustrated radical acceptors have been four or five carbon atoms large. Some workers have added anomeric radicals to fairly large olefin acceptors. By adding anomeric radicals to suitable "sugar acceptors", Giese[22] has efficiently constructed *C*-dissacharides (See Chapter 8 for a full discussion).   Equation 16 highlights the salient features. The addition is stereoselective and this is reflected in the $\alpha:\beta$ ratio of 10:1, but reduction of the adduct radical by tin hydride exhibits only low selectivity (60:40).

eq. 16

Vogel[23] has added the 2,3,4,6-tetra-*O*-acetyl-D-glucopyranosyl radical to a 3-methylidene-7-oxabicyclo[2.2.1]heptan-2-one system to give *C*-disaccharide precursor **52**. It is interesting to note that the anomeric bromide is attacked preferentially over the phenyl seleno function. No trace products arising from reduction of the phenyl seleno group were detected. This selectivity is in accordance with the order of reactivity put forward by Beckwith.[24] The addition is, as expected, highly stereoselective. The α-anomer **52** was formed in 48.5% while a small amount (6%) of β-isomer (not shown) was also isolated from the reaction mixture. The intermolecular radical addition proceeds efficiently because it is a Michael type addition. Transfer of hydride to the adduct radical also occurred with very high *exo* face selectivity.

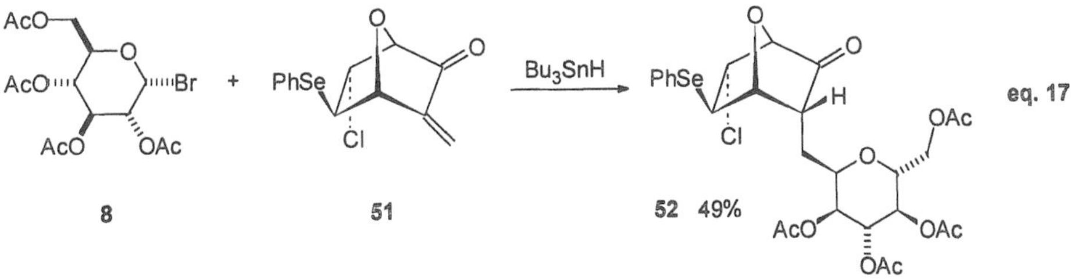

eq. 17

B. RADICAL ADDITION TO EXOMETHYLENIC SUGARS

In the cases presented thus far, all have the common feature of anomeric radicals adding to radical acceptors. Motherwell has reversed this by adding carbon based radicals to difluoro-exomethylenic sugars to give C-glycosides.[25] Intermolecular addition to electron deficient alkenes is the usual strategy and addition onto simple enol ethers would seem to be feasible in an intermolecular fashion. Calculations on the difluoroenol ethers showed that the LUMO energy is lowered considerably as compared to the simple enol ethers. This is significant since the SOMO of the free radical will interact with the LUMO of the olefin and by lowering the LUMO one increases the rate of addition relative to simple radical reduction.

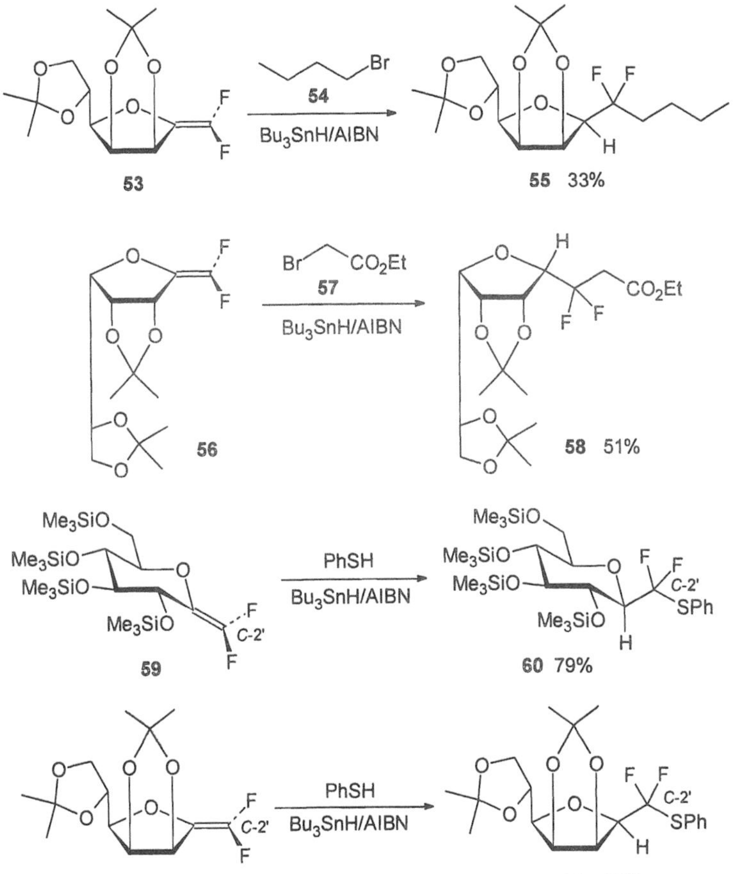

Scheme 12

Reproduced with permission from ref. 25, Copyright 1989, pg. 68-69,
courtesy Georg Thieme Verlag, Stuttgart, FRG.

The radical derived from 1-bromobutane adds to compound **53** to give an anomeric radical that is reduced from the bottom face to provide the *C*-glycoside **55**. The yield was low (33%), but 23% of **53** was recovered from the reaction mixture. More success was achieved with the electrophilic radical derived from **57**. It added to compound **56** regioselectively, on the least hindered difluoromethylene terminus, and the resulting anomeric radical was reduced from the less encumbered top face to give the *C*-glycoside **58** in 51% yield. Keck allylation was also used to generate *C*-glycosides (*vide infra*), but a radical precursor at *C*-2' is required first. Radical addition of thiophenol to difluoroenol ether **59** and **61** gives compounds **60** and **62**, respectively (Scheme 12).

C.   KECK ALLYLATION

Keck allylations have been utilized for forming carbon-carbon bonds at the anomeric center. The reactions exhibit similar selectivity as additions to Michael acceptors and have the advantage that a double bond is left in place on the side chain poised for further manipulation. Keck has assembled *C*-glycosides using this allylation methodology.[26] Both furanoses and pyranoses were used as the starting materials. In the gluco series the selectivity of the addition was moderate (*ca.* 1:1). In contrast the lyxose derivative **64** gave a high level of selectivity (α to β ratio 90 to 10). When compound **66** was exposed to photolytic allylation conditions, α-anomer **67** was formed as the major product. Thermal additions were also examined in all the cases shown below, but photolytic conditions generally gave better yields.

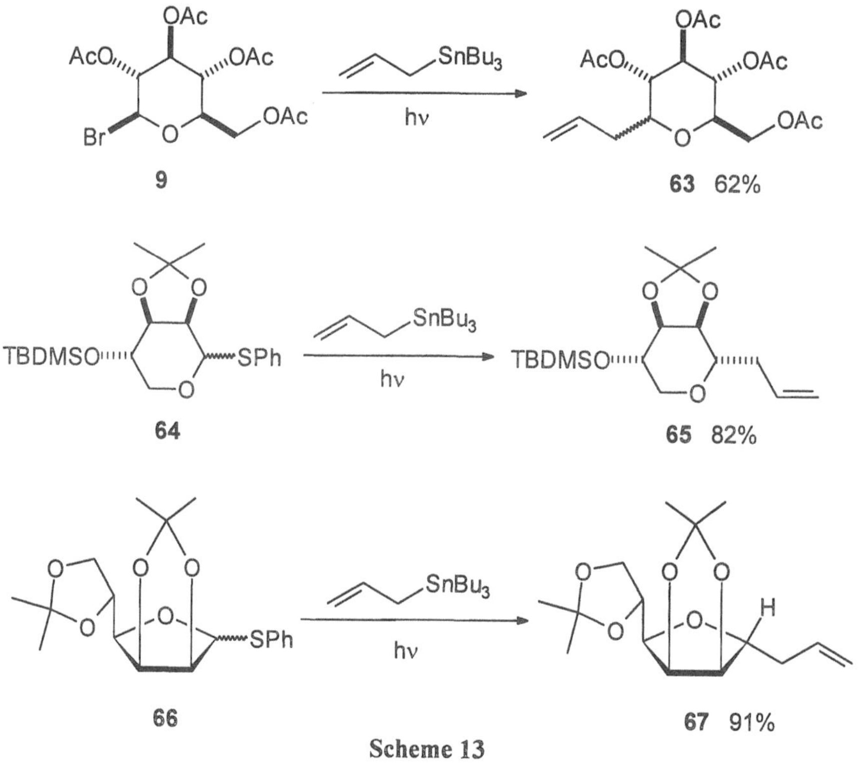

Scheme 13

Paulsen and Matschulat[27] have applied the Keck reaction to the *N*-acetyl Neuraminic acid derivative **68** to obtain a mixture of allyl *C*-glycosides **70** and **69** in a 1.8:1 ratio in 92% combined yield (eq. 18).

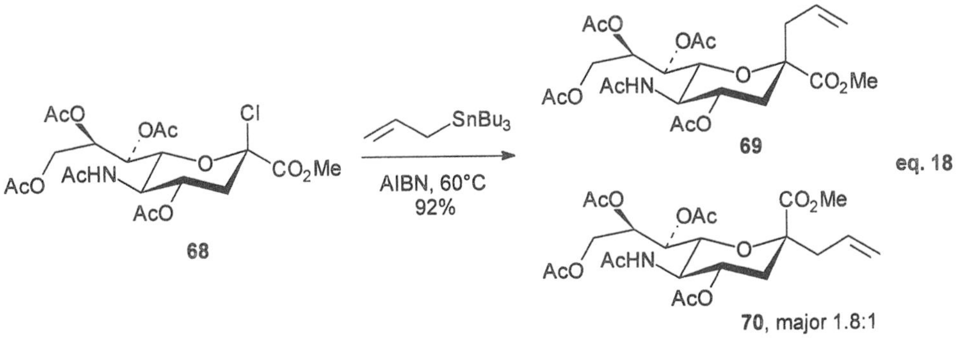

eq. 18

Compound **71** also served as a suitable precursor since deoxygenation then gave the required compounds (Scheme 14).

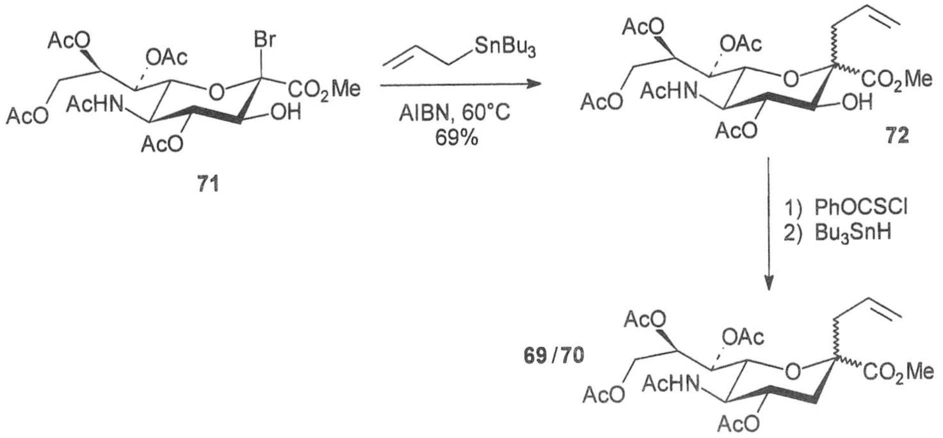

Scheme 14

Bednarski and Nagy[28] used a combination of chemical and enzymatic methods to synthesize a *C*-glycoside of *N*-acetyl neuraminic acid. Enzyme catalyzed aldol reaction between *N*-acetyl mannosamine **73** and sodium pyruvate **74** gave the intermediate acid that was esterified and converted to chloride **75**. Allylation under photolytic conditions then gave a mixture of allyl *C*-glycosides **76** that were deprotected and separated (Scheme 15).

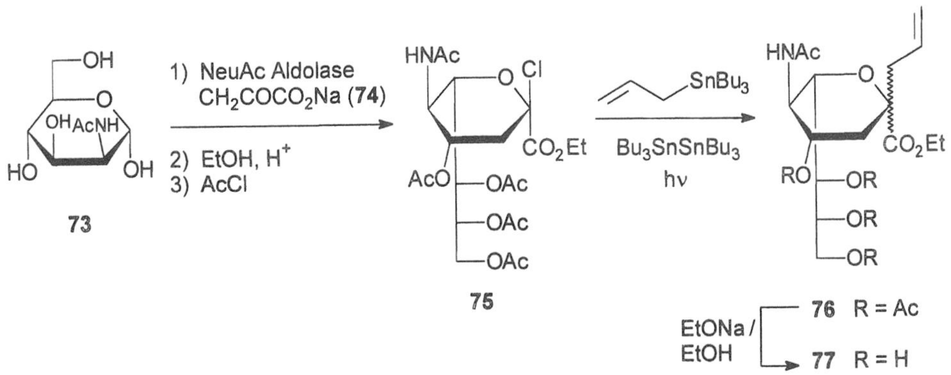

Scheme 15

The allylation of the KDO derivative **78** has also been examined.[29] Reaction of **78** under standard conditions gave a 1:1 mixture of anomers **79** and **80** in 30% combined yield. The diacetonide **81** underwent a more stereoselective reaction and afforded a 90:10 mixture of **82** and **83**, respectively. Presumably the locked conformation and the two β-acetonide groups shield the top face thereby impeding the approach of the allylstannane (Scheme 16).

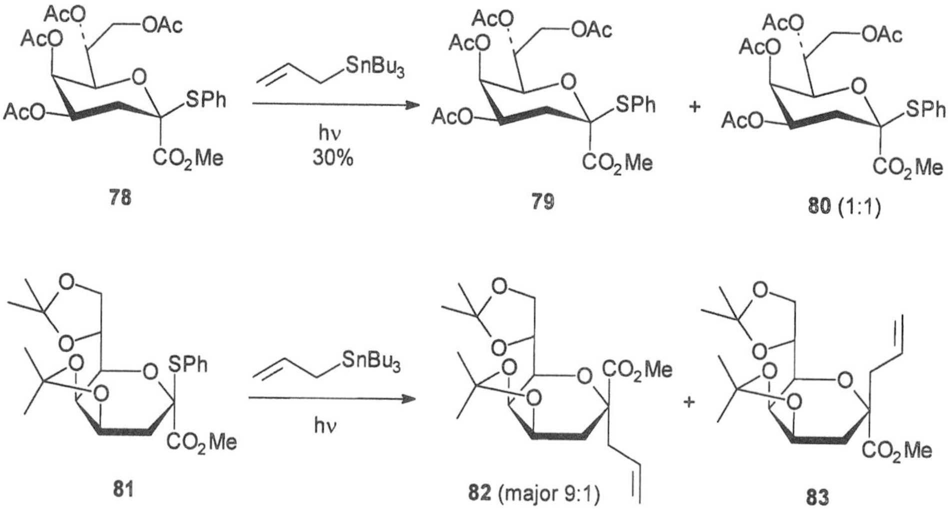

Scheme 16

Giese has also used the Keck allylation reaction to construct *C*-glycosides (Scheme 17).[30] The glucopyranosyl bromide **8** reacts with allyl tin compound **84** in an axial fashion to give the acrylate ester **85**. Exposure of this compound to ozone then furnishes the keto-ester *C*-glycoside **86**. It is interesting to note that the radical addition is highly stereoselective and this result is in contrast to the first entry in Scheme 13, *vide supra*.

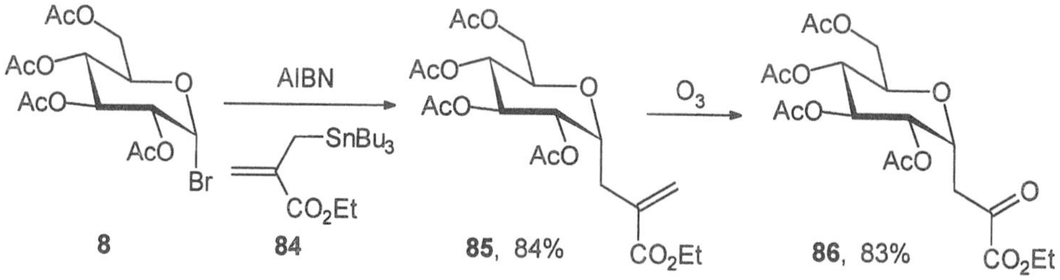

Scheme 17

The Keck allylation has been applied to chain extend an existing *C*-glycoside. Sulfide **60** made from the corresponding exomethylenic difluoroenol ether underwent Keck allylation to give **87** in acceptable yield (Scheme 18).[25]

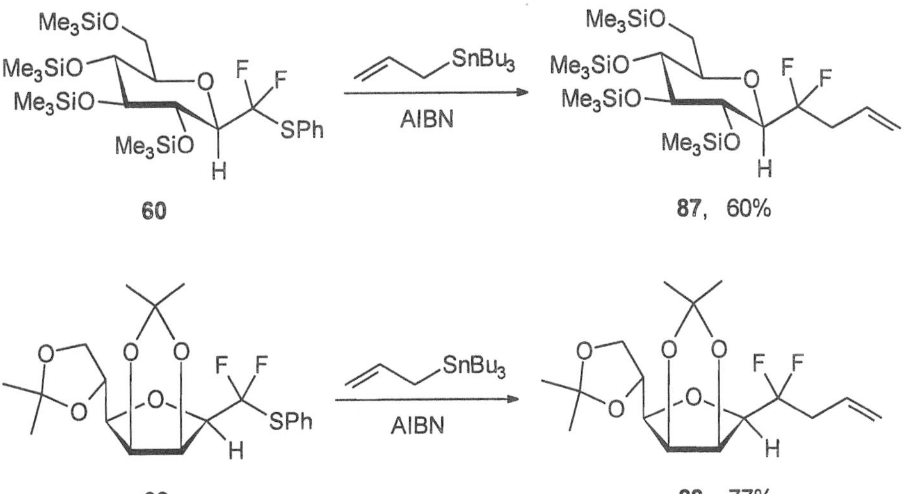

Scheme 18

D. REDUCTION TO β-C-PYRANOSIDES

Due to the stereoselective nature of additions to glucopyranosyl radicals, both the α- and β-C-glycosides can be accessed by simply alteration in the strategy. To construct α-C-glycosides the new carbon-carbon bond must be constructed last so that the radical addition is forced to occur from the axial direction as shown in eq. 19.

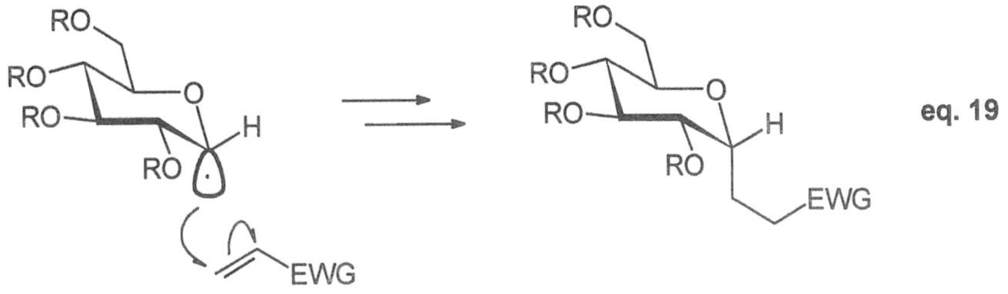

eq. 19

If, however, an anomeric carbon-carbon bond already exists, then generation of an anomeric radical followed by axial reduction with tin hydride should lead to the β-C-glycoside with the anomeric carbon-carbon bond now in an equatorial position (eq. 20).

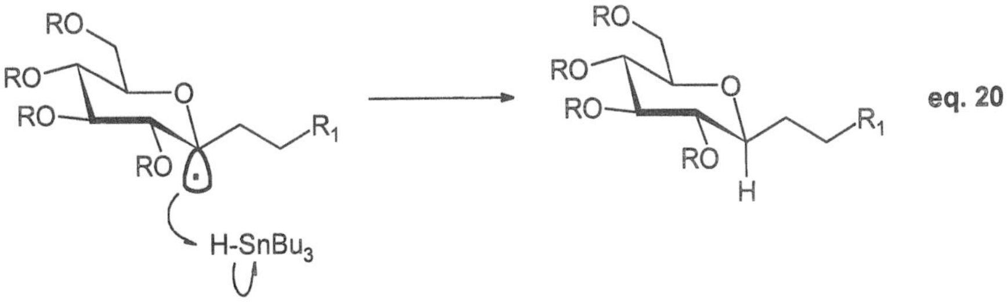

eq. 20

Crich has utilized the latter strategy to construct 2-deoxy-β-C-glycosides.[31] The radical precursor is made from the anomeric sulfones **89** by applying technology developed by Sinaÿ.[32] These are deprotonated with LDA to give an anomeric anion that is quenched with dimethyl carbonate. The sulfone group is reductively removed to give the enolate **90**.

With the enolate in hand, the carbon aglycone can be introduced to provide the *C*-glycoside **91**. Saponification and treatment of the acid **92** with triethylamine and the heterocycle **93** furnishes the radical precursor, and this is photolyzed (tungsten light) in the presence of a tertiary thiol to give the product β-*C*-glycoside **94** (Scheme 19).

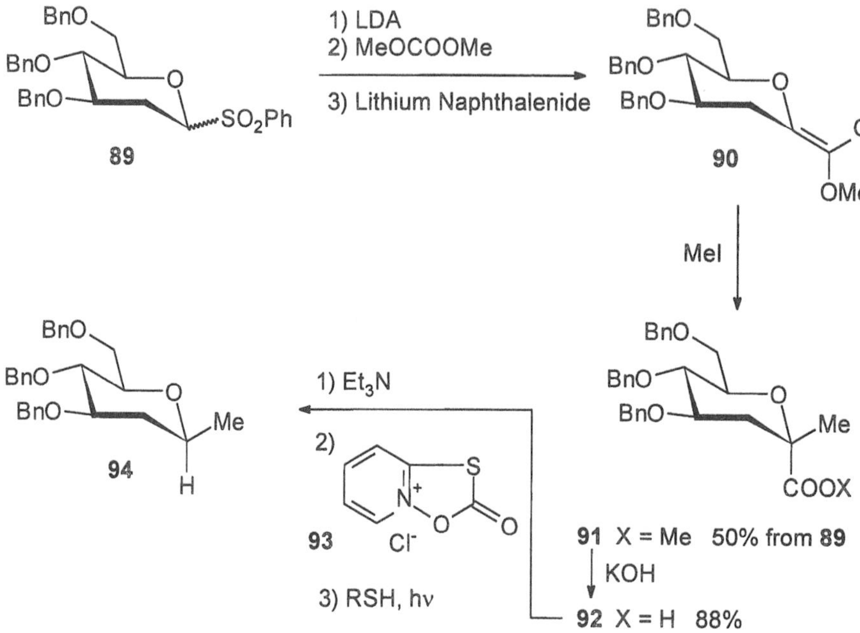

Scheme 19

Earlier, Vasella had utilized a similar strategy using nitrosugars as the radical precursors.[33] Base catalyzed Henry aldol condensation of **95** with paraformaldehyde followed by acetylation gave a mixture of separable compounds **47** and **96**. Separate treatment of these compounds with tributyltin hydride gave the same product, the β-*C*-glycoside **97** (Scheme 20).

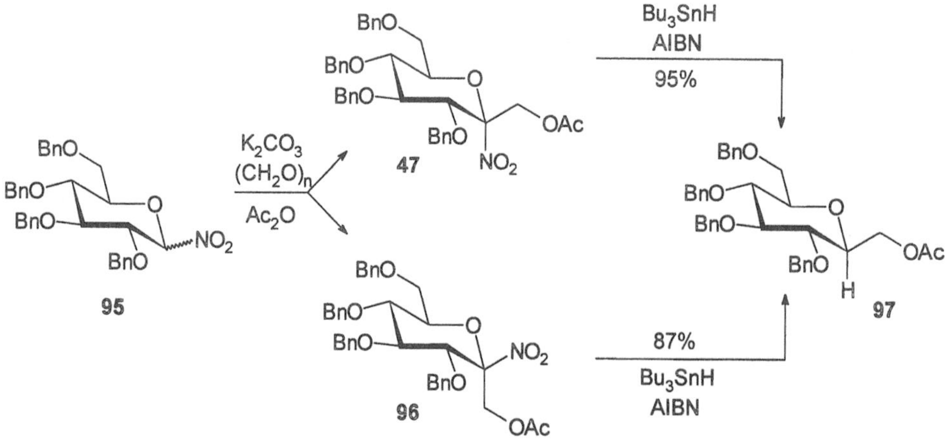

Scheme 20

Two furanose sugars were also examined.[33] The D-manno derived nitrosugar **98**, available from an ionic Michael addition with acrylonitrile, gave **99** as the main product of reduction when treated with tributyltin hydride (eq. 21).

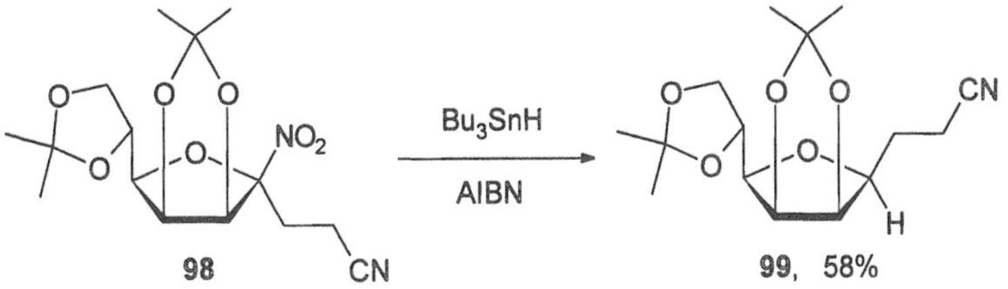

Presumably, the radical is reduced from the more accessible bottom face. One would therefore expect that reduction of the D-ribo derivative **100** would also occur from the sterically less encumbered side, now the top face, but a mixture of compounds (47:53) was obtained in favor of **104**.

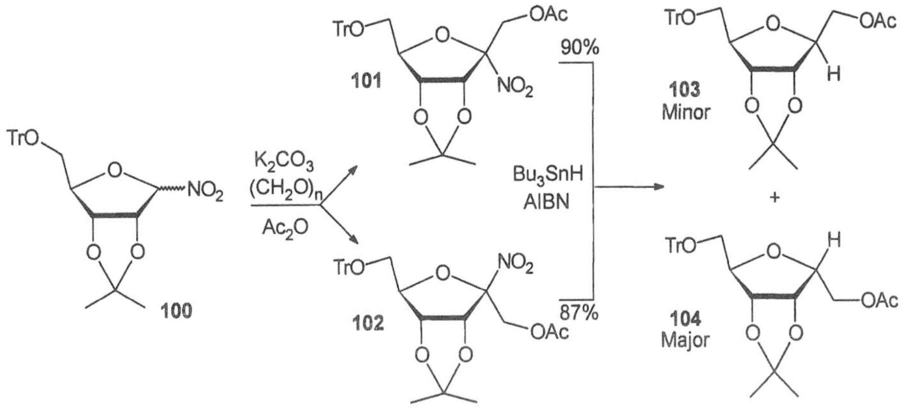

Scheme 21
Reproduced with permission from ref. 33, Copyright 1983 courtesy Verlag Helvetica Chimica Acta.

It would seem that electronic factors are opposing the steric factors in the above case. Vasella has suggested an explanation for these observations.[33] If we assume that the radicals are in an envelope type conformation with the side chains in a pseudo equatorial position, then a stereoelectronic effect is now possible for intermediates **105** and **106**. The LUMO of the β-axial C-O bond also stabilizes the SOMO of the anomeric β-radical **105**. Steric factors also favor addition from the bottom face.

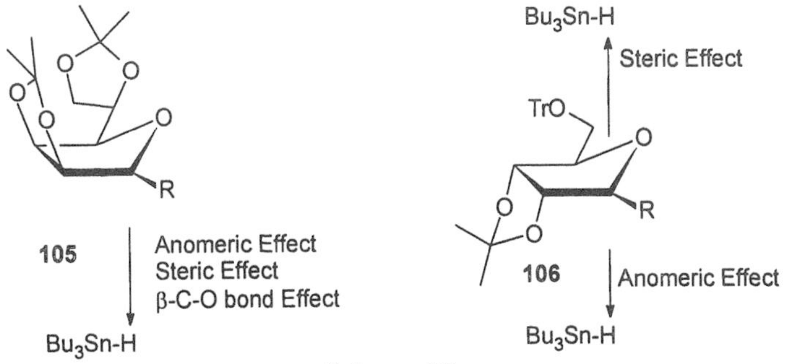

Scheme 22

For the radical **106**, derived from the D-ribo sugar, the radical anomeric effect also favors reduction from the bottom face while the steric factor favors top face reduction. In this example the radical anomeric effect is at least as strong as the steric effect. The above arguments are, of course, only valid if there is conformational stability around the radical center.

In the introduction it was stated that the concentration of radicals in solution is kept low to avoid dimerization. Giese has purposely conducted free radical reactions in concentrated solution, in the absence of trapping agents, to dimerize anomeric radicals (Scheme 23).[34] Three compounds were isolated in the amounts shown. The *C*-2 hydroxyl was methylated to prevent rearrangement (*vide infra*).

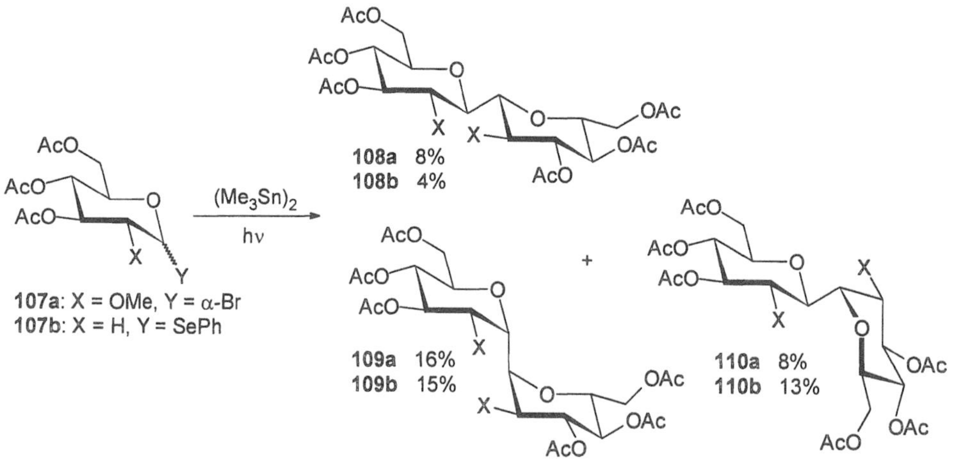

107a: X = OMe, Y = α-Br
107b: X = H, Y = SePh

108a 8%
108b 4%

109a 16%
109b 15%

110a 8%
110b 13%

Scheme 23

The 2-deoxy sugar **107b** exhibited different behavior in its dimerization. The ratio of products differed from **107a**. The axial-axial coupling product **109b** still dominated, but now a greater proportion of axial-equatorial coupling product **110b** was formed. This would seem to indicate that the rate of equatorial and axial attack on the glucosyl radical is equal, while attack on the 2-deoxy sugar is faster from the axial direction by a factor of 1.8. ESR measurements show that while the glucosyl radical adopts a boat conformation, the two deoxy sugar exists in chair conformation that favors axial attack.

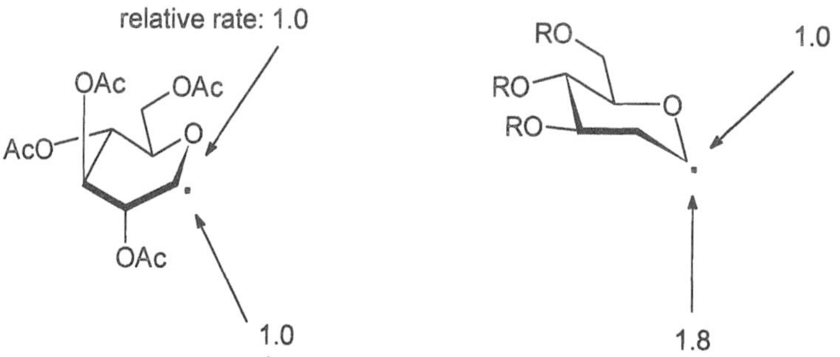

relative rate: 1.0

1.0

1.0

1.8

Scheme 24

If an acceptor is absent when the anomeric radical is generated and a 2-*O*-acetyl group is present, an interesting rearrangement occurs.[35] The radical attacks the ester carbonyl oxygen transferring the O-Ac onto the the anomeric carbon and the radical formed at *C*-2 is then reduced by tributyltin hydride to give the 2-deoxy sugar (eq. 22).

eq. 22

III. INTRAMOLECULAR APPROACHES

There have been several approaches to *C*-glycoside synthesis that involve intramolecular free radical addition. To be included in this category the radical and acceptor must be part of the same molecule. The connection joining the acceptor and radical source can be temporary or permanent. The approaches that utilize a temporary connection may do so to control stereochemistry or to obtain greater selectivity. Once carbon-carbon bond formation is complete the connection may then be cleaved to furnish a free *C*-glycoside. When a more permanent connection is used, the new ring that is formed is preserved in the product *C*-glycoside.

Work at Ciba-Geigy, led by De Mesmaeker, has shown that intramolecular cyclization of an anomeric radical onto a radical acceptor is an efficient way of making *C*-glycosides. The initial idea (eq. 23) was to attach the acceptor via a suitable connector that could be cleaved at a later stage, if required.

eq. 23

Initial efforts began with compound **112** using a methylene group (X = CH$_2$) as the connector (eq. 24).[36] Cyclization gave the products of *cis* ring fusion, with compound **113** predominating. Curiously, compound **115** was also isolated from the reaction mixture.

eq. 24

Compound **115** arises from an intramolecular hydrogen abstraction reaction. The adduct radical abstracts the α-alkoxy hydrogen atom, via a six-membered ring transition state, to give the free radical **117** which can be reduced from the α or β face. Reduction from the α face leads to compound **114** while β face reduction gives the L-ido compound **115**.[37]

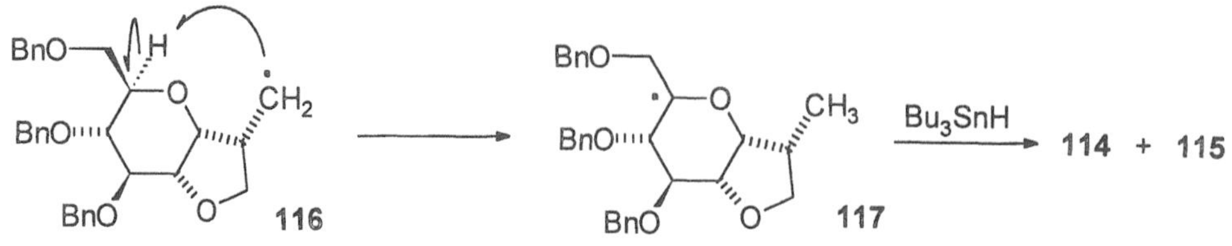

BnO—, H, O, CH₂ **116** → BnO—, O, CH₃, Bu₃SnH → **114 + 115**, **117**

Scheme 25

The amount of hydrogen abstraction product could be suppressed by conducting the reaction with a higher concentration of tributyltin hydride (0.2M). It should be noted that no reduction product was detected at these higher concentrations of tin hydride indicating that these cyclizations are indeed very fast. The workers also showed that the β-*C*-glycoside could be obtained by using a sugar precursor with the D-manno configuration compound **118**.[35] Cyclization proceeded as expected to give the expected products, but now the configuration at the anomeric center is β due to the geometric requirement for *cis* ring fusion.[38]

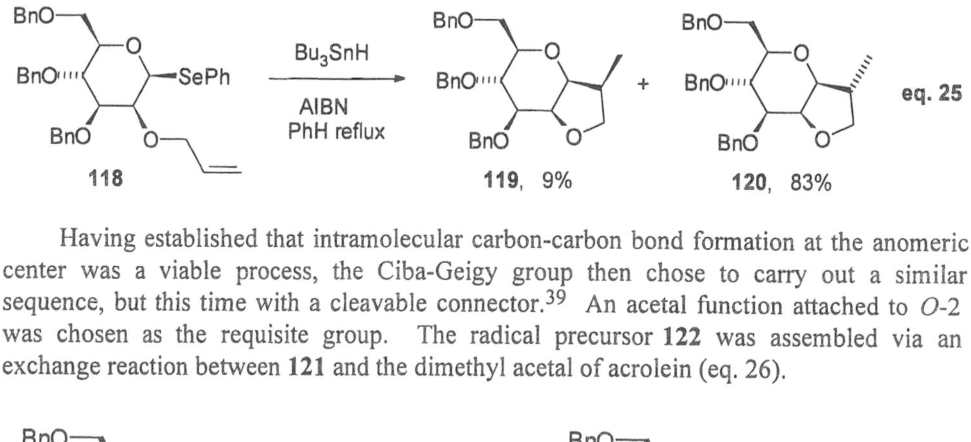

BnO—, O, SePh, **118** →[Bu₃SnH / AIBN / PhH reflux] BnO—, O, **119, 9%** + BnO—, O, **120, 83%**   eq. 25

Having established that intramolecular carbon-carbon bond formation at the anomeric center was a viable process, the Ciba-Geigy group then chose to carry out a similar sequence, but this time with a cleavable connector.[39] An acetal function attached to *O*-2 was chosen as the requisite group. The radical precursor **122** was assembled via an exchange reaction between **121** and the dimethyl acetal of acrolein (eq. 26).

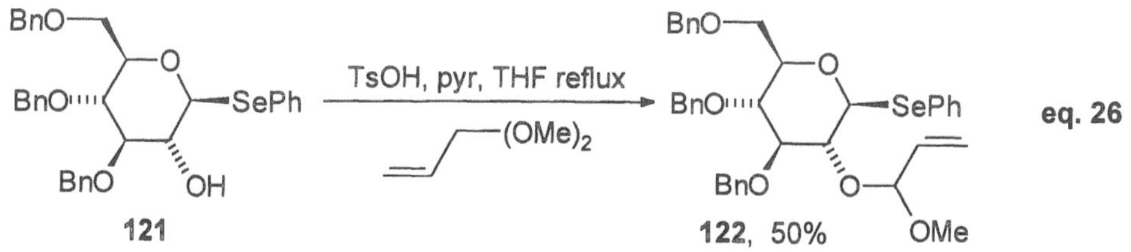

BnO—, O, SePh, OH, **121** →[TsOH, pyr, THF reflux / (OMe)₂] BnO—, O, SePh, O, OMe, **122, 50%**   eq. 26

Radical cyclization of **122** at high tributyltin hydride concentration (0.3M) gave a mixture of compounds **123** and **124** (1:1) in ≥ 89% yield. The mixture of compounds was oxidized to the isomeric lactones **125** and **126**. The lactones were separated and **126** was carried on via standard manipulations to the β-*C*-glycoside **127**, Scheme 26. The above reactions show that depending on the stereochemical disposition of the radical acceptor either the α or β-*C*-glycoside can be obtained in a highly stereoselective fashion. It is noteworthy that the stereochemistry at the anomeric center is dictated by stereochemical disposition of the hydroxyl group at *C*-2.

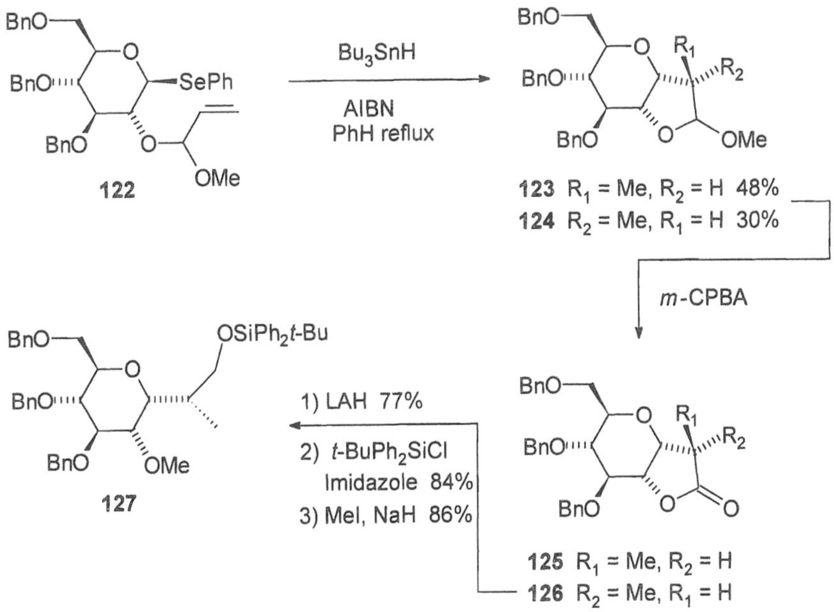

Scheme 26

The stereoselectivity of the above cyclizations was also examined.[40] It was found that increased selectivities could be achieved by performing these reactions at low a temperature (-78°C) using a photochemical initiation and that the nature of the group at *C*-4 strongly affects the stereochemistry at *C*-8. When a benzylidene is used to protect *O*-4 and *O*-6, as in compound **128**, the stereoselectivity increases in favor of the α-*C*-8 epimer **129**.

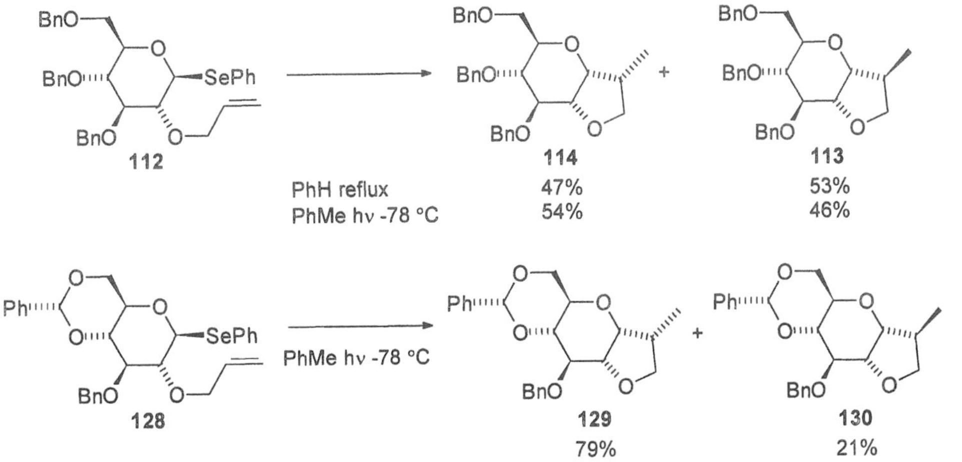

Scheme 27
Reproduced with permission from ref. 40, Copyright 1990,
pg. 688-689, Georg Thieme Verlag, Stuttgart, FRG.

Removal of the substituent at *C*-4 as in **131**, or inversion to the galactosyl compound **134**, increases the selectivity further, Scheme 28. When the galactosyl compound **137** is tied up in the *O*-4, *O*-2 benzylidene only the α-*C*-8 anomer **138** is formed. The mannosyl derivative **140** also exhibits high selectivities due to the fact that the radical exists in a chair conformation, Scheme 28. This will put the side chain axial and place the methyl group *exo*, away from the ring.

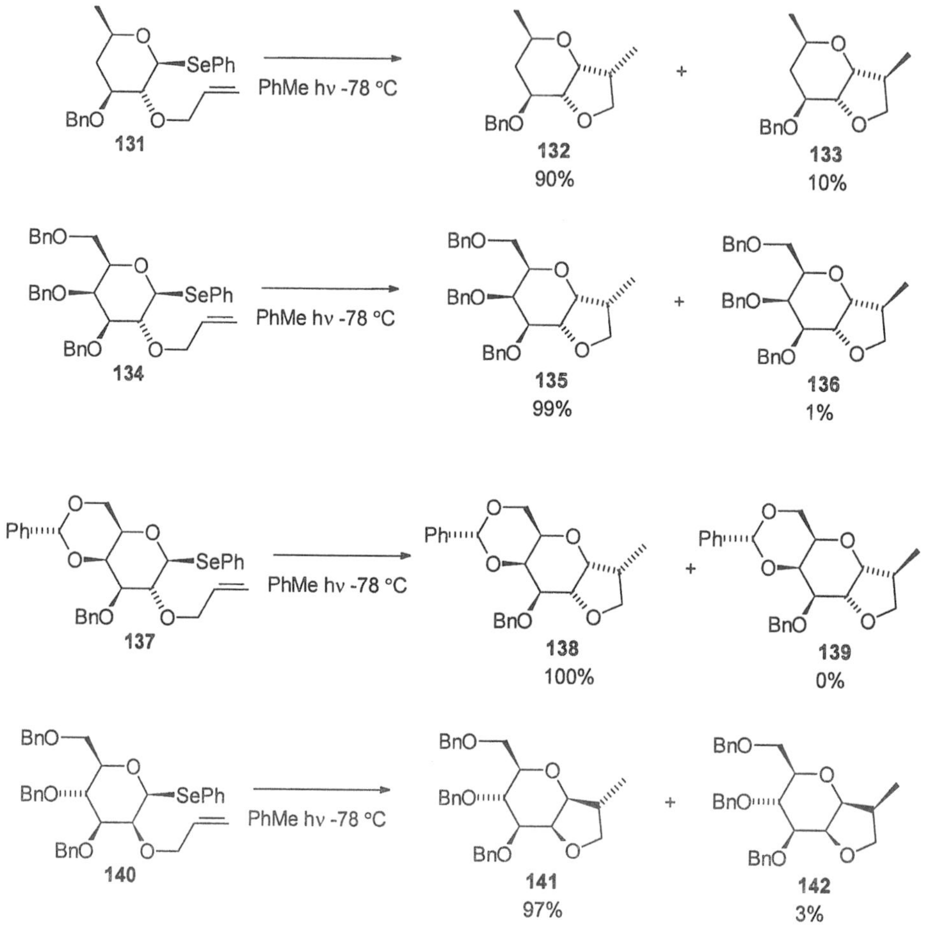

Scheme 28
Reproduced with permission from ref. 40, Copyright 1990,
pg. 688-689, Georg Thieme Verlag, Stuttgart, FRG.

These results would seem to indicate that high selectivities are obtained when the radical exists in a half-chair conformation as it would for compounds **134** and **137**. The steric bias, a more pronounced effect in the chair (mannosyl radical **144**) and half-chair (galactosyl radical **145**) conformations (as opposed to the boat conformation, glucosyl radical **143**), and the orientation of the SOMO are the likely factors responsible for these selectivities.

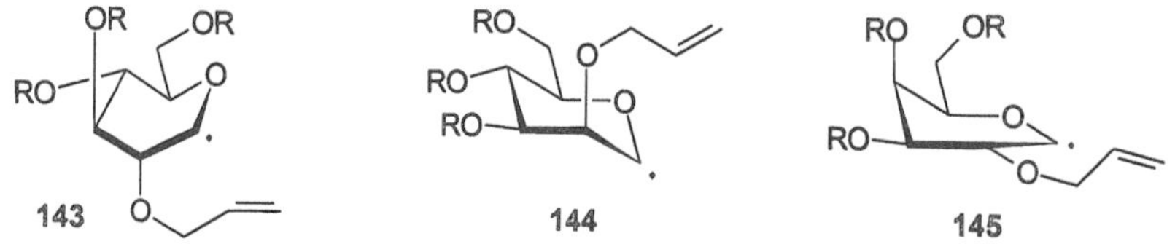

Scheme 29

In order to establish the preferred conformation of the glucosyl radical **146**, Giese and co-workers attempted to trap the radical in an intramolecular fashion.[41] The workers favor the twist boat conformation **146** over the inverted chair **147**. This is based on the observation that the ESR spectrum of the radical **146** is similar to that of the radical **148**, which cannot easily invert and form a chair with axial substituents (Scheme 30).

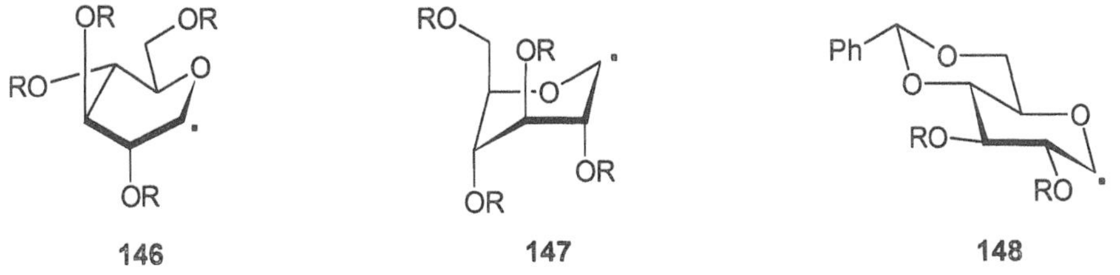

Scheme 30

Their work is shown below in Scheme 31.[41] Keck allylation of **149** gave a mixture of compounds and the major **151** was treated with trimethylsilyl iodide to give the radical precursor **152**. Exposure of **152** to standard conditions gave two compounds, of which the major **153** (90:10) had the methyl group *exo* to the ring. Formation of these two compounds shows that the radical must adopt a twist boat or inverted chair conformation; otherwise, only reduced product would be expected.

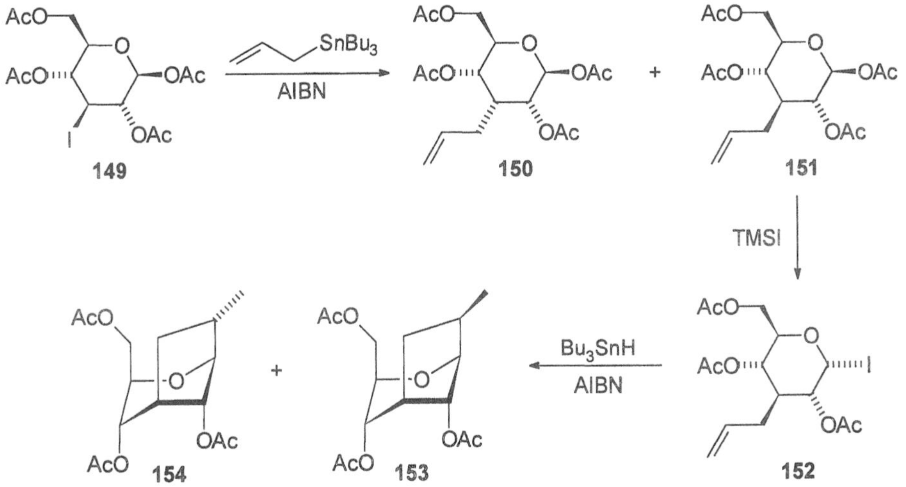

Scheme 31

In order to finalize this work Giese attached a radical acceptor to *C*-6 in order to determine if the preferred conformation is a twist boat or inverted chair.[41] The reasoning being that if the radical adopts a twist boat conformation, the pendant would now be too far away from the anomeric radical to have cyclization occur. Alternatively, if the preferred conformation was an inverted chair, now the pendant would be axial and cyclization should occur. The precursor **156** was easily made by Keck allylation of primary iodide **155**, and successive treatment of **156** with trimethylsilyl iodide and tributyltin hydride gave only the product of reduction **157** (Scheme 32). This result seems to indicate that the preferred conformation of the glucosyl radical is the twist boat.

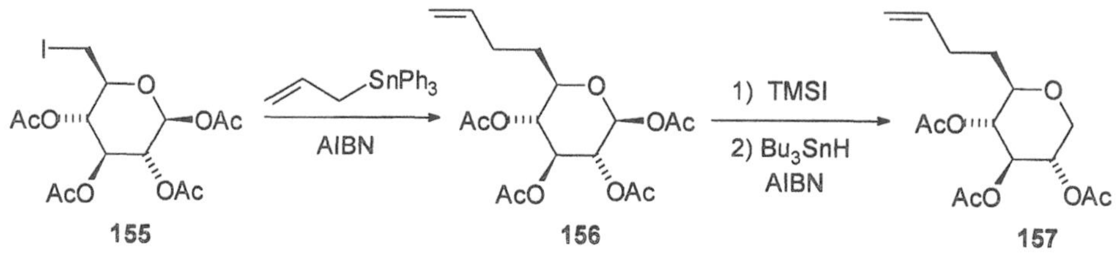

Scheme 32

Work by Stork[42] relied on the previous explorations by De Mesmaeker. The Columbia group used intramolecular radical cyclization onto an acetylene connected via a silicon atom to a suitably located oxygen atom. The reactions are most efficiently carried out by performing the cyclizations thermally. The reaction mixtures are treated with TBAF and this serves to desilylate the tethered oxygen atom, cleave the silicon acetylene bond, and presumably also aid in the removal of tin residues. The free styryl *C*-glycoside is then obtained. Both *O*-2 and *O*-3 of glucose can be derivatized to give the α- and β-*C*-glycosides respectively (Scheme 33).

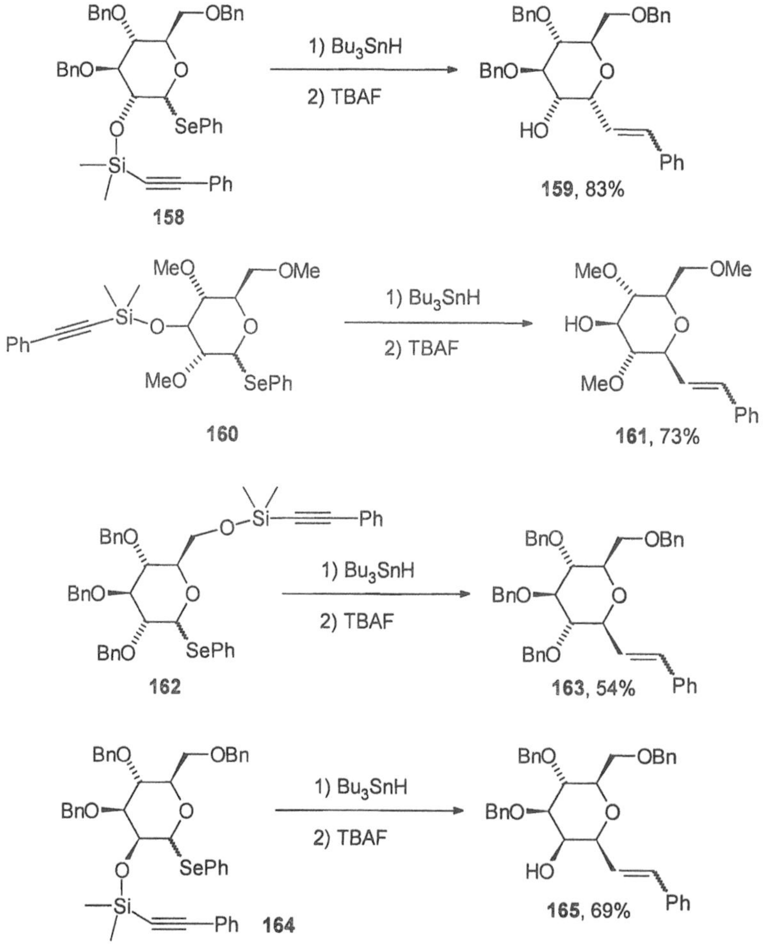

Scheme 33

It is noteworthy that the first cyclization proceeded in good yield (83%) and the second cyclization, a 6-*exo* closure, occurred in slightly lower yield (73%). Even with the tether on *O*-6, now a 7-exo closure, 54% of β-*C*-glycoside **163** was isolated using syringe pump techniques to affect cyclization. The mannoside **164**, with the tether on *O*-2, gave a 69% yield of β-*C*-glycoside **165**. The reaction is also applicable to furanoses and both anomers of D-ribose can be obtained by judicious choice of hydroxyl group (Scheme 34). The styryl glycosides can be used as such or they can be cleaved to the corresponding aldehydes by ozonolysis. The methodology is more versatile than shown. Once cyclization occurs and the tether is removed the now free hydroxyl can be derivitized and that center inverted to give a new sugar. For example compound **165** could be converted to the gluco derivative by inversion at *C*-2. Thus the methodology can give access to both anomers in selected cases.

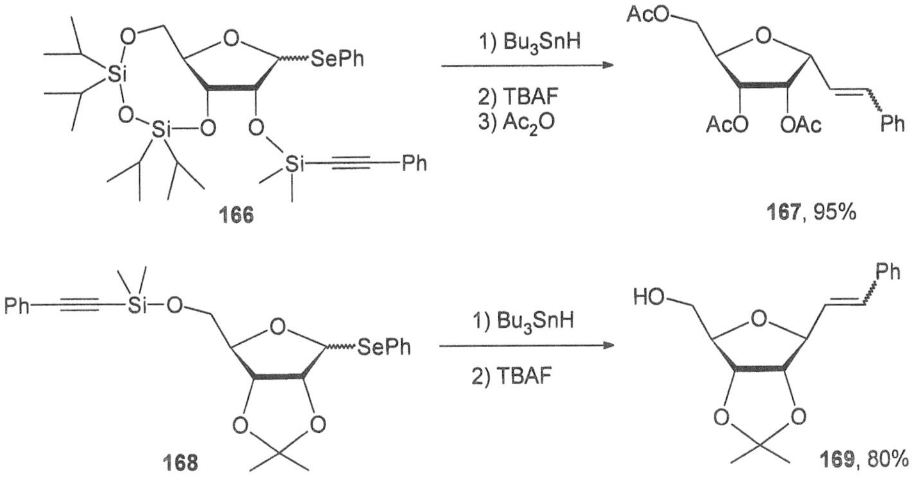

Scheme 34
Reprinted with permission from ref. 42, Copyright 1991 American Chemical Society.

Lee and Park cyclized bromides **170** and **172** to give the *C*-furanosides **171** and **173** in 84% and 81% yield, respectively (Scheme 35). The starting bromides were built up from suitable tartrate esters.[43]

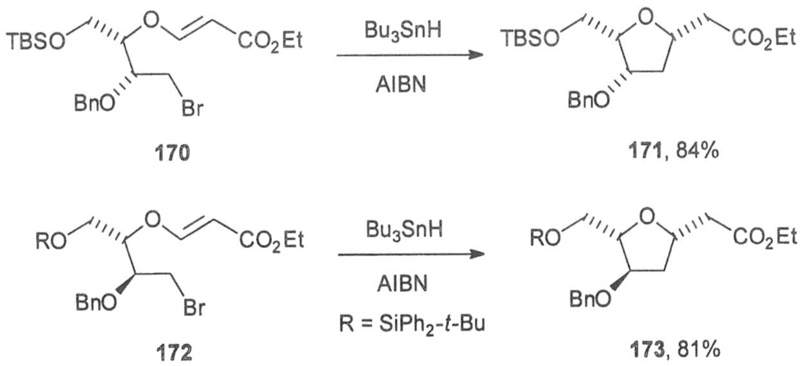

Scheme 35

Fraser-Reid has used a serial radical process to construct *C*-glycosides (Schemes 36 and 37).[44] The first cyclization is an intramolecular closure onto an enol ether to give an anomeric radical that can be trapped intermolecularly with a suitable radical acceptor to give an α-*C*-glycoside. Glycal **174** was converted to **165** in the usual way and radical cyclization under the standard conditions in the presence of acrylonitrile gave **177**. Alternatively, when cyclization was carried out in the presence of *t*-butylisocyanide **176** and **178** were obtained in a 1.4:1 ratio.

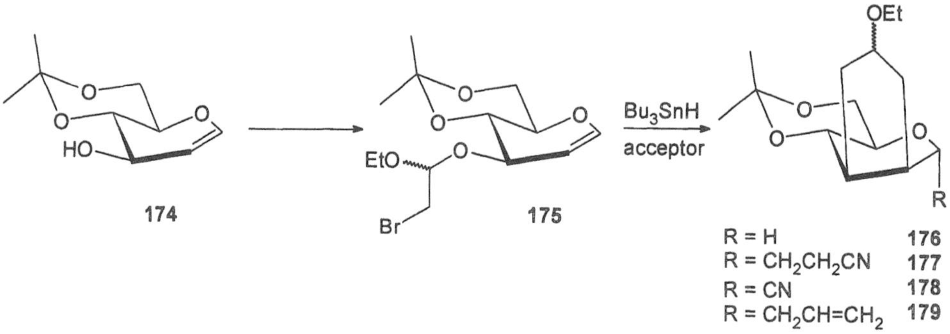
OEt
O
O
O
HO
174
EtO
O
Br
175
Bu$_3$SnH
acceptor
R
R = H 176
R = CH$_2$CH$_2$CN 177
R = CN 178
R = CH$_2$CH=CH$_2$ 179

Scheme 36

Exposure of **175** to Keck allylation conditions gave **179** in 70% yield. The selectivity of *C*-glycoside formation for the derivative **180** was lower than that of **175** and this is probably due to the effect of the *syn* carbon bridge on *C*-2. In compound **175** the radical anomeric effect and the steric effect both favor the formation of the α-anomer. This is another example that illustrates the delicate balance between the influence of the electronic and steric effects in stereocontrol at the anomeric center during carbon-carbon bond formation. In the reaction of **180** the α-anomer, **181**, was still the major product being formed in a 3:1 ratio over the β-anomer **182**, Scheme 37.

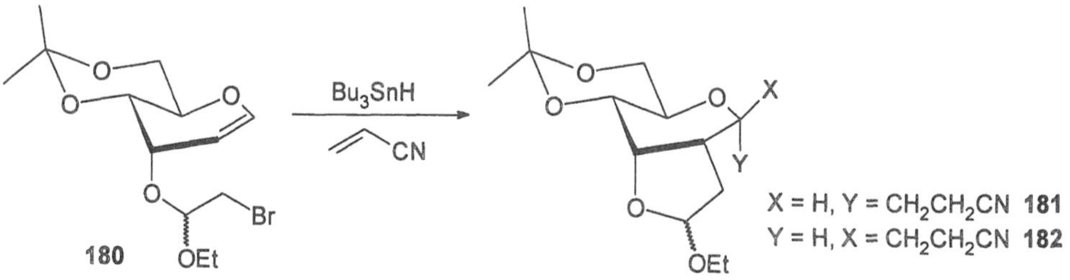
O
O
O
O
Br
180
OEt
Bu$_3$SnH
CN
X
Y
OEt
X = H, Y = CH$_2$CH$_2$CN 181
Y = H, X = CH$_2$CH$_2$CN 182

Scheme 37

Sinaÿ has used intramolecular radical cyclization to efficiently construct the *C*-disaccharide corresponding to methyl α-maltoside.[45] The free hydroxyls on each of the sugars were tethered together via a silicon atom and the product **185** exposed to standard free radical conditions to give a 40% yield of cyclized product **186** as a single diastereomer. The selectivity of this reaction is excellent and this can be contrasted to the modest selectivities that Giese observed during reduction of his *C*-disaccharide adduct radicals. Sinaÿ observed much lower selectivities when the acceptor was tethered via its *O*-3.

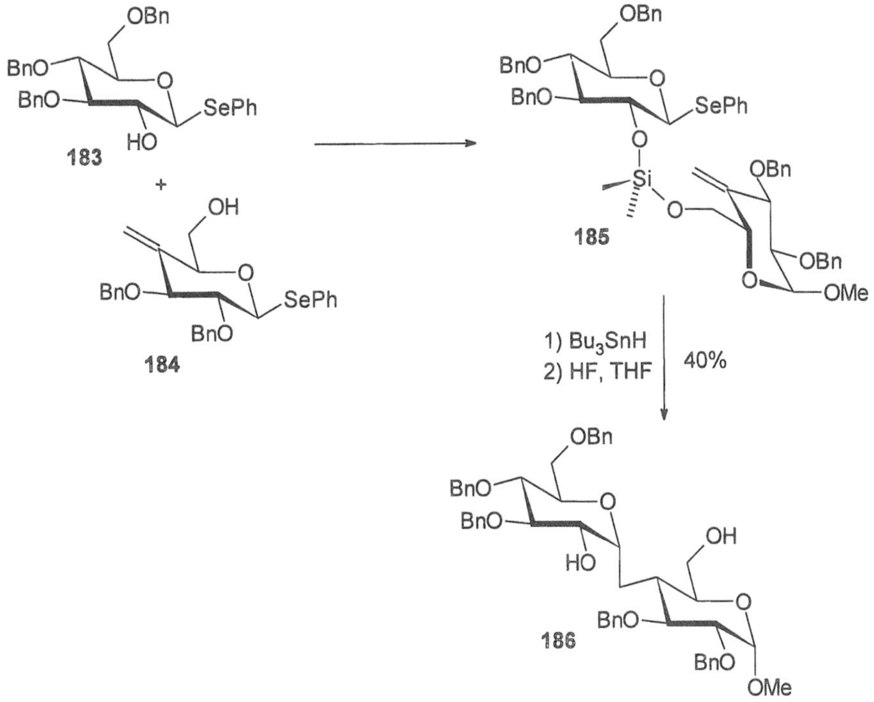

Scheme 38

Spiroketals can also be considered *C*-glycosides. The discovery of the papulacandin class of antibiotics[46] has sparked interest in glycosidic spiroketal formation.[47] Radical methods are well suited for stereoselective spiroketal formation. Sinaÿ[48] began with the exomethylenic gluco derivative **187** and treatment of this compound with *N*-iodo-succinimide and allyl alcohol gave **188** in 77% yield (eq. 27).

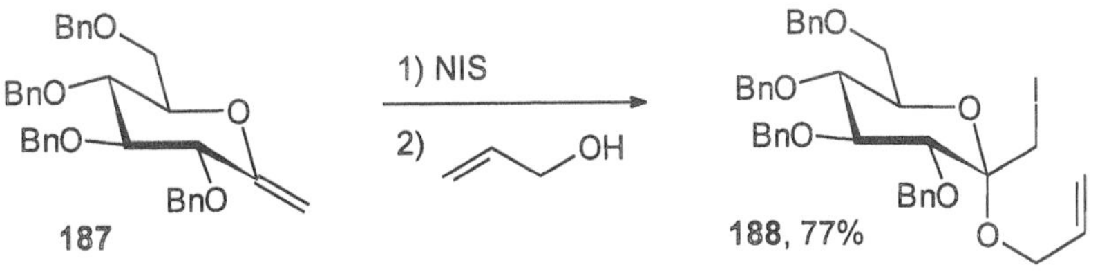

When this compound was exposed to standard radical conditions spiroketal **189** was produced as a single isomer. The formation of **189** was expected on the basis of the anomeric effect and the chair transition state model. Stereochemical conformation was provided by transformation of **189** into **191** (Scheme 39). This was accomplished by anomeric reduction to give the expected β-*C*-glycoside which was oxidized to aldehyde **190** via the method of Swern. Catalytic hydrogenolysis removed the benzyl protecting groups which allowed hemiacetal to occur between the aldehyde function and *O*-2. Acetylation gave **191**, in which the methyl group is now axial and its stereochemistry could be unambiguously assigned by ^{1}H NMR.

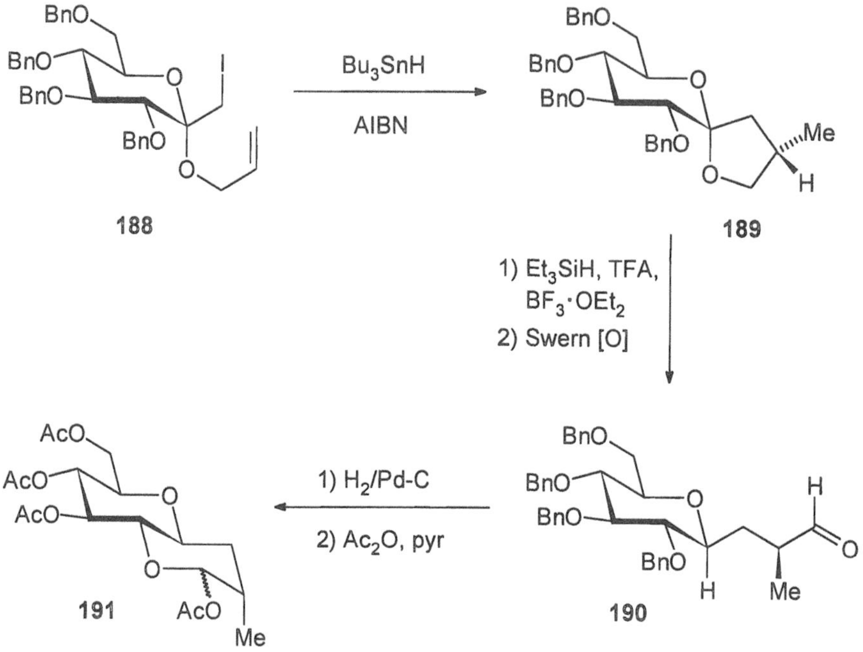

Scheme 39

Work by Ferrier used hydrogen abstraction to generate an anomeric radical that could then cyclize onto a suitably poised tether.[49] Epoxidation of the β-glucopyranoside **192** with *m*-CPBA afforded a 1:1 mixture of epoxides. Treatment of the separated epoxides with iodine/triphenylphosphine/imidazole gave the iodo epoxides **194** and **196**.

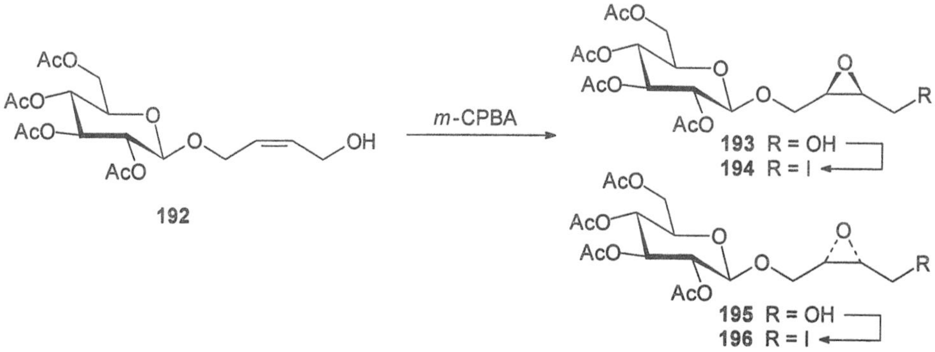

Scheme 40

Compound **196** was then exposed to tributyltin hydride under UV irradiation to give compound **200** in 68% yield. Its proposed mechanism of formation is shown in Scheme 41. The formed primary radical causes epoxide ring opening to give a reactive, oxygen based radical which is poised to abstract the axial anomeric hydrogen via a 6-membered ring transition state to provide the anomeric radical **199** which then cyclizes onto the double bond to give the product **200**. The β-linked spiroketal **201** is easily generated from **200** by treatment with a Lewis acid.

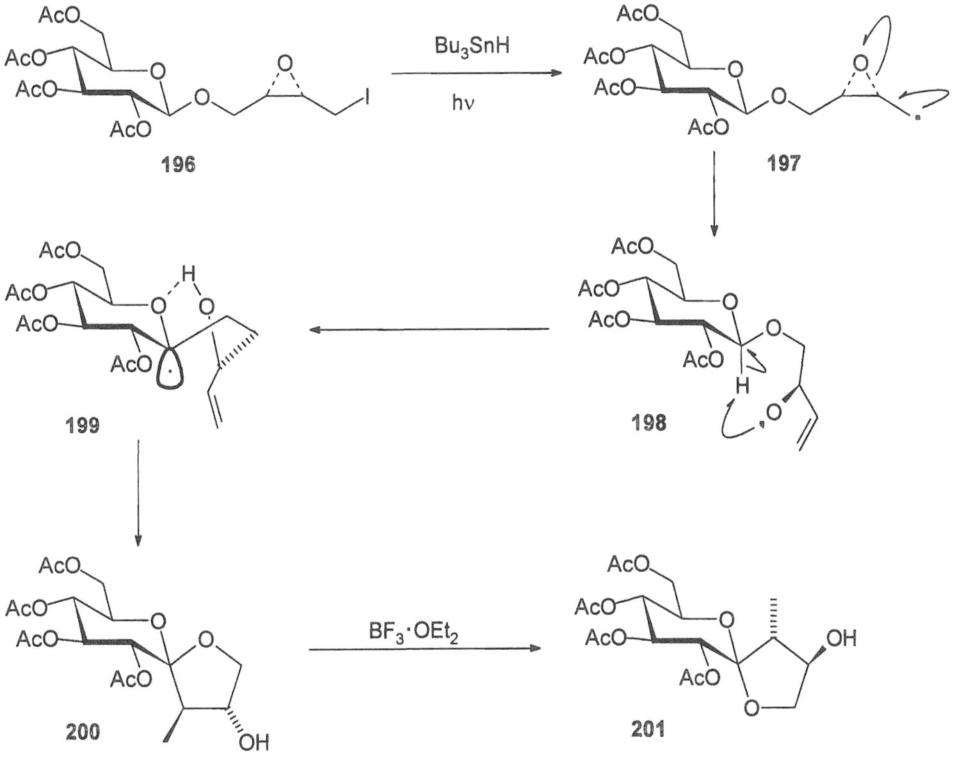

196 197 199 198 200 201

Scheme 41

The observed selectivity is due to the stabilization of conformer **199** by hydrogen bonding which occurs between the anomeric oxygen atom and the allylic hydroxyl. This selectivity is contrasted to that obtained when **194** is exposed to the same conditions as above. The reaction now gives a mixture of isomers. The conformer **202**, stabilized by hydrogen bonding, has the double bond away from the axial radical thus making cyclization difficult. An equatorial radical, though, can cyclize and this explains the appearance of α-linked spiroketals in the product distribution.

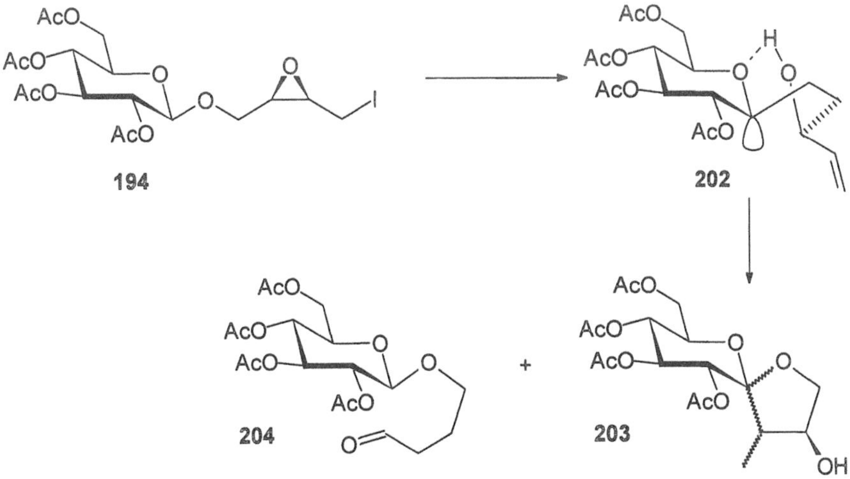

194 202 204 203

Scheme 42

IV. ANOMERIC CARBENES

Although not free radicals, anomeric carbenes are electron deficient species and their inclusion here seems appropriate. There have been a few recent reports in which anomeric carbenes react with olefins to provide bis *C,C*-glycosides.

The requisite precursors are most easily accessed from sugar lactones (eq. 28). Treatment of **205** with trimethyl silyl azide followed by exposure to boron trifluoride etherate gave the anomeric diazide **206** in 45% yield.[50]

Photolysis of the anomeric diazide in the presence of acrylonitrile gave a mixture of isomers **208** in 65% combined yield, eq. 29.[51]

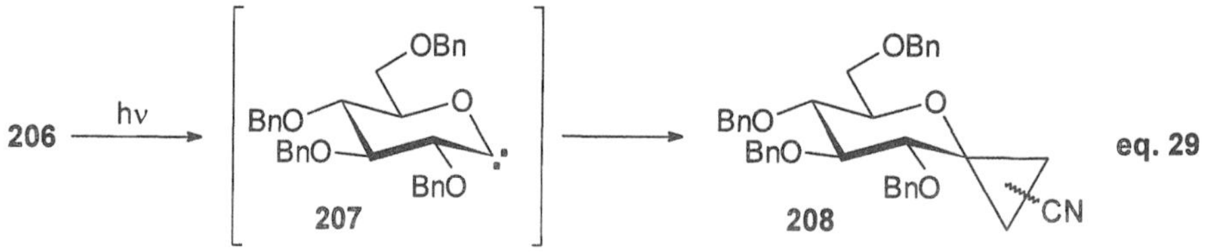

Vasella has reacted the glucopyranosyl carbene, derived from the anomeric diazirene **209**,[52] with C_{60} to obtain a spiro linked bis*C,C*-glycoside of C_{60} **210**. When deprotected a water soluble C_{60} derivative can be obtained (Scheme 43).[53]

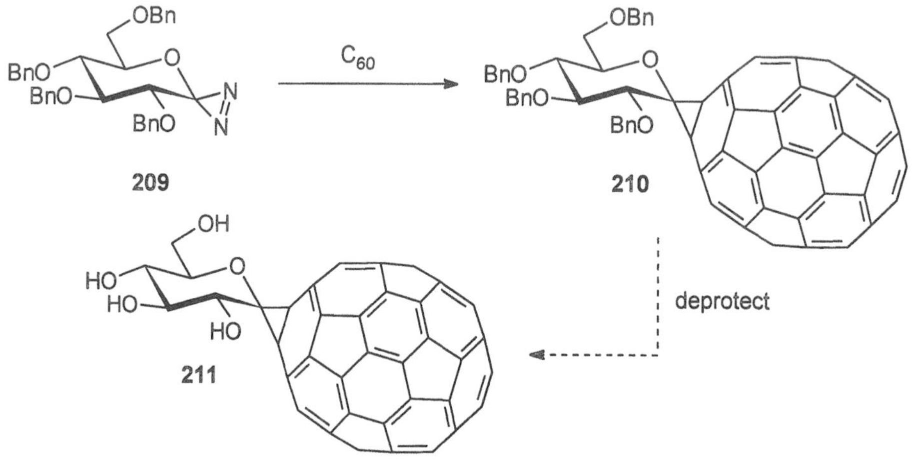

Scheme 43

V. CONCLUSION

This chapter has summarized the radical chemistry of *C*-glycoside formation. In general terms the α-anomer for *C*-pyranosides is favored in intermolecular additions. Both Michael additions and Keck fragmentations have proven useful in these additions. The formation of *C*-furanosides is governed by a combination of steric and electronic factors. Several initiators and radical processes have been developed. If access to the β-*C*-pyranoside is desired then generation of an anomeric *C*-glucosyl radicals will provide the requisite compound, after reduction Intermolecular additions tend to give the product of *cis*-ring fusion and therefore the anomeric configuration is controlled by the configuration of the tethered oxygen atom.

VI. REFERENCES

1. For some recent reviews see: (a) Curran, D.P. Radical Cyclizations and Sequential Radical Reactions, in *Comprehensive Organic Synthesis*, Vol. 4, Trost, B.M., and Fleming, I., Eds., Pergamon Press, New York, 1991, Chapter 4.2 p. 779. (b) Larid, E.R. and Jorgensen, W.L. *J. Org. Chem.* **1990**, *55*, 9. Jasperse, C.P., Curran, D.P.; Fevig, T.L. *Chem. Rev.* **1991**, *91*, 1237. (c) Curran, D.P. *Synthesis* **1988**, 417 and 489. (d) Ramamiah, M. *Tetrahedron* **1987**, *43*, 3541. (e) Giese, B., *Radicals in Organic Synthesis*, Pergamon Press, 1986.

2. Giese, B. *Angew. Chem. Intl. Ed. Engl.* **1983**, *22*, 753.

3. Keck, G.E.; Yates, J.B. *J. Am. Chem. Soc.* **1982**, *104*, 5829.

4. Giese, B.; Dupuis, J.; Nix, M. *Org. Syn.* **1987**, *65*, 236.

5. Giese, B.; Dupius, J. *Angew. Chem. Intl. Ed. Engl.* **1983**, *22*, 622.

6. Giese, B.; Heuck, K. *Chem. Ber.* **1979**, *112*, 3759.

7. Dupius, J.; Giese, B.; Rüegge, D.; Fischer, H.; Korth, H.G.; Sustmann, R. *Angew. Chem. Intl. Ed. Engl.* **1984**, *23*, 896.

8. Giese, B.; Dupuis, J. *Tetrahedron Lett.* **1984**, *25*, 1349. See also: *Substituent Effects in Radical Chemistry*, Viehe, H.G.; Janousek, Z.; Merengi, R. Eds., D. reidel Publishing Co., Derdrecht, Holland, 1986, p. 283-296 and Rychnovosky, S.D.; Powers, J.P.; LePage, T.J. *J. Am. Chem. Soc.* **1992**, *114*, 8375.

9. Giese, B.; Dupuis, J.; Leising, M.; Nix, M.; Linder, H.J. *Carbohydr. Res.* **1987**, *171*, 329.

10. Araki, Y.; Endo, T.; Tanji, M.; Nagasawa, J.; Ishido, Y. *Tetrahedron Lett.* **1987**, *28*, 5853.

11. Adlington, R.M.; Baldwin, J.E.; Basak, A.; Kozyrod, R.P. *J. Chem. Soc., Chem. Commun.* **1983**, 944.

12. Hart, D.J.; Seely, F.L. *J. Am. Chem. Soc.* **1988**, *110*, 1631.

13. Nicolaou, K.C.; Dolle, R.E.; Chucholowski, A.; Randall, J.L. *J. Chem. Soc., Chem. Commun.*, **1984**, 1153.

14. Ferrier, R.J.; Haines, S.R.; Gainsford, G.J.; Gabe, E.J. *J. Chem. Soc., Perkin Trans. I* **1984**, 1683.

15. Blattner, R.; Ferrier, R.J.; Renner, R. J. *Chem. Soc., Chem. Commun.* **1987**, 1007.

16. Barton, D.H.R.; Ramesh, M. *J. Am. Chem. Soc.* **1990**, *112*, 891.

17. Thoma, G.; Giese, B. *Tetrahedron Lett.* **1989**, *30*, 2907.

18. Giese, B.; Gobel, T.; Ghosez, A. *Chem. Ber.* **1988**, *121*, 1807 and Giese, B. *Pure and Appl. Chem.* **1988**, *60*, 1655.

19. Abrecht, S.; Scheffold, R. *Chimia* **1985**, *39*, 211.

20. Ono, N.; Miyake, H.; Tamura, R.; Kaji, A. *Tetrahedron Lett.* **1981**, *22*, 1705.

21. Dupuis, J.; Giese, B.; Hartung, J.; Leising, M. *J. Am. Chem. Soc.* **1985**, *107*, 4332.

22. Giese, B.; Witzel, T. *Angew. Chem. Intl. Ed. Engl.* **1986**, *25*, 450. For a study dealing with the selectivity of the reduction step see: Giese, B.; Damm, W.; Witzel, T.; Zeitz, H.-G. *Tetrahedron Lett.* **1993**, *34*, 7053.

23. Vogel, P.; Bimwala, R.M. *Tetrahedron Lett.* **1991**, *32*, 1429.

24. Beckwith, A.J.; Pigou, P.E. *Aust. J. Chem.* **1986**, *39*, 77.

25. Motherwell, W.B.; Ross, B.C.; Tozer, M.J. *Synlett* **1989**, 68.

26. Keck, G.E.; Enholm, E.J.; Yates, J.B.; Wiley, M.R. *Tetrahedron* **1985**, *41*, 4079.

27. Paulsen, H.; Matschulat, P. *Liebigs Ann. Chem.* **1991**, 487.

28. Nagy, J.O.; Bednarski, M.D. *Tetrahedron Lett.* **1991**, *32*, 3953.

29. Wåglund, T.; Claesson, A. *Acta. Chem. Scand.* **1992**, *46*, 73.

30. Giese, B.; Linker, T.; Muhn, R. *Tetrahedron* **1989**, *45*, 935.

31. Crich, D.; Lim, L.B.L. *J. Chem. Soc. Perkin Trans I* **1991**, 2205.

32. Beau, J.-M.; Sinaÿ, P. *Tetrahedron Lett.* **1985**, *26*, 6193.

33. Baumberger, F.; Vasella, A. *Helv. Chim. Acta* **1983**, *66*, 2210.

34. Giese, B.; Ruckert, B.; Gröninger, K.S.; Muhn, R.; Lindner, H.J. *Liebigs. Ann. Chem.* **1988**, 997.

35. Korth, H.G.; Sustmann, R.; Gröninger, K.S.; Leisung, M.; Giese, B. *J. Org. Chem.* **1988**, *53*, 4364 and references cited therein.

36. De Mesmaeker, A.; Hoffmann, P.; Ernst, B.; Hug, P.; Winkler, T. *Tetrahedron Lett.* **1989**, *30*, 6307.

37. For a study on this reaction see: De Mesmaeker, A.; Waldner, A.; Hoffmann, P.; Hug, P.; Winkler, T. *Synlett* **1992**, 285.

38. Beckwith, A.L.J. *Tetrahedron* **1981**, *37*, 3073.

39. De Mesmaeker, A.; Hoffmann, P.; Ernst, B.; Hug, P.; Winkler, T. *Tetrahedron Lett.* **1989**, *30*, 6311.

40. De Mesmaeker, A.; Waldner, A.; Hoffmann, P.; Mindt, T.; Hug, P.; Winkler, T. *Synlett* **1990**, 687.

41. Gröninger, K.S.; Jäger, K.F.; Giese, B. *Liebigs. Ann. Chem.* **1987**, 731.

42. Stork, G.; Suh, H.S.; Kim, G. *J. Am. Chem. Soc.* **1991**, *113*, 7054.

43. Lee, E.; Park, C.M. *J. Chem. Soc., Chem. Commun.* **1994**, 293.

44. Lopez, J.C.; Fraser-Reid, B. *J. Am. Chem. Soc.* **1989**, *111*, 3450.

45. Xin, Y.C.; Mallet, J.-M.; Sinaÿ, P. *J. Chem. Soc., Chem. Commun.* **1993**, 864.

46. Traxler, P.; Tosch, W.; Zak, O. *J. Antibiot.* **1987**, *40*, 1146 and references therein.

47. Perron, F.; Albizati, K.F. *Chem. Rev.* **1989**, *89*, 1617 and see also Boivin, T.L.B. *Tetrahedron* **1987**, *43*, 3309.

48. Haudrechy, A.; Sinaÿ, P. *Carbohydr. Res.* **1991**, *216*, 375.

49. Ferrier, R.J.; Hall, D.W. *J. Chem. Soc., Perkin Trans. I* **1992**, 3029.

50. Praly, J.-P.; El Kharraf, Z.; Descotes, G. *J. Chem. Soc., Chem. Commun.* **1990**, 431.

51. Praly, J.-P.; El Kharraf, Z.; Descotes, G. *Tetrahedron Lett.* **1990**, *31*, 4441.

52. Briner, K.; Vasella, A. *Helv. Chim. Acta* **1989**, *72*, 1371.

53. Vasella, A.; Uhlmann, P.; Waldraff, C.A.A.; Diederich, F.; Thilgen, C. *Angew. Chem. Intl. Ed. Engl.* **1992**, *31*, 1388.

Chapter 8

C-DISACCHARIDE SYNTHESIS

I. INTRODUCTION

A C-disaccharide occurs when the linking oxygen in a disaccharide is replaced with a carbon atom as shown in Figure 1.

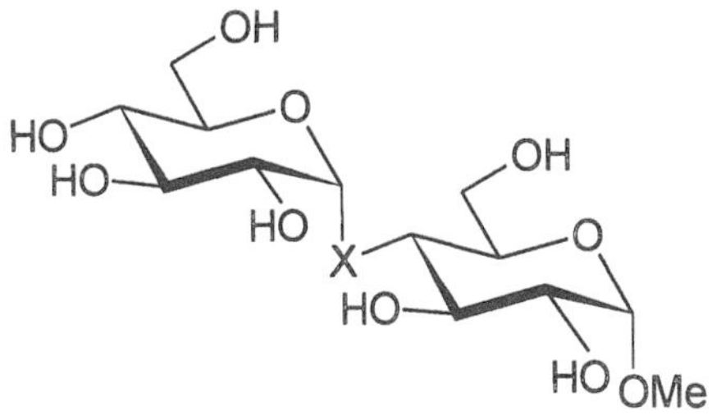

X = O,   methyl α-maltoside   **1**
X = CH₂, methyl α-C-maltoside **2**

Figure 1

There has been some confusion in the literature on the proper use of the term C-disaccharide. Strictly speaking, this term is only applicable when the anomeric carbon of one sugar is bonded via **one** carbon atom to any position on another sugar. Sugars that are linked by more than one carbon atom are not C-disaccharides, but a closely related class of compounds. Also, if the carbon linkage is not anomeric in nature, the compound is only a branched carbon sugar.

This chapter will address the chemical synthesis of C-disaccharides and non C-disaccharidic compounds, since these compounds may also possess biological activity making synthetic efforts toward these compounds worthwhile.

Although C-disaccharides are not commonly found in nature, this structural type of compound does exist in some natural products. The natural product palytoxin can be considered as being comprised of numerous C-disaccharide sub-units.[1] There has been considerable interest in this compound class due to its potential for enzyme inhibition. Disaccharides are prone to hydrolytic cleavage by various enzymes and the replacement of the interglycosidic oxygen atom with a carbon atom leads to a non-hydrolyzable function. This chapter will deal with the synthetic methods that have been developed for C-disaccharide synthesis. Some are extensions of C-glycoside synthesis, while others are tailored exclusively for C-disaccharide construction.

II. SYNTHETIC APPROACHES TO *C*-DISACCHARIDES

A.   ANIONIC APPROACHES

1.   Carbanionic Methods

Sinaÿ and Rouzad were the first to synthesize a *C*-disaccharide.[2] The workers chose to construct a β-(1→6')-*C*-disaccharide through the use of anionic chemistry. The plan was to link an aglyconic *C*-6' anion with an appropriate anomeric acceptor. Their work is shown in Scheme 1. Swern oxidation of the primary hydroxyl gave the aldehyde **4** which was directly converted to the dibromo-olefin **5** (70% overall). Exposure of **5** to *n*-BuLi gave the acetylenic anion which was immediately reacted with the benzylated gluconolactone **7** to give a mixture of anomers **8**. This mixture was of no consequence since stereoselective reduction (Et$_3$SiH, BF$_3$·OEt$_2$) furnished the β-anomer **9** exclusively. The synthesis was completed by catalytic hydrogenation which served to reduce the triple bond and also remove the benzyl protecting groups to give *C*-disaccharide **10**.

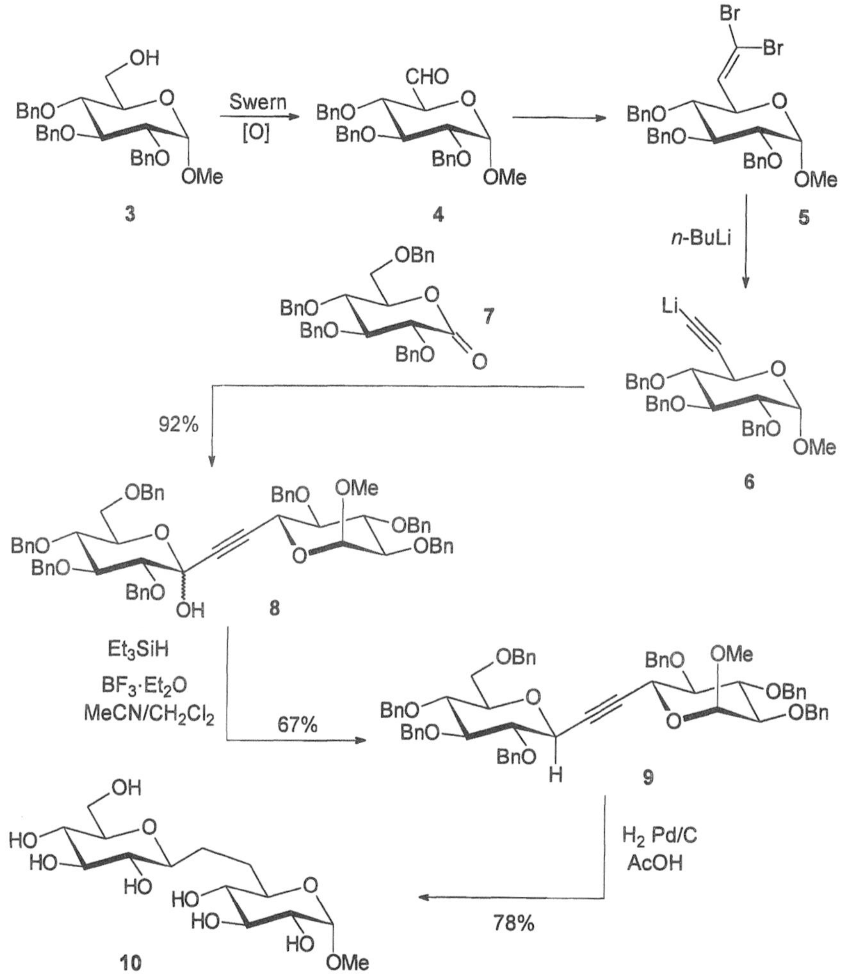

Scheme 1

Sinaÿ has also used tin based chemistry, in this case to generate an anomeric anion, and in doing so has reversed the natural polarity of this electrophilic center.[3] The vinylic stannane **12**, available from the thioglycoside **11** in 3 steps (for a full discussion of anionic approaches to *C*-glycosides see Chapter 3), underwent tin-lithium exchange to give the lithiated species **13** which when treated with dimethyl carbonate gave two compounds. The expected product **14** was formed in 36% yield along with the bis-adduct, **15**, which arises from the coupling of two anions with one molecule of dimethyl carbonate. Compound **15** can be envisaged as a precursor to a (1→1')-*C*-disaccharide.

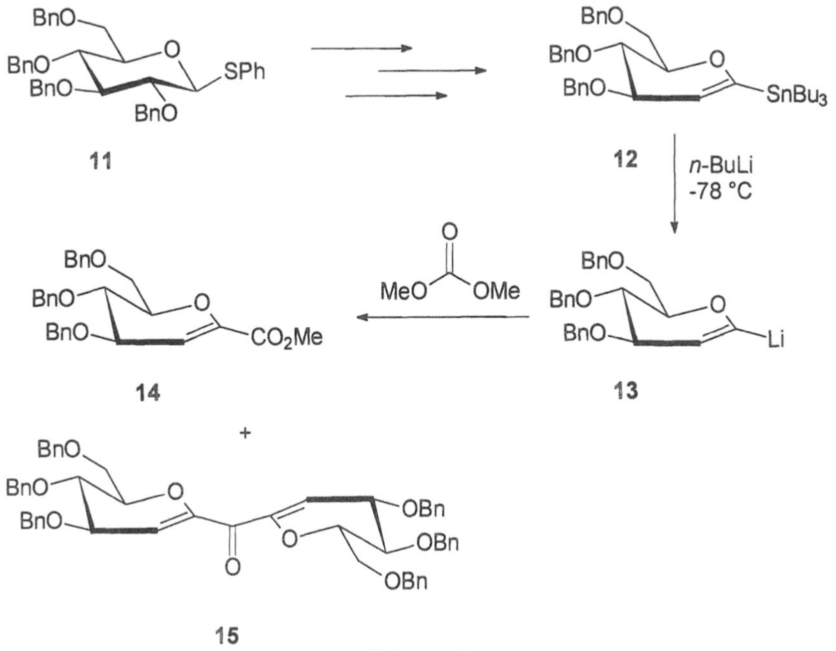

Scheme 2

Further work by Sinaÿ, using anomeric anions, also led to β-(1→6')-*C*-disaccharides.[4] In this instance, the anomeric anion was generated from the anomeric sulfones **16**. The anion was then reacted with aldehyde **4** to give a mixture of four compounds, which were not isolated, but treated directly with lithium naphthalide followed by quenching with methanol. This served to desulfonylate the adducts and led exclusively to the β-anomers **19**. Presumably, the reaction proceeded via radical intermediates which might explain the high level of stereoselectivity observed.[5]

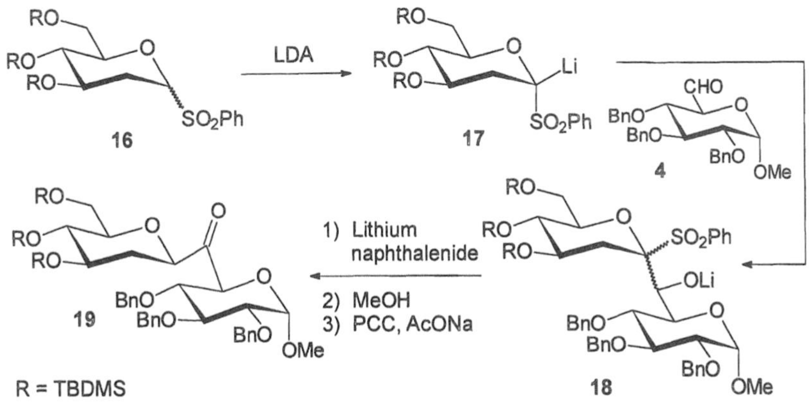

Scheme 3

Schmidt and Preuss[6] also utilized an anionic approach in their synthesis of a β-(1→4') linked *C*-disaccharide. The aglycone was to serve as the nucleophile and a protected gluconolactone as the electrophile. Moffatt oxidation of the 1,6-anhydro glucopyranose **20** provided ketone **21** which was further transformed into the olefin **22** via standard Wittig olefination. The lithiated compound **24** was obtained via a four step protocol: stereoselective hydroboration, oxidation, iodination, and lithium iodide exchange. Compound **24** was then coupled with the protected gluconolactone **7** to give an anomeric mixture which was stereoselectively reduced to furnish **26**. Exposure of **26** to trifluoroacetic acid (TFA) and acetic anhydride served to cleave the anhydro bridge. This was followed by hydrogenolysis of the benzyl groups and acetylation to give the compounds **27**. The same sequence was carried out using the galactolactone **28** to provide the corresponding *C*-disaccharide **29**.

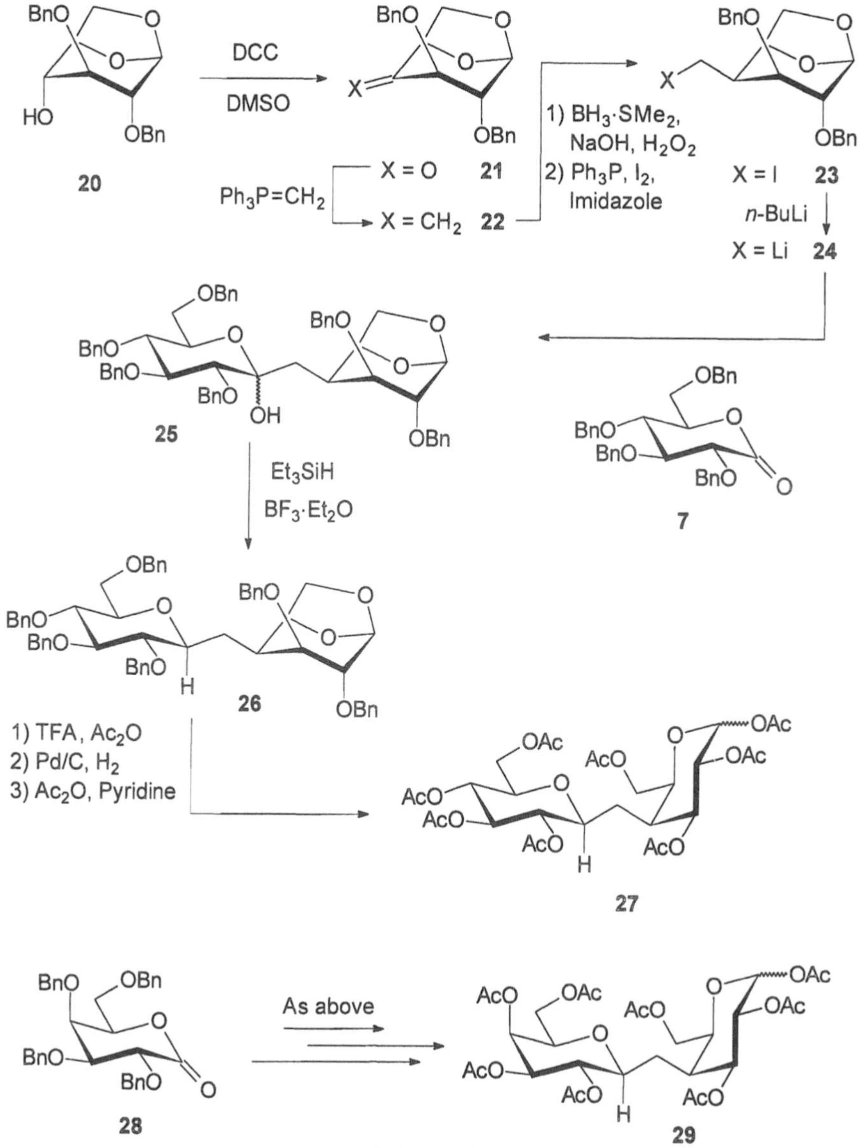

Scheme 4

There are a few noteworthy points in Scheme 4. The hydroboration of **22** is stereoselective and B-H is delivered from the opposite face of the anhydro bridge and the adjacent benzyl ether function. It is this step which determines the configuration of the aglyconic nucleophile as galacto.

Earlier work by Schmidt and Preuss[7] reversed the nature of the key coupling step. In the following case (Scheme 5), the aglycone was to serve as the electrophile in the reaction with an anomeric anion. The required nucleophile was readily available from previous work. Sequential treatment of the benzylated D-glucal **30** with sulfenyl chloride, DBU, and m-CPBA provided **31** as a mixture of isomers. The required electrophile was made from the gluco derivative **33** as follows: Oxidation and Wittig reaction gave **35**. Hydroboration and oxidation gave the primary alcohol **36** which was oxidized to give the aldehyde **37**.

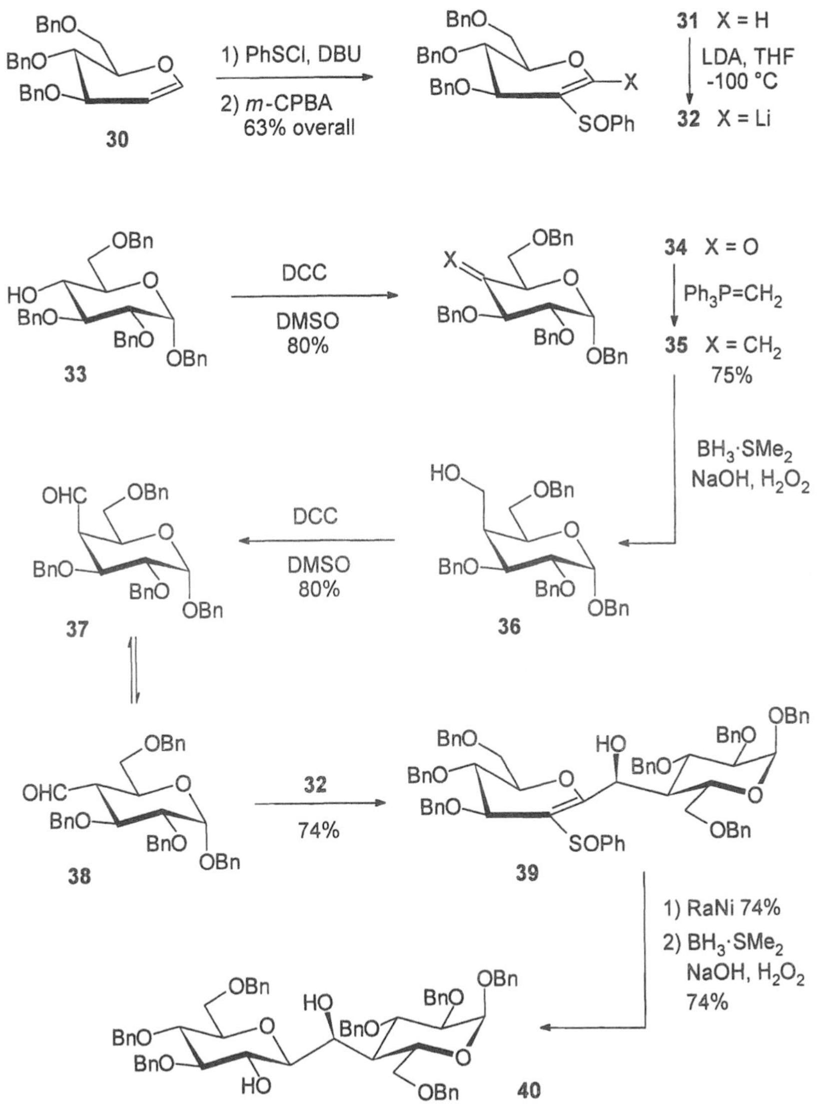

Scheme 5

Compound **37** was isomerized to the more stable equatorial compound **38** and this was reacted with the vinyl lithium species **32** (derived from **31**) to give C-disaccharide **39** (10:1

d.s.) as the major compound. The sequence was completed by Raney nickel induced cleavage of the carbon-sulfur bond followed by hydroboration and oxidation of the intermediate borane to afford the Glc-β-(1→4')-Glc *C*-disaccharide **40**.

A *C*-disaccharide is characterized by having a carbon atom linker instead of the normal oxygen atom. The oxygen atom is divalent and possesses two lone pairs of electrons. A methylene (CH_2) linker has a very different electronic character. Some workers have constructed *C*-disaccharides with heteroatoms on the linking carbon atom to mimic the electronic effect of the oxygen lone pair. The above product **40** is a case in point. Although not illustrated, the above sequence was also carried out with the galacto aldehyde **37** to give the corresponding Glc-β-(1→4')-Gal *C*-disaccharide.

Daly and Armstrong[8] have used acetylenic anions to connect two sugar units via sequential addition to a lactone and a ketone. Strictly speaking, the product is not a *C*-disaccharide since two carbon atoms are linking the sugars together. Initial attempts focused on the use of a bulky acetylene equivalent, the cobalt complex **43**. It was hoped that reduction of the carbocation would favor the equatorial conformation of this group and give the compound with the gluco configuration, but all attempts were unsuccessful and led to mixtures of products in low yield.

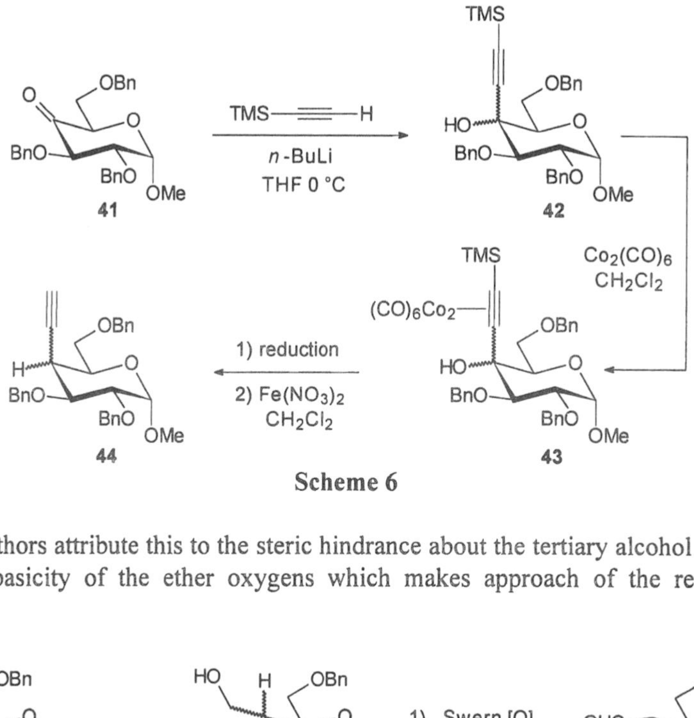

Scheme 6

The authors attribute this to the steric hindrance about the tertiary alcohol coupled with competing basicity of the ether oxygens which makes approach of the reducing agent difficult.

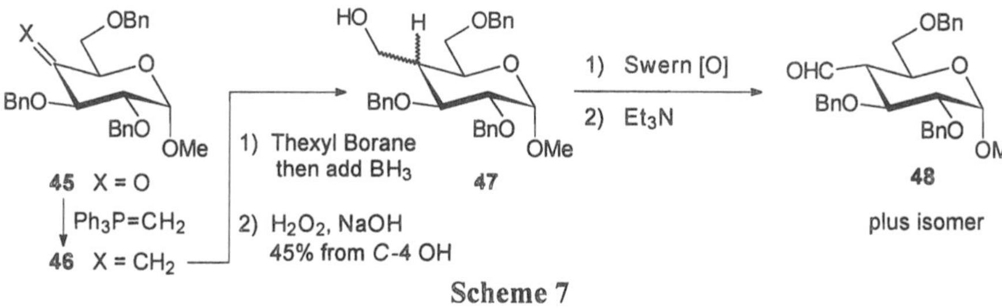

Scheme 7

They were then forced to rely on more conventional means to introduce the acetylene stereoselectively at *C*-4. Wittig reaction on **45** followed by hydroboration, oxidation and equilibration to the more stable equatorial aldehyde gave **47** (equatorial:axial ratio of 2:1).

The equatorial to axial ratio was increased during formation of the dibromo olefin 50 (> 80%) as per the method of Corey and Fuchs.[9] The acetylide anion was generated by addition of butyllithium to 50 and the resulting anion was then quenched with the lactone 7 to give a mixture of hemiketals 51. Stereoselective reduction then gave the fully protected product 52.

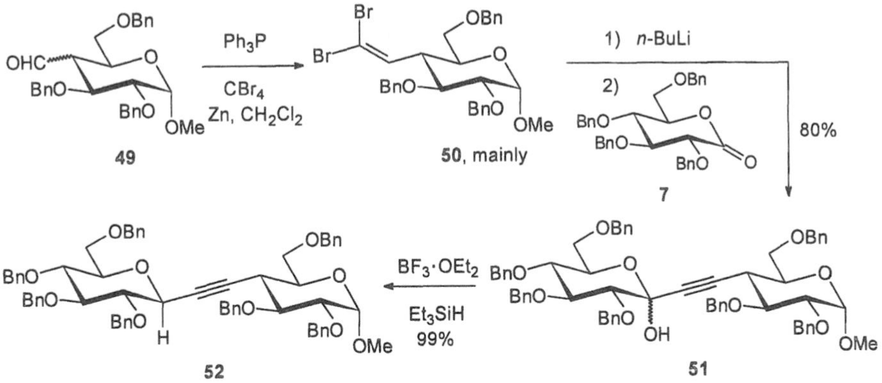

Scheme 8

In order to assemble sugars in a "linear" manner, Armstrong and Sutherlin[10] have been working at constructing linear C-oligosaccharides. These water soluble compounds have potential for use as cross-linking agents to covalently link bipolymers or smaller molecules. The sequence began with the benzylidene 53. Selective reduction of the benzylidene ring gave the liberated primary alcohol (95%) which was oxidized to furnish aldehyde 55. Condensation of 55 with the borane 54 gave compound 56 with >95% diastereoselectivity. Protective group manipulation of 56 provided 57 and ozonylitic cleavage of the double bond gave a lactol which was oxidized to provide the lactone 58.

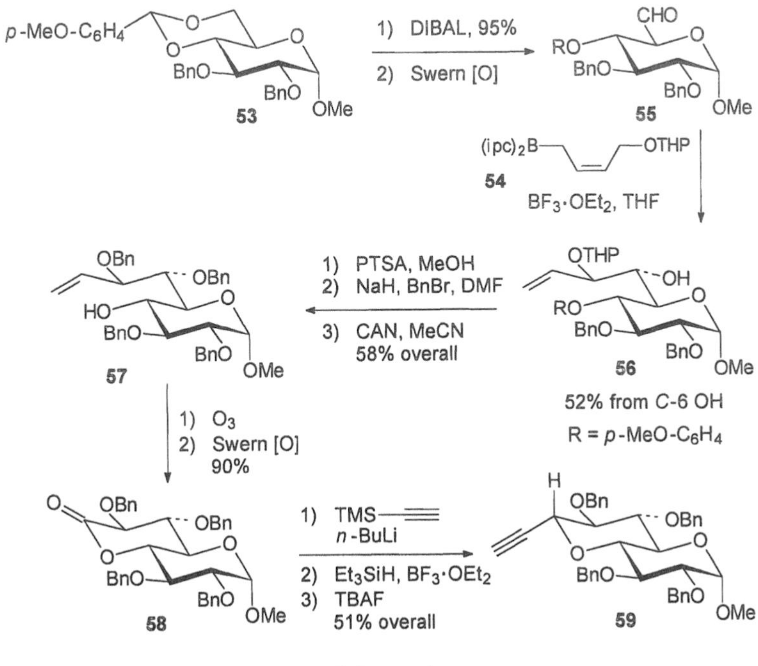

Scheme 9

The sequence was completed (**59→60**) by acetylide anion addition, and stereoselective reduction. Treatment of **59** with *n*-BuLi and addition of the formed anion to **58** gave, after hemiketal reduction, **60** in 54% overall yield (eq. 1).

Kishi has been very active in the area of both *C*-glycoside synthesis and *C*-disaccharide synthesis. His *C*-disaccharide program has been concerned not only with the assembly of *C*-disaccharides, but also with the study of their conformation in solution[11] and their comparision to the solution conformation of *O*-disaccharides. For the *C*-disaccharides to be biologically useful, their conformation must be similar to that of the corresponding naturally occurring disaccharides.

The anionic approach used by the Kishi group relied heavily on the Ni(II)/Cr(II) mediated coupling of vinyl iodides and aldehydes[12] to assemble the carbon framework of the *C*-disaccharide.[13] Further manipulation followed by eventual cyclization would give the requisite compound. The vinyl iodide **61** (available from the corresponding bromide) was coupled with the aldehyde **62** (available from 2,3-*O*-dibenzyl-4,5-isopropylidene-D-arabinose in three steps) to give a 10:1 ratio of products, the major having the erythro configuration. Mitsunobu inversion then gave the desired threo alcohol **63**.

Directed epoxidation using *m*-CPBA gave a 1:3 ratio of *syn* and *anti* epoxides. Cyclization was affected by exposing the major epoxide **64** to CSA and wet CH_2Cl_2 causing acetonide cleavage and cyclization of the secondary alcohol onto the quaternary carbon of the epoxide with preservation of the stereochemical integrity to give **65**. The sequence was completed by deprotection of the triol to give *C*-sucrose **66** in ~30% overall yield from **61**.

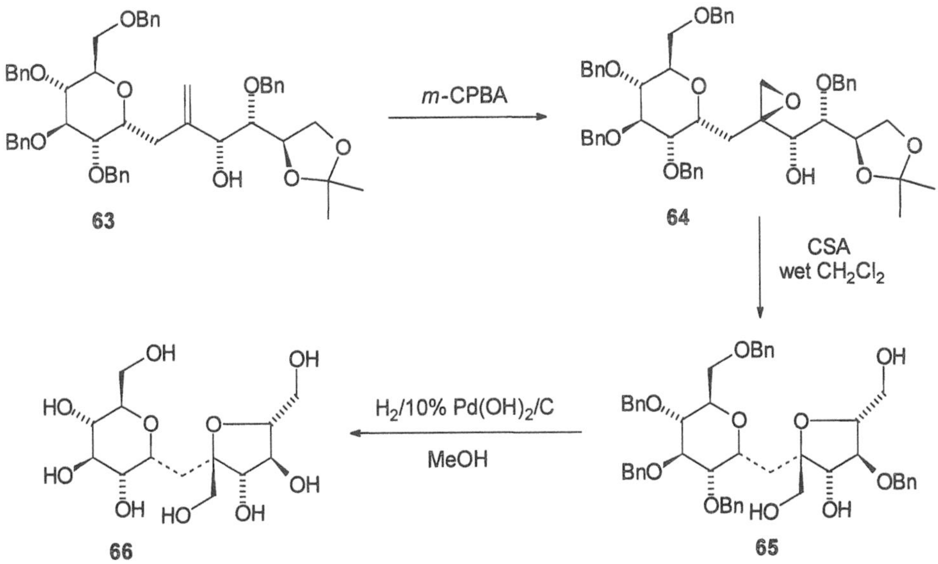

Scheme 10

A similar approach was also utilized in the synthesis of the carbon analogue of isomaltose.[14] Coupling of the vinyl iodide **67** with the aldehyde **68** gave **69**. Hydrogenation of the double bond was followed by protection of the secondary alcohol as its THP ether. Desilylation (TBAF) was followed by Swern oxidation to bring the sequence as far as aldehyde **71**.

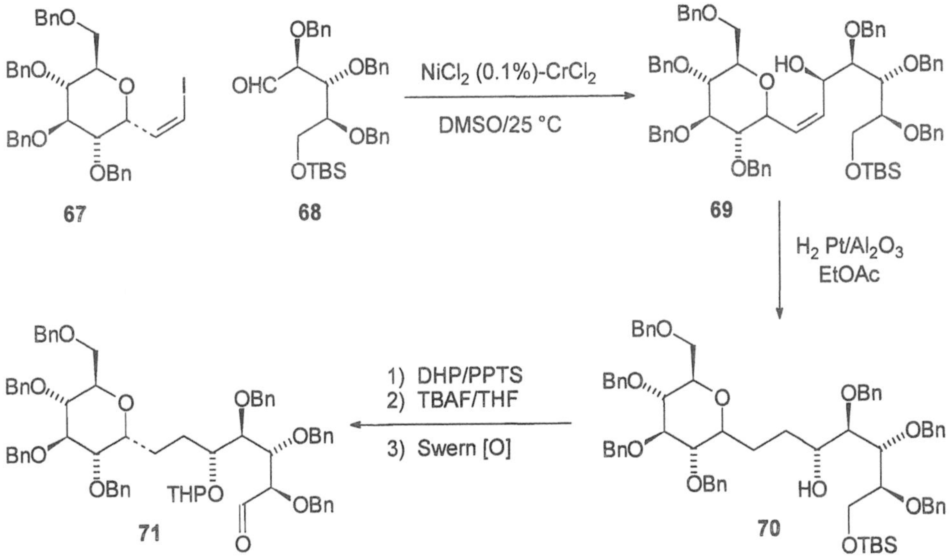

Scheme 11

Removal of the THP ether under acidic conditions preceded cyclization to the lactol. Methylation of the anomeric hydroxyl followed by debenzylation then gave **72**, the carbon analog of methyl isomaltose, eq. 3.

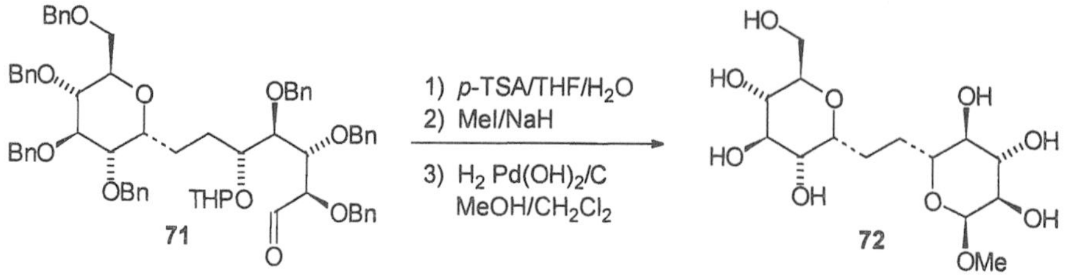

eq. 3

2.   Use of Nitrosugars

Vasella used an anionic approach with anomeric nitrosugars as the key intermediates to construct bis (1,1) linked sugars[15] and also *C*-disaccharides.   The lithium salt of the nitrosugar **74** was reacted with the bromo nitro sugar **73** to provide the bis α,α-(1,1) linked sugar **75** in 76% yield.   Exposure of **75** to Na$_2$S gave the olefinic compound **76** which underwent hydrogenation to give **77** in 79% yield.

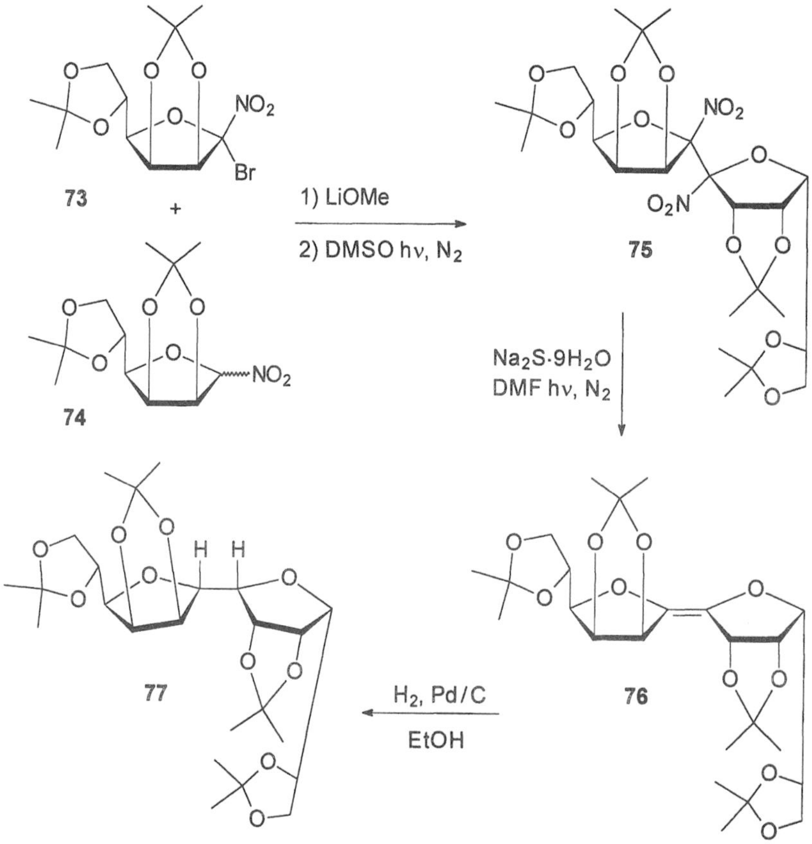

Scheme 12

Reprinted with permission from ref. 15, Copyright 1984 courtesy Verlag Helvetica Chimica Acta.

Vasella's *C*-disaccharide[16] synthesis involved nitro aldol chemistry (Scheme 13).   The anion, generated from an anomeric nitrosugar **78**, condensed with the protected galacto aldehyde derivative **79** to give, after acetylation, the α-(1→6')-*C*-disaccharide **80**.   Reductive denitration with tributyltin hydride gave an anomeric radical which was reduced from the more accessible top face to give the β-(1→6')-*C*-disaccharide **81**.

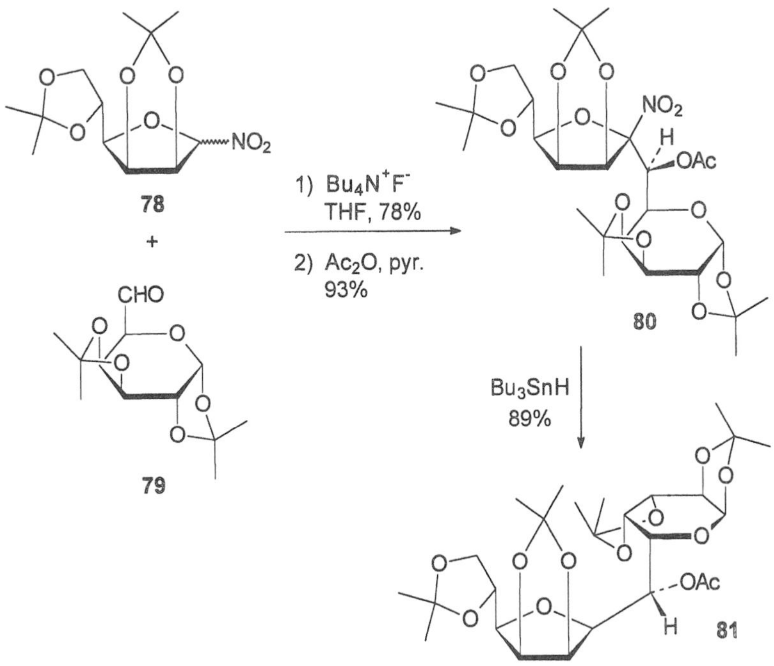

Scheme 13

Martin has also used nitro aldol chemistry to construct *C*-disaccharides.[17] The key step is shown below in eq. 4. The sequence began with fluoride ion mediated nitro aldol condensation of **82** with the aldehyde **79** to give **83** as a mixture of isomers.

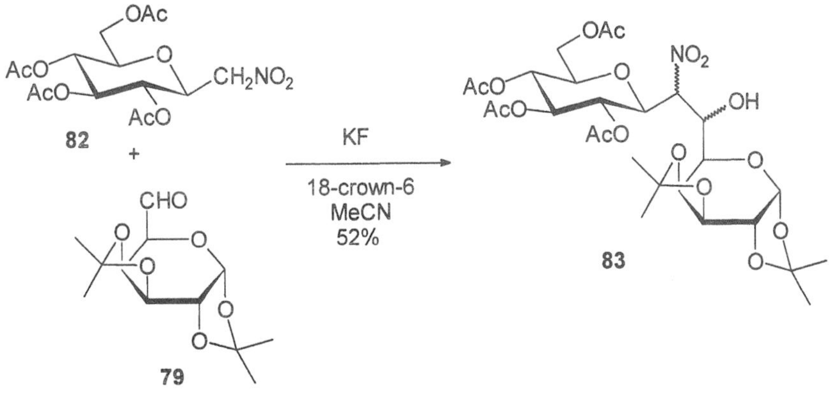

eq. 4

Acetylation and elimination gave the nitroalkene **84**. Selective double bond reduction with sodium borohydride gave a mixture of epimers and reduction of the secondary nitro group with tributyltin hydride went smoothly to give **86**. Finally, deacetylation and hydrolysis gave the free Glc-β-(1→6')-*C*-disaccharide **87**, Scheme 14. This synthetic approach differs from Sinaÿ's first *C*-disaccharide synthesis in the sense that Martin began with a *C*-glycoside that was coupled with a *C*-6 aldehyde. No anomeric carbon-carbon bond was formed in this *C*-disaccharide synthesis, and the β-linkage in the product is a consequence of the starting nitrosugar **82**.

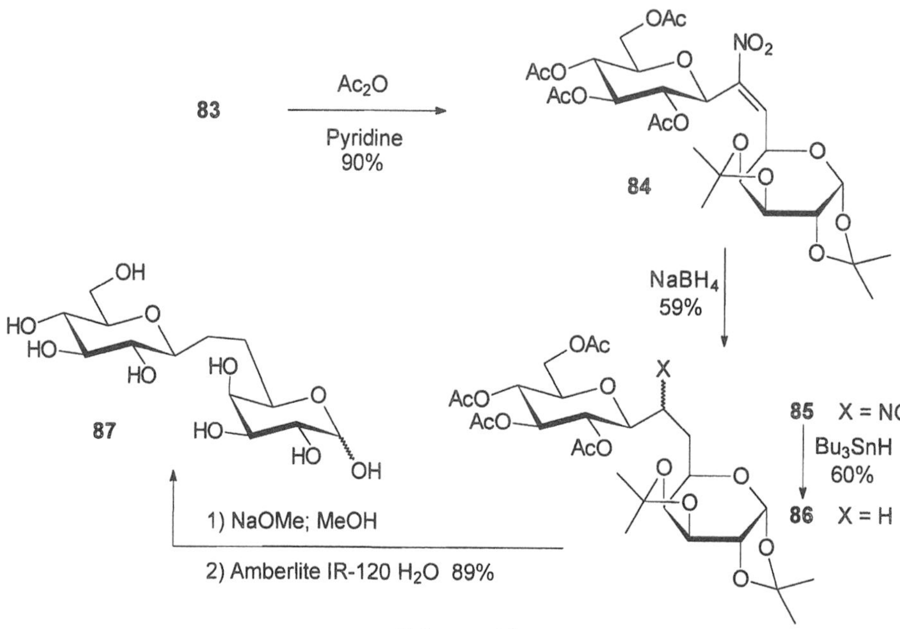

Scheme 14

A similar sequence[17] (Scheme 15) was applied with **82** using the open chain gluco derivative **88**. Here the carbon-carbon bond forming step occurs between *C*-1 of the aglycone to give the β,β-(1→1') linked *C*-disaccharide **93** as the final product, Scheme 15.

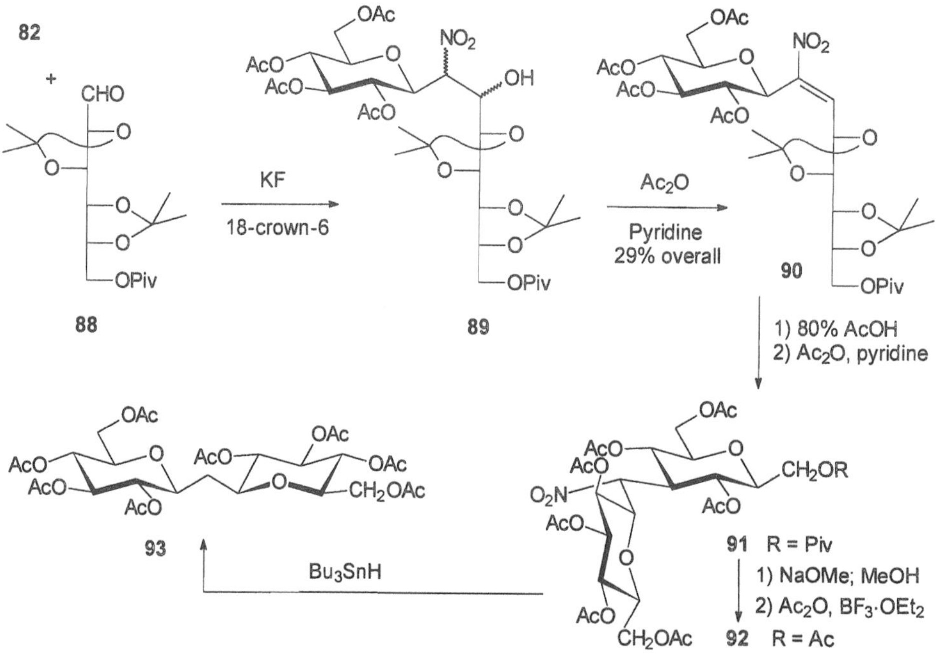

Scheme 15

Hydrolysis of the isopropylidene groups allows a 5- or 6-*exo* closure of the appropriate hydroxyl onto the nitro olefin to give the *C*-disaccharides. It should be noted that a mixture of *C*-pyranosyl/*C*-furanosyl (55:45, 49.5% overall yield) disaccharides were obtained and that only the major (*i.e.* **91**) is shown above in Scheme 15. The work also demonstrates how the electrophilic nature of the anomeric carbon was used to advantage (in the sugar's open form) in the key carbon-carbon bond forming reaction.

B. FREE RADICAL APPROACHES
1. Intermolecular Additions

Free radical methods are particularly well suited for carbon-carbon bond formation at the anomeric center and in many cases the stereoselectivity can occur with good levels of predictability. In order to use this tactic for the construction of *C*-disaccharides, suitable sugar acceptors must be available.

Giese and Witzel[18] were the first to apply radical methods to *C*-disaccharide synthesis. The requisite anomeric radical was easily generated from the corresponding bromide, **94** and addition of the radical to the carbohydrate olefin **95** (available form D-glyceraldehyde) gave a mixture of compounds **96**. The stereoselectivity of the addition at the anomeric center was high and favored formation of the α-anomer (α:β ratio of 10:1). This is not unexpected and is explained on the basis of the radical anomeric effect. Reduction of the adduct radical, however, exhibited poor selectivity (60:40), but one isomer **97** could be isolated in pure form by precipitation from ether and then reduction followed by acetylation gave the α-(1→2') linked *C*-disaccharide **98** in fair overall yield.

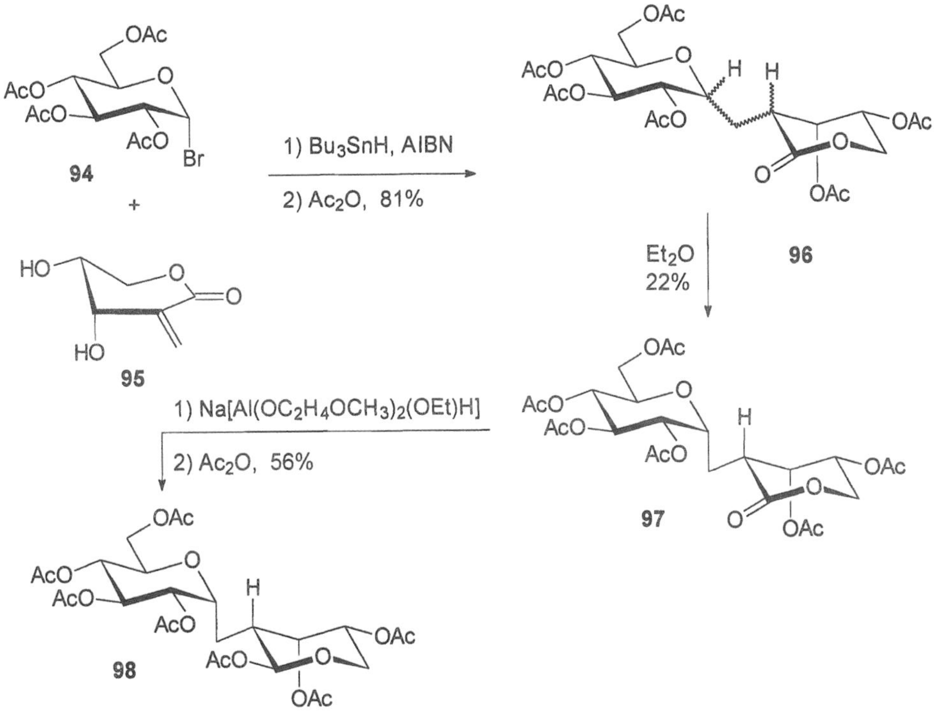

Scheme 16

This methodology was extended to the preparation of the carbon linked sugars **100** and **102** in 35% and 54% yield, respectively. The linked pseudo *C*-linked sugar **100** (eq. 5) arises from radical addition of *C*-4 gluco radical derived from the iodide **99** to the olefin **95**. The addition proceeds from the α-face opposite the two adjacent groups (OBn and CH$_2$OBn) to give an adduct radical which captures hydride opposite from the two OAc groups to give the product **100**.

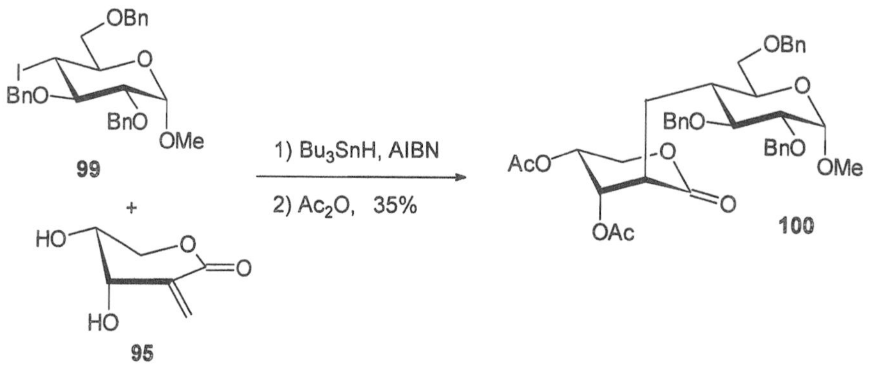

eq. 5

The stereoselectivity of the reduction of the adduct radical, associated with the formation of the 2,6 linked sugar **102** (eq. 6), is rather low and **102** is obtained as a mixture of isomers.

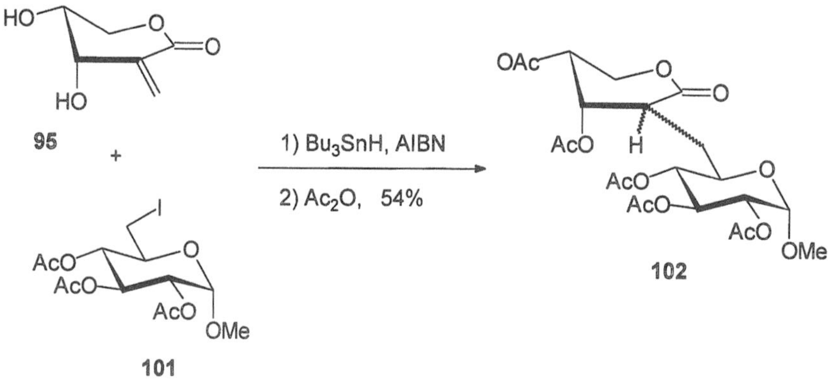

eq. 6

The above methodology is limited by the availability of the corresponding acceptor sugar. Collaborative work between Giese and Schmidt[19] relied on anionic chemistry to construct the requisite olefin sugar (Scheme 17) and radical chemistry (Scheme 18) to assemble the *C*-disaccharides.

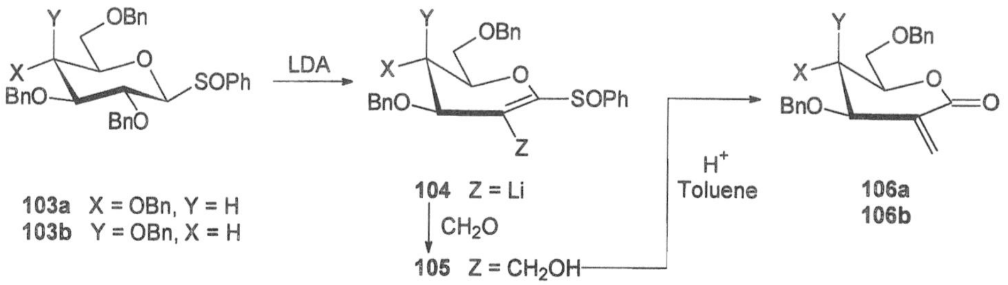

103a X = OBn, Y = H
103b Y = OBn, X = H

104 Z = Li
      CH$_2$O
105 Z = CH$_2$OH

106a
106b

Scheme 17

With the requisite radical acceptors in hand (**106a-b**), the radical additions could now be carried out. Three sugar radicals were examined: glucosyl, fucosyl, and mannosyl and their reactions with the acceptors are shown in Scheme 18.

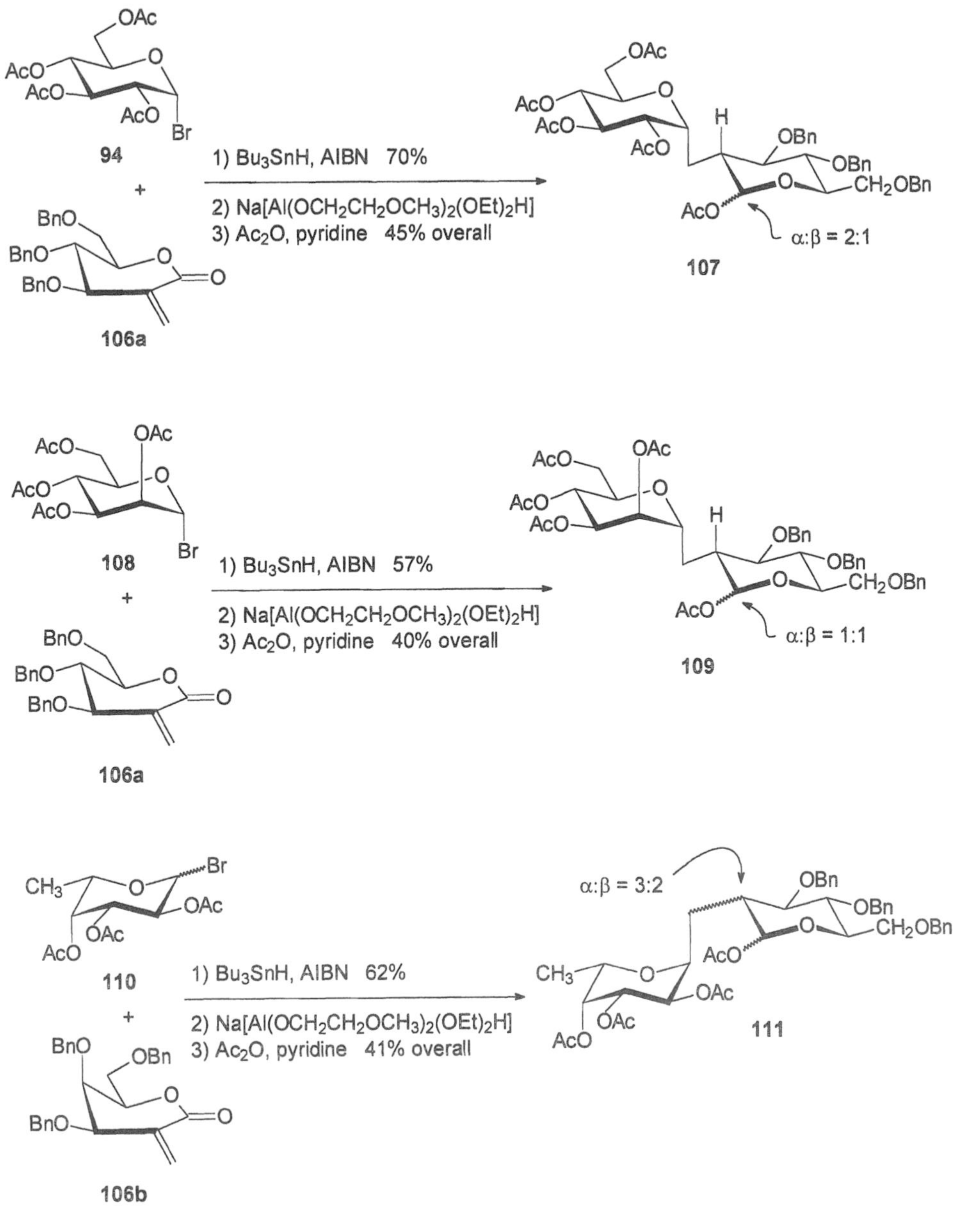

Scheme 18

The first entry (Scheme 18) shows that both the radical addition and hydrogen donation by tributyltin hydride to the intermediate adduct radical occurred with high levels of stereoselectivity. The product is the methylene analogue of kojibose 107. Reduction and acetylation occurred with moderate stereoselectivity (α:β ratio of 2:1). Addition of the mannosyl radical to lactone 106a also proceeded with high selectivity and reduction, and acetylation in this instance, occurred with no selectivity to give 109. The motivation for the third entry arises from the frequent occurrence of the α-L-fucopyransoyl-(1→2')-D-galactose moiety in glycoconjugates. The addition is highly selective, but reduction of the adduct radical with tributyltin hydride exhibits low selectivity. The above examples illustrate that although the stereoselectivity of the radical addition is predictable, the corresponding reduction of the secondary adduct radical is not always so predictable.

Vogel and Bimwala[20] have also assembled *C*-disaccharides by intramolecular radical addition (Scheme 19 & 20). Their approach involves the addition of an anomeric radical to a suitable olefin acceptor. The acceptor is in fact a masked sugar and the adduct is easily transformed into a *C*-disaccharide in a few steps after radical addition. The radical acceptor **113** is readily available from the optically pure oxabicyclo compound **112**.[21] Addition of the glucosyl radical derived from **94** proceeds from the α face (α:β ratio of 48.5% : 6.0%) and adduct reduction is also highly stereoselective. It is noteworthy that the phenylseleno group survives the reductive conditions. *Exo* face delivery of hydride from $NaBH_4$ gave the alcohol **115**. Oxidation of the selenide and selenoxide fragmentation was followed by acetylation to provide the olefin **116**. Hydroxylation of the olefin from the *exo* face gave, after acetylation, **117**.

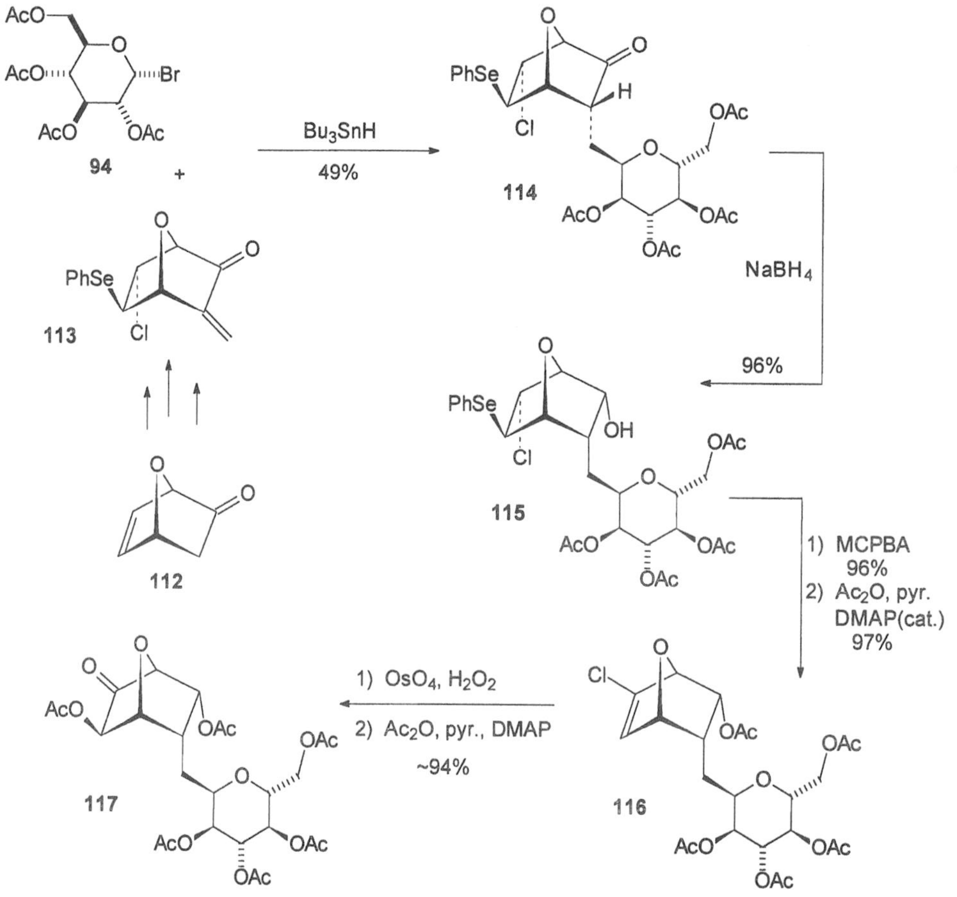

Scheme 19
Reprinted with permission from ref. 20, Copyright 1992 American Chemical Society.

Bayer Villager oxidation went as predicted and furnished the lactone **118** in 94% yield. Treatment of the lactone **118** with K_2CO_3 in methanol gave the pyranosyl/furanosyl *C*-disaccharide **119**. Reduction and treatment of **119** with acid followed by acetylation gave the peracetylated derivative **121** (Scheme 20). Under acidic conditions **118** gave a mixture of two partially acetylated sugars that were reacetylated and separated to provide **122** and **123**. Reduction followed by reacetylation then gave the protected *C*-disaccharides **124** and **125**, α-D-Glc*p*-*C*-(1→3)-β-L-Man*p*-OMe and α-D-Glc*p*-*C*-(1→3)-β-L-Man*f*-OMe, respectively.

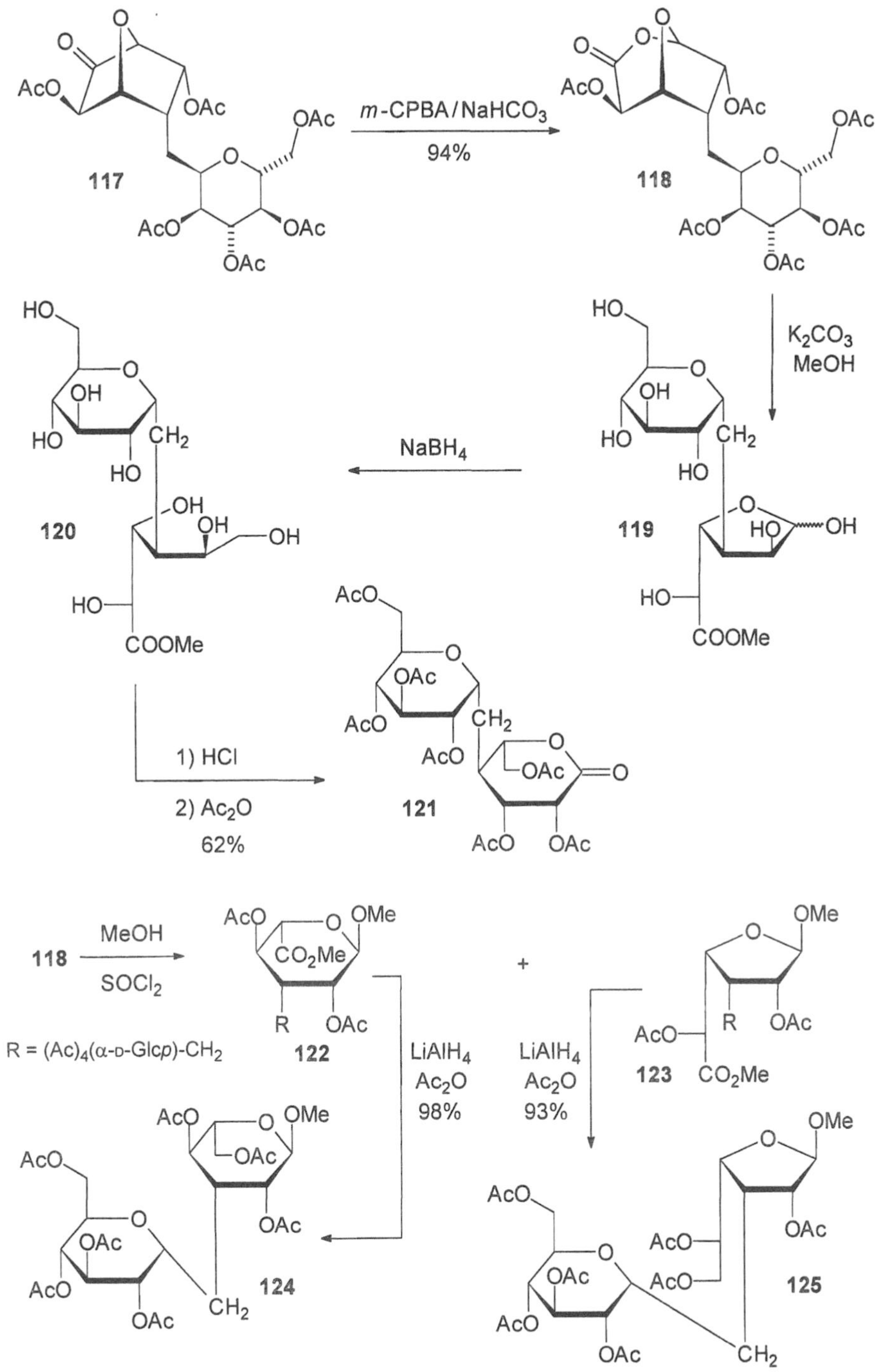

Scheme 20

Reprinted with permission from ref. 20, Copyright 1992 American Chemical Society.

Motherwell[22] has constructed a $(1\to6')$-C-disaccharide by adding the C-6 radical derived from **127** to the difluoroexomethylenic sugar **126**. The addition occurs in low yield, but with a high level of stereoselectivity. The adduct radical then abstracts hydride from tributyltin hydride from the more accessible top face to give the C-disaccharide **128**, eq. 7.

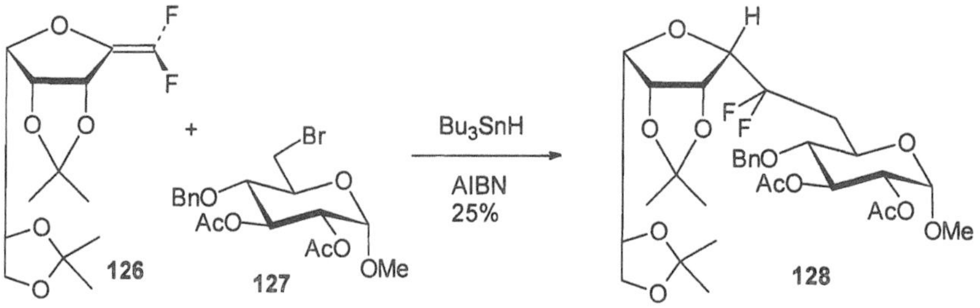

Martin[23] has applied the $S_{RN}1$ coupling reaction between a suitable free radical and nitronate anion to quickly assemble *C*-disaccharides. Initial attempts focused on using an organomercurial in the $S_{RN}1$ coupling but this met with failure. Satisfactory results were realized when the alkylcobaloxime was used. Its preparation is shown in Scheme 21. Compound **129** underwent mercuriocyclization/iodo-demercuration in 62% yield to give **130**. The glycosylmethylcobaloxime **131** was accessed by displacement in 88% yield. Coupling with the 6-nitrosugar **132** gave a mixture of **134** and **136** after acetylation. Reductive denitration of **134** then furnished the *C*-disaccharide corresponding to isomaltoside **135**.

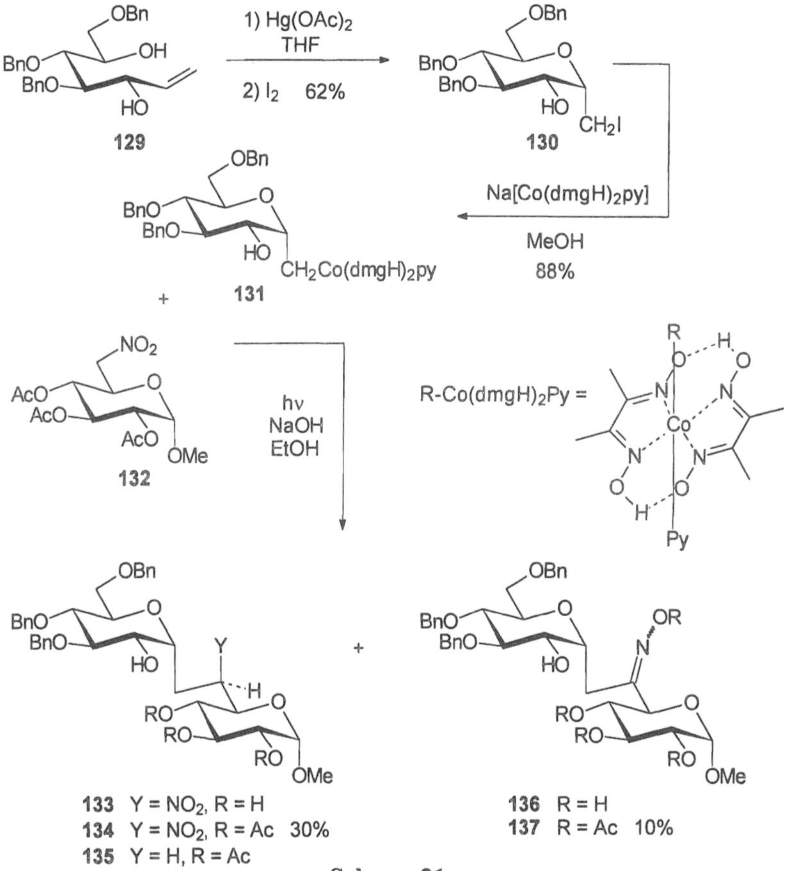

133 Y = NO₂, R = H
134 Y = NO₂, R = Ac 30%
135 Y = H, R = Ac

136 R = H
137 R = Ac 10%

Scheme 21

The corresponding β-(1→6')-linked *C*-disaccharides were made by a strategic reversal in the coupling step, with the cobaloxime group now on the aglyconic sugar as shown below in Scheme 22. Both the gluco and galacto derivatives were made by this methodology.

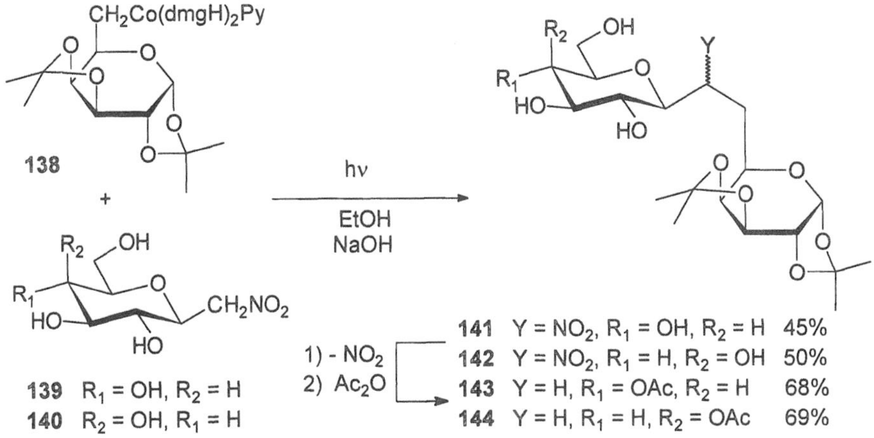

Scheme 22

The coupling steps were carried out in fair yield and acetylation and reductive denitration then furnished the *C*-disaccharides **143** and **144**. It is relevant to note that the coupling steps are carried out with unprotected sugars and this minimizes the need for elaborate protection-deprotection schemes. The β-D-*C*-Glc-(1→6')-α-D-Glc-OMe disaccharide was quickly made by a coupling reaction between the equatorial cobaloxime **145** and the 6-nitrosugar **146** to give, after reacetylation, **148** (51%), a suitable precursor to *C*-gentiobiose.

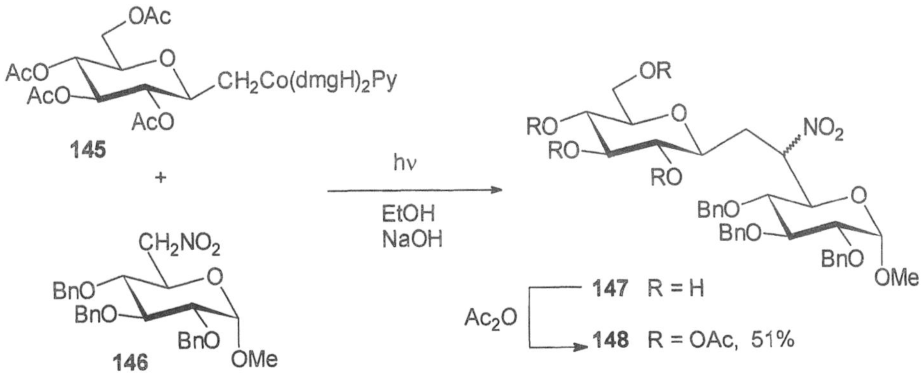

Scheme 23

The above methodology works well when the nitro sugar is used in excess. This is acceptable due to the fact that the nitrosugars are readily available and that the cobaloximes are easily made from suitable iodides by simple displacement reactions. Although the yields in the key carbon-carbon bond forming reactions are only ~50%, the lack for elaborate protection coupled with the ready availability of starting materials makes the above synthesis of (1→6')-*C*-disaccharides rather efficient.

2.   Intramolecular Approaches

There are only two true intramolecular free radical approaches to *C*-disaccharides synthesis both by Sinaÿ. An example by Fraser-Reid which deals with a tandem radical addition approach to a branched *C*-sugar is also included in this section.

Exposure of the bromide **149** to free radical conditions in the presence of the radical acceptor **150** gave a 70% yield of the bis *C*-sugar **151** (eq. 8).[24]

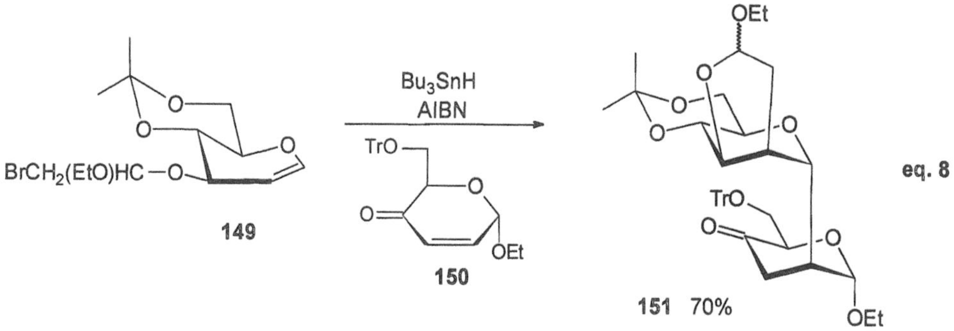

eq. 8

Sinaÿ *et al.* have been the only ones who have synthesized a true *C*-disaccharide via intramolecular radical cyclization. Two different tethers were examined. In one case the two fragments were tethered together via a silicon atom[25] by treatment of the suitably protected sugar **152** with butyllithium followed by quenching with dimethylchlorosilane. Then, treatment with **153** and imidazole gave the tethered intermediate **154**. Slow addition of tributyltin hydride and AIBN in benzene to **154** then furnished the protected (1→4')-*C*-disaccharide **155** as a single isomer. It is interesting to note that the cyclization is a 9-*endo* trigonal one and that reduction of the adduct radical by tin hydride was highly stereoselective, Scheme 24.

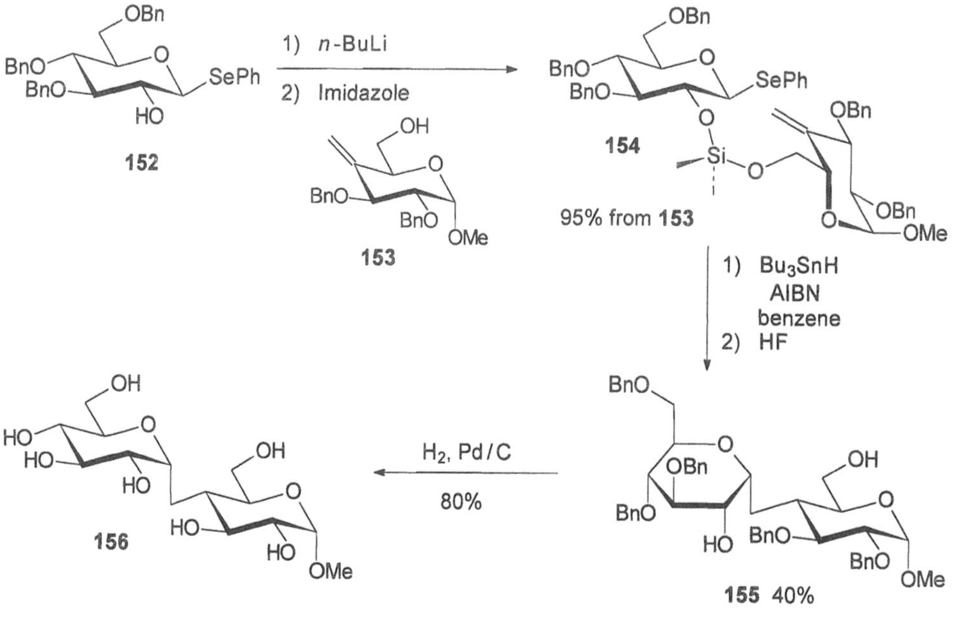

Scheme 24

Reproduced with permission from Ref. 25, Copyright 1993, pg. 865, courtesy the Royal Society of Chemistry, Thomas Graham House, Science Park, Milton Road, Cambridge, CB4 4WF, UK.

The highly observed selectivity is not a general trend and this is supported by the stereochemistry of the product resulting from cyclization when the tether is between *O*-2 and *O*-3. Cyclization of **157** gave a mixture of *C*-disaccharides **158**.

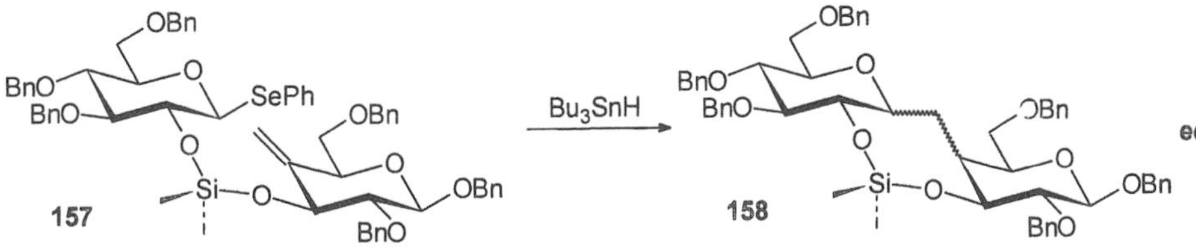

The second approach utilized a ketal function[26] as the temporary connector. Treatment of the orthoester **159** with benzeneselenol and mercuric bromide gave the anomeric selenide **160**. Exposure of the acetate **160** to Tebbe's reagent then gave **161** which was coupled with the olefin **162** to give the ketal **163**. 9-*Endo* trigonal radical cyclization furnished **164** and the sequence was completed by exposure to TFA and water, diacetylation, debenzylation and acetylation to give the methyl α-D-man-*C*-(1→4')-α-D-glc-OMe disaccharide **165**.

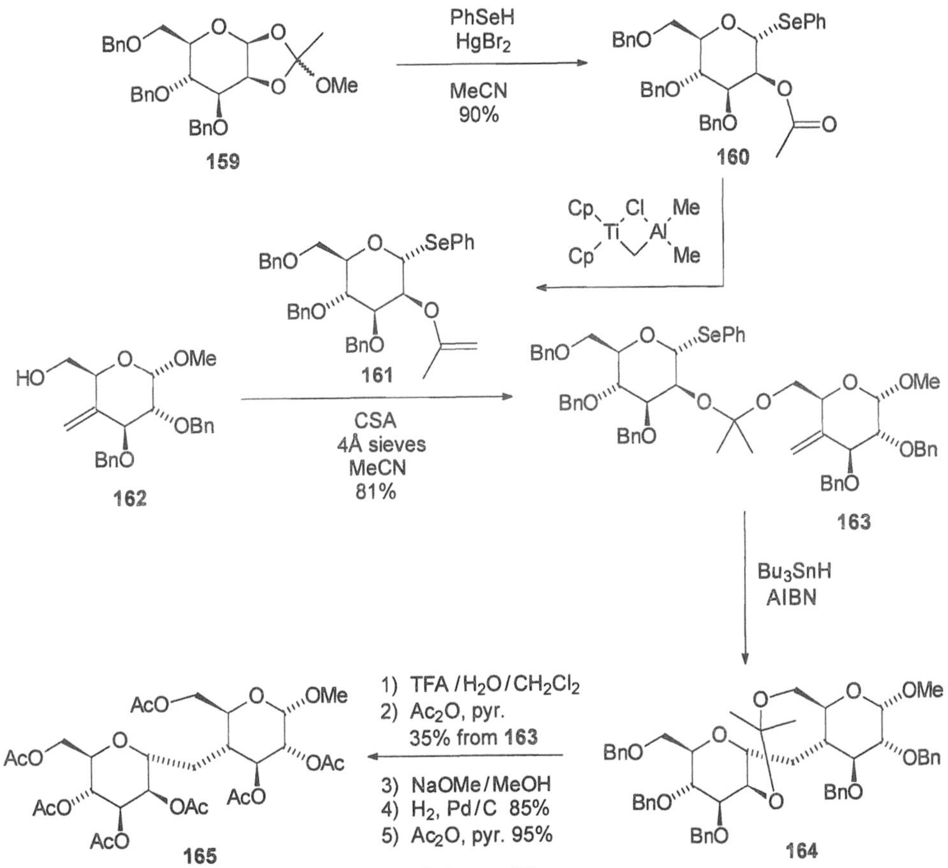

Scheme 25

A small amount (10:1 ratio) of the β-*C*-manno isomer **166** was also isolated from the reaction mixture (Fig. 2).

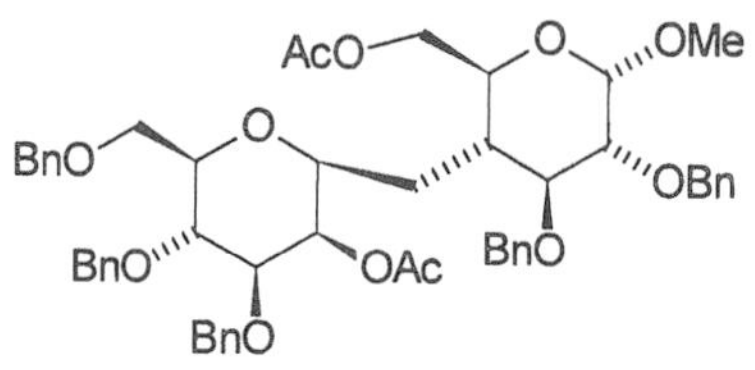

166 minor

Figure 2

C. CYCLOADDITION APPROACHES

Danishefsky[27] has used a cyclocondensation reaction between the diene 167 and the protected aldehydo sugar **79** under Lewis acid conditions to give, after desilylation and elimination, compound **168** exclusively. The observed stereoselectivity is probably due to an alignment (implied by **79**) as shown in which the C-O aldehyde bond is *anti* to the *C*-5-*O* bond. The resulting compound can be envisaged as possessing the galacto configuration, assuming that further transformations could successfully oxygenate the ring.

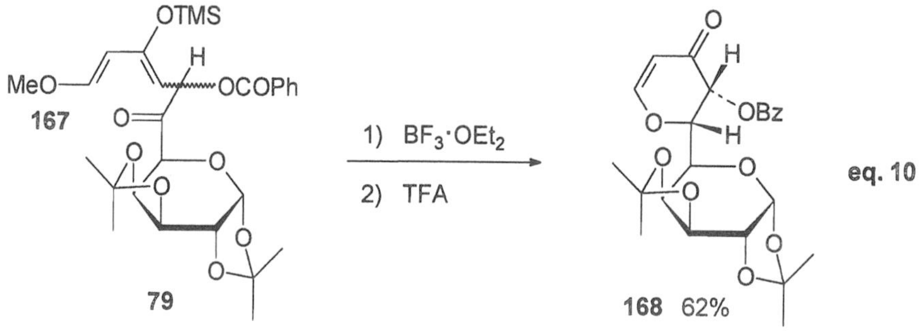

eq. 10

Cyclocondensation of the same diene with the aldehyde **169** derived from D-ribose gave **170** after treatment with trifluoroacetic acid. In this case the cycloaddition occurs in a similar sense, but the disposition of the OBz group is now such that the configuration is L-gluco.

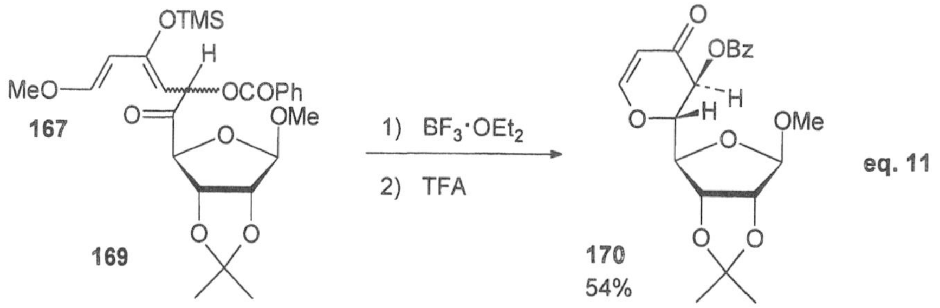

eq. 11

Jurczak *et al.* used high pressure conditions in similar cycloadditions (Scheme 26) to achieve very high levels of asymmetric induction.[28] Cycloaddition at 20 Kbar gave **172** as the major compound. The stereochemical sense of the cycloaddition is opposite to that observed by Danishefsky. The reaction proceeded via *endo* addition and the configuration of the OMe group could be inverted by treatment with pyridinium *p*-toluenesulfonate in acetone to give **173**. The absolute configuration of **172** was established via chemical degradation and comparison of the product with an authentic sample.

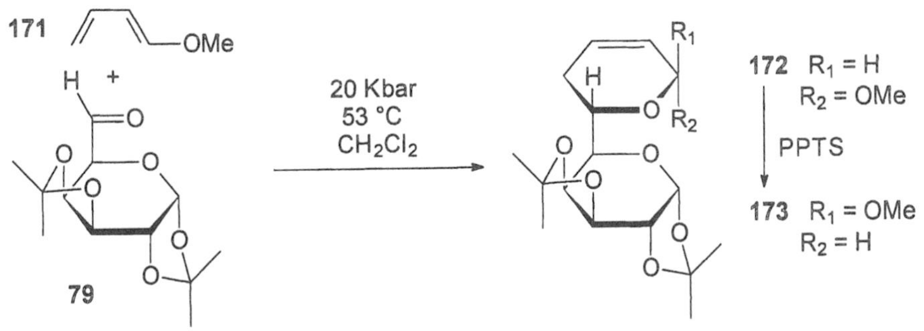

Scheme 26

C-Disaccharides have also been assembled by the use of a 1,3-dipolar addition reaction.[29] The D-xylo derivative 174 was dehydrated *in situ* with tolyluene di-isocyanate to give the nitrile oxide 175 which underwent cycloaddition to provide two isomers (total of four possible isomers) 177 and 179 in 33% yield each.

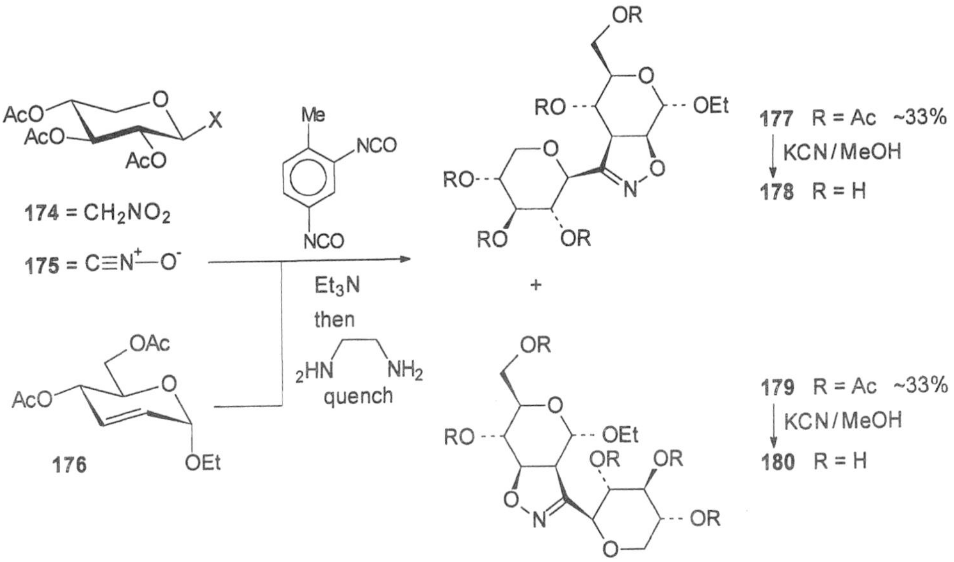

Scheme 27

Reductive hydrolysis of 178 gave after reacetylation the D-xyl-C-(1→3)-α-D-glc-OEt linked C-disaccharide 182 (in 97% yield for 182 from 177). The linker is now a carbonyl group in lieu of the interglycosidic oxygen atom. When 180 was exposed to identical conditions compound 183 was formed in 90% yield and characterized as its hexa-acetate 184. ^{1}H NMR analysis revealed that the configuration at *C*-2 was gluco and not manno as in the cycloadduct 180. Presumably epimerization occurs under the reductive hydrolytic conditions to give 183 with the carbonyl group in a more favorable equatorial position.

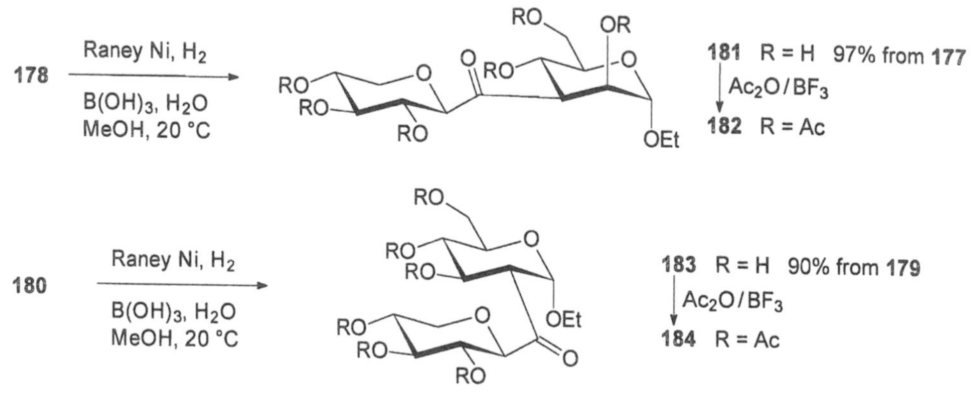

Scheme 28

The cycloadditions occurred exclusively from the top β-face opposite the substituents at *C*-4 and *C*-1. Variation came into play only as far as the regiochemistry was concerned. Presumably, cycloaddition on the corresponding lactones would exhibit more regioselectivity. Alternatively, attachment of the nitrile to the olefinic sugar via a temporary connection may increase the regioselectivity by forcing the cycloaddition to proceed in an intramolecular fashion. Nevertheless, the above cycloaddition strategy is very efficient and *C*-disaccharides are quickly assembled from accessible precursors.

D.   CATIONIC APPROACHES

Cationic approaches to anomeric carbon-carbon bond formation have been widely utilized in *C*-glycoside synthesis. The natural electrophilic nature of this center makes this approach an obvious choice.

Isobe and co-workers have used glycals and acetylenes in their preparation of *C*-glycosides.[30] In an extension of this methodology they wished to assemble (1→1') linked sugars spaced by one or more acetylene groups.[31] Reaction of the glycals **185** and **188** with the trimethylsilyl acetylene **186** gave **187** and **189**, respectively. Further attempted reaction of **187** with another molecule of glycal gave no coupled product.

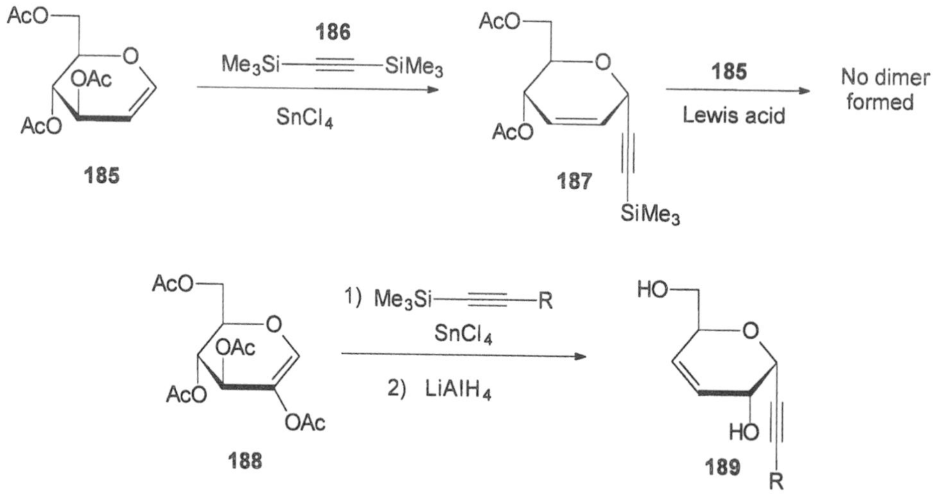

Scheme 29

When the 1,4-bis(trimethylsilyl)-1,3-butadiene **190** was used, the first glycosidation proceeded in excellent yield and the second glycosidation occurred in 55% along with recovered **191** (19%) to give the bis-sugar **192**. Alternatively, the bis-sugar could be obtained directly from **185** in good overall yield by performing the reaction without isolation of the intermediate **191**.

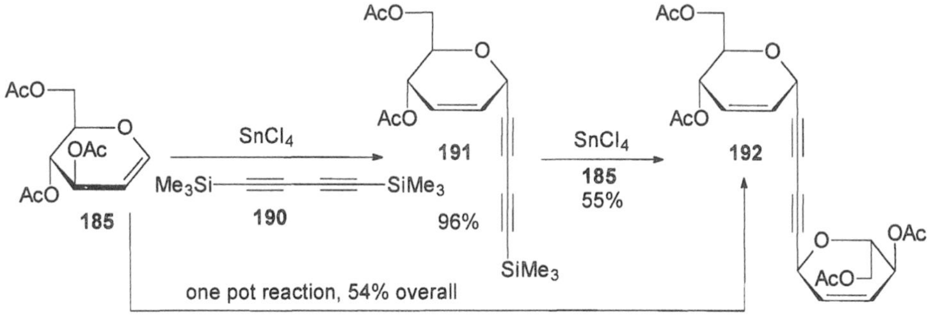

Scheme 30

Instead of utilizing the 1,4-bis(trimethylsilyl)-1,3-butadiene **190**, the workers decided to link the sugars via an ene-diyne connector that is a familiar sub-unit in several natural products. Reaction of 1,6-bis-(trimethylsilyl)-hex-1,5-diyn-3-ene **193** with glycal **185** gave **194** in 79% and further reaction along with the dimer **195** (10%). When an excess of **193** was used the dimer **195** was formed in 68% yield without isolation of the mono-glycoside.

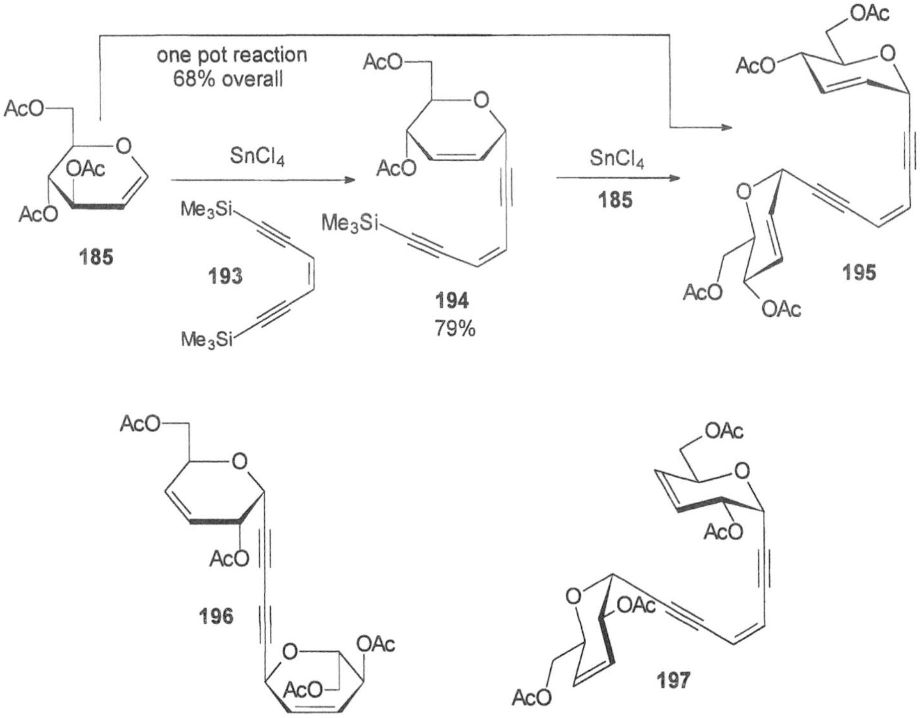

Scheme 31

Compounds **196** and **197** arise from a similar sequence of reactions. Treatment of **188** (See Scheme 29) with SnCl₄ and subsequent exposure to the appropriate bis-silyl compound, NaBH₄ reduction, and reacetylation (Ac₂O) then gave the bis-sugar derivatives **196** and **197** in 54% and 78% overall yield, respectively (Scheme 31).

Stütz and de Raadt[32] extended the Ferrier allylation to *C*-disaccharide synthesis. Reaction of the allylic silanes **198** and **199** with **185** under boron trifluoride etherate catalysis gave the *C*-disaccharides **199** and **201**, respectively (Scheme 32).

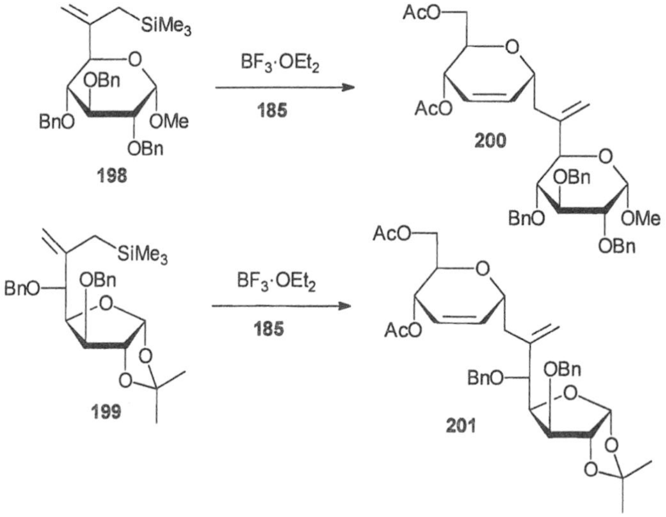

Scheme 32

Nicotra[33] has observed that treatment of **202** with a Lewis acid leads to the formation of a *C*-disaccharide **205** as a (1:1) mixture of compounds.

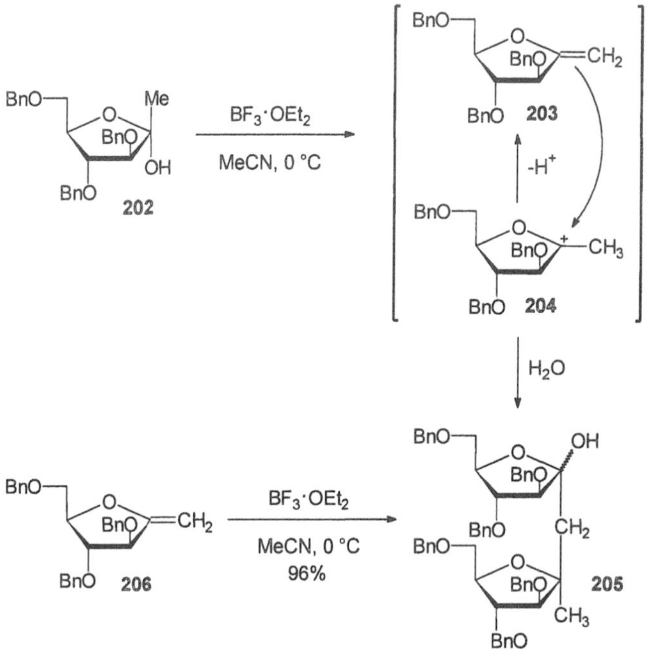

Scheme 33

Presumably, the reaction goes through the enol ether intermediate **203** which is formed under the Lewis acid conditions. This is supported by the fact that the exomethylenic sugar **206** undergoes *C*-disaccharide formation under identical conditions.

The situation with the pyranose sugar (Scheme 34) was slightly different. No *C*-disaccharide could be obtained when **207** was exposed to the reaction conditions, but **208** dimerized easily (66% yield) to give **209** as the sole compound.

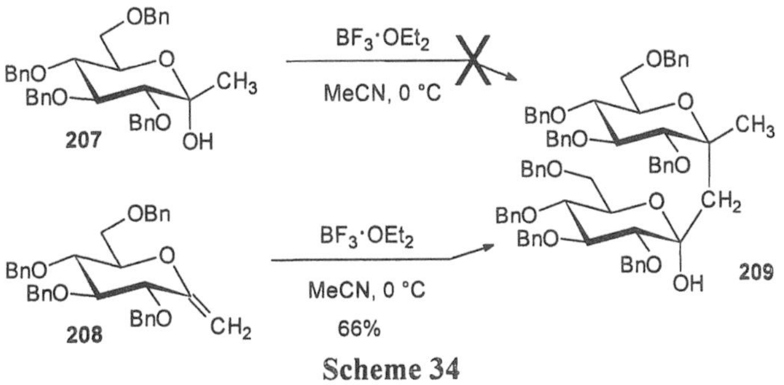

Scheme 34

E. WITTIG APPROACHES

Nicotra's approach to *C*-disaccharide assembly relied on a Wittig coupling step between the *C*-glycoside **212** and a protected D-glyceraldehyde, Scheme 35.[34]

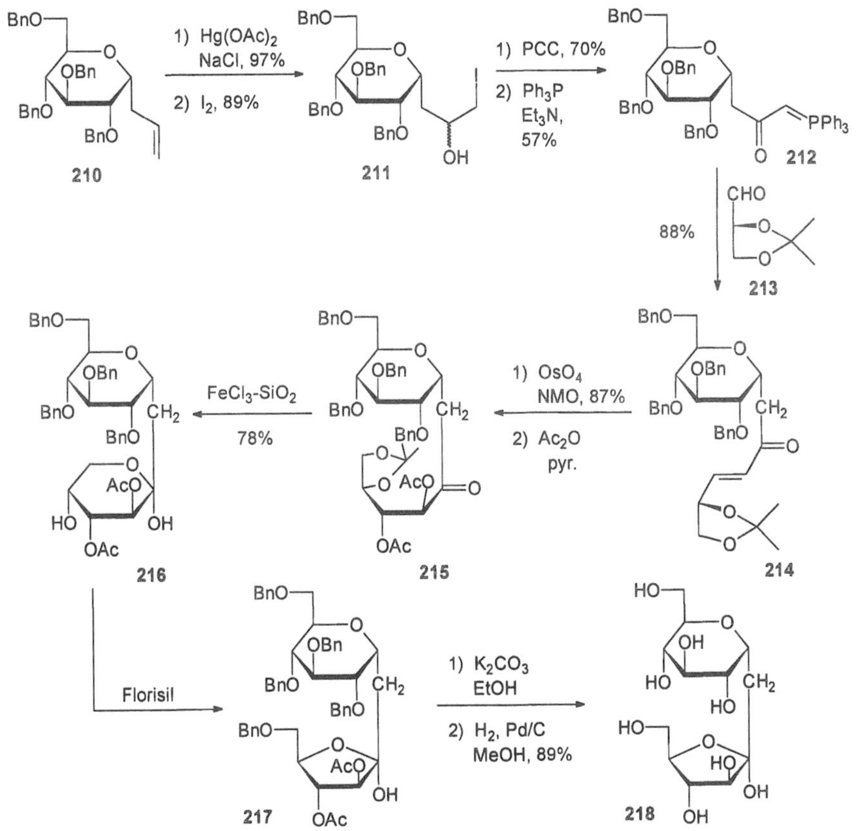

Scheme 35

Reproduced with permission from ref. 34, Copyright 1989 courtesy the Royal Society of Chemistry.

The α-*C*-glycoside **210** was subjected to hydroxymercuration to give, after treatment with iodine, the iodo alcohol **211**. PCC oxidation and displacement of iodide with triphenyl phosphine in the presence of triethylamine then gave the ylide **212**. Exposure of the ylide **212** to 2,3-O-isoproplyidene-D-glyceraldehyde (**213**) afforded the olefin **214**. Stereoselective osmylation was followed by acetylation to give **215**. Deisopropylidenation using FeCl$_3$-SiO$_2$ then gave the *C*-disaccharide **216**. Exposure to florisil gave the fructofuranosidic form **217** and finally, deprotection gave **218** which exists as a mixture of pyranosidic and furanosidic forms (^{13}C NMR).

Kishi and co-workers[35] have utilized a conceptually similar strategy as that outlined in Scheme 35. Their sequence begins (Scheme 36) with a Wittig reaction between the anhydro sugar ylide **220** and the aldehyde **219** to give the *cis*-olefin **221**, exclusively. Osmylation gave a 6:1 mixture of diols with **222** predominating. Selective benzylation then gave **223** in good yield. Cyclization to the hemiketal **224** was affected by Swern oxidation, acetonide hydrolysis and ring closure, and finally stereoselective silane reduction (7:1). Deprotection and acid catalyzed cleavage of the anhydro bridge then gave **226**.

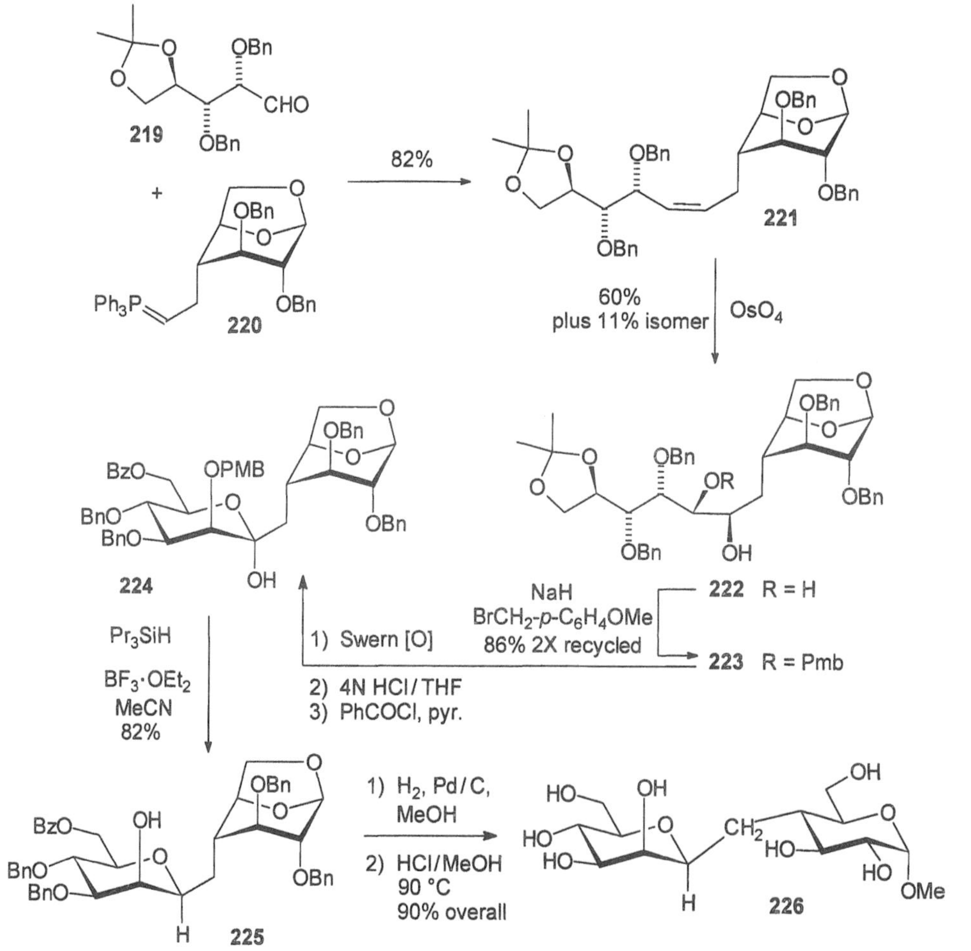

Scheme 36

A similar sequence was also carried out on the minor diol **227** to eventually give the gluco isomer **228**. Alternatively, compound **225** could be converted into **229** by a two step protocol: Swern oxidation followed by stereoselective reduction with $BH_3 \cdot Et_3N$.

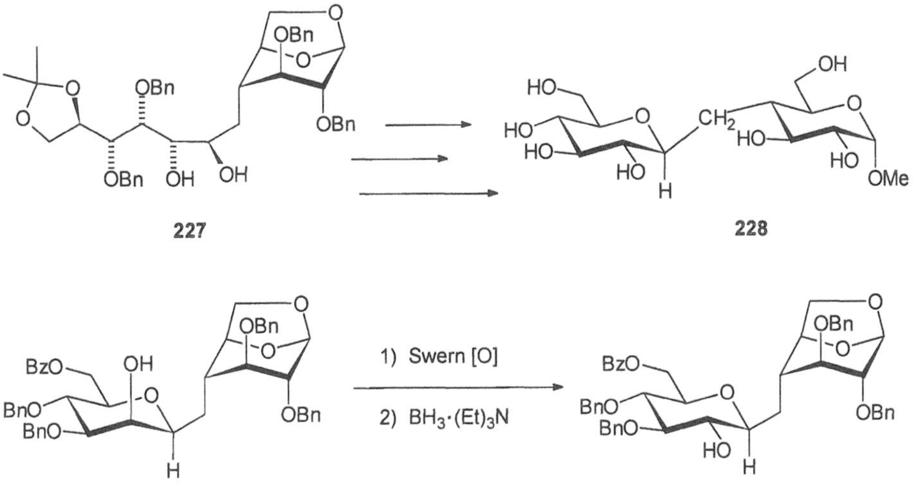

Scheme 37

The α-linked *C*-disaccharides[35] could be accessed by alteration of the cyclization step. Compound **230** was exposed to acetic acid-water to cleave the ketal. Lead tetratacetate served to oxidatively cleave the diol to give an aldehyde which cyclizes to give a lactol which is then benzoylated to give **231**α (51%) and **231**β (46%). The stage was now set to introduce the hydroxymethyl function at which would eventually become *C*-5. Treatment of **231**β with $BF_3 \cdot OEt_2$ and propargyl trimethylsilane gave the product **232** (10:1) as the major product. Exposure of **232** to the action of ozone followed by sodium borohydride reduction gave the hydroxy methyl function at *C*-5 and hydrogenation and bridge cleavage then afforded the *C*-disaccharide **233** (Scheme 38).

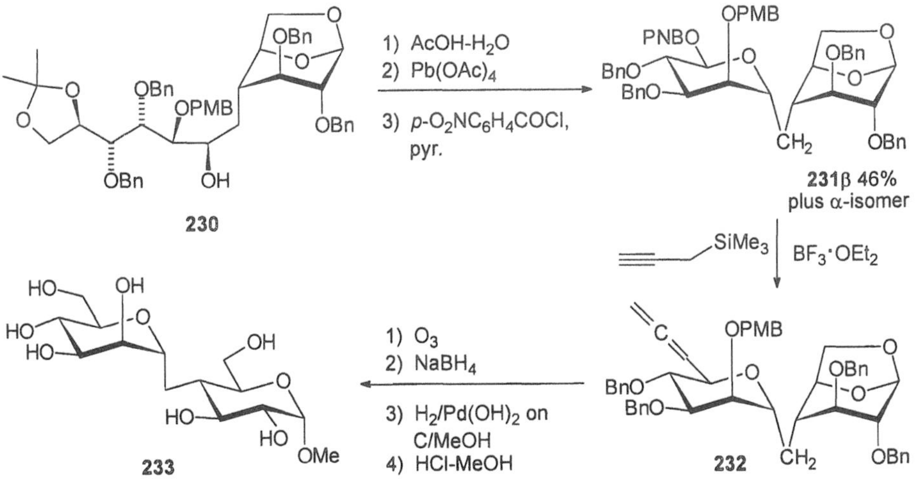

Scheme 38

Kishi's approach to the above *C*-disaccharide is complementary to that found in Scheme 36. In the Scheme 36 silane reduction of the hemiketal **224** (derived from cyclization of *O*-5 on *C*-1 ketone) dictates the stereochemistry as β. In Scheme 38, the stereochemistry of the linkage comes about as a result of the stereochemical disposition of the hydroxyl at *C*-1. Cyclization on a *C*-5 aldehyde gives compound **231** in which the hydroxymethyl group at *C*-5 is lacking. The hydroxymethyl function was then introduced by a stereoselective *C*-glycosidation as shown in Scheme 38.

Armstrong and Teegarden[36] used a combination of Wittig and cyclization strategy to assemble a (1,4') linked sugar. Wittig reaction of the ylide generated from **234** with the aldehyde **235** gave a 1 to 3 ratio of olefins **237** and **236**, respectively. The silyl groups were removed with TBAF and the primary alcohol **239** was selectively silylated to give **240**. Exposure of **240** to NBS and a trace of bromine then gave **241** selectively in 32% yield. Compound **241** is the result of a 6-*endo* cyclization. The sequence was completed by simultaneous debenzylation and debromination with Na/NH₃. Finally, desilylation and acetonide hydrolysis gave **243**, Scheme 39.

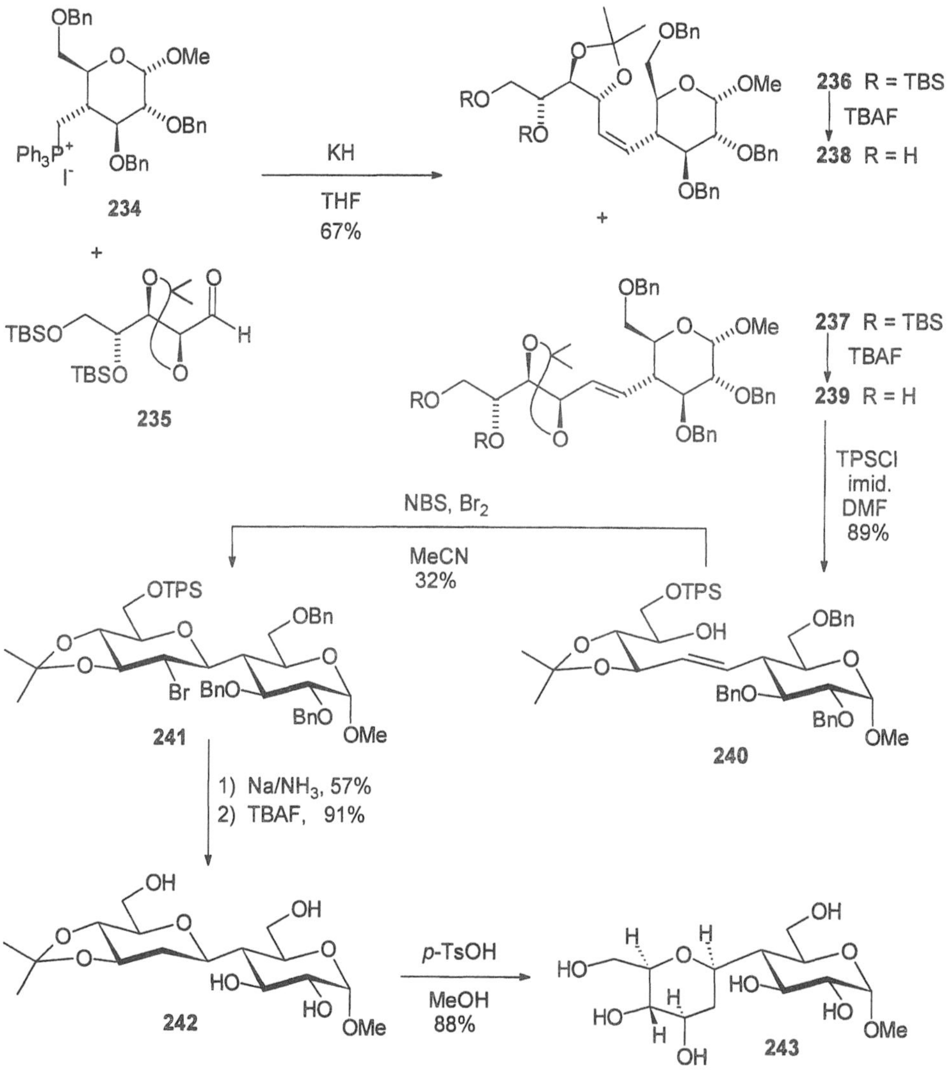

Scheme 39

Fraser-Reid has joined the primary alcoholic carbon of pyranose and furanose sugars via Wittig chemistry to produce linked sugars. Scheme 40 shows the work. The acid **244** was converted to the imidazolide **245** which was then reacted with methylenetriphenyl-phosphorane to give the Wittig reagent **246**. The ylide **246** was then exposed to the aldehyde **79** to give the linked sugar **247**. Alternatively, the ylide **246** could be reacted with the aldehyde **249** to give the 5-6 linked sugar.[37]

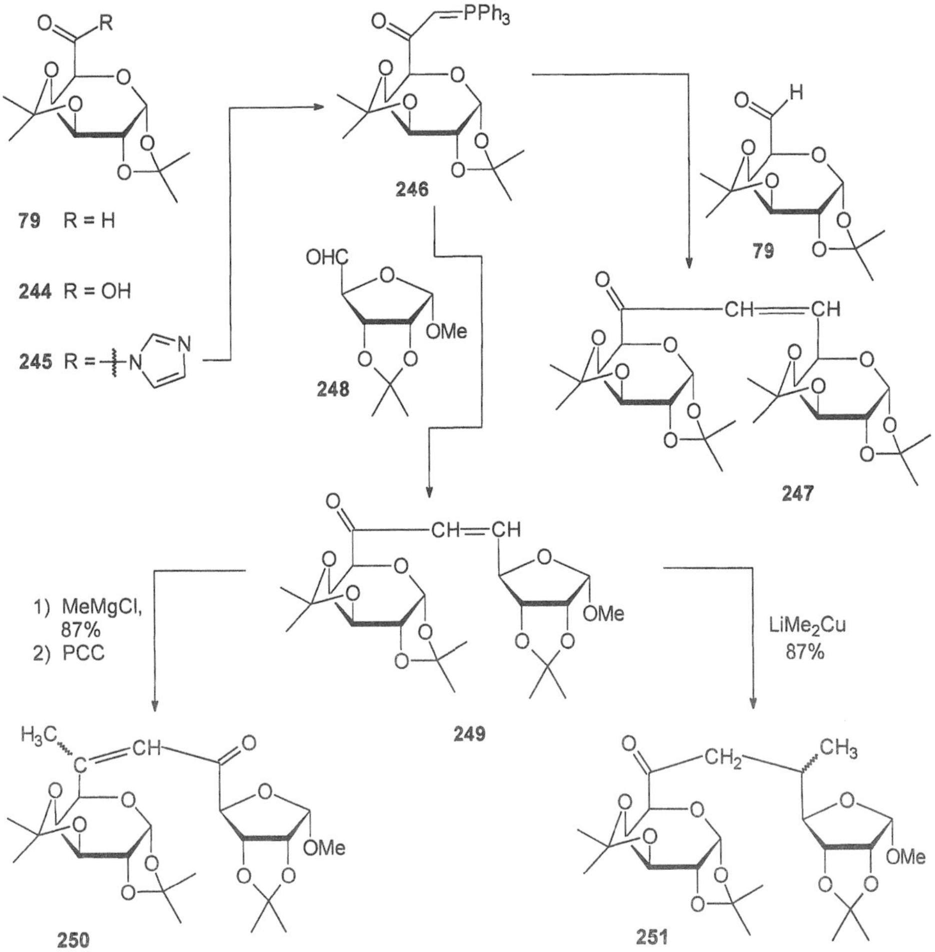

Scheme 40

The α,β unsaturated aldehyde **249** could be reacted with lithium dimethyl cuprate to give a 3:1 mixture of isomers **251**. Alternatively, **249** when exposed to methyl magnesium chloride gave after treatment with PCC the product of carbonyl transposition **250** as a mixture of isomers about the terminal olefin.

III. SYNTHETIC APPROACHES TO C-TRISACCHARIDES

A. ALDOL APPROACH TO A C-TRISACCHARIDE

Kishi et al.[38] used an aldol approach in the construction of a C-trisaccharide that was an analogue related to the Type II O(H) blood group determinant. The work is shown below in Schemes 41-44. The C-4 branched sugar **252** undergoes a carbon-carbon bond forming aldol condensation with the aldehyde **253** to give a mixture of aldols **254**. Desilylation with

TBAF affected cyclization to the hemiketals which when exposed to a Lewis acid and methanethiol gave the thioketals 255. Stereoselective reduction with tributyltin hydride was followed by Swern oxidation to furnish the 2'-deoxy-C-disaccharidic ketone 256.

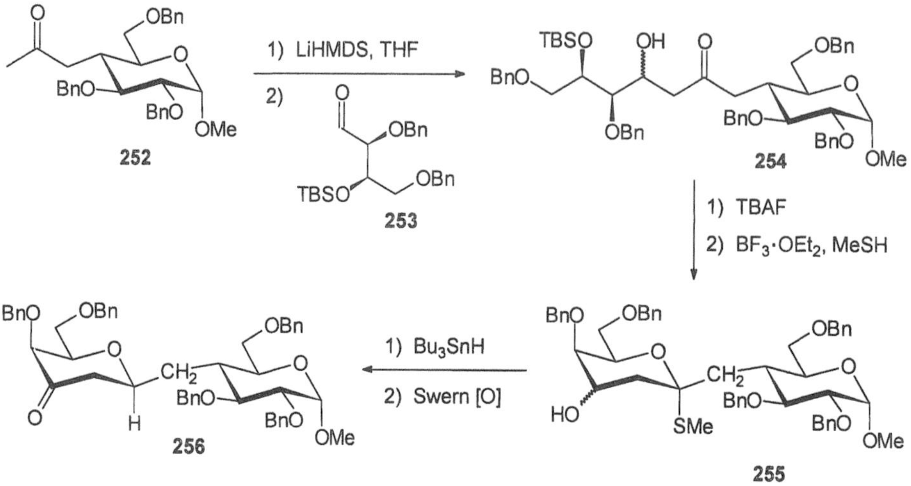

1) LiHMDS, THF
2)

253

1) TBAF
2) BF₃·OEt₂, MeSH

1) Bu₃SnH
2) Swern [O]

Scheme 41

The stage was now set to couple the C-disaccharide with an appropriate acceptor to give a C-trisaccharide. Regioselective enolate formation of 256 using LiHMDS gave the kinetic enolate which was treated with TMEDA and transmetallized with MgBr₂ and then exposed to aldehyde 257 to give a mixture of equatorial aldol products 257 (1:2 ratio) which were then dehydrated to give the isomeric olefins 259.

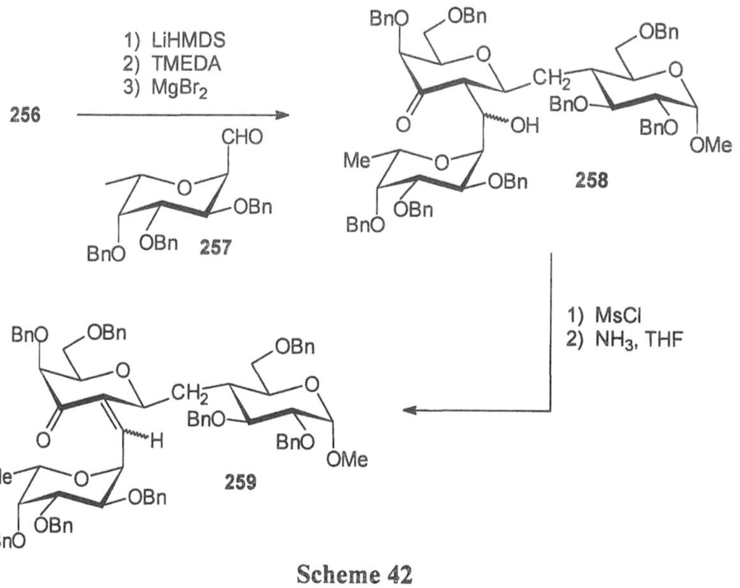

256

1) LiHMDS
2) TMEDA
3) MgBr₂

1) MsCl
2) NH₃, THF

Scheme 42

1,4-Addition of hydride to the enone system (Scheme 43) gave an enolate which was reduced from an axial direction to give the equatorial product 260. The sequence was completed by sodium borohydride reduction of the ketone and deprotection to give the C-trisaccharide 262.

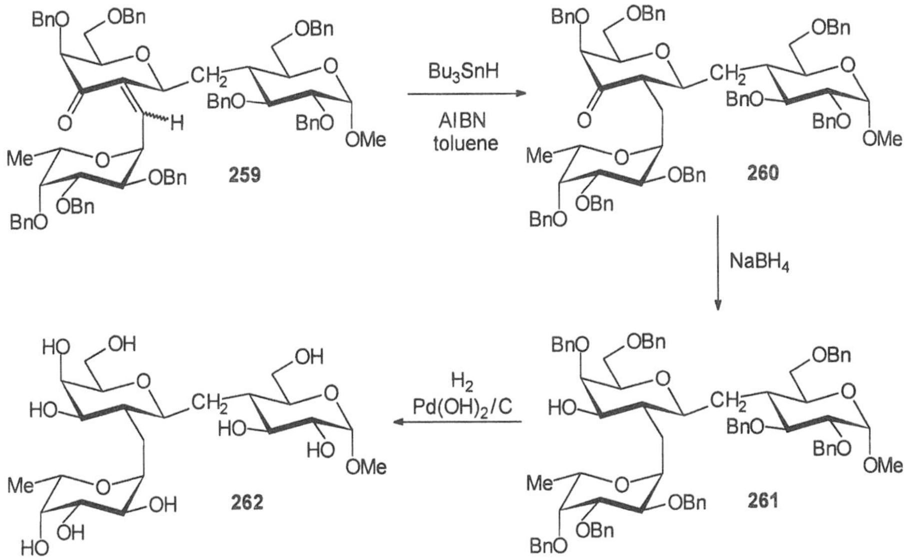

Scheme 43

A similar sequence (Scheme 44) was carried out starting with **263** to give compound **264**. The sequence differed slightly from that above since the ketone moiety was not only reduced, but the resulting hydroxy deoxygenated via a radical pathway to give 3-deoxy-*C*-trisaccharide **264**.

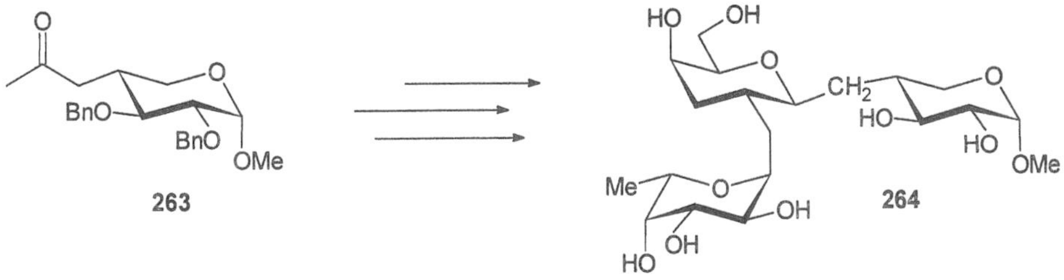

Scheme 44

IV. CONCLUSION

There have been several different approaches to *C*-disaccharide construction. Anionic approaches are well suited for connecting a carbon atom to the anomeric center, especially when the anomeric center exists as a lactone, since stereoselective reduction will then lead to a β-*C*-disaccharide. *C*-1 Nitro glycosides have also been used to gain access to the β-*C*-disaccharides since the starting equatorial nitro sugars are readily available. Radical methods complement the anionic approaches nicely because the intermolecular addition of anomeric radicals to olefins occurs almost exclusively in an axial sense, the only requisite being that the radical acceptor must be easily transformable into another sugar moiety. Cycloadditions have also seen some utility in *C*-disaccharide synthesis and cationic approaches have seen limited use, although some interesting branched bis-(1,1) sugars have been quickly assembled through the use of this tactic.

V. REFERENCES

1. Armstrong, R.W.; Beau, J.-M.; Cheon, S.H.; Christ,,W.J.; Fujioka, H.; Ham, W.-H.; Hawkins, L.D.; Jin, H.; Kang, S.H.; Kishi, Y.; Martinelli, M.J.; McWhorter, Jr., W.W.; Mizuno, M.; Nakata, M.; Stutz, A.E.; Talamas, F.X.; Taniguchi, M.; Tino, J.A.; Ueda, J.-i.; White, J.B.; Yonaga, M. *J. Am. Chem. Soc.* **1989**, *111*, 7525 and references cited therein.

2. Rouzad, D.; Sinaÿ, P. *J. Chem. Soc., Chem. Commun.* **1983**, 1353.

3. Lesimple, P.; Beau, J.-M.; Jaurand, G.; Sinaÿ, P. *Tetrahedron Lett.* **1986**, *27*, 6201.

4. Beau, J.-M.; Sinaÿ, P. *Tetrahedron Lett.* **1985**, *26*, 6189.

5. If the desulfinylation proceeds via radical intermediates then the axial radical would be formed preferentially, see Chapter 7 for additional discussion.

6. Preuss, R.; Schmidt, R.R. *J. Carbohydr. Chem.* **1991**, *10*, 887.

7. Schmidt, R.R.; Preuss, R. *Tetrahedron Lett.* **1989**, *30*, 3409.

8. Daly, S.M.; Armstrong, R.W. *Tetrahedron Lett.* **1989**, *30*, 5713.

9. Corey, E.J.; Fuchs, P.L. *Tetrahedron Lett.* **1972**, 3769.

10. Sutherlin, D.P.; Armstrong, R.W. *Tetrahedron Lett.* **1993**, *34*, 4897.

11. Wang, Y.; Babirad, S.A.; Kishi, Y. *J. Org. Chem.* **1992**, *57*, 468.

12. See: Jin, H.; Uenishi, J.; Christ, W.J.; Kishi, Y. *J. Am. Chem. Soc.* **1986**, *108*, 5644 and Takai, K.; Tagashira, M.; Kuroda, T.; Oshima, K.; Utimoto, K.; Nozakai, H. *J. Am. Chem. Soc.* **1986**, *108*, 6048.

13. Dyer, U.C.; Kishi, Y. *J. Org. Chem.* **1988**, *53*, 3383.

14. Goekjian, P.G.; Wu, T.-C.; Kang, H.-Y.; Kishi, Y. *J. Org. Chem.* **1987**, *52*, 4823.

15. Aebischer, B.; Meuwly, R.; Vasella, A. *Helv. Chim. Acta* **1984**, *67*, 2236.

16. Baumberger, F.; Vasella, A. *Helv. Chim. Acta* **1983**, *66*, 2210.

17. Martin, O.R.; Lai, W. *J. Org. Chem.* **1990**, *55*, 5188.

18. Giese, B.; Witzel, T. *Angew. Chem. Intl. Ed. Engl.* **1986**, *25*, 450.

19. Giese, B.; Hoch, M.; Lamberth, C.; Schmidt, R.R. *Tetrahedron Lett.* **1988**, *29*, 1375.

20. Bimwala, R.M.; Vogel, P. *J. Org. Chem.* **1992**, *57*, 2076.

21. For an overview on the chemistry of naked sugars see: Vogel, P.; Fattori, D.; Gasparini, F.; Le Drian, C. *Synlett* **1990**, 173.

22. Motherwell, W.B.; Ross, B.C.; Tozer, M.J. *Synlett* **1989**, 68.

23. Martin, O.R.; Xie, F.; Kakarla, R.; Benhamza, R. *Synlett* **1993**, 165.

24. Lopez, J.C.; Fraser-Reid, B. *J. Am. Chem. Soc.* **1989**, *111*, 3450.

25. Xin, Y.C.; Mallet, J.-M.; Sinaÿ, P. *J. Chem. Soc., Chem. Commun.* **1993**, 864.

26. Vauzeilles, B.; Cravo, D.; Mallet, J.-M.; Sinaÿ, P. *Synlett* **1993**, 522.

27. Danishefsky, S.J.; Maring, C.J.; Barbachyn, M.R.; Segmuller, B.E. *J. Org. Chem.* **1984**, *49*, 4564.

28. Jurczak, J.; Bauer, T.; Jarosz, S. *Tetrahedron Lett.* **1984**, *25*, 4809.

29. Dawson, I.M.; Johnson, T.; Paton, R.M.; Rennie, R.A.C. *J. Chem. Soc., Chem. Commun.* **1988**, 1339.

30. Ichikawa, Y.; Isobe, M.; Konobe, M.; Goto, T. *Carbohydr. Res.* **1987**, *171*, 193.

31. Tsukiyama, T.; Peters, S.C.; Isobe, M. *Synlett* **1993**, 413.

32. de Raadt, A.; Stütz, A.E. *Carbohydr. Res.* **1991**, *220*, 101.

33. Lay, L.; Nicotra, F.; Panza, L.; Ruso, G.; Caneva, E. *J. Org. Chem.* **1992**, *57*, 1304.

34. Carcano, M.; Nicotra, F.; Panza, L.; Russo, G. *J. Chem. Soc., Chem. Commun.* **1989**, 642.

35. Babirad, S.A.; Wang, Y.; Kishi, Y. *J. Org. Chem.* **1987**, *52*, 1370.

36. Armstrong, R.W.; Teegarden, B.R. *J. Org. Chem.* **1992**, *57*, 915.

37. Jarosz, S.; Mootoo, D.; Fraser-Reid, B. *Carbohydr. Res.* **1986**, *147*, 59.

38. Haneda, T.; Goekjian, P.G.; Kim, S.H.; Kishi, Y. *J. Org. Chem.* **1992**, *57*, 490.

Chapter 9

SYNTHESIS OF ALKYL *C*-GLYCOSIDE NATURAL PRODUCTS

I. INTRODUCTION

This chapter will deal with the synthetic approaches towards natural products that are in their own right *C*-glycosides or contain structural features that are highly reminiscent of *C*-glycosides. In nature these compounds can exist as *C*-pyranosides with either the α or β configuration.

Figure 1

The corresponding *C*-furanosides have also been found in several natural products as well and they also can exist in either configuration. The subject of naturally occurring *C*-glycosides has been reviewed many years ago.

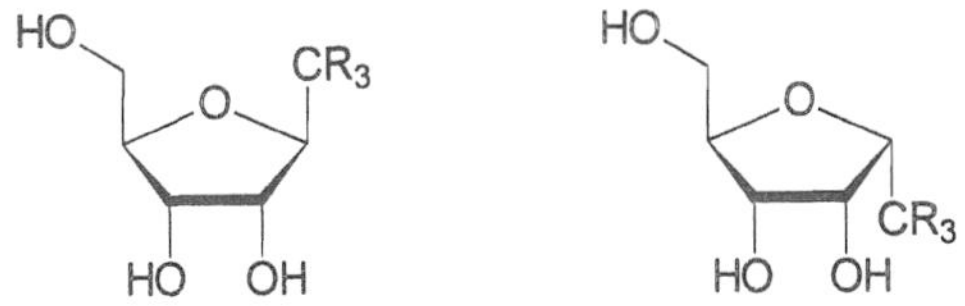

Figure 2

The aglycone can be aliphatic or aromatic in nature and many biologically active naturally occurring *C*-glycosides possess an aromatic aglycone. This subject is dealt with in a separate chapter.

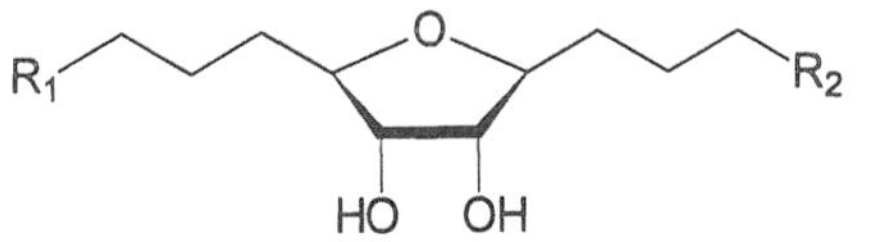

Figure 3

Some natural products that contain a *C*-furanoside also have carbon functionality on both *C*-1 and *C*-4. This bis-carbon branched furan occurs in several natural products and the topic of selected synthetic approaches toward this class of compounds has been recently reviewed.[1] The organization of the chapter is such that approaches to *C*-pyranosides will be dealt with first and this will be followed by approaches to *C*-furanosides.

II. ALKYL *C*-PYRANOSIDE NATURAL PRODUCTS

A.   AMBRUTICIN
1.   Isolation and Structure Determination

Workers from Warner-Lambert isolated ambruticin[2] from *Polyangium cellulosum fulvum* in 1977 and it was shown to possess interesting antifungal properties.[3] Exact mass measurement gave a molecular formula of $C_{28}H_{42}O_6$ and further spectral elucidation pointed to the structure (1) shown below.   Single X-ray determination then further supported the structure and absolute configuration.[4] Degradation studies[5] have also been carried out on ambruticin in order to determine its absolute configuration. The salient structural features include a polyene aglycone chain that has a cyclopropane in the array of double bonds. The anomeric C-C bond is β and the glycosidic group is a 4-deoxy gluco unit.   The *C*-6 hydroxyl, though, has been homologated by one carbon and at the other end of the chain is a cyclic enol ether with two stereocenters.   Epiambruticin was also isolated; it differs only from ambruticin in the configuration of the *C*-3 hydroxyl group.

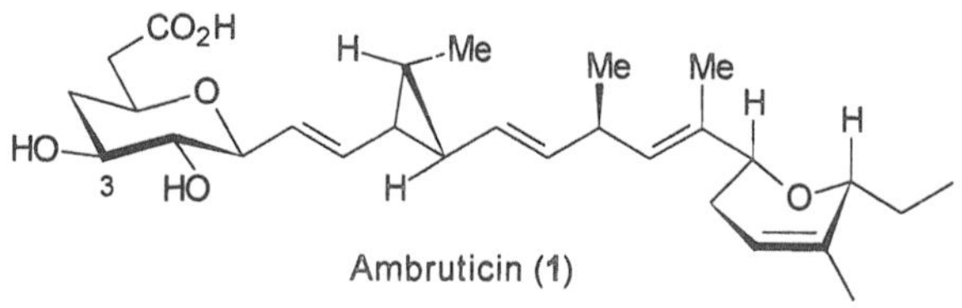

Ambruticin (1)

Figure 4

2.   Total Synthesis of Ambruticin

In 1990 Kende *et al.* were the first to synthesize naturally occurring ambruticin.[6] They chose to disconnect the molecule at the anomeric carbon and on the other side of the cyclopropane as shown below.

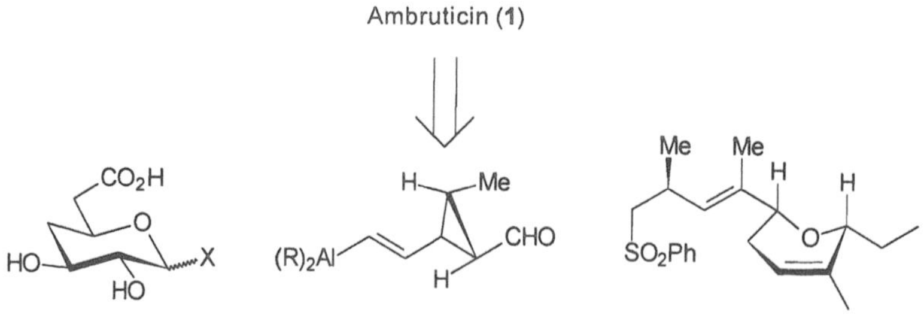

Ambruticin (1)

Scheme 1

A key point in their strategy was an aluminum based coupling of an anomeric fluoride. The right hand side was attached by the use of a Julia coupling to give the intact carbon skeleton.   The synthesis began with the preparation of the sugar portion.   Diol 2 was selectively silylated and the *C*-4 oxygen removed via a Barton type process. Deprotection and oxidation gave an acid which underwent Arndt-Eistert homologation to 5. Anomeric hydrolysis, via a two-step protocol, gave the lactol which upon treatment with DAST yielded the requisite anomeric fluorides 7.

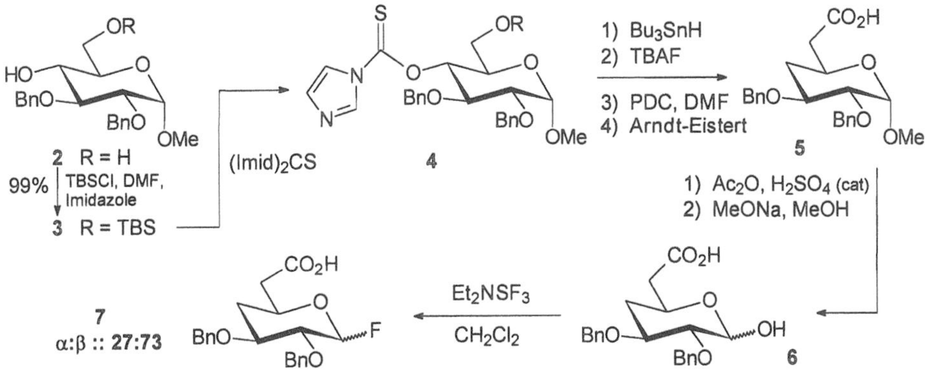

Scheme 2

The cyclopropane moiety was assembled starting from the optically pure diester **8**. Double deprotonation and subsequent reaction with 1,1-bromochloroethane then provided **9**. Selective hydrolysis was followed by conversion of the acid to the aldehyde by a two step procedure to provide **10**. The formed aldehyde then served as a handle for introduction of an acetylene unit. Olefination gave the ester **11** and reduction, tritylation, and treatment with *n*-butyllithium gave the requisite acetylene **12**.

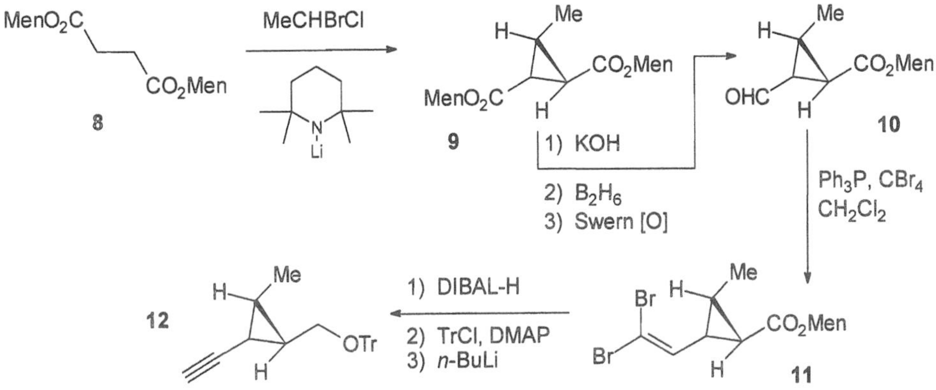

Scheme 3

With the central and left hand side of the molecules in hand they now set out to assemble these two pieces.

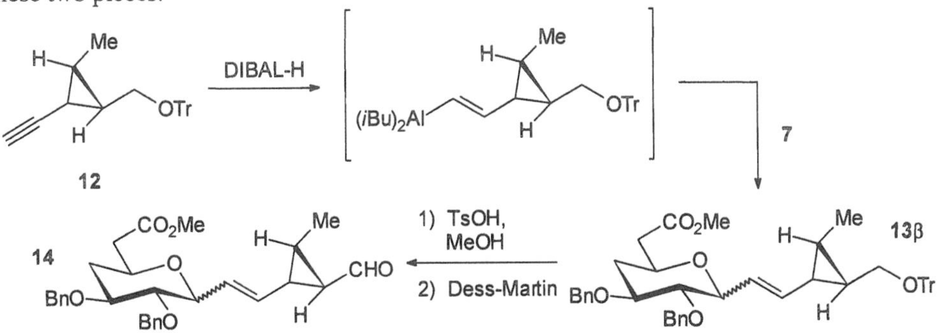

Scheme 4

Hydroalumination of the triple bond gave an *E*-aluminate which was then coupled with the anomeric fluorides **7** to give a 77% yield of anomers **13** (α:β of 1:1.8). Detritylation and oxidation then furnished **14**.

The right hand side pyran ring was constructed by the use of cycloaddition chemistry. Hetero Diels-Alder reaction of the diene **15** with methyl glyoxylic acid provided a mixture of isomers of which the desired *cis*-isomer **17** was formed in 65% yield. The mixture of isomers was derivatized and separated. The *cis*-isomer was then resolved with (+)-α-phenylethylamine to give optically pure **18**. Exposure of **18** to *trans*-propenyltrimethyltin in the presence of Pd(0) catalysis then gave the unsaturated ketone **19**. Grignard reaction gave a tertiary alcohol which was acetylated to provide **21**. Enolate formation followed by trapping with *t*-butyldimethylsilyl chloride gave the silyl enol ether **22** which underwent Claisen rearrangement to give, after hydrolysis, acid **24**. Ester formation was followed by enolate formation and reaction with *p*-toluenesulfonyl fluoride then produced sulfone **26**, after decarboxylation.

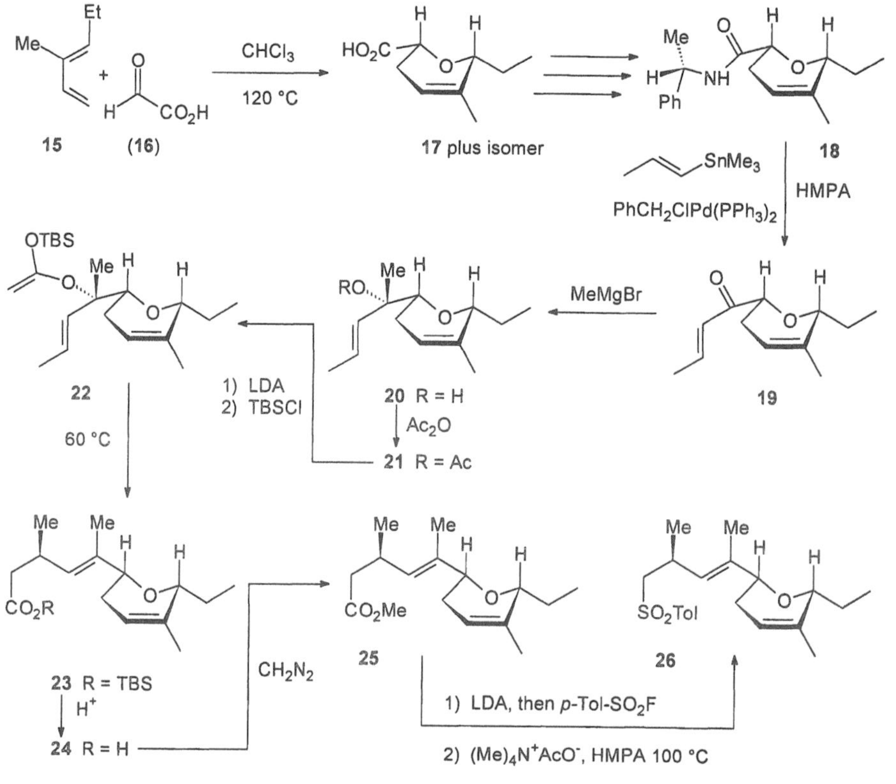

Scheme 5

Coupling of the two fragments was accomplished by condensing the anion derived from **26** with aldehyde **14** to give a diastereomeric mixture of hydroxy sulfones. Reductive elimination then gave the required *E*-olefin **27**. The Julia coupling proceeded in good overall yield. The next step was deprotection and saponification of the ester to liberate the acid group and Birch reduction caused deprotection thereby furnishing natural ambruticin (**1**). The above synthesis has the merit of being highly convergent. Key steps involve aluminum mediated stereoselective β-*C*-glycoside formation and Claisen rearrangement to setup both the olefin geometry and configuration of the methyl group on the left hand portion of the molecule.

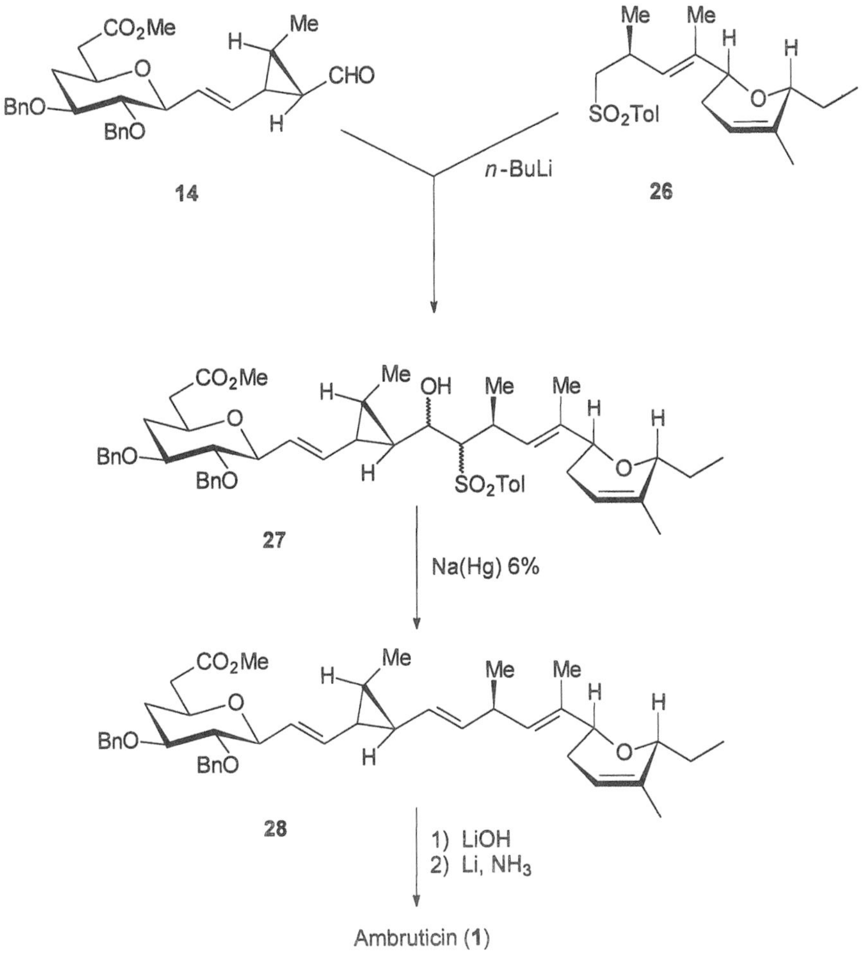

Scheme 6

Davidson has also made inroads towards the synthesis of ambruticin.[7] Their approach involved the use of Wittig chemistry on the 6-cyano lactol **29** with the phosphonate-sulfone **30** to provide a mixture of anomers **31** with an α:β of 3:2. The cyano group was then converted to the ester **32**, a piece corresponding to the left side of ambruticin.

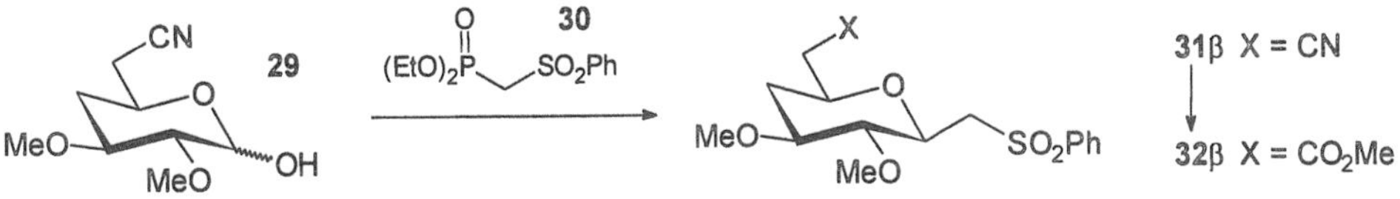

Scheme 7

In 1991, just as Kende's total synthesis[6a] was completed, Davidson and co-workers[8] published a strategy for assembly of the pyran ring based on the Claisen rearrangement. Their model studies began with the tertiary alcohol **33**. Acetylation with the anhydride **34** gave **35** and enolate formation followed by trapping with trimethylsilyl chloride at low temperature gave **36**. Rearrangement occurred at room temperature to give, after hydrolysis, **37**. Heating of **37** with sodium hydrogen carbonate in DMF then yielded the requisite sulfone **38**. To further their model they then coupled the anion of **38** with the aldehyde **39** to provide a mixture of isomers **40**. Treatment with sodium amalgam finally gave the expected diene **41**. Their model study bears a striking resemblance to the total synthesis recently described by Kende.

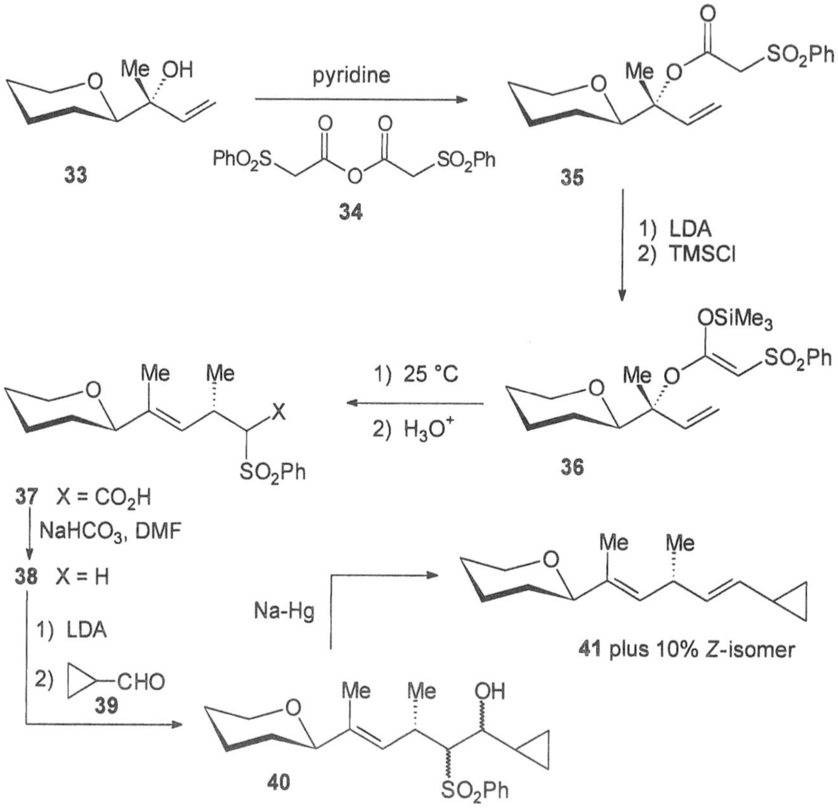

Scheme 8

The two above Schemes show how the Davidson group constructed the left hand and a model of the right hand ring of ambruticin. Scheme 9 shows their approach[9] toward the central fragment of ambruticin, the cyclopropane moiety **45**. Their approach relied on the acid **42**, which is available in optically pure form for either isomer. Iodolactonization with iodine in acetonitrile gave predominantly the *trans* lactone **43**. Ring opening with lithium *t*-butoxide then gave the epoxide **44** which results when the free alkoxide cyclizes onto the iodine bearing carbon. The cyclopropane ring was formed by treating the ester with LDA to furnish the *trans* cyclopropane exclusively in about 50% overall yield from acid **42**.

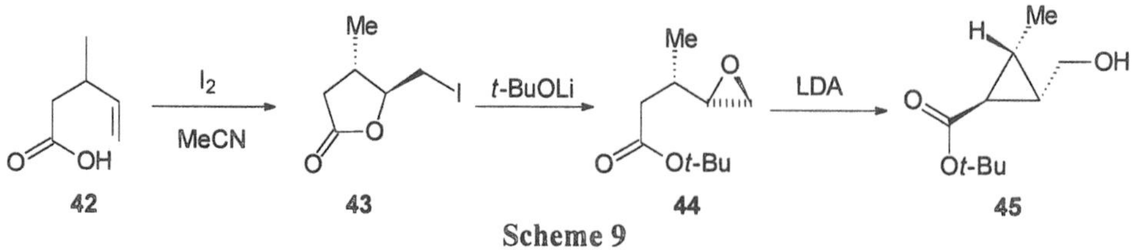

Scheme 9

Sinaÿ has also made progress towards the synthesis of ambruticin.[10] His work began with the assembly of the western fragment and as a result they developed a general method for constructing β-E-vinyl-C-glycosides as shown in Scheme 10. The initial carbon-carbon bond formation relies on the condensation of an acetylenic anion with a sugar lactone to provide a mixture of isomers **47**. This mixture is of no consequence since stereoselective reduction then furnishes the requisite β-alkynyl C-glycoside **48**. Reduction with lithium aluminum hydride in dimethoxyethane then provides access to the desired *trans* olefin **49**.

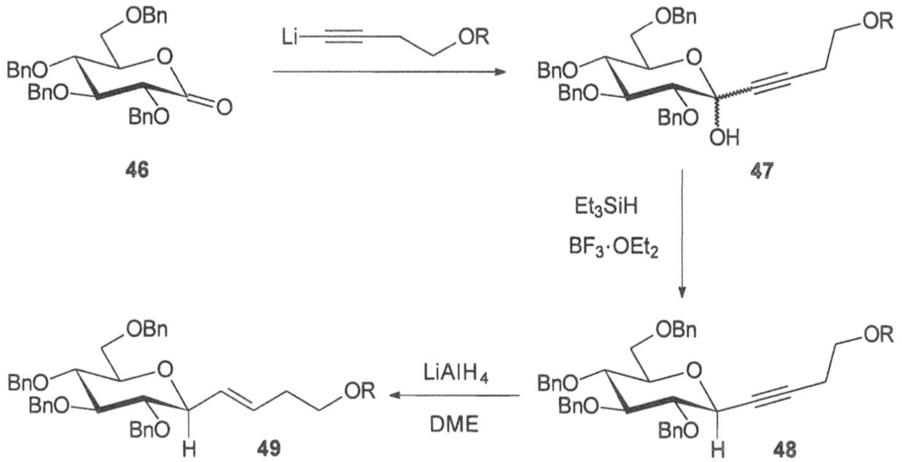

Scheme 10

A similar sequence was carried out using the dibromide **50** as the acetylene precursor. Treatment of such olefins with *n*-butyllithium leads directly to the lithium acetylide anion which can then be condensed with the lactone **46** to give after suitable manipulations the C-glycoside **51**.

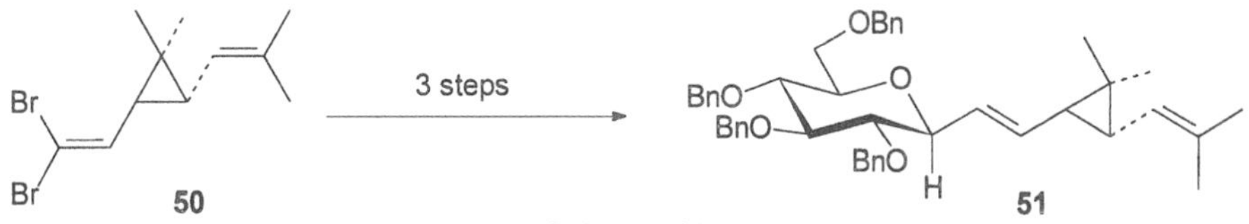

Scheme 11

B.   PSEUDOMONIC ACIDS

1.   Isolation and Structure

The family of pseudomonic acids is comprised of several members. Pseudomonic acid A and B are highly oxygenated naturally occurring C-glycosides. The central pyran ring has two carbon chains, one containing an ester and the other an epoxide. The structure is reminiscent of a sugar and this is reinforced by the presence of the two *syn* hydroxyl groups on the ring. It is not surprising that synthetic approaches towards this molecule have utilized a strategy based on C-glycosides. The pseudomonic acids A (**52**) and B (**53**) are produced by *Pseudomonas fluorescens* (NCIB 10586).[11] Pseudomonic acid A has good activity against several strains of bacteria.[12] Recently, pseudomonic acid C (**54**), a compound missing the epoxide function, was isolated[13] and showed good biological activity under acidic to basic conditions. This is contrasted to pseudomonic acid A which undergoes rearrangement to an inactive species under these conditions.[14] Much of the synthetic activity has been directed towards pseudomonic acid C (**54**) and its methyl ester (**55**).

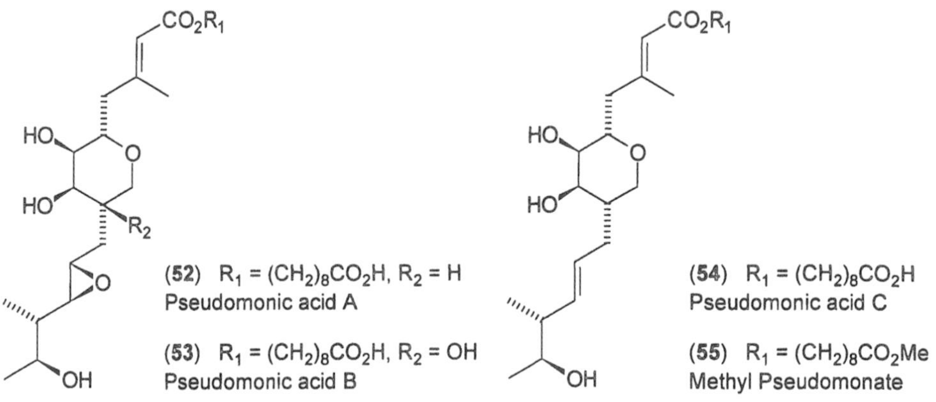

(52) $R_1 = (CH_2)_8CO_2H$, $R_2 = H$
Pseudomonic acid A

(53) $R_1 = (CH_2)_8CO_2H$, $R_2 = OH$
Pseudomonic acid B

(54) $R_1 = (CH_2)_8CO_2H$
Pseudomonic acid C

(55) $R_1 = (CH_2)_8CO_2Me$
Methyl Pseudomonate

Figure 5

2.   Synthesis of Pseudomonic Acids

Kozikowski *et al.* began their synthesis[15] with the diester **56**. Decarboxylation and reduction then furnished **57**. Hydrolysis of the ketal gave an intermediate hemiacetal that was dehydrated to **58**. Benzoloxyselenation furnished the expected product **59** and oxidative fragmentation then provided **60**. Osmylation was followed by ketalization, hydrogenolysis of the benzyl groups, and silylation to give **61** (Scheme 12).

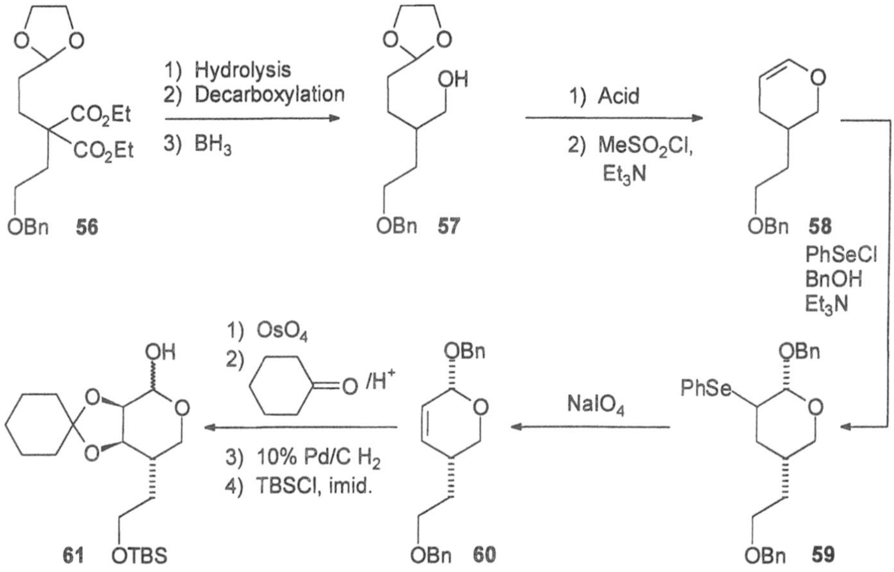

Scheme 12

Wittig reaction of **61** in a sealed tube in acetonitrile with **62** gave a mixture of *C*-glycosides of which the major product was the desired one, **63**. The stage was now set for introduction of the northern side chain and this was accomplished by Horner-Emmons reaction with **64** to give a 4:1 mixture of separable *E:Z* isomers. The southern side chain was introduced in a similar fashion. Desilylation, PCC oxidation, and Wittig reaction with ylide **67** produced **68** along with the corresponding *Z*-olefin (*E:Z* of 60:40). Acid hydrolysis of the ketal then gave racemic pseudomonic acid C (**54**), Scheme 13.

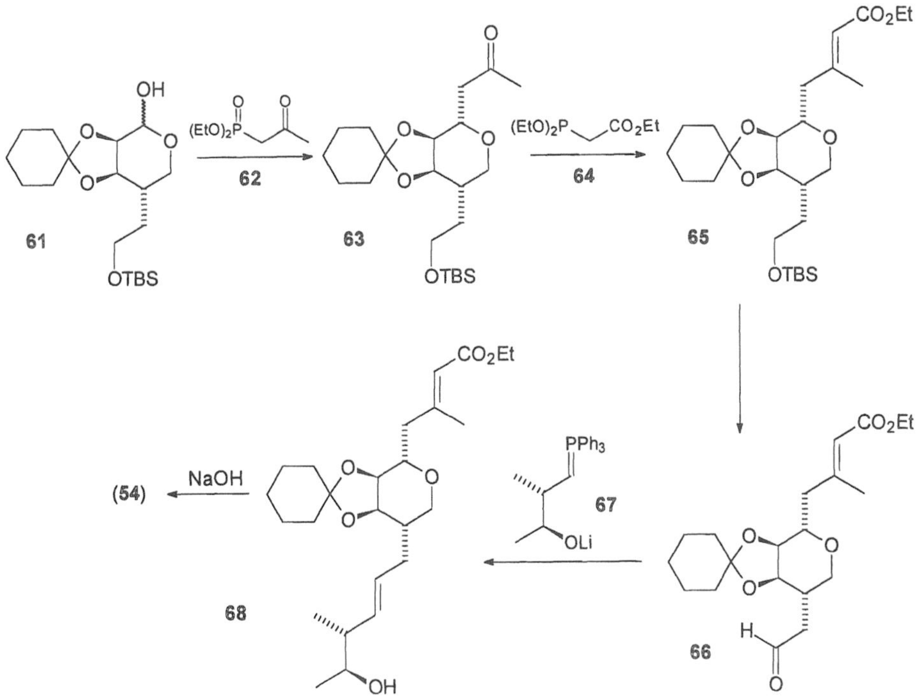

Scheme 13

Kozikowski[16] also used the chiron approach to synthesize the methyl ester of pseudomonic acid B. Lewis acid catalyzed allylation of the anomeric acetate **69** gave a mixture of C-glycosides with **70** predominating in a 80:1 ratio. It is noteworthy that the silyl ether did not survive the reaction conditions. Resilylation and oxymercuration/demercuration gave, after protection, **71**. Treatment with TBAF and Collins oxidation yielded the ketone **72**. Addition of allyl magnesium bromide occurred from the less hindered face to give **73** exclusively. Silylation and oxidative cleavage of the remaining double bond then furnished the aldehyde **74**.

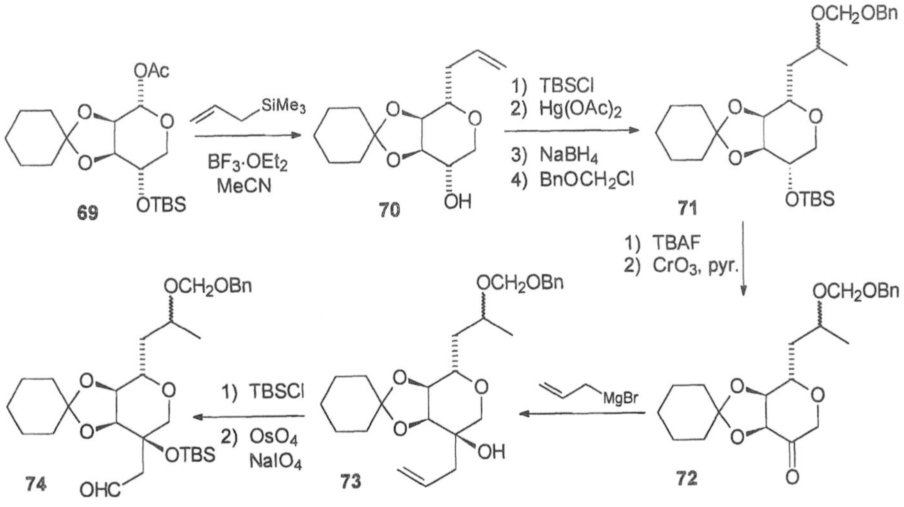

Scheme 14

The southern *E*-olefin was installed using Julia coupling technology to give 76. Lithium ammonia cleavage of the benzyloxy group was followed by oxidation to the ketone 77 which underwent Wittig olefination with the phosphonate 78. Desilylation and ketal hydrolysis then yielded the product methyl pseudomonate B (79).

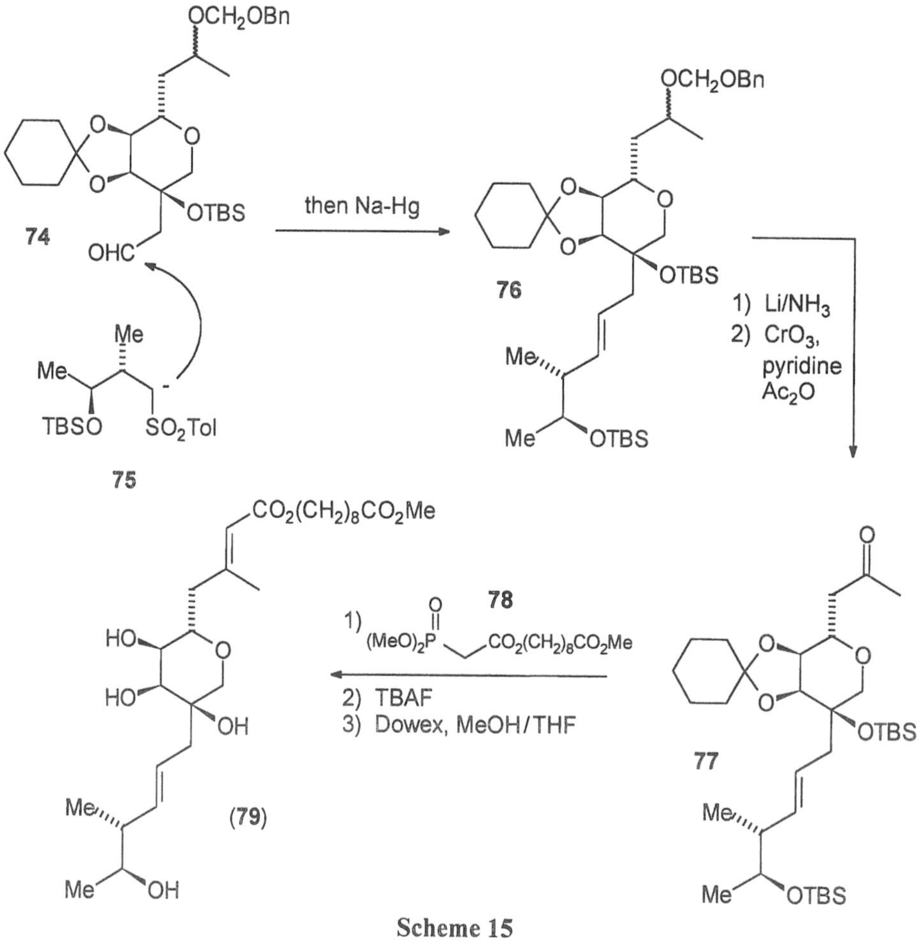

Scheme 15

Keck and co-workers[17] also employed the chiron approach in their synthesis of pseudomonic acid C and one of the main features of their synthesis is a free radical based key carbon-carbon bond forming reaction. The sequence began with the conversion of L-lyxose into 81 via a two step protocol. The free hydroxyl was converted into the xanthate and exposure to free radical allylation conditions gave 83 which was then converted to aldehyde 84. The aldehyde was reduced and the primary hydroxyl converted to the sulfone 85 in the usual way. The sulfone and the aldeyhde 86 were condensed in a Julia type coupling and elimination of the formed mesylate then gave the *trans* olefin 89. With the southern tail in position they were now poised to install the northern chain. After some experimentation they found that Wittig methodology was well suited. After a lithium ammonia deprotection of the anomeric benzyl group, reaction of the lactol with the ylide 90 gave a mixture of α and β isomers (6:1) in favor of the desired isomer. Concomitant deprotection of the silyl and dioxolane groups was achieved with 80% acetic acid and saponification of the methyl ester then furnished optically pure pseudomonic acid C (54).

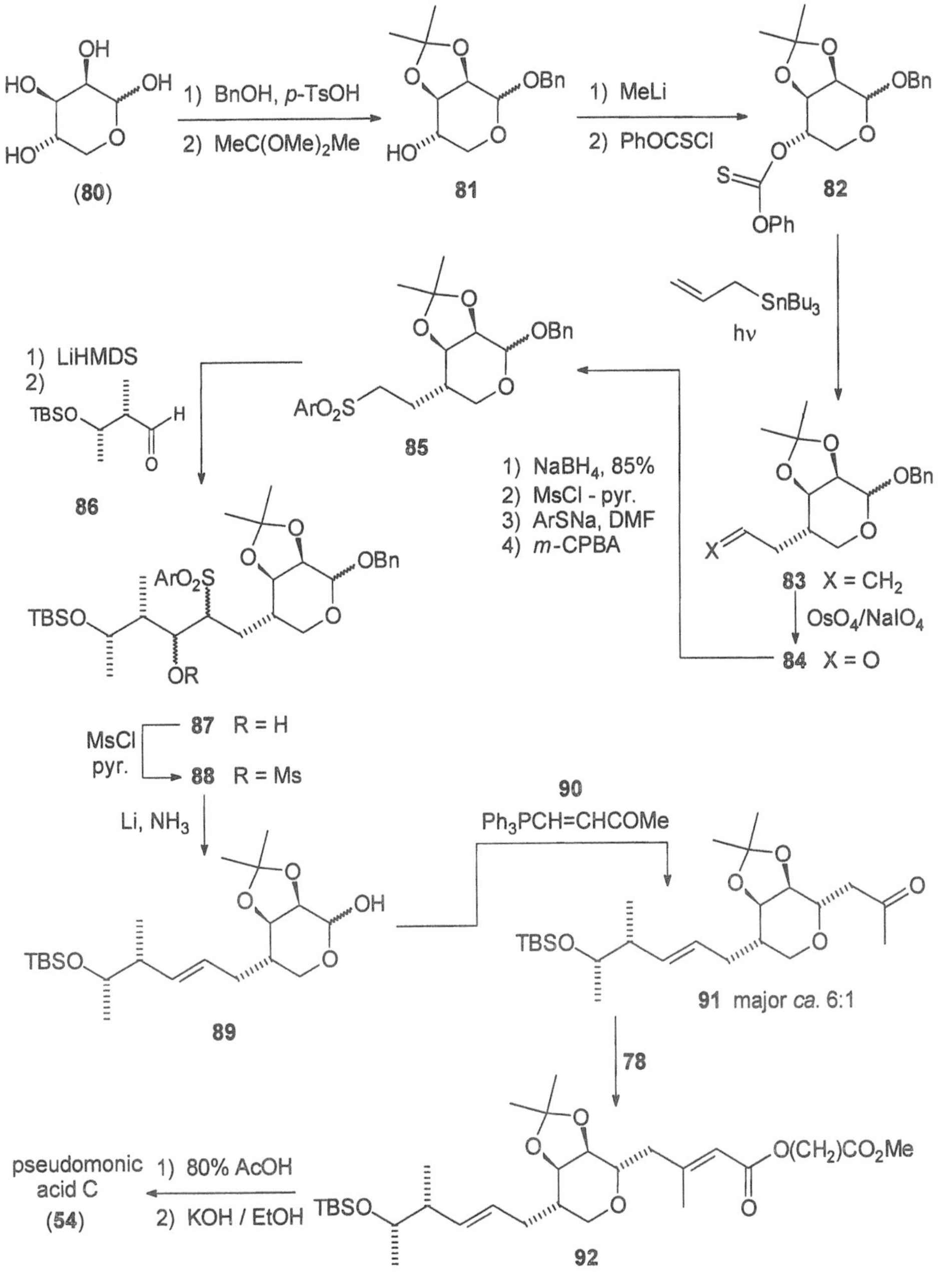

Scheme 16

Curran and Suh[18] used a combination of the Ireland ester enolate rearrangement and palladium mediated coupling to install the two carbon chains of pseudomonic acid C. Their approach begins with the glycal 93. Acetylation and silyl enol ether formation led to 94 and exposure to mild heat (60°C) followed by treatment with potassium fluoride gave acid 95. It is worthy to note that the acid at C-1 is poised for selective manipulation in the presence of the acetate at C-4 which itself is ready to be replaced by an appropriate nucleophile. The acid was converted into the methyl ketone and π-allyl palladium mediated displacement, with retention of configuration, with the anion 97 furnished 99, after reductive cleavage of the sulfone. With the beginning of the two chains in place Curran now decided to introduce the vicinal hydroxyl function. *Syn* hydroxylation and ketalization now provided 100 and Horner-Emmons olefination gave the ester 101. Selectivity in the reduction between the two esters was achieved with the use of the bulky "ate" complex 102 (which is derived from treatment of DIBAL with *n*-BuLi) to provide alcohol 103 which was then oxidized to aldehyde 104. Compound 104 is a key intermediate in the Kozikowski's synthesis[15] of pseudomonic acid C and therefore this work constitutes a formal total synthesis of pseudomonic acid C.

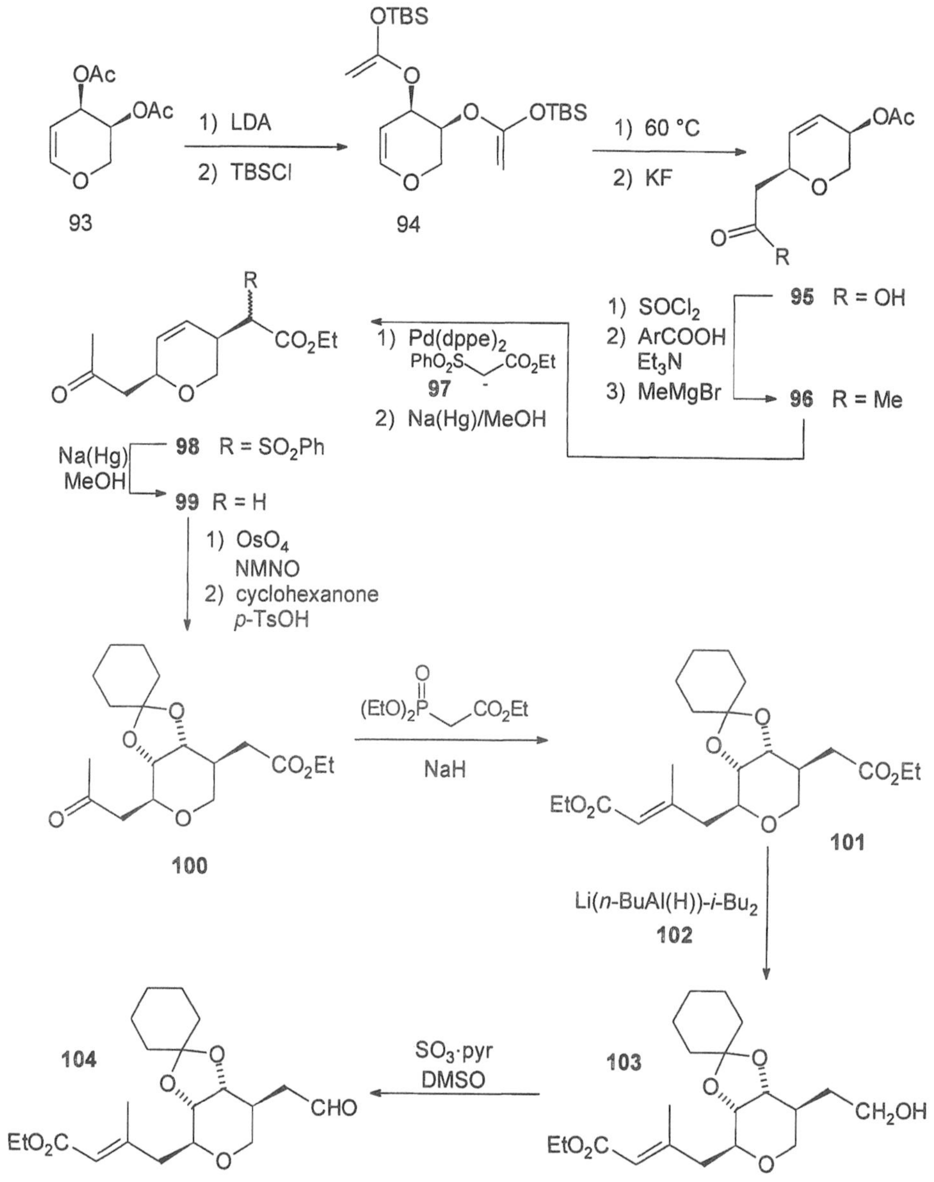

Scheme 17

Fleet has also used the chiron approach to synthesize pseudomonic acid C. His strategy relied on coupling two optically active fragments via Wittig chemistry. The side chain moiety is derived from L-arabinose while the ring portion is derived from D-arabinose. The salt **112** was synthesized as shown in Scheme 18.[19] The methyl furanoside **105** was selectively derivatized at *C-5* and the resulting sulfonate displaced with iodide which was then reduced to give the *C-5* deoxy sugar **107**. Treatment of the diol under Mitsunobu conditions gave the epoxide **108**, stereospecifically. The epoxide was then regioselectively opened with cuprate **109** to give **110**. The sequence was completed by anomeric hydrolysis, reduction, and conversion of the primary alcohol to the iodide **111**. Treatment of this iodide with triphenylphosphine then gave the salt **112**.

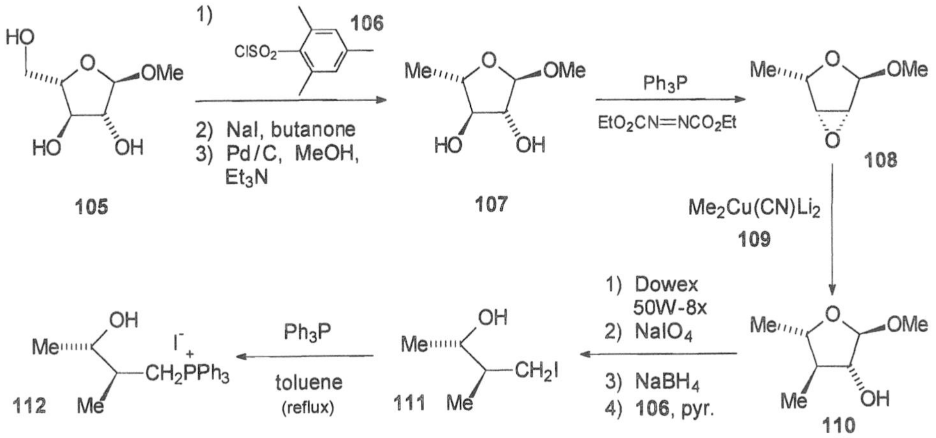

Scheme 18

The pyran portion[20] of pseudomonic acid C was itself derived from D-arabinose (**113**). The anomeric position was deoxygenated via a two step protocol and the triol was pyrolyzed via its orthoformate to allylic alcohol **115**. Treatment of **115** with the dimethyl acetal of *N,N*-dimethyl acetamide and heating caused Claisen rearrangement to the amide **116** in 82% yield. With the beginnings of one of the carbon chains in place Fleet now used the existing functionality to introduce a second allylic alcohol to affect another Claisen rearrangement to install the second requisite chain. Iodocyclization of **116** occurred in the presence of iodine and exposure to DBU then furnished the lactone **118**.

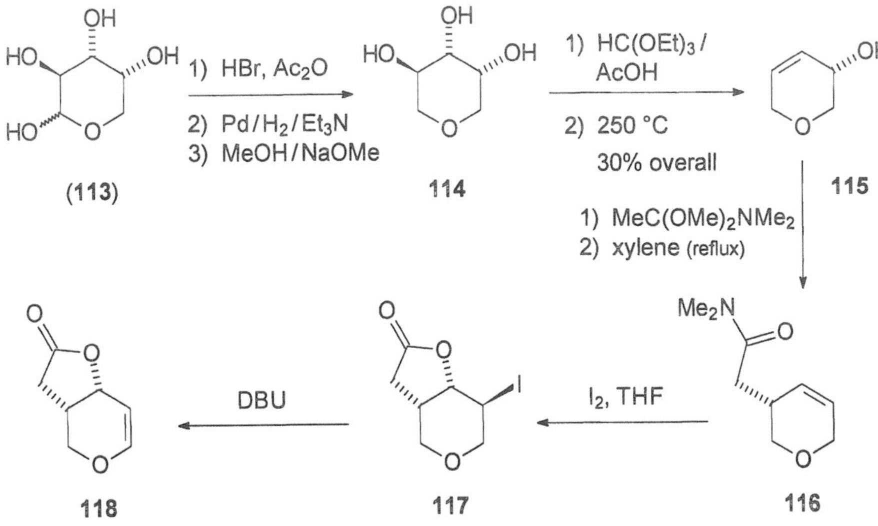

Scheme 19

The lactone was reduced (Scheme 20)[21] with sodium borohydride and the primary alcohol selectively silylated. Claisen rearrangement proceeded smoothly and the treatment of the amide with one equivalent of methyllithium gave the ketone **120**. Osmylation and ketalization then provided **121** which underwent olefination to **122**. The silyl protecting group was removed with TBAF and the alcohol oxidized to aldehyde **123** with PCC. A second Wittig reaction with the optically pure ylide **124** was followed by an acetic acid quench to give **125** in 25% yield. The cyclohexylidene was hydrolyzed to give crystalline ethyl monate **126** which has been previously converted into pseudomonic acid C.[22]

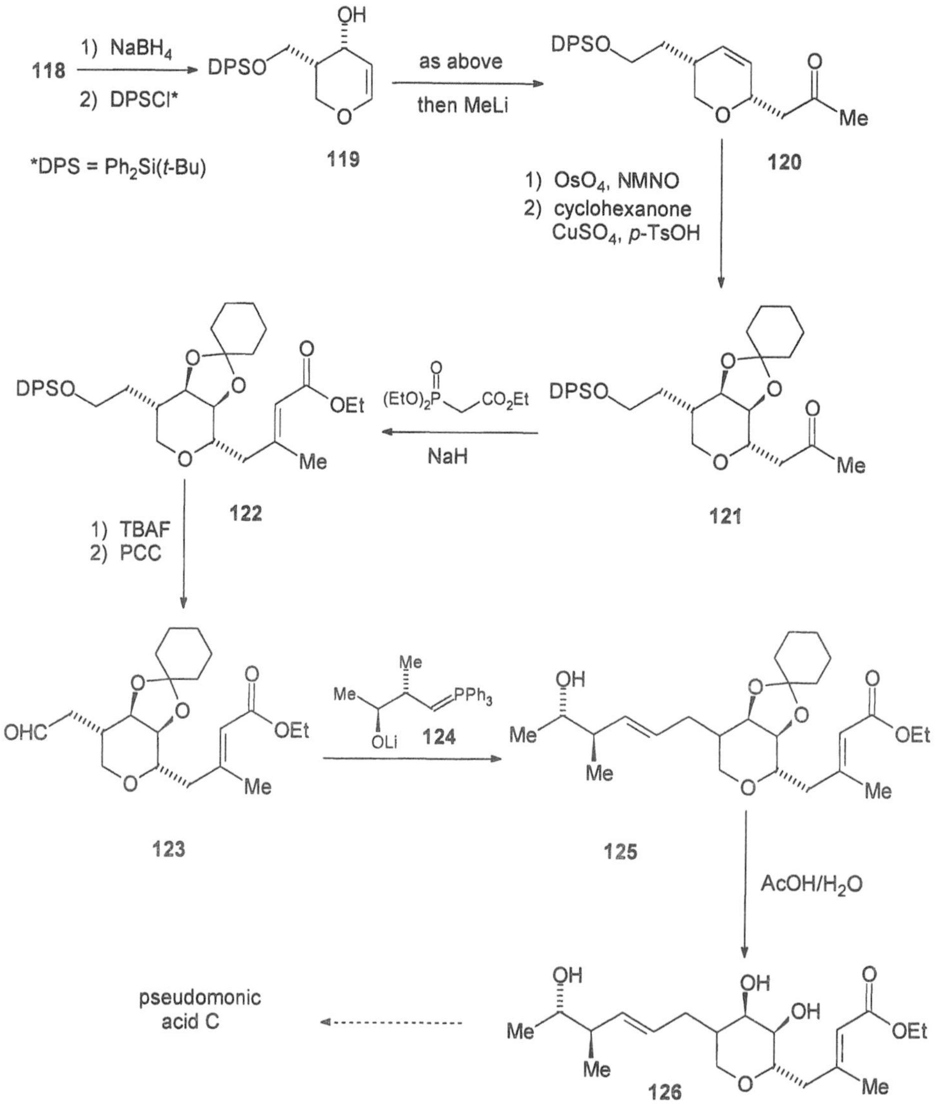

Scheme 20

Snider *et al.* used a very ingenious approach in the assembling of the pyran ring with the two requisite side chains by the use of three tandem reactions.[23] 1,5-Hexadiene was exposed to Lewis acid catalyzed ene reaction to give, after acetylation, **129**. A second ene reaction, this time with ethyl aluminum dichloride and three equivalents of formaldehyde, gave a 32% yield of **132** and **133** in a 16:1 ratio. The second ene reaction gives the intermediate **130** which loses ethane to give **131** which is then poised to direct the cycloaddition to furnish **132** as the major product, Scheme 26.

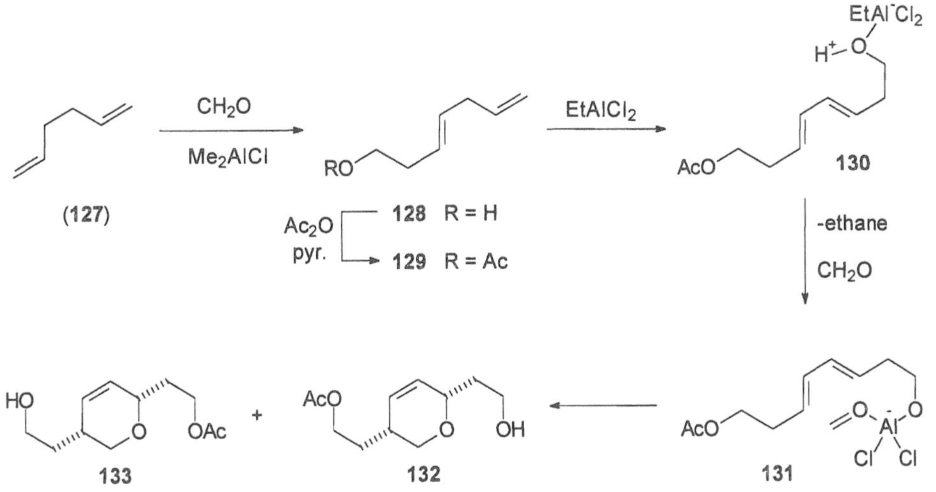

Scheme 21

With the differentially protected alcohol in hand Snider now proceeded to synthesize an advanced intermediate in Kozikowski's pseudomonic acid C synthesis.[15] The alcohol was oxidized to the aldehyde and treated with an excess of methyl magnesium chloride to give **134** as a mixture of diastereomers. Silylation was followed by oxidation to **135**. The sequence was completed by *syn* hydroxylation and ketalization to furnish Kozikowski's advanced intermediate **63**.

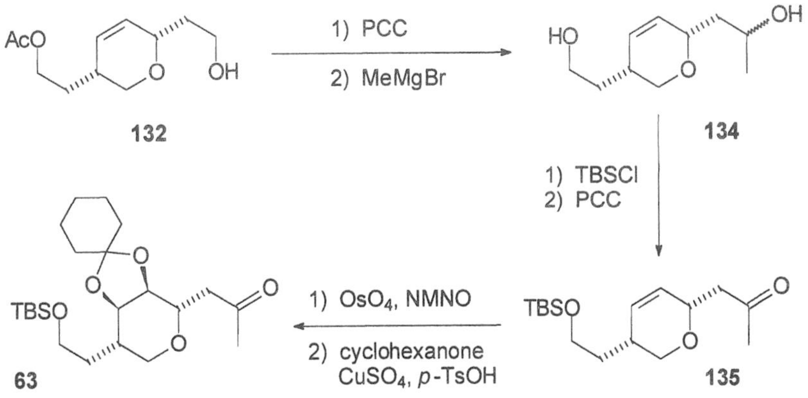

Scheme 22

Williams' approach[24] relied on a highly stereoselective aldol condensation between the aldehyde **137** (derived from the corresponding alcohol by Swern oxidation) with the thiol ester **136** to give a 79% yield of **138**. Reduction gave **139** which was tosylated, the acetonide cleaved, and the resulting triol exposed to base to afford the product of cyclization **140**. With the carbon framework of the tetrahydropyran ring in place side chain elaboration was now the issue at hand.

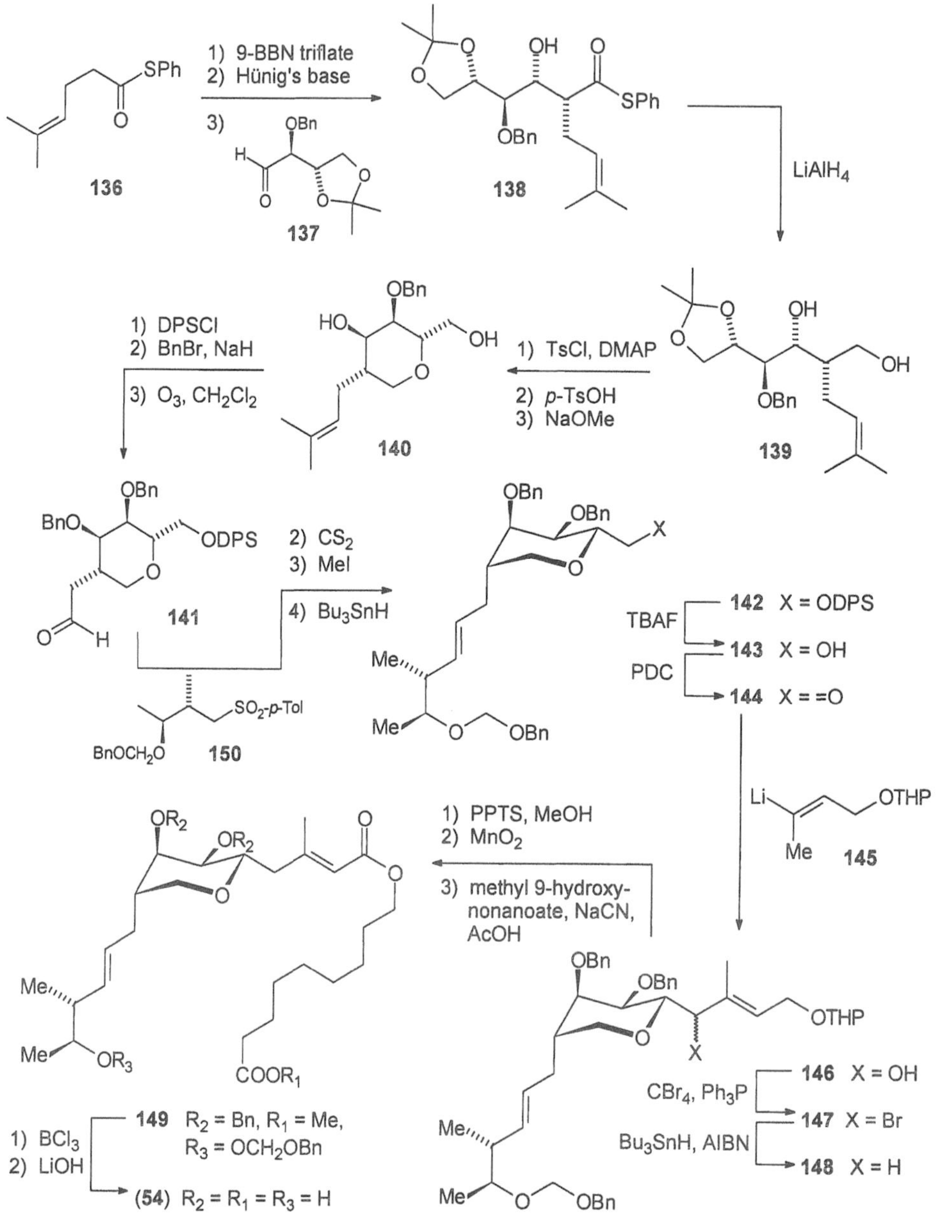

Scheme 23

Silylation and benzylation were followed by the exposure to the action of ozone to furnish the aldehyde **141**. Julia coupling with the optically active sulfone **150** provided **142** in which the double had the desired *trans* configuration. Conversion of the silyl ether to the aldehyde was accomplished in two steps and condensation with the anion **145** gave a mixture of alcohols which were deoxygenated via their corresponding bromides by tin hydride mediated reduction to **148**. The THP ether was removed in the normal way and oxidation followed by esterification then gave the fully protected compound **149** which was deblocked to yield pseudomonic acid C (**54**).

Raphael and co-workers[25] used an S_N2' addition to an α,β-unsaturated lactone to set up the carbon framework of pseudomonic acid C. Lactol **151** underwent olefination to **152** which was then cyclized in basic methanol to **153**. The double bond was introduced by selenoxide fragmentation and saponification then gave acid **154**. Phenylselenolactonization with PhSePF$_6$ gave a lactone which underwent selenoxide fragmentation to give the olefinic lactone **155**. S$_N$2' reaction with di-t-butyl sodiomalonate was catalyzed by Pd(Ph$_3$P)$_4$ to give the addition product **156** which can easily be envisaged as an intermediate for the preparation of pseudomonic acid C. It is worthy to note that coordination of the palladium probably occurs from the less hindered face, opposite the lactone, and addition of the carbon based nucleophile then is forced to proceed *syn* to the lactone to give a net *cis* addition after β-elimination of PdL$_n$-H.

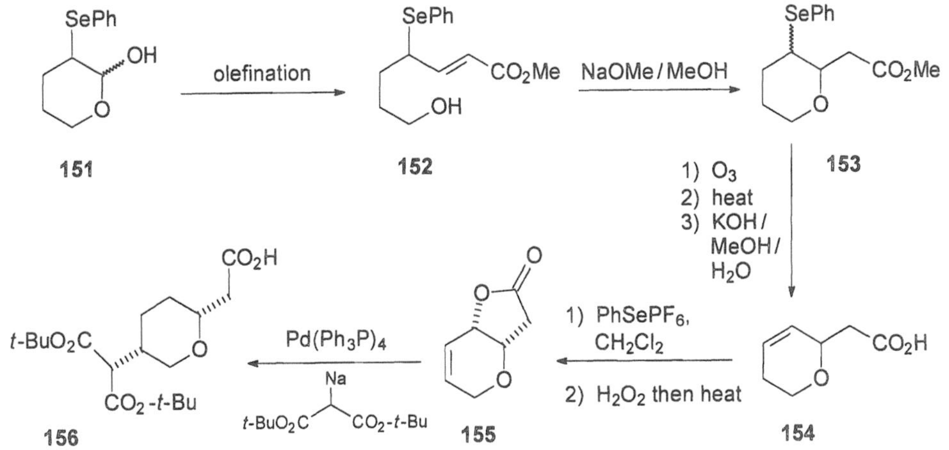

Scheme 24

The sequence was completed by conversion of **156** into the advanced intermediate **63**. The acid **156** was converted into the methyl ester by cuprate addition of the corresponding acid chloride and this was followed by thermal decarboxylation to give **157**. The acid was reduced to the diols **158** and silylation followed by oxidation furnished a ketone. The formal total synthesis was completed by *syn* hydroxylation and ketalization to provide **63**.

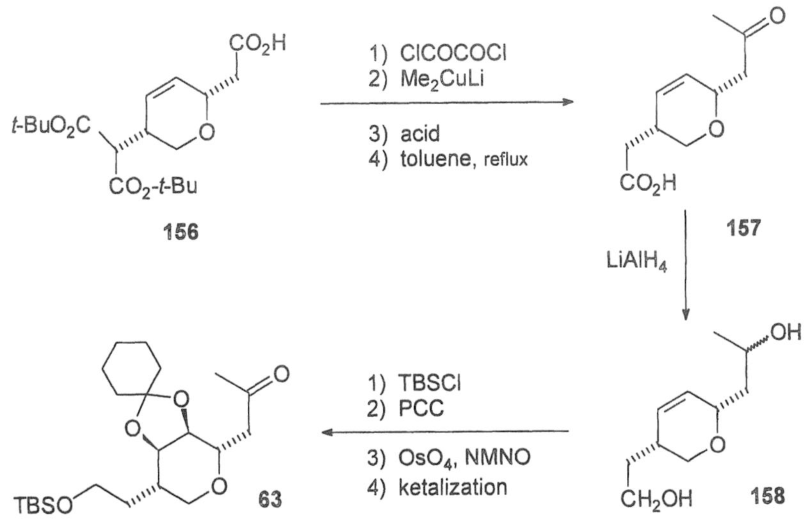

Scheme 25

Sinaÿ and co-workers[22] used a chiron approach to assemble pseudomonic acid C in optically pure form. D-Glucose was converted into the methyl glycoside **159** via the literature method. Protection and regioselective ring opening then brought the sequence as far as **161**. Treatment of the bromide with activated zinc and subsequent reduction of the formed aldehyde then gave the olefin **162**. The primary hydroxyl was removed via reduction of the corresponding iodide and the benzoate cleaved in the usual fashion to provide **163**. Exposure of the secondary allylic alcohol to thionyl chloride in the presence of pyridine caused rearrangement to furnish the desired primary allylic chloride **164**.

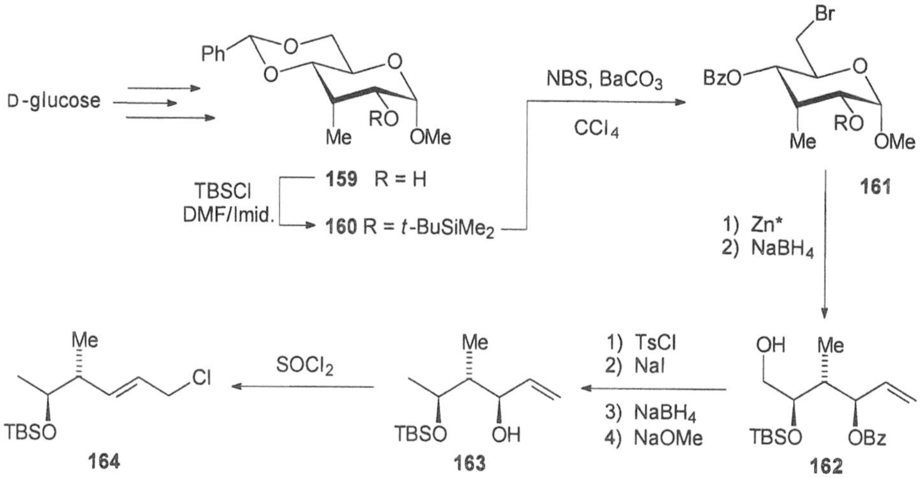

Scheme 26

The pyran ring system of pseudomonic acid C was accessible from the available cyanide **165**. The nitrile was converted to the primary alcohol in a four step protocol and the formed tetrol was then acetalized to the diol **166**. Ditosylation was followed by treatment with base to give epoxide **168**.

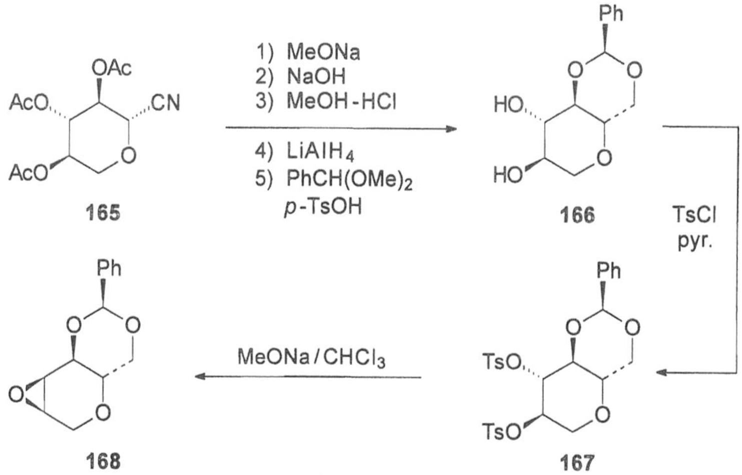

Scheme 27

The stage was now set for coupling of the allylic chloride (which would serve as a nucleophile) with the electrophilic epoxide **168**. Copper catalyzed ring opening of the epoxide proceeded smoothly with the Grignard reagent derived from **164** to give the regioisomer shown. The reaction also proceeded with the anion on the primary carbon and with *trans* stereoselectivity. Hydrolysis removed the protecting groups and the *syn* hydroxyls were ketalized to give **171**. Tosylation of the primary alcohol was followed by displacement with cyanide to furnish **172**. Nickel catalyzed addition of trimethyl aluminum proceeded smoothly to give, after hydrolysis of the protecting groups, ketone **173**. The synthesis was completed along previous lines by Wittig extension of the ketone and ester formation to give methyl pseudomonate (**55**).

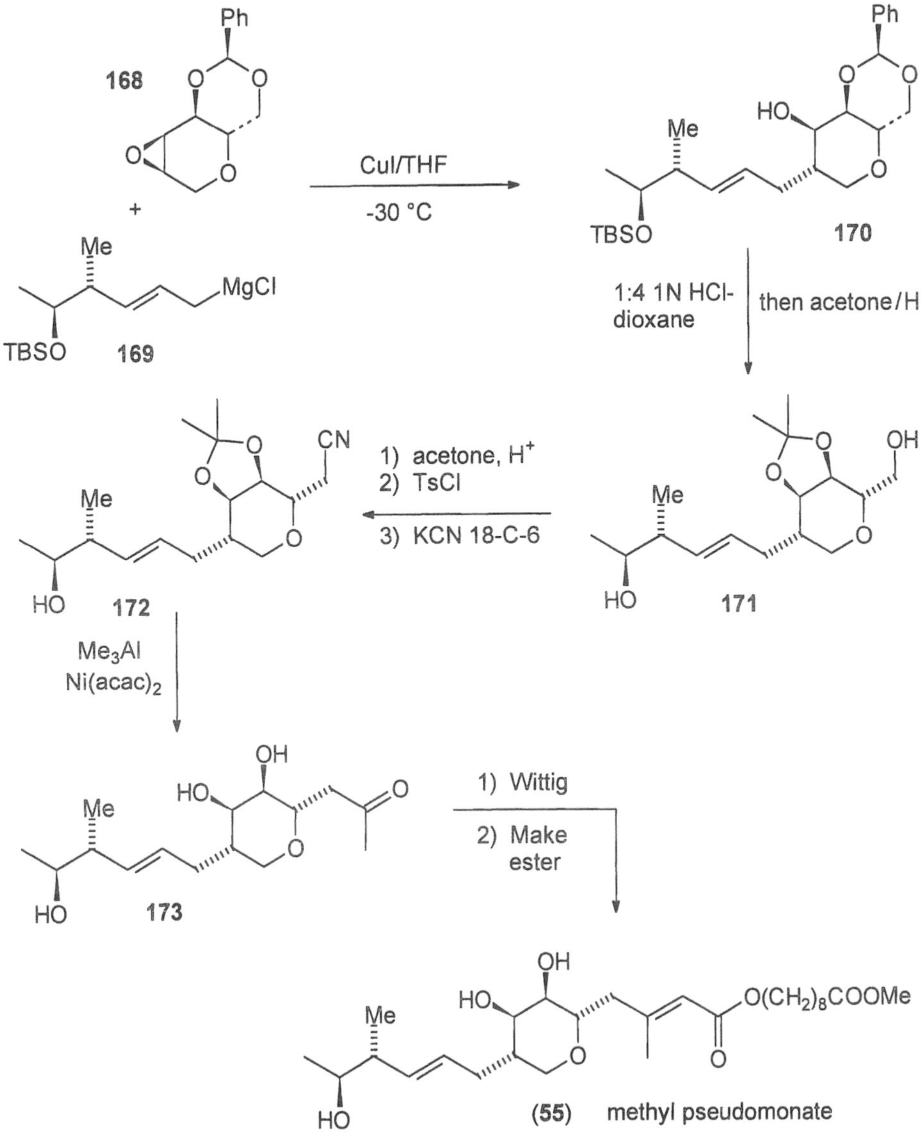

Scheme 28

Swiss workers also relied on the ready availability of carbohydrates as the starting point for their synthetic studies towards the pseudomonic acids.[26] The acetal **174** was subjected to Horner olefination to give a mixture of open chain isomers which were then cyclized with sodium methoxide to *C*-glycoside **176**. A second Wittig reaction then also provided an isomeric mixture of unsaturated esters which could be photochemically isomerized and then separated by fractional crystallization to furnish pure 177 in about 80% yield over two cycles. When 177 was exposed to acid in the presence of 2,2-dimethoxy propane a 1:1 mixture of acetonides 178 and 179 (not shown) was obtained. Compound 178 was then oxidized to the ketone and imine aldol condensation gave aldehyde 180 which was subsequently protected and reduced to afford 180. Compound 180 possesses most of the features of the pseudomonic acids with the exception of an extra hydroxyl at *C*-8 which is well suited for the preparation of pseudomonic acid B.

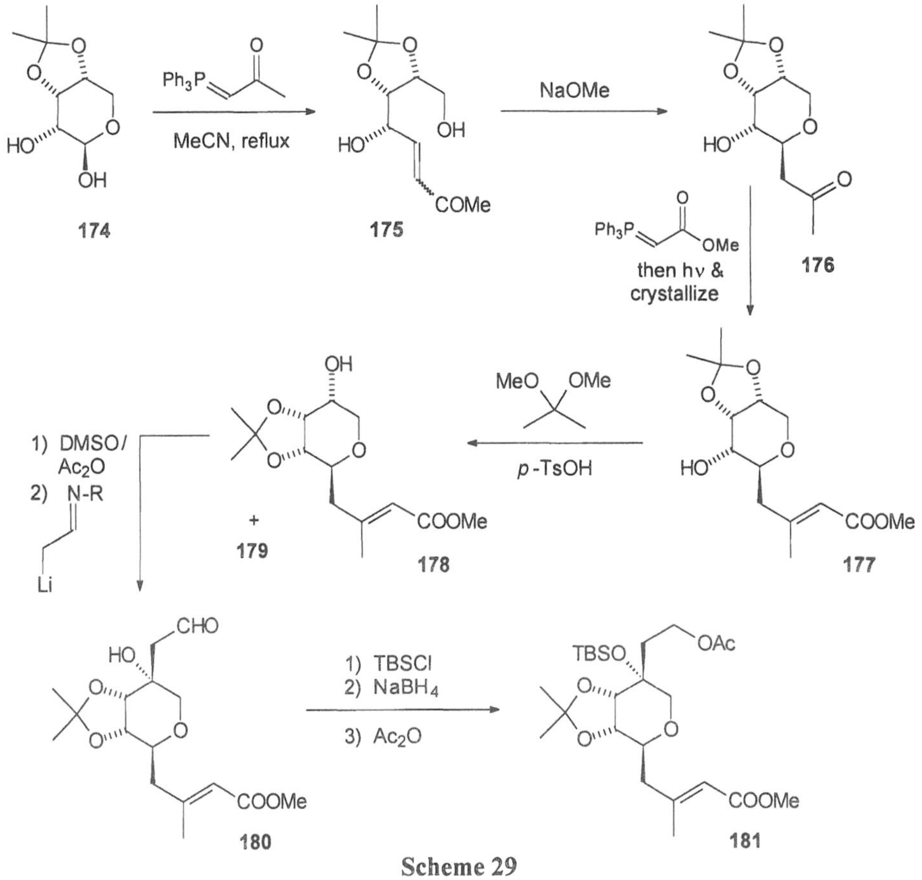

Scheme 29

Both Nagarajan and Keck used a free radical based method in their approaches to the syntheses of pseudomonic acid C. Nagarajan and Vaman Rao[27] began by converting D-xylose into its benzyl glycoside by the standard method and selective benzoylation then gave the diol 182. Corey-Winter olefination gave 183 and bromohydrin formation followed by benzoylation then afforded 184. After considerable experimentation, it was found that the vinyl sulfone 185 served as an optimal radical acceptor to give compound 186, with radical addition occurs opposite the adjacent OBz group. The anomeric protecting group was removed to provide the lactol 187 and Horner olefination then gave the *C*-glycoside 188 as the minor isomer (1:5). It is not difficult to envisage coupling of the sulfone with an appropriate aldehyde (189) and further elaboration of the ketone should eventually give pseudomonic acid C.

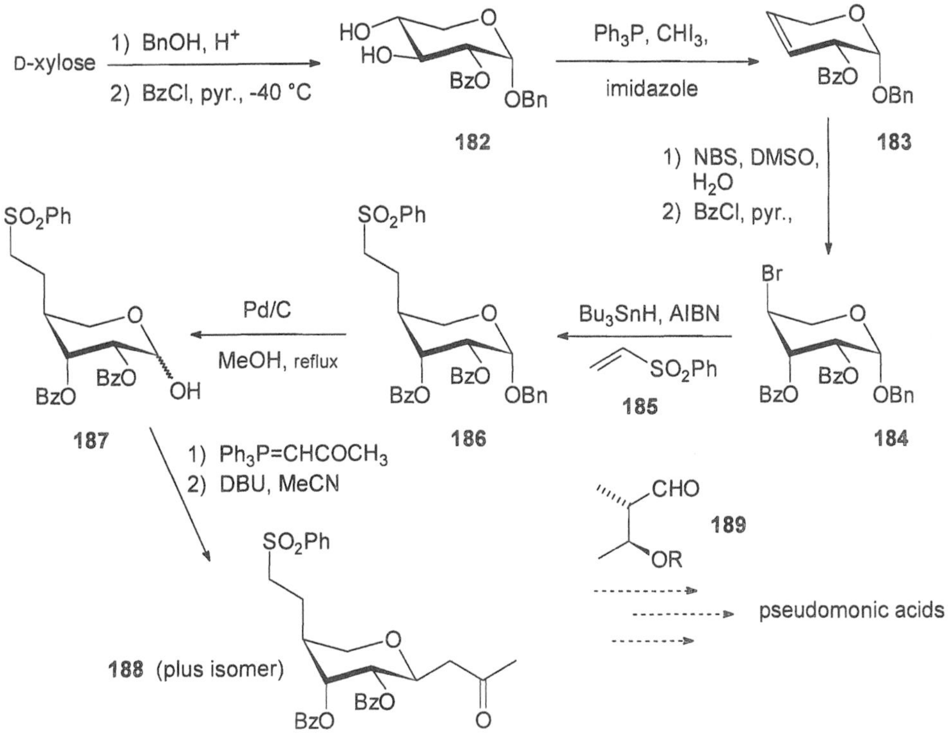

Scheme 30

Keck's approach[28] also relied on radical chemistry to introduce the southern chain. In this synthesis Keck wished to introduce the fully functionalized chain in a one step procedure by coupling **192** and **196** (Scheme 32).

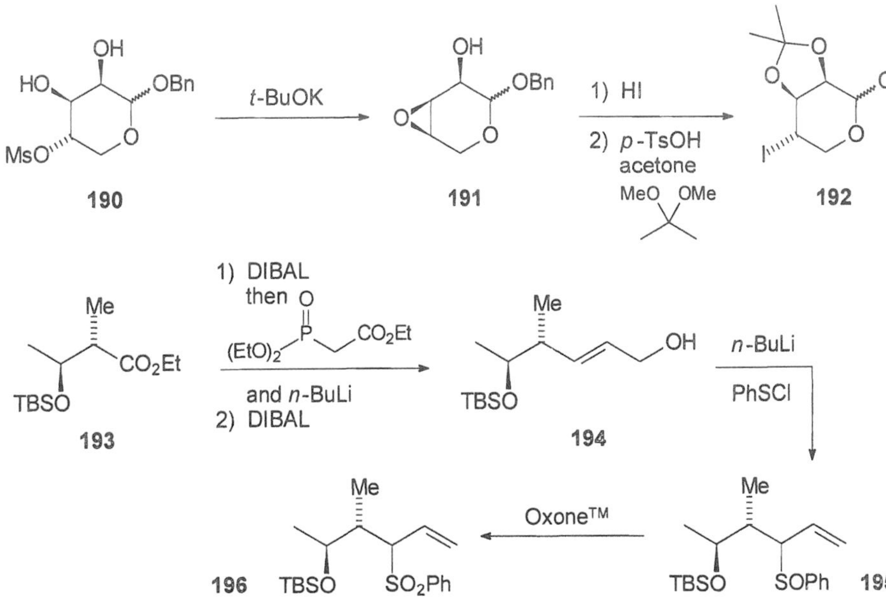

Scheme 31

Iodide **192** was prepared from the epoxide **191** via acid catalyzed ring opening with hydroiodic acid and ketalization (Scheme 31). Keck chose to use a sulfone group in the fragmentation reaction and this is probably due to the ease of incorporation of sulfur as opposed to tin. Wittig chain elongation on the ester **193** proceeded uneventfully to yield the allylic alcohol **194**. Sulfenate rearrangement then positioned the sulfur moiety at the correct location and oxidation then gave the requisite sulfones **196**. The stage was now set for the critical coupling (Scheme 32) and standard conditions failed to give any product and after some experimentation it was found that slow addition of the radical precursor **192** to a THF solution of the sulfones **196** and hexabutylditin in the presence of light gave a 74% yield of the adduct **197**. The formal total synthesis was completed by removal of the anomeric protecting group to give the lactol **198** which has been previously converted[17] into pseudomonic acid C.

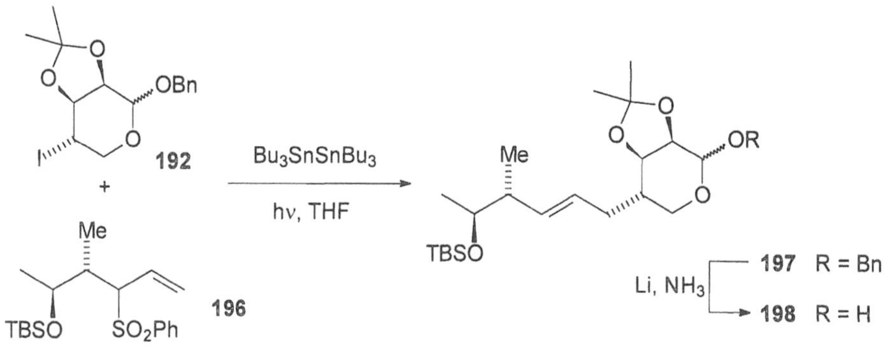

Scheme 32

Barrish et al.[29] used elegant methodology to construct pseudomonic acid C. In their first generation synthesis of the pyran ring they utilized the hydroxy ester as their starting point with the intention to expand a five membered ring lactone to a δ-lactone by the use of Bayer-Villager type chemistry. The ester **199** was saponified and halolactonization of the acid gave the bromolactone **200**.

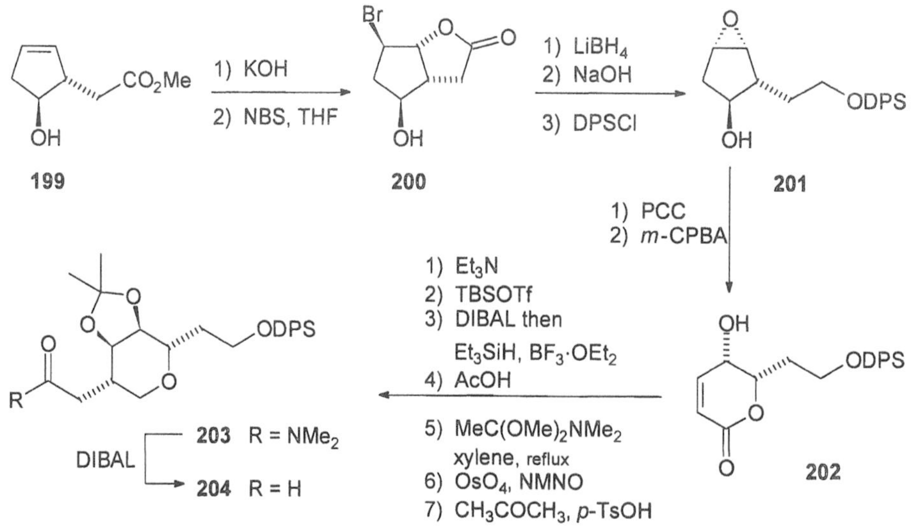

Scheme 33

The lactone (Scheme 33) was reduced to a diol that, when exposed to base, gave the epoxide **201** after silylation of the primary alcohol. The remaining secondary alcohol was oxidized with PCC and then further oxidized with *m*-CPBA to give after base induced oxirane opening lactone **202**. The ketone was reduced down via the intermediacy of a lactol (with protection of the secondary alcohol) and this was followed by Claisen type chemistry to introduce the southern side chain to give, after hydroxylation, ketalization, and partial reduction of the amide, the aldehyde **204**. The workers were now poised (Scheme 34) to apply the Ireland ester rearrangement so as to properly set up the required stereochemistry in the side chain. Reaction of **204** with the Grignard reagent **205** in the presence of zinc bromide gave a 4:1 ratio of isomers. The major (**206**) was acylated with acid **207** under dehydrating conditions to **208**. The requisite *E*-enolate was generated by deprotonation in the presence of trimethylsilyl chloride to give, after rearrangement and work-up, a 4:1 ratio of the desired threo (12*S*, 13*R*) product. Changing the base to *t*-octyl-*t*-butylamide improved the ratio up to 7:1 but gave a low yield. The formed ester was reduced and deoxygenated via an iodide by the use of stannane chemistry to provide **209**.

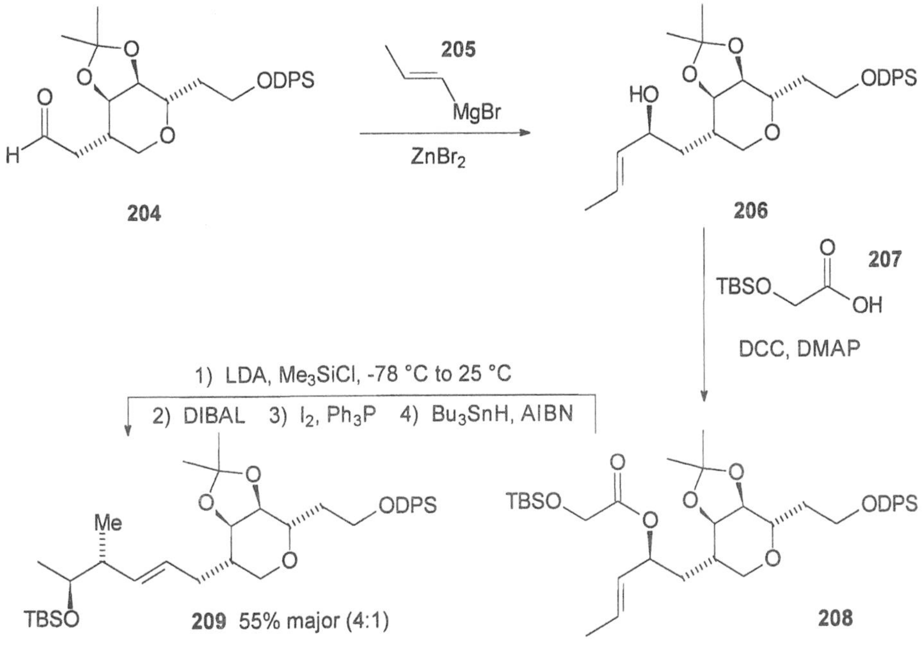

Scheme 34

Although the above work is elegant, the workers felt that they could improve the modest stereoselectivity they encountered. They decided to use the Midland procedure to stereoselectively reduce an acetylenic ketone. Addition of propynyl lithium to the amide **203** gave ketone **210** and reduction with (*R*)-alpineborane then gave the *R* alcohol with excellent selectivity (20:1). It is interesting to note that the direction of the reduction with the borane is completely controlled by the reagent and not the local chirality of the molecule. Esterification was followed by controlled hydrogenation to now afford the *cis* olefin **211**. In this case the workers now only needed to generate a *Z*-enolate, by using a more chelating controlling group (OBn), which (after silylation) would then undergo rearrangement to give the threo (12*S*, 13*R*) compound. This was indeed found to be true and rearrangement gave **212**, after hydrolysis and esterification. The sequence was then continued along previous lines (Schemes 34 & 35).

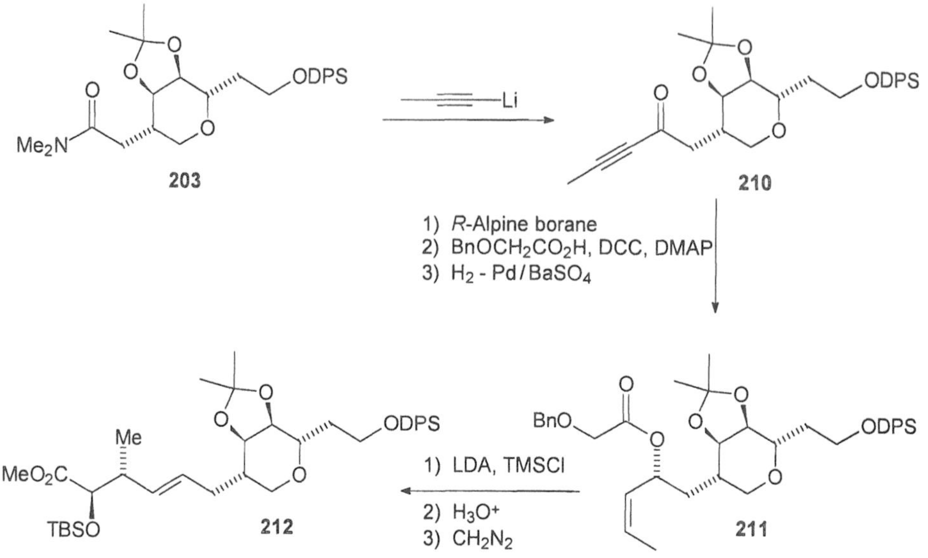

Scheme 35

The ketone **209** was accessed by desilylation, oxidation, addition of methyllithium and oxidation (Scheme 36). Wittig elongation and deprotection then furnished pseudomonic acid C (**54**).

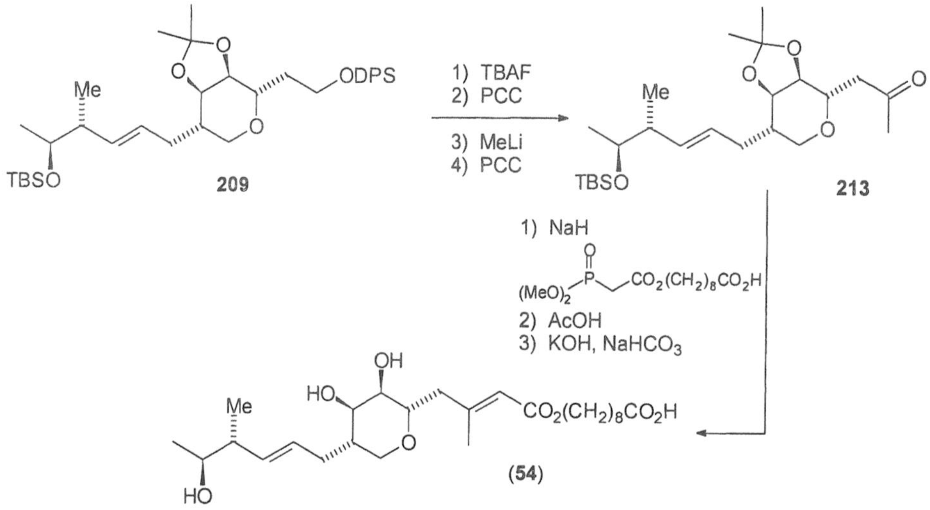

Scheme 36

The workers also found that Diels-Alder chemistry could be used to assemble the pyran ring portion and the reader is encouraged to consult the original paper[29] for more details.

White and co-workers[30] also employed the Claisen rearrangement to install the southern side chain in their enantiospecific synthesis of monic acid C. Dihydropyran (**214**) was condensed with (-)-borneol (**215**) in the presence of bromine to give **216** along with the corresponding isomer. The mixture was treated with DBU to give the diastereomeric pair of isomers **217**. Epoxidation with *m*-CPBA then gave the epoxides **218** and **219** which underwent ring opening with sodium phenyl selenide to give the isomeric alcohols **220** and **221**, respectively. Both alcohols were eventually converted to the intermediate lactone **224**. Compound **220** underwent selenoxide fragmentation to give an allylic alcohol which was converted to **222**, in the usual way, and when exposed to high temperature **222** underwent Claisen rearrangement to **223**.

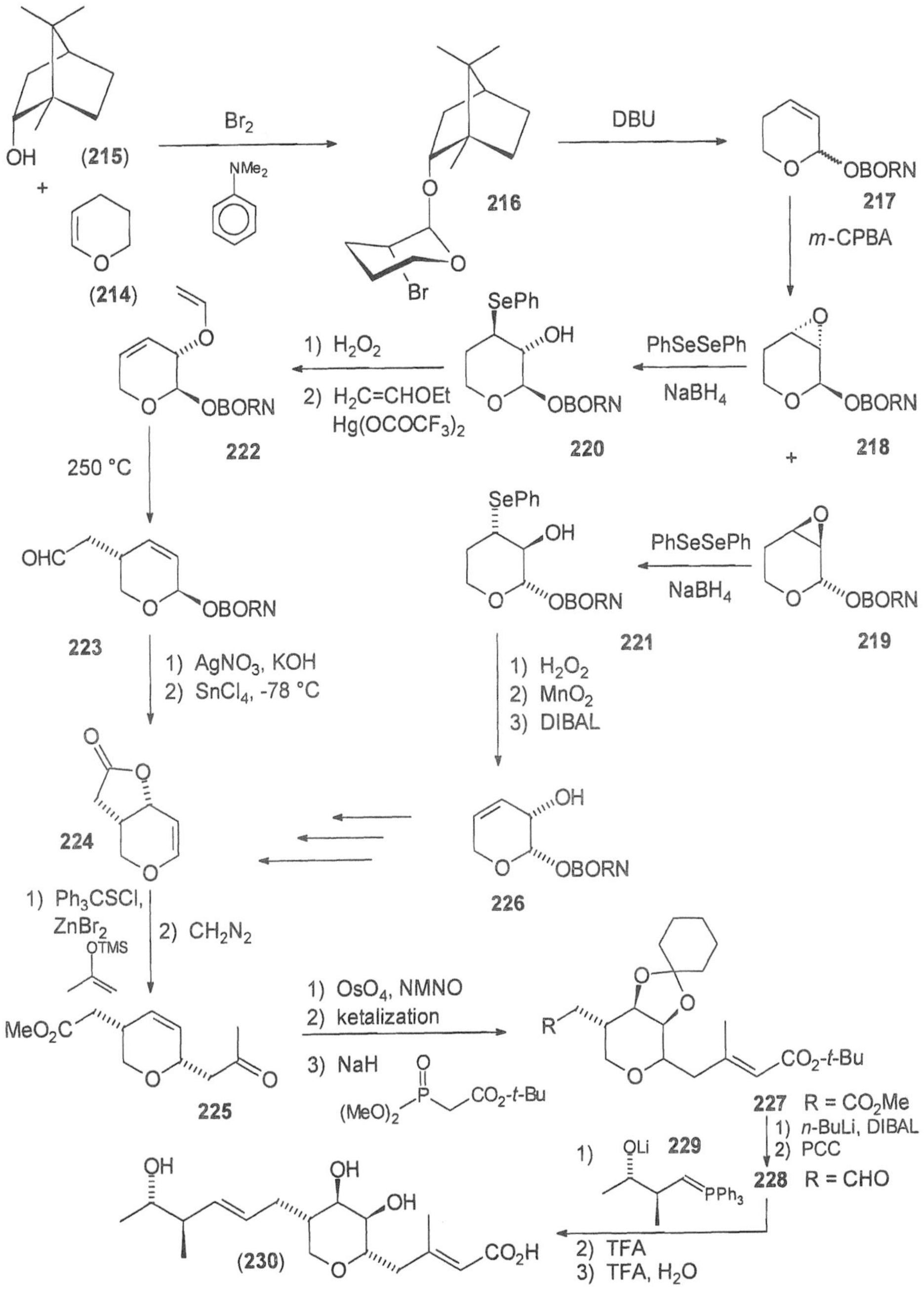

Scheme 37

Oxidation to the acid was followed by cyclization to regenerate (-)-borneol and lactone **224**. Isomer **221** was oxidized and selectively reduced to give alcohol **226** which was converted to lactone **224** by a similar sequence to that above. Allylic displacement on **224**, under Lewis acid catalysis with the silyl enol ether of acetone, gave, after esterification, **225**. *Syn* hydroxylation and ketalization was followed by Wadsworth-Emmons reaction to give the diester **227** which was transformed to aldehyde **228**. The southern side chain was installed using the optically active ylide **229** (available from *trans*-2-butenol by Sharpless epoxidation, regioselective epoxide ring opening with lithium dimethyl cuprate and conversion to the ylide), and hydrolysis of the *t*-butyl ester then provided optically pure Monic acid C (**230**).

Work led by Bates[31] was centered around acid catalyzed cyclization of a 1,5 diol to construct the pyran ring with a good level of stereoselectivity (Scheme 38). Coupling of the acetylide anion **231** (available from diethyl malonate and butyl vinyl ether in several steps) with the aldehyde **232** gave **233**. Exposure of **233** to dilute HCl caused deprotection and cyclization was affected during distillation of the product to provide **235** in 70% yield plus 15% of the *trans* isomer. Reprotection and conversion of the primary alcohol to the bromide then afforded **236** which was hydroxylated and ketalized to furnish **237**. The sequence was completed by a one carbon elongation to give **238**, after some functional group manipulations.

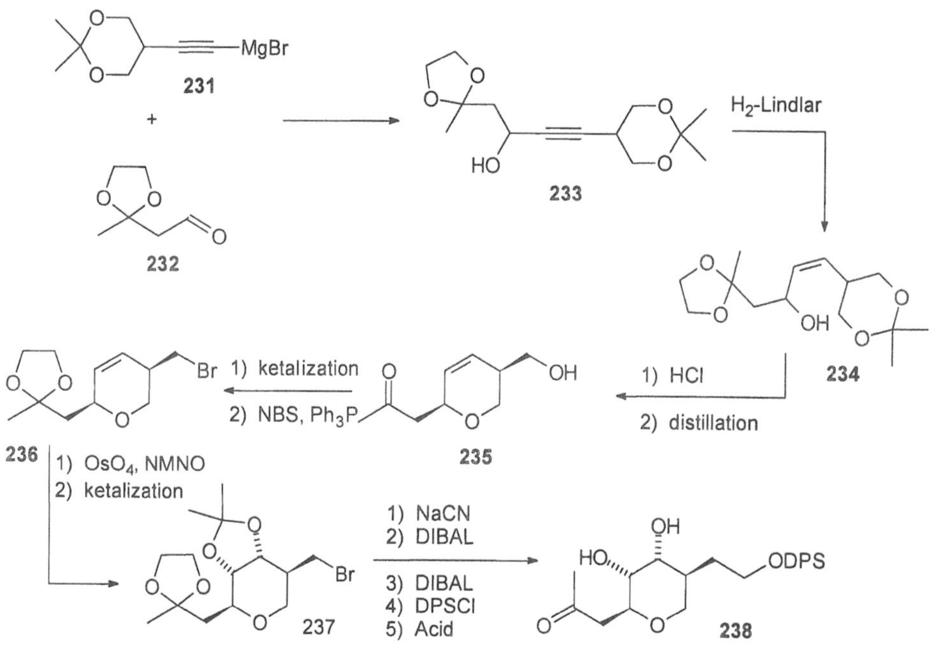

Scheme 38

Paterson and Alexander[32] used a Lewis acid mediated condensation of an enol ether with an α-halo ether in a model study directed towards the preparation of the pseudomonic acids (Scheme 39). Condensation of **239** with the chlorosulfide **240** gave compound **241** with good selectivity. Oxidation to the sulfoxide and elimination then provided **242** which possesses one of the required side chains of the pseudomonic acids. Psedomonic acid C has been converted to pseudomonic acid A,[33] and therefore all total syntheses of pseudomonic acid C must be considered as formal total syntheses of pseudomonic acid C.

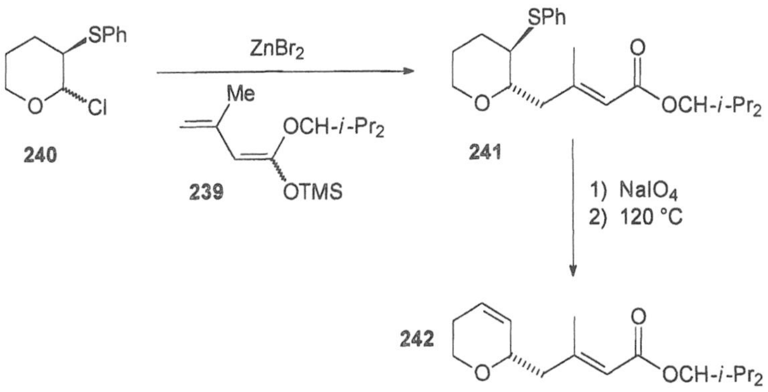

Scheme 39

III. ALKYL C-FURANOSIDE NATURAL PRODUCTS

A. MUSCARINES

1. Isolation

L-(+)-Muscarine (243), an alkaloid first isolated[34] from *Amanita muscaria*, has been an intensely investigated molecule from a chemical and physiological point of view. As a result, several strategies have been developed for the synthesis of the natural product as well as analogs. In the body muscarine behaves as a selective antagonist of acetylcholine on several smooth muscles.[35] These include the smooth muscle found in the exocrine glands, heart and the gastrointestinal tract. More recently,[36] it has been found that muscarine has an involvement with the secondary messengers such as inositol phospholipid metabolites. Several analogs that are closely related to muscarine are shown below in Fig. 6. These include the epi and allomuscarines. Synthesis of all these compounds has been accomplished. The structure of muscarine is characterized by an anomeric carbon-carbon bond, with a hydroxyl group at C-3 and a methyl group at C-5. Natural muscarine is dextrorotatory and the X group is OH, although many syntheses have focused on making the corresponding muscarine halides. Muscarine can be considered a 3,5-dideoxy-C-glycoside and several "C-glycosidic like" synthetic approaches have been applied to its construction. Alternatively, muscarine can be viewed as having a structure made of a 1,5-disubstituted tetrahydrofuran and some of the characteristic approaches to these molecules have naturally been applied to the synthesis of muscarines.

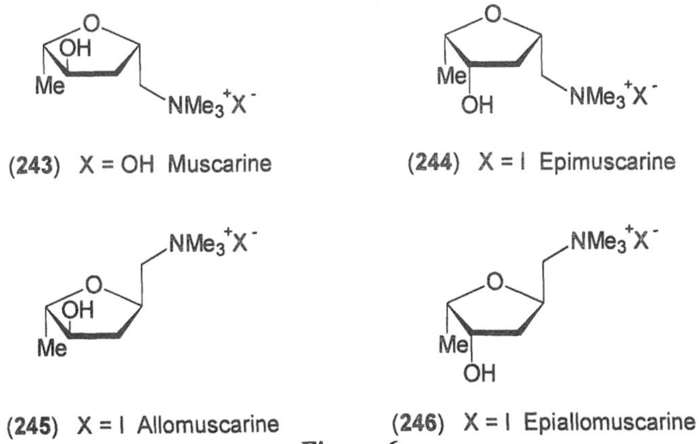

(243) X = OH Muscarine   (244) X = I Epimuscarine

(245) X = I Allomuscarine   (246) X = I Epiallomuscarine

Figure 6

2.   Synthesis of Muscarines

a.   *From Carbohydrates*

Joullié and Wang[37] used the gluco derivative 247 (available from D-glucose in four steps) as their starting point in their synthesis of epiallomuscarine (246). Treatment of 247 with sodium hydride gave an intermediate epoxide that was reduced to give a regioisomeric mixture of alcohols (3:2). The major 248 was acylated, deprotected, oxidized, and converted to the amide 250. Further reduction and quaternization then gave (246) epiallomuscarine.

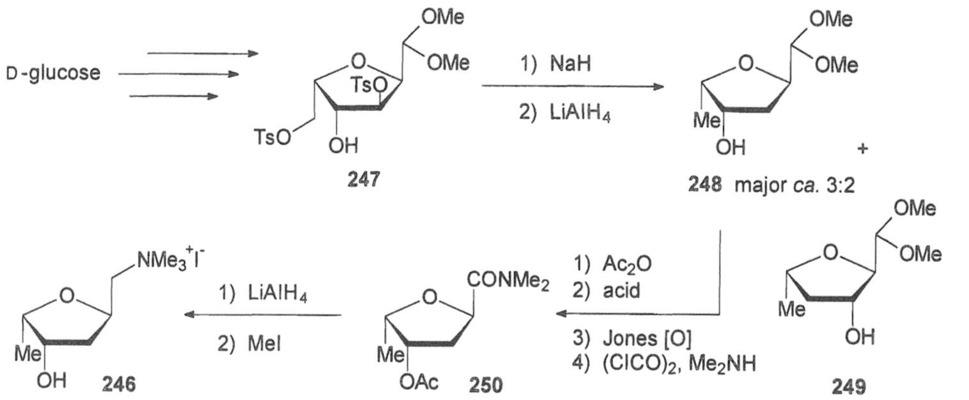

Scheme 40

The cyanide 251 (Scheme 41)[38] was treated with an excess of methanolic hydrochloric acid to give a mixture of three compounds: the β-ditoluoyl ester 252 and two mono tolyl esters 253 and 254 in approximately 63% overall yield. The esters were converted to the amides 255 and reduction followed by quaternization then gave D-(-)-muscarine iodide (256).

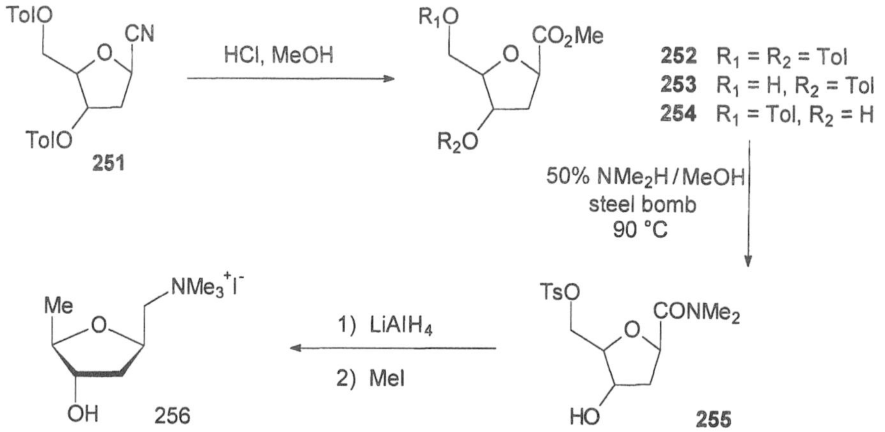

Scheme 41

Fleet *et al.*[39] have also employed a chiron approach to their synthesis of (+)-muscarine from L-rhamnose without having to resort to a protection-deprotection strategy. Exposure of L-rhamnose to aqueous buffered bromine solution furnished the lactone 258 after acid hydrolysis. Selective mesylation gave 259 and this diol was reacted with TFAA in the presence of triethylamine to give a bis(trifluoroacetate) which then underwent elimination to a vinyl mesylate. The remaining triflouroacetate group was removed by treatment with acetic acid in methanol to then provide **260**. Hydrogenation from the less hindered face furnished **261** and reduction of the lactone was followed by a base induced cyclization between O-5 and the mesyl group to give the tetrahydrofuran **262**. The sequence was completed by tosylation and displacement with trimethylamine to give the muscarine salt **263**.

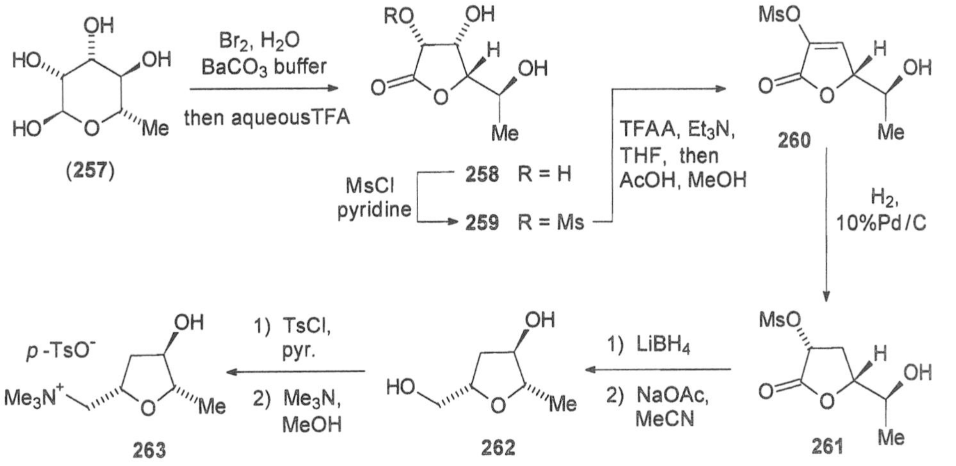

Scheme 42

In further work Fleet and co-workers[40] applied a similar strategy to the preparation of the muscarine analog **266** which contains an extra hydroxy group at the three position, Scheme 43. Exposure of lactone **264** to triflic anhydride in THF/pyridine followed by work up in the presence of methanol furnished ester **265**. Presumably the triflic anhydride esterifies O-2 and further reaction then causes tetrahydrofuran formation. The sequence was completed by reduction of the ester, tosylation, and derivitization to the muscarine analog **266**.

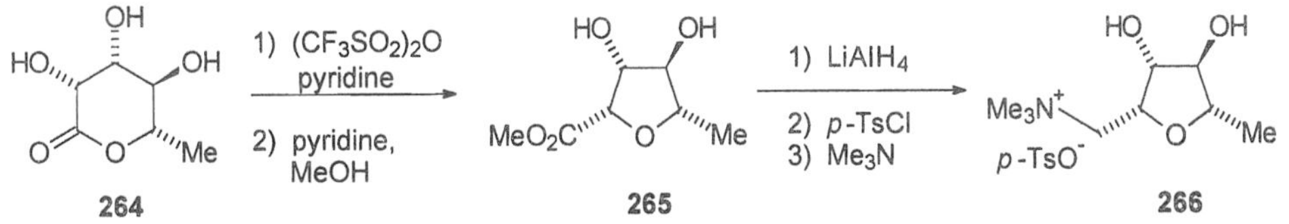

Scheme 43

Chapleur and Bandzouzi[41] have applied methodology developed in their laboratories[42] to the synthesis of L-(+)-muscarine from the lactone **267**. Olefination of the lactone as shown below in Scheme 44 led to the dichloroolefin **268**. When **268** was treated with LDA an elimination reaction ensued and ketone **269** was formed. Reduction with Raney nickel gave a mixture of compounds of which **271** was to be used in the synthesis of muscarine. The free hydroxyl was inverted to the required configuration by Mitsunobu chemistry and acid hydrolysis then gave the diol **272**. The diol was cleaved with periodate and the resulting aldehyde reduced to **273**. Ester cleavage, selective tosylation, and displacement of the formed tosylate with trimethylamine then gave the L-(+)-muscarine salt (**274**).

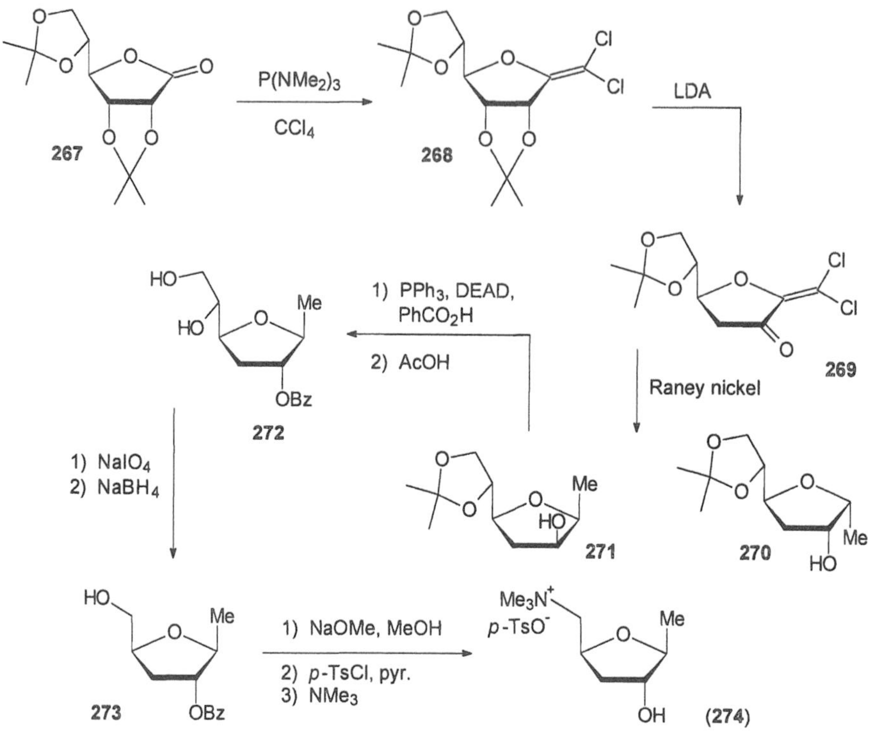

Scheme 44

b.   Cyclofunctionalization Approaches

Both halocyclization and selenocyclization are viable methods for the construction of tetrahydrofurans and several workers have applied this strategy to muscarine synthesis. The fundamental differences between their approaches lie in the methodology used to assemble the precursor to cyclofunctionalization.

Mulzer and co-workers[43] used an organometallic addition/epoxide ring opening sequence to make their precursor to iodocylization.

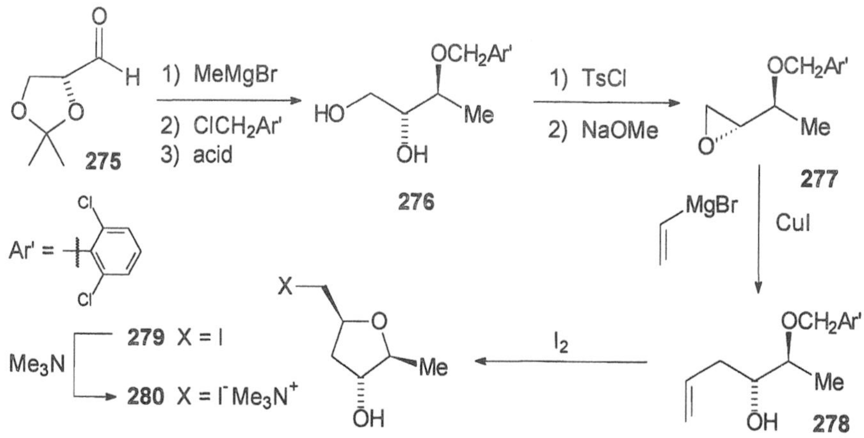

Scheme 45

Addition of methyl magnesium bromide to the aldehyde **275** gave, after benzylation, and acetonide hydrolysis, **276** as the major product. The ratio of addition was 68:32 in favor of the *anti* isomer **276**. Tosylation and base treatment of **276** produced epoxide **277** which was allowed to react with vinyl magnesium bromide under copper iodide catalysis to provide **278**. Exposure of **278** to iodine gave **279** (plus isomer). It is noteworthy that the benzyl protecting group is removed under these conditions. The sequence was then completed in the usual way by displacement of iodide with trimethylamine to give the muscarine salt **280**, Scheme 45.

Joullié and co-workers[44] used selenocyclization during the assembly of the muscarine analog **286** (Scheme 46). Alcohol **281** was treated with phenylselenyl chloride in the presence of triethylamine to give the selenide **282**. Reduction with sodium borohydride gave a 4:1 mixture of *trans* and *cis* isomers, respectively and acetylation, selenoxide fragmentation, and ozonolysis yielded aldehyde **284** which was then converted to the muscarine analog **286** via the intermediacy of the amide **285** by sequential reduction and methylation.

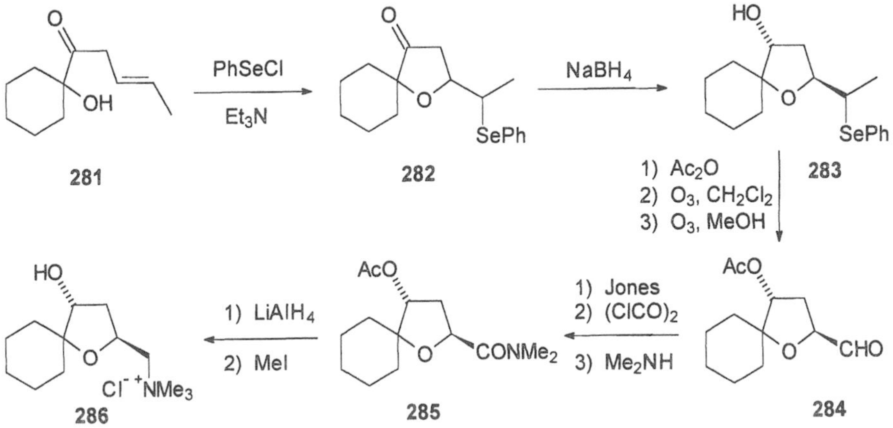

Scheme 46

Chan and Li[45] found that aqueous zinc mediated allylation of the aldehyde **287** showed a reversal of selectivity to that found when the reaction is performed in organic media. The non-chelation *anti* product **288** was formed (71:29 of *anti:syn*) in 60% yield. To complete the synthesis **288** was cyclized to **290** in the presence of iodine and conversion to muscarine iodide (**280**) was routinely accomplished as shown below in Scheme 47.

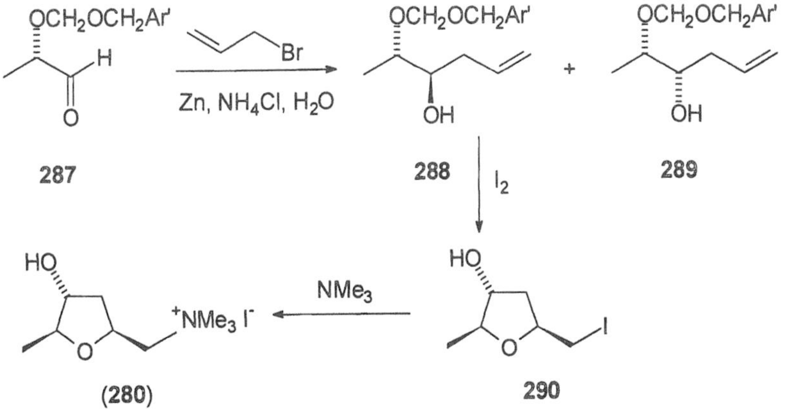

Scheme 47

The need for muscarine analogs prompted Shapiro and co-workers[46] at Sandoz to develop a general route to such compounds. The ester 291 (available from L-lactic acid) was converted to the Wadsworth-Emmons reagent 292 by treatment with the anion derived from methyl phosphonate. Olefination with the ketone 293 produced olefin 294 which was reduced with L-selectride® to the *syn* product 295. The silyl group was removed and mercuriocyclization followed by reductive work-up then provided the tetrahydrofuran 296. The sequence was completed by Mitsunobu inversion to give the muscarine analog 297 with the proper relative and absolute stereochemistry, Scheme 48.

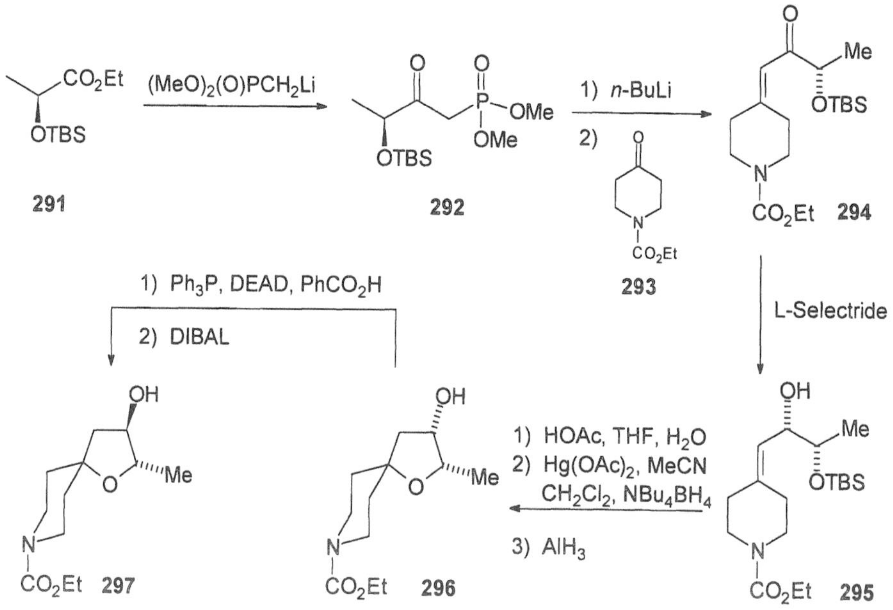

Scheme 48

c.   *Direct Cyclization*

Direct cyclization approaches to formation of the tetrahydrofuran ring have also been investigated. Still[47] began with the olefin 298 which when exposed to ozone gave a hydrate that was acetylated to provide the diacetate 299.

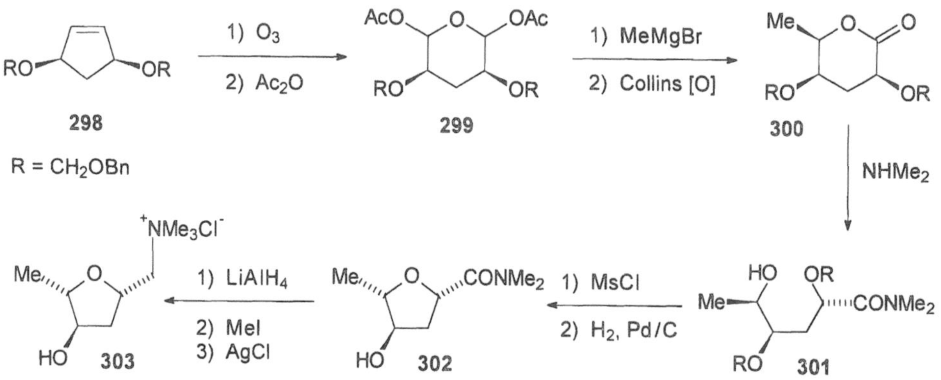

Scheme 49

Chelation controlled addition of methyl Grignard gave a lactol which was oxidized to lactone **300**. The lactone was opened and the resulting free hydroxyl mesylated, and the benzyl group cleaved to affect cyclization to **302**. Conversion to muscarine was accomplished by quaternization and halide exchange.

The work shown below in Scheme 50 illustrates the utility of 5-*endo* cyclization onto an acetylene for tetrahydrofuran formation.[48] Alkylation of the acetylene **304** gave **305** and exposure of this compound to potassium *t*-butoxide in DMSO produced the dihydrofuran **306**. Hydroboration and oxidation then furnished **307**. Conversion to (+)-muscarine iodide was carried out via a four step protocol. The above sequence is short and has the advantage that the starting acetylene is readily available from (*R*)-*O*-benzylglycidol.

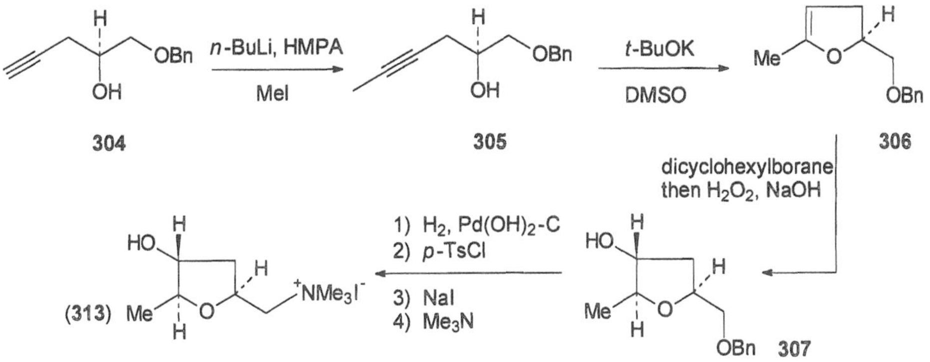

Scheme 50

Chmielewski and Guzik[49] used a tandem ene–5-*endo* tetragonal closure to synthesize allomuscarine. Ene reaction between butyl gloxylate and 1-butene gave the *trans* alcohol **310** which underwent epoxidation with *m*-CPBA to **311**. Cyclization was affected by treatment with stannic chloride at -40°C in dichloromethane to produce the tetrahydrofuran **312**. The synthesis of allomuscarine was completed by transformation of the ester group into the quaternary amine **313** in the normal fashion, Scheme 51.

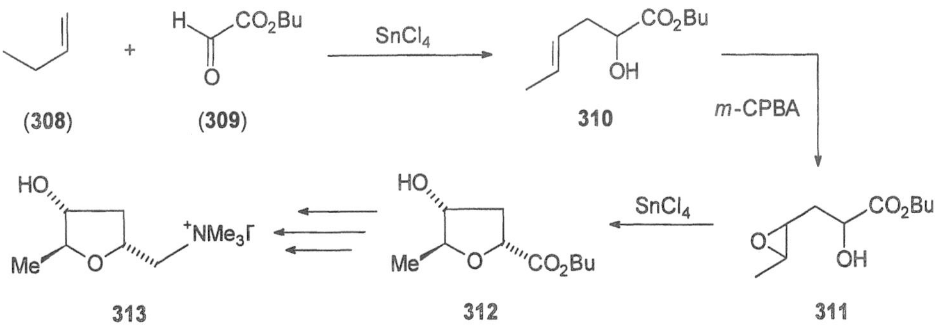

Scheme 51

Transition metal mediated reactions are starting to become standard methods in the arsenal of the synthetic organic chemist. Workers at Bio-Mega used a rhodium based carbenoid cyclization as the key step in their synthesis of (+)-muscarine.[50] The optically pure bromo acid **314** (available from D-alanine) underwent a displacement reaction to afford **315**. Conversion to the diazo ketone **316** was uneventful and exposure to rhodium acetate gave the tetrahydrofuran **317**. Deprotection and reduction then gave **318** as the major product. Standard methods then provided (+)-muscarine chloride, Scheme 52.

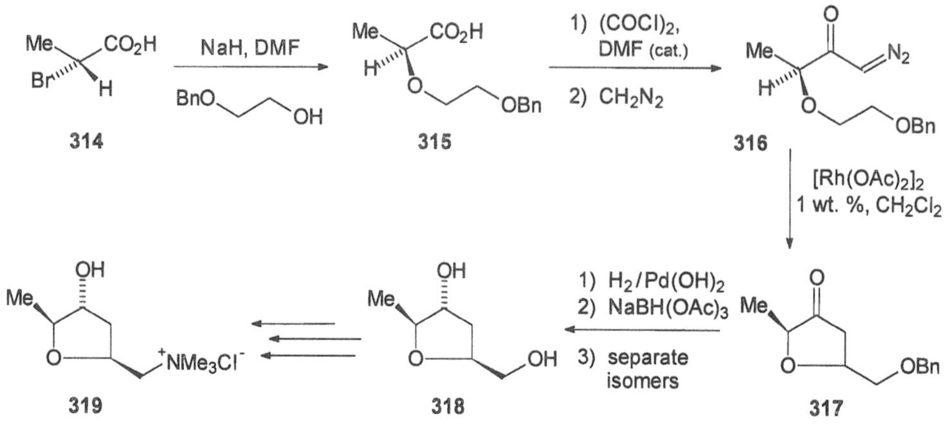

Scheme 52

d.   Ring Expansion

Ring cleavage approaches have also found their way into muscarine synthesis. Maan and co-workers[51] used the cyclocondensation of furan with the tetrabromide in the presence of zinc dust to give an unstable cycloadduct that was hydrogenated to the ketone **322**. Oximation and Beckman rearrangement then afforded lactam **323**. Acid hydrolysis was followed by exhaustive methylation to provide an analog of muscarine chloride, **325**, Scheme 53.

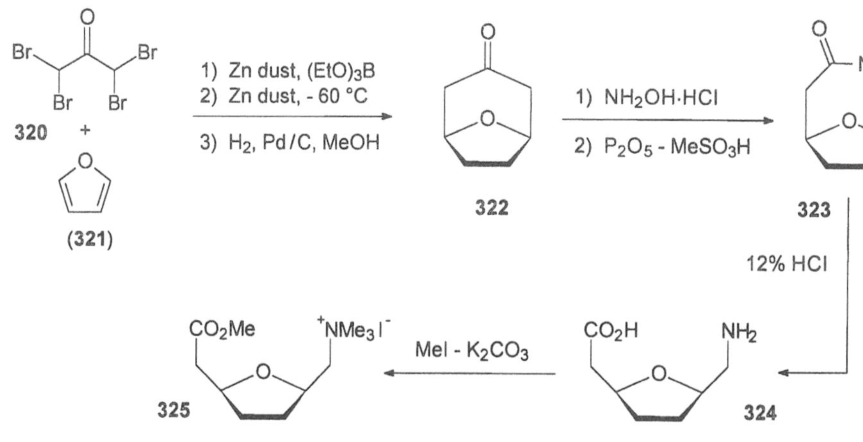

Scheme 53

Pirrung and DeAmicis[52] also used a cycloaddition approach, in the making of a cyclobutane that was then ring expanded to provide the requisite tetrahydrofuran ring. Ketene cycloaddition (Scheme 54) gave the chloroketone **327** which underwent dechlorination when exposed to zinc copper couple to give **328** as the major (3:1) compound. Oxacarbene formation (MeOH, hʋ) gave the acetals **329** in 55% combined yield. *C*-Glycosidation, under Lewis acid catalysis, with trimethylsilyl cyanide then yielded a 1:1 mixture of cyanides **330** and **331**. The mixture was separable and the undesired isomer **330** was converted to allomuscarine in an analogous manner to which **332** was converted to muscarine.

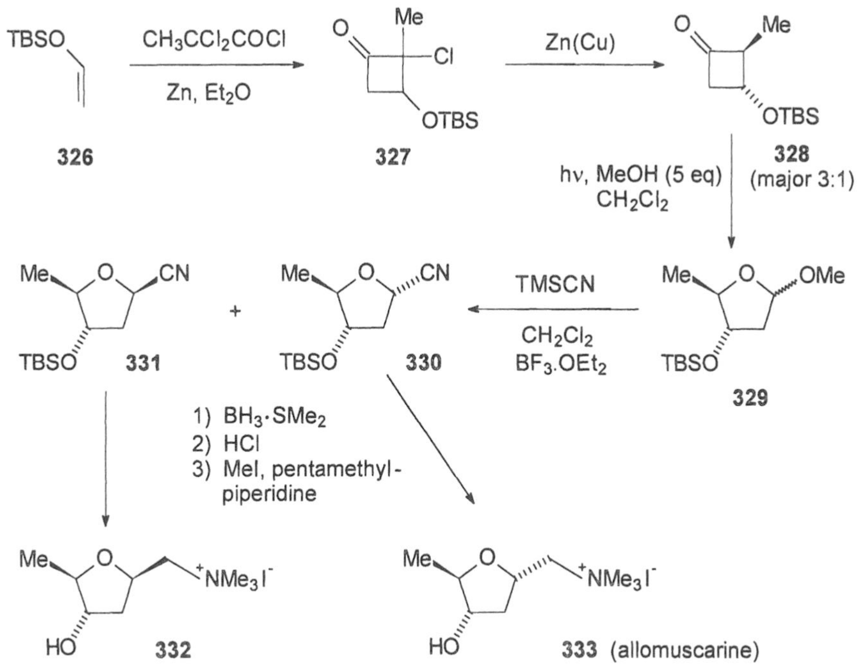

Scheme 54

IV. CONCLUSION

The synthetic approaches to ambruticin have essentially relied on the stereoselective formation of a β-*C*-alkenyl glycoside. Efficient disconnections have been performed at the double bond and the anomeric center. Julia coupling has been a reliable method for olefin formation. The use of Claisen rearrangement has met with success for installation of the side chains in the right hand side ring.

The approaches to the pseudomonic acids have been varied, yet very similar in how the side chains are installed. Many workers have used the chiron approach to assemble the pyran portion of the molecule and Wittig type olefination to address the issue of side chain construction while others have applied free radical chemistry to address the issue of side chain assembly. Sigmatropic rearrangements have also seen utility in the synthesis of pseudomonic acids. Direct cyclization and ene-Diels Alder strategies have also been applied with success.

The muscarines are a much less complex group of natural products but a rich variety of chemistry has been applied in their synthesis. Chiron approach, cyclization, transition metal mediated processes, cyclofunctionalization, and ring expansion strategies are all examples of some of the more interesting tactics that have been applied to the construction of these molecules.

V. REFERENCES

1.   Harmange, J.-C.; Figadère, B. *Tetrahedron: Asymmetry* **1993**, *4*, 1711.
2.   Ringel, S.M.; Greenough, R.C.; Roemer, S.; Connor, D.; Gutt, A.L.; Blair, B.; Kanter, G.; von Strandtmann, M. *J. Antibiot.* **1977**, *30*, 371.
3.   (a) Shadomy, S.; Utz, C.; White, S. *Antimicrob. Agents Chemother.* **1978**, *14*, 95. (b) Levine, H.B.; Ringel, S.M.; Cobb, J.M. *Chest* **1978**, *73*, 302.
4.   Connor, D.T.; von Strandtmann, M. J. *J. Org. Chem.* **1978**, *43*, 4606.
5.   (a) Just, G.; Potvin, P. *Can. J. Chem.* **1980**, *58*, 2173. The workers ozonized ambruticin and compared the fragments with synthetic material made in optically pure form.

Ambruticin

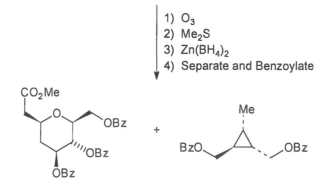

1) O$_3$
2) Me$_2$S
3) Zn(BH$_4$)$_2$
4) Separate and Benzoylate

6.   (a) Kende, A.S.; Fujii, Y.; Mendoza, J.S. *J. Am. Chem. Soc.* **1990**, *112*, 9645. (b) Kende, A.S.; Mendoza, J.S.; Fujii, Y. *Tetrahedron* **1993**, *49*, 8015.
7.   Barnes, N.J.; Davidson, A.H.; Hughes, L.R.; Procter, G. *J. Chem. Soc., Chem. Commun.* **1985**, 1292.
8.   Davidson, A.H.; Eggleton, N.; Wallace, I.H. *J. Chem. Soc., Chem. Commun.* **1991**, 378.
9.   Barnes, N.J.; Davidson, A.H.; Hughes, L.R.; Procter, G. and Rajcoomar, V. *Tetrahedron Lett.* **1981**, *22*, 1751.
10.   Sinaÿ, P.; Beau, J.-M.; Lancelin, J.-M. in Organic Synthesis, An interdisciplinary Challenge; Blackwell: London, 1985.
11.   (a) Pseudomonic acid A: Banks, G.T.; Barrow, K.; Chain, E.B.; Fuller, A.T.; Mellows, G.; Woolford, M. *Nature (London)* **1971**, *234*, 416. (b) Pseudomonic acid B: Chain, E.B.; Mellow, G.J. *J. Chem. Soc., Chem. Commun.* **1977**, 318.
12.   (a) Basker, M.J.; Comber, K.R.; Clayton, J.P.; Hannan, P.C.T.; Mizen, L.W.; Rogers, N.H.; Slocombe, B.; Sutherland, R. *Curr. Chemother. Infect. Dis., Proc. Int. Congr. Chemother. 11th*, **1979**, *1*, 471. (b) For studies on the competitive inhibition of isoleucyl-t-RNA synthetase see: Hughes, J.; Mellows, G.; Soughton, S. *FEBS Lett.* **1980**, *122*, 322. Hughes, J.; Mellows, G. *Biochem. J.* **1980**, *191*, 209.
13.   (a) Pseudomonic acid C: Clayton, J.P.; O'Hanlon, P.J.; Rogers, N.H. *Tetrahedron Lett.* **1980**, *21*, 881. (b) Pseudomonic acid D differs only from pseudomonic acid A in the ester side chain; it contains a *trans* double bond between *C*-4 and *C*-5 in the 9-hydroxynonenoic acid side chain: O'Hanlon, P.J.; Rogers, N.H.; Tyler, J.W. *J. Chem. Soc., Perkin Trans I* **1983**, 2655.
14.   Pseudomonic acid A rearranges to *trans*-ring fused bicyclic structures. Clayton, P.J.; Oliver, R.S.; Rogers, N.H.; King, T.J. *J. Chem. Soc., Perkin Trans. I* **1979**, 838.

OH
HO
Me
Me
OH
O
Me
$CO_2(CH_2)_7CH_2CO_2H$

Pseudomonic acid A

pH < 4

Me
OH
Me
O
OH
H
OH
OH
H
O
Me
$CO_2(CH_2)_7CH_2CO_2Me$

15. Kozikowski, A.P.; Schmiesing, R.J.; Sorgi, K.L. *J. Am. Chem. Soc.* **1980**, *102*, 6577.
16. Kozikowski, A.P.; Sorgi, K.L. *Tetrahedron Lett.* **1984**, *25*, 2085.
17. Keck, G.E.; Kachensky, D.F.; Enholm, E.J. *J. Org. Chem.* **1985**, *50*, 4317.
18. Curran D.P.; Suh Y.-G. *Carbohydr. Res.* **1987**, *171*, 161.
19. Fleet, G.W.J.; Shing, T.K.M. *Tetrahedron Lett.* **1983**, *24*, 3657.
20. Fleet, G.W.J.; Gough, M.J. *Tetrahedron Lett.* **1982**, *23*, 4509.
21. Fleet, G.W.J.; Gough, M.J.; Shing, T.K.M. *Tetrahedron Lett.* **1983**, *24*, 3661.
22. Beau, J.-M.; Aburaki, S.; Pougny, J.-R.; Sinaÿ, P. *J. Am. Chem. Soc.* **1983**, *105*, 621.
23. Snider, B.B.; Phillips, G.B. *J. Am. Chem. Soc.* **1982**, *104*, 1113 and Snider, B.B.; Phillips, G.B.; Cordova, R. *J. Org. Chem.* **1983**, *48*, 3003.
24. Williams, D.R.; Moore, J.L.; Yamada, M. *J. Org. Chem.* **1986**, *51*, 3916.
25. Jackson, R.F.W.; Raphael, R.A.; Stibbard, J.H.A.; Tidbury, R.C. *J. Chem. Soc., Perkin Trans. I* **1984**, 2159 and Raphael, R.A.; Stibbard, J.H.A.; Tidbury, R. *Tetrahedron Lett.* **1982**, *23*, 2407.
26. Schönenberger, B.; Summermatter, W. ; Ganter, C. *Helv. Chim. Acta.* **1982**, *65*, 2333.
27. Vaman Rao, M.; Nagarajan, M. *J. Org. Chem.* **1988**, *53*, 1432.
28. Keck, G.E.; Tafesh, A.M. *J. Org. Chem.* **1989**, *54*, 5845.
29. Barrish, J.C.; Lee, H.L.; Baggiolini, E.G.; Uskokovic, M.R. *J. Org. Chem.* **1987**, *52*, 1372 and Barrish, J.C.; Lee, H.L.; Mitt, T.; Pizzolato, G.; Baggiolini, E.G.; Uskokovic, M.R. *J. Org. Chem.* **1988**, *53*, 4282.
30. Kuroda, C.; Theramongkol, P.; Engebrecht, J.R. ; White, J.D. *J. Org. Chem.* **1986**, *51*, 956.
31. Bates, H.A., Farina, J.; Tong, M. *J. Org. Chem.* **1986**, *51*, 2637.
32. Alexander, R.P.; Paterson, I. *Tetrahedron Lett.* **1983**, *24*, 5911.
33. Kozikowski, A.P.; Schmiesing, R.J.; Sorgi, K.L. *Tetrahedron Lett.* **1981**, *22*, 2059.
34. Eugster, C.H.; Waser, P.G. *Experentia*, **1954**, *10*, 298.
35. Wang, P.C.; Joullié, M.M. *The Alkaloids*. Vol. XXIII. Ed. Brossi, A., Academic Press, New York, **1984**, p. 327-380.
36. Horwitz, J.; Anderson, C.H.; Perlman, R.L. *J. Pharmacol. Exp. Ther.* **1986**, *237*, 312 and Horwitz, J.; Perlman, R.L. *J. Pharmacol. Exp. Ther.* **1985**, *233*, 235.
37. Wang, P.-C.; Joullié, M.M. *J. Org. Chem.* **1980**, *45*, 5359.
38. Pochet, S.; Huynh-Dinh, T. *J. Org. Chem.* **1982**, *47*, 193.
39. Mantell, S.J.; Fleet, G.W.J.; Brown, D. *J. Chem. Soc., Chem. Commun.* **1991**, 1563.
40. Mantell, S.J.; Ford, P.S.; Watkin, D.J.; Fleet, G.W.J.; Brown, D. J. *Tetrahedron Lett.* **1992**, *33*, 4503.

41. Bandzouzi, A.; Chapleur, Y. *J. Chem. Soc., Perkin Trans. I* **1987**, 661.

42. Bandzouzi, A.; Chapleur, Y. J. *Carbohydr. Res.* **1987**, *171*, 13.

43. Mulzer, J. Angermann, A.; Münch, W.; Schlichthörl, G.; Hentzschel, A. *Liebigs Ann. Chem.* **1987**, 7.

44. Lysenko, Z.; Ricciardi, F.; Semple, J.E.; Wang, P.C.; Joullié, M.M. *Tetrahedron Lett.* **1978**, *30*, 2679.

45. Chan, T.H.; Li, C.J. *Can. J. Chem.* **1992**, *70*, 2726.

46. Shapiro, G.; Buechler, D; Hennet, S. *Tetrahedron Lett.* **1990**, *31*, 5733.

47. Still, W.C.; Schneider, J.A. *J. Org. Chem.* **1980**, *45*, 3375.

48. Takano, S.; Iwabuchi, Y.; Ogasawara, K. *J. Chem. Soc., Chem. Commun.* **1989**, 1371.

49. Chmielewski, M.; Guzik, P. *Heterocycles* **1984**, *22*, 7.

50. Adams, J.; Poupart, M.A.; Grenier, L. *Tetrahedron Lett.* **1989**, *30*, 1753.

51. Cowling, A.P.; Mann, J.; Usmani, A.A. *J. Chem. Soc., Perkin Trans. I* **1981**, 2116.

52. Pirrung, M.C.; DeAmicis, C.V. *Tetrahedron Lett.* **1988**, *29*, 159.

Chapter 10

SYNTHESIS OF NATURALLY OCCURRING ARYL *C*-GLYCOSIDES

I. INTRODUCTION

The contents of this chapter will address the synthetic approaches to naturally occurring aryl *C*-glycosides. As in Chapter 9, the focus will be on specific natural products and the strategies used in their synthesis. The natural products that have been selected for this chapter are nogalamycin and its congeners, the vineomycins (with emphasis on vineomycinone B_2), the papulacandins (including chaetiacandin), the gilvocarcin class of antibiotics, and a few miscellaneous aryl *C*-glycosides that have also been synthesized. Although their have been numerous synthetic methods developed to construct anomeric aryl carbon-carbon bonds the bulk of these methods will not be presented here. The strategies used to assemble aryl *C*-glycosides can be found throughout the text depending on the method used to assemble the compound (*i.e.* Lewis acid mediated approaches, nucleophilic approaches, etc.). Some of the natural products listed here are biologically important compounds, but the emphasis of the chapter will be on the synthetic strategies applied to their preparation and not on structure-activity relationships or relative potency studies.

II. ARYL *C*-GLYCOSIDE NATURAL PRODUCTS

A. NOGALAMYCIN AND ITS CONGENERS
1. Structure and Properties
Nogalamycin (1) is a carbon linked anthracycline antibiotic isolated in 1968[1] that is active against Gram positive bacteria and it also possesses antitumor activity.[2]

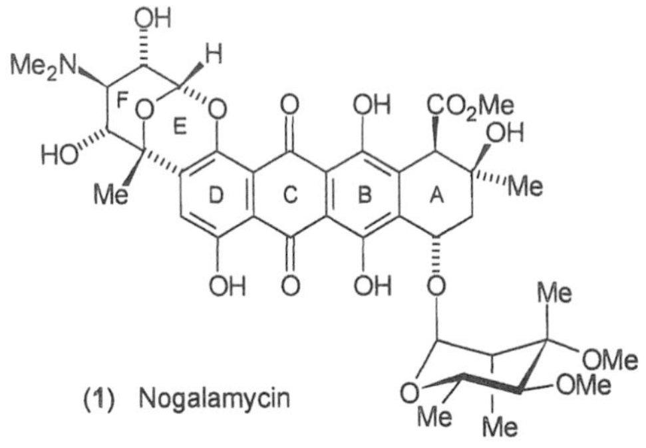

(1) Nogalamycin

Figure 1

Although no total synthesis of (1) has appeared several approaches to the DEF ring system have been published. If the sugar unit is removed then one obtains a nogarol (2), a compound which still possesses significant biological activity (Fig. 2).[2] The semi-synthetic analog menogaril (**3**),[3, 4] which is in clinical trials and holds promise in the treatment of breast carcinoma, is also a popular synthetic target and two total syntheses of this compound have been reported. It differs from nogalamycin in the sense that it is missing the ester function and the sugar residue has been replaced by a methyl group.

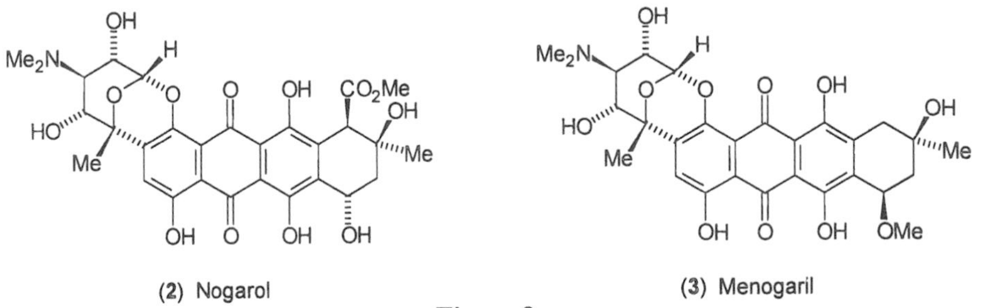

(**2**) Nogarol   (**3**) Menogaril

Figure 2

The unique and interesting structural feature of all these compounds is the DEF ring system. It is comprised of a *C*-glycosidic linkage to an amino sugar which is further linked to the anthracycline to form the E ring. It is more appreciated when drawn as in Fig. 3.

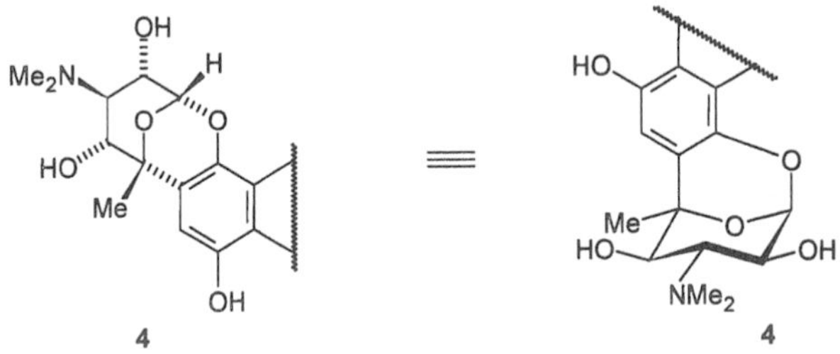

Figure 3

2.   Synthesis of the Nogalamycins

Bates' successful racemic approach[5] to the DEF ring system of nogalamycin began with furan **5** which was oxidized and methylated to the pyrans **6** and **7**.

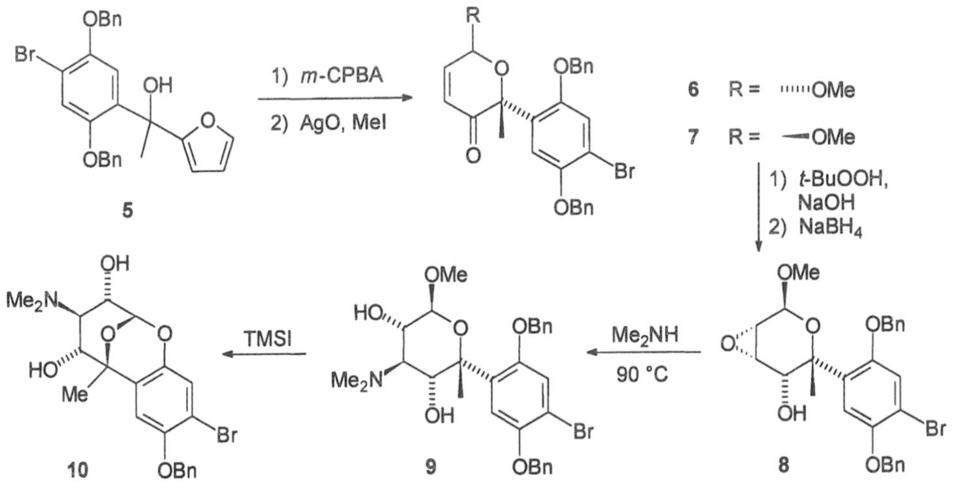

Scheme 1

267

The major compound **7** was epoxidized to **8**. It was found to be better if the ketone was first reduced and then the amino functionality introduced by ring opening the epoxide, away from the anomeric carbon to give **9**. Exposure of **9** to trimethylsilyl iodide caused selective debenzylation and cyclization to then furnish the tricyclic compound **10** which is well matched to the DEF ring system found in nogalamycin, Scheme 1.

Terashima and co-workers successfully prepared menogaril in optically pure form. Their strategy is illustrated in Scheme 2. The right hand side was to be assembled by the use of Diels-Alder chemistry between a suitable diene and the quinone **12**. The DEF ring system was envisaged as coming from addition of a suitable aryl nucleophile to the ketone **15** which could then be cyclized to afford the target ring system. The strategy is flexible and allows for variation in the substituents, especially on the A ring.

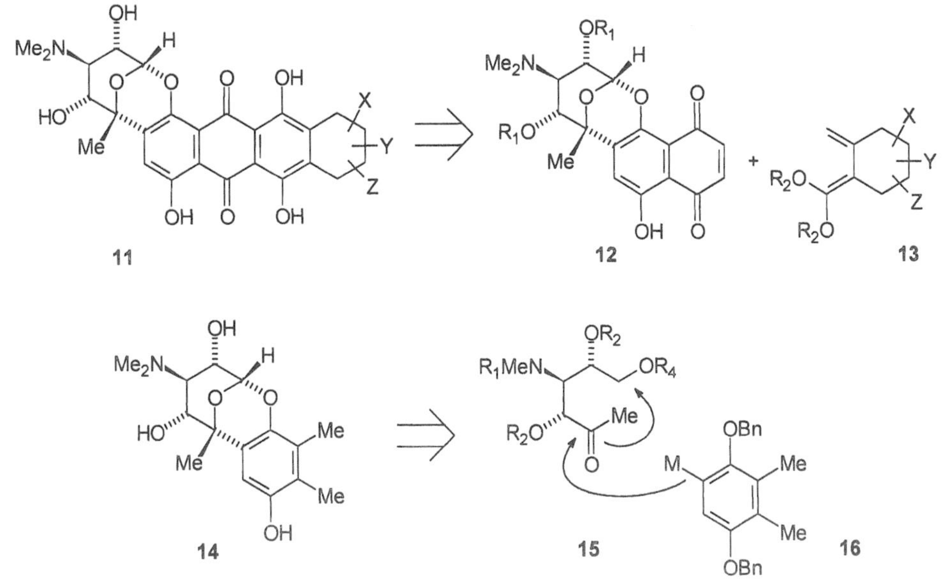

Scheme 2

Ketone **20** was available in optically pure form from D-arabinose.[6] Diol **17** was made by sequential glycosidation, ketalization, mesylation, and hydrolysis of D-arabinose.

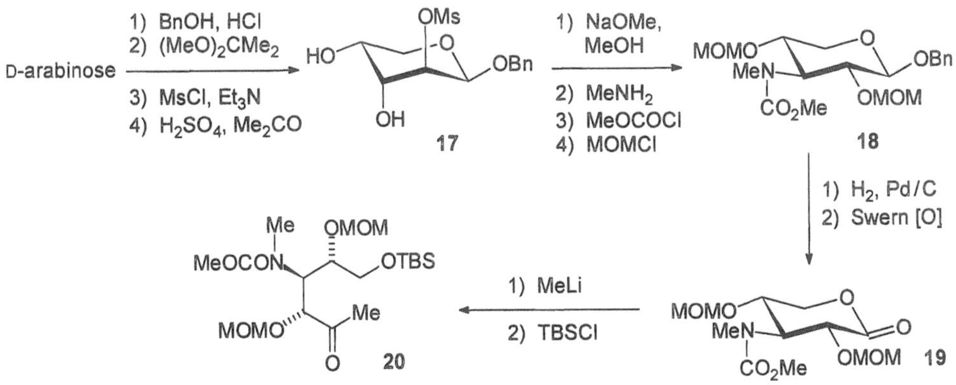

Scheme 3

Exposure of diol **17** to sodium methoxide in methanol gave an epoxide that was ring opened with methylamine and protected to give **18**. Removal of the anomeric protecting group gave a lactol which was oxidized and treated with methyllithium to deliver the pivotal ketone **20**.

The CDEF ring system of menogaril was assembled as shown in Scheme 4.[7] The 2-lithionaphthalene **21** was condensed with ketone **20** to give **22** as the major isomer (14:1) in good combined yield. A multistep sequence then provided **23** and exposure to CAN gave quinone **24** regioselectively. The naphthoquinone was reduced and cyclization was brought about by treatment with trimethylsilyl bromide to give **25** (TMSBr also caused removal of the MOM protecting groups). Selective acylation gave a diacetate which was demethylated and oxidized to the naphthoquinone **26**.

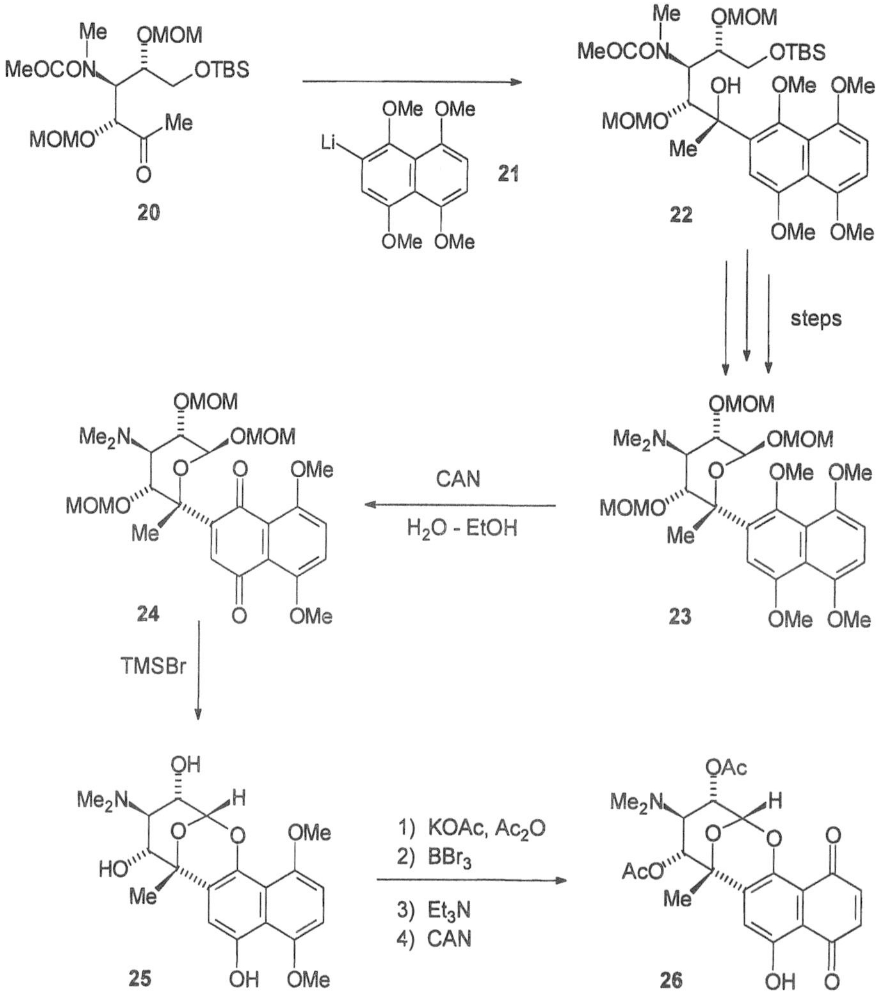

Scheme 4

With the chiral naphthoquinone in hand the workers were now poised to synthesize a few nogalamycin congeners. They chose to begin by preparing nogarene (**31**) which is a nogalamycin congener that possesses no stereochemistry in the A ring.[8] Diene **29** (available from **28** by double deprotonation and silylation of the anion derived from acid **28**) underwent Diels-Alder reaction at 60°C to give, after air oxidation and acid work-up, adduct

30. The synthesis was completed by oxidation of **30** with CAN in the presence of CSA to provide (+)-nogarene (**31**), Scheme 5.

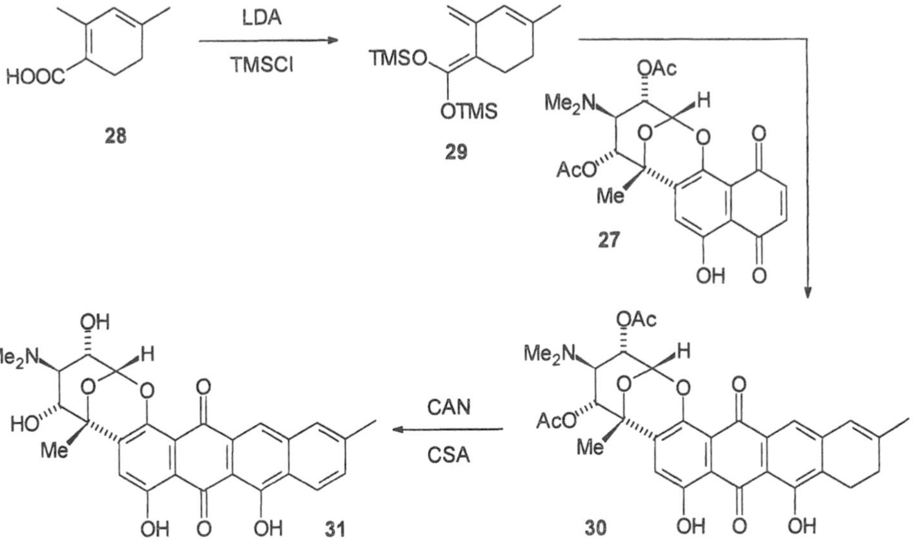

Scheme 5

They next chose to prepare (+)-7-deoxynogarol by utilizing a similar strategy.[8] The diene **34** was prepared from the intermediacy of epoxide **32** by reduction and stepwise oxidation to an acid which was then converted to the corresponding trianion and silylated. Diels-Alder reaction of **34** with **27** gave a mixture of isomers in which the desired one was minor (**35**:**36** :: 1:4). Deacetylation then gave the target (**37**).

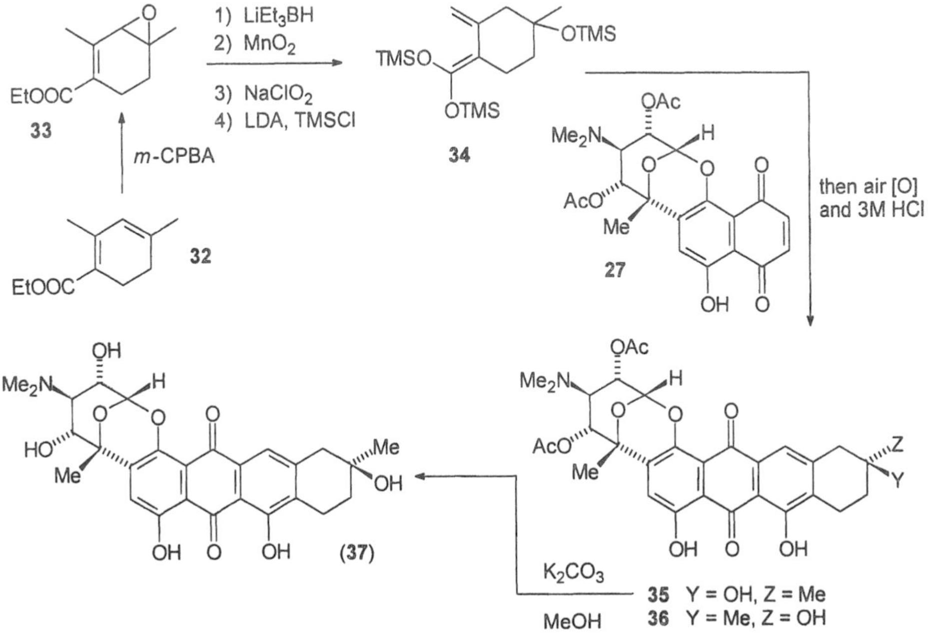

Scheme 6

Terashima and co-workers[8] were now poised to apply the same strategy to be the first to synthesize menogaril. In order to use the Diels-Alder approach they needed the diene **40**. Its preparation was not facile, but was accomplished as shown in Scheme 7. The diester **38** was hydrolyzed and decarboxylated to the keto-acid **39**. Reduction was followed by trianion formation and trapping with trimethylsilyl chloride to provide **40**. Diels-Alder reaction with **27** then provided the expected ratio of **41** and **42** of which **41** was converted to menogaril. The workers did make several analogs[9] by variations in the A ring and they concluded from these studies that the *C*-7 methoxy and the absolute stereochemistry at *C*-9 are important for activity. Also, aromatization of the A ring causes a severe loss in activity.

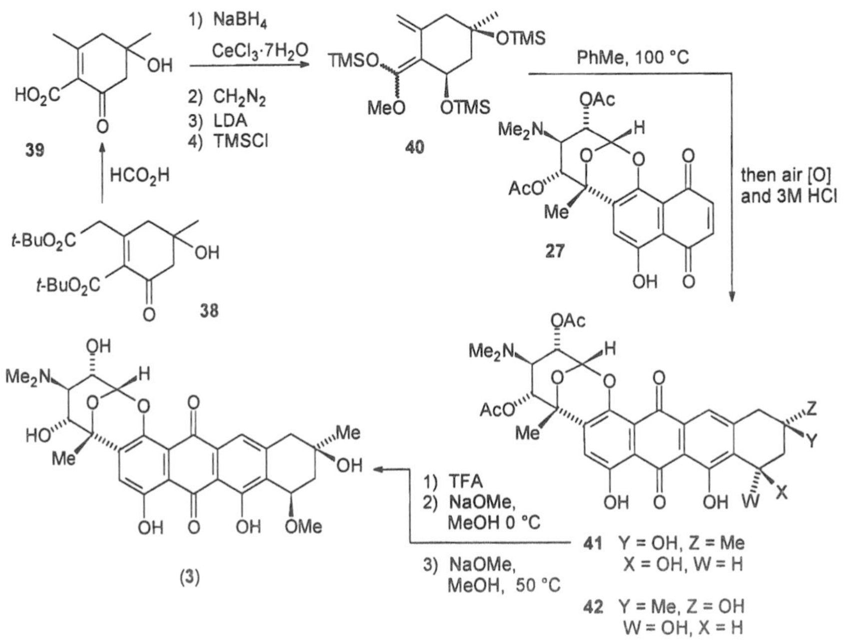

Scheme 7

DeShong's approach[10] also relied on cycloaddition chemistry, but he employed a 1,3-dipolar addition to assemble an advanced model of the DEF ring system of nogalamycin.

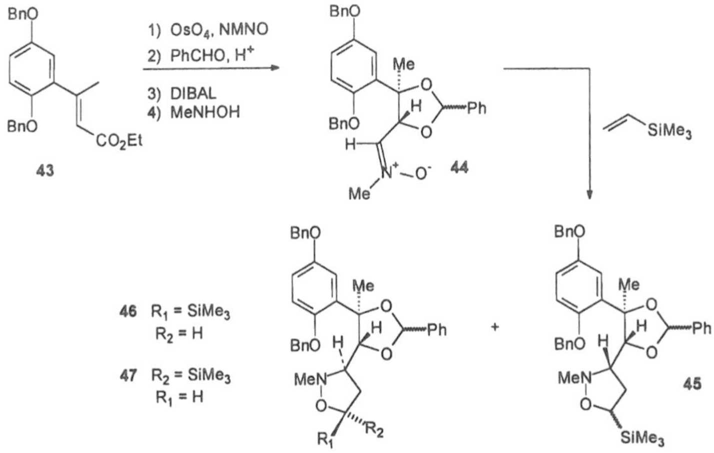

Scheme 8

Ester **43** was hydroxylated and protected as its benzylidene acetal and DIBAL reduction was followed by condensation with methylhydrazone to give the nitrone **44**. Cycloaddition with vinyltrimethylsilane gave 3 cycloadducts in a 60:20:20 ratio, Scheme 8. Compounds **46** and **47** were fragmented with fluoride ion to give the α,β-unsaturated aldehyde which when hydrogenated and exposed to acid cyclized to the tricyclic system **49**, Scheme 9.

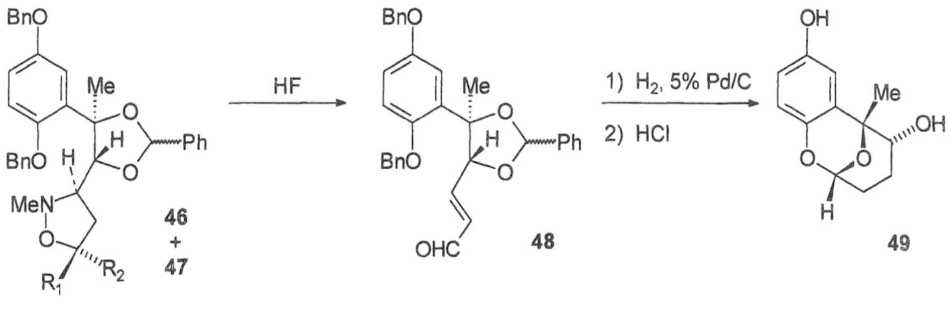

Scheme 9

The minor isomer, **45**, was fragmented to **50** with acetyl chloride and hydrogenation gave an intermediate tetrol **51** that cyclized to give the skeleton corresponding to the DEF ring system (this ring system is missing one hydroxyl group at *C*-2') of nogalamycin.

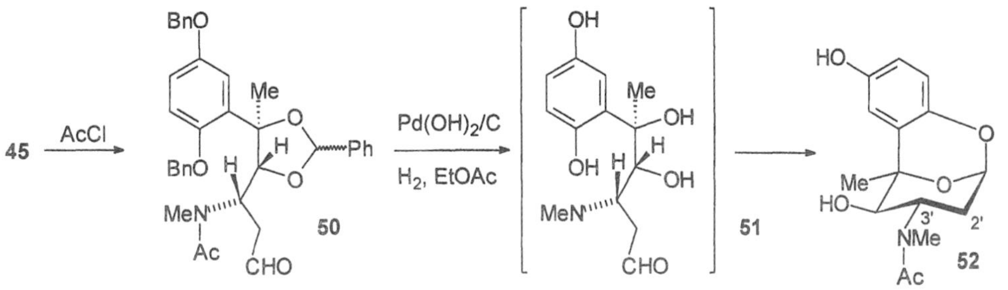

Scheme 10

Hauser *et al.*[11] used the oxidative conversion of furans to pyrones in their preparation of model compounds of the DEF ring system of nogalamycin. Fries rearrangement of **53** gave ketone **54** which was then treated with methyllithium (**54→55**), Scheme 11.

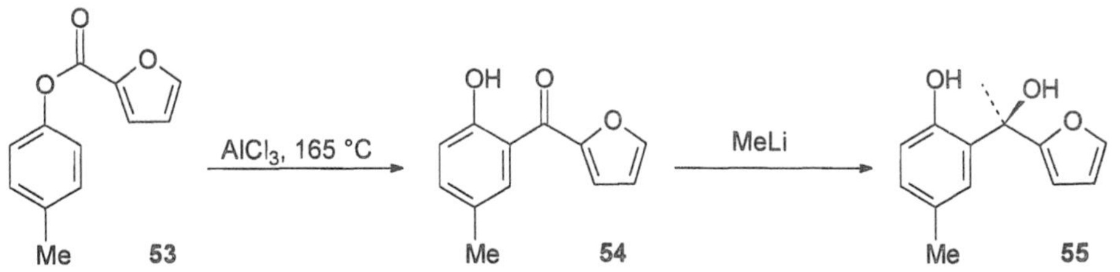

Scheme 11

Exposure of furan **55** to bromine in methanol followed by acid treatment furnished the benzoxocin **57** in good yield. Transformation of the pyran portion into a suitable "sugar" derivative was straightforward. Reduction was followed by *syn* hydroxylation to furnish the model compound **59** that possesses the manno configuration (Scheme 12). Other sugar configurations were also accessible from **57** such as **60**.

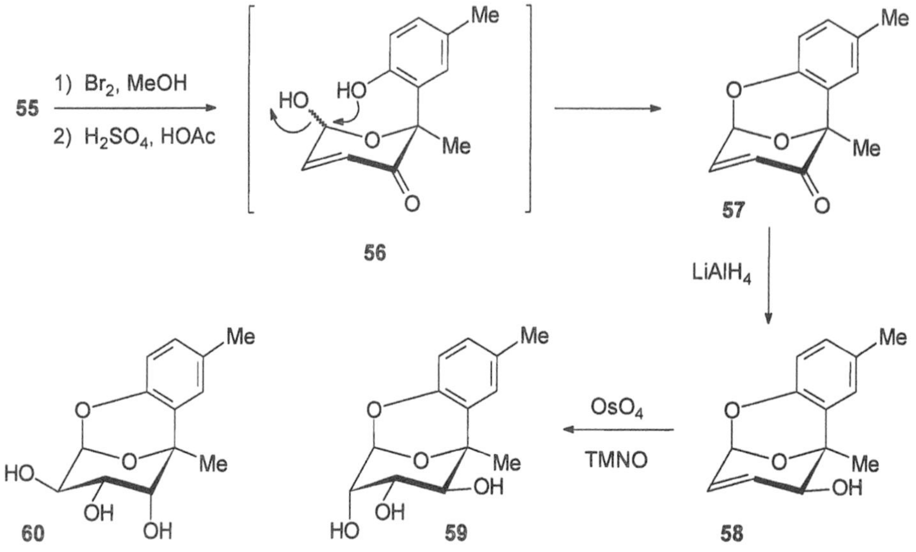

Scheme 12

After subsequent experimentation[12] the workers found that **57** could be used to gain access to deoxyamino sugar analogs of the key ring system. The double bond was reduced and the amino function introduced via the intermediacy of oxime **61** by reduction to **62**. Alternatively, the amino group could be inverted by epimerization. Accordingly, alcohol **62** was oxidized via the Swern method and base induced epimerization (triethylamine in dichloromethane) furnished the desired product **66**, after reduction.

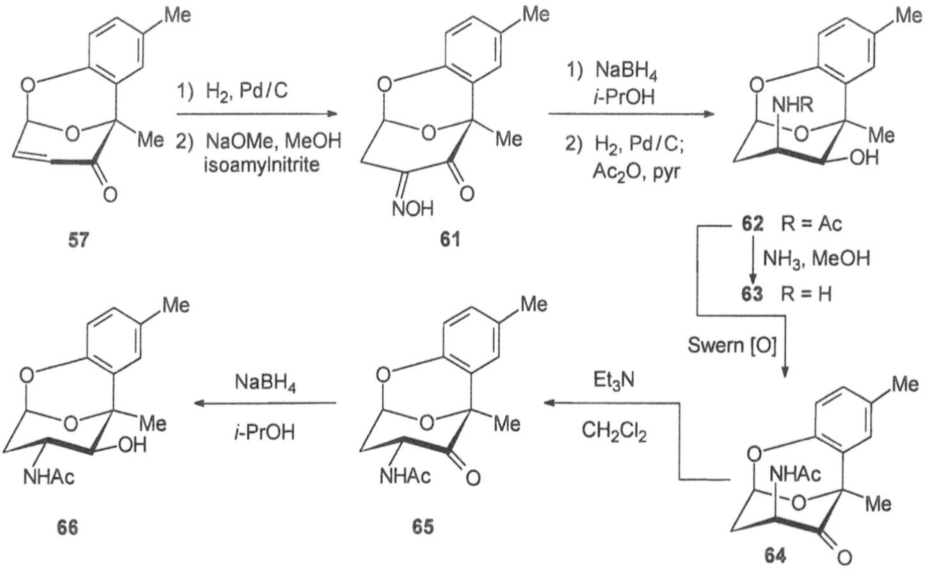

Scheme 13

In further work Hauser and Adams[13] used the addition of an organometallic to a sugar ketone to setup the stereochemistry of C-1' in the EF ring juncture of nogalamycin. Both ketones (68 and 71) were accessible from the aldehyde 67 by addition of the appropriate organometallic followed by Collins oxidation (Scheme 14). Careful hydrogenation of 69 and 72 gave the expected diols 70 and 73 respectively.

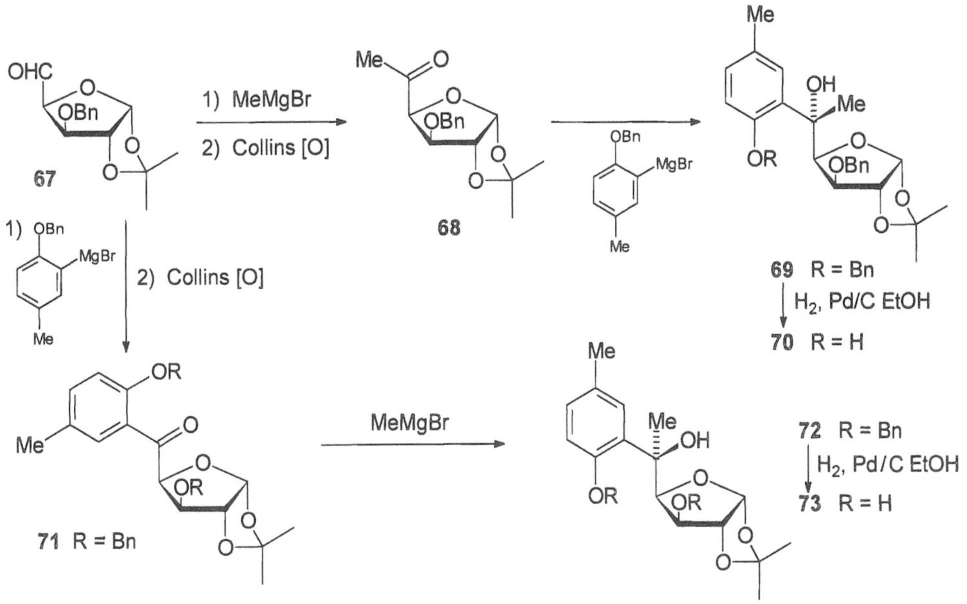

Scheme 14

Acid treatment of 70 gave, after acetylation, 74 (16%) and 75 (17%). When 73 was exposed to the same conditions an acceptable 57% yield of the L-ido compound 75 was obtained along with 4% of the gluco isomer 74 (Scheme 15).

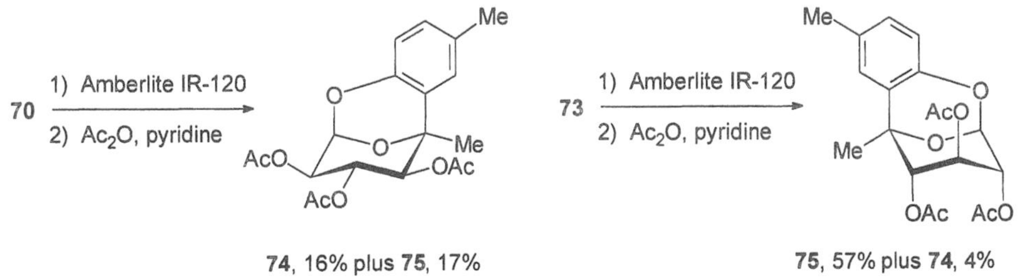

74, 16% plus 75, 17%   75, 57% plus 74, 4%

Scheme 15

In 1991 Hauser, Chakrapani, and Ellenberger[14] reported the total synthesis of racemic 7-con-O- methylnogarol (3). The construction of the DEF ring system follows their earlier work in which the keto-amide is converted to the hexenulose via oxidative ring expansion of a suitable furan. Keto-amide 76 was condensed with 2-furyllithium to give the tertiary alcohol 77 that when exposed to m-CPBA rearranged to pyran 78. Glycosidation gave a 1:5 mixture of isomers and the major (79) was epoxidized (t-BuOOH, triton B, CH_2Cl_2) and the keto function selectively reduced to 81. Ring opening with dimethylamine was followed by selective demethylation to then furnish the product of ring closure 83. The alcohols were protected as their MEM ethers and orthometallation followed by DMF quench gave a aldehydo-amide that when treated with trimethylsilyl cyanide and subsequent exposure to acid gave the 3-cyanoisobenzofuranone **85**, Scheme 16.

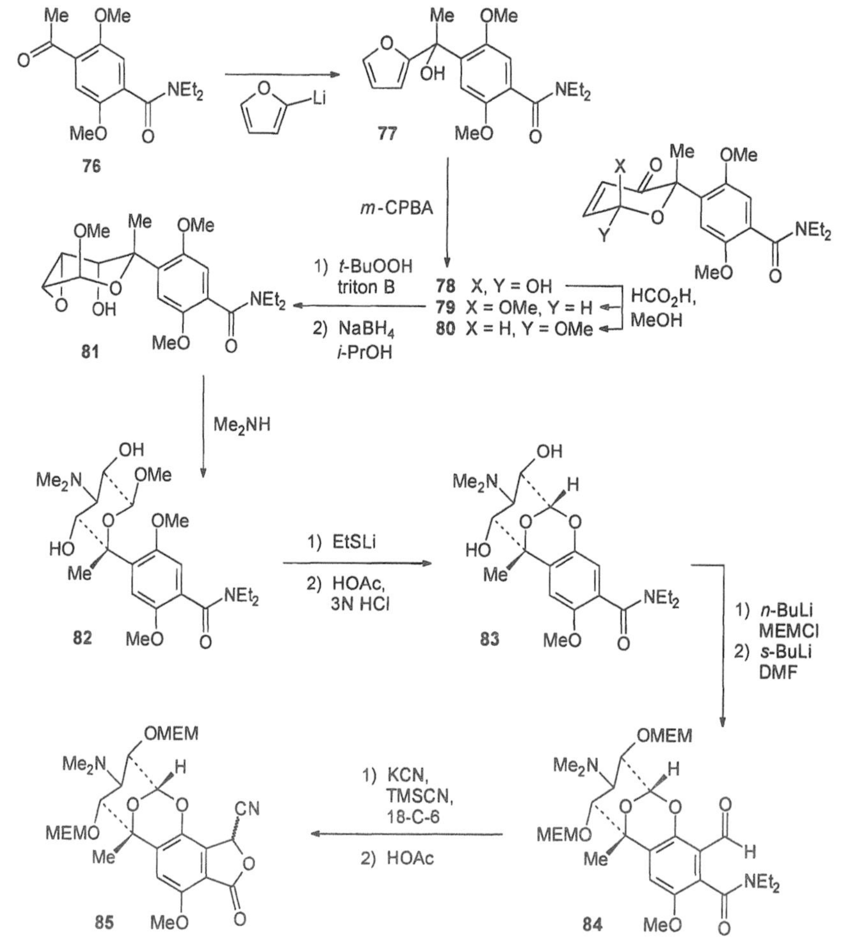

Scheme 16

Cycloaddition of the anion derived from **85** with the enone **86** gave a 63% yield of adduct **87** after oxidation, Scheme 17.

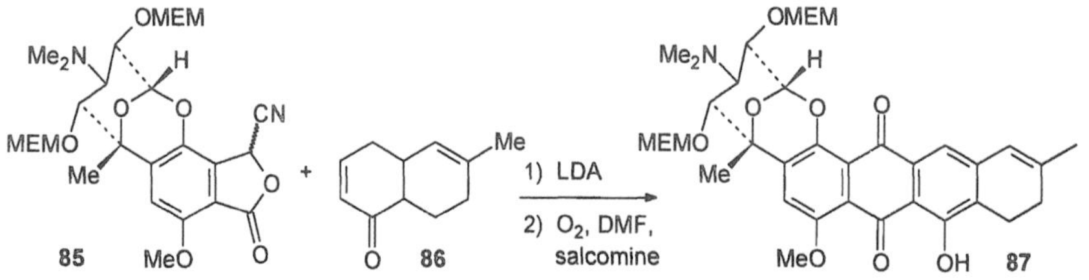

Scheme 17

Protective group manipulations gave an intermediate that was epoxidized to provide a mixture of oxiranes **88**. The epoxides were reductively cleaved to furnish a mixture of tertiary alcohols of which the major (6:1) was the desired one. Further protective group manipulations and benzylic oxidation gave **90** as a mixture of epimers, but this mixture could be converted to the desired one by treatment with TFAA and sodium methoxide. Conversion into the target (3) was completed by deprotection and methylation, Scheme 18.

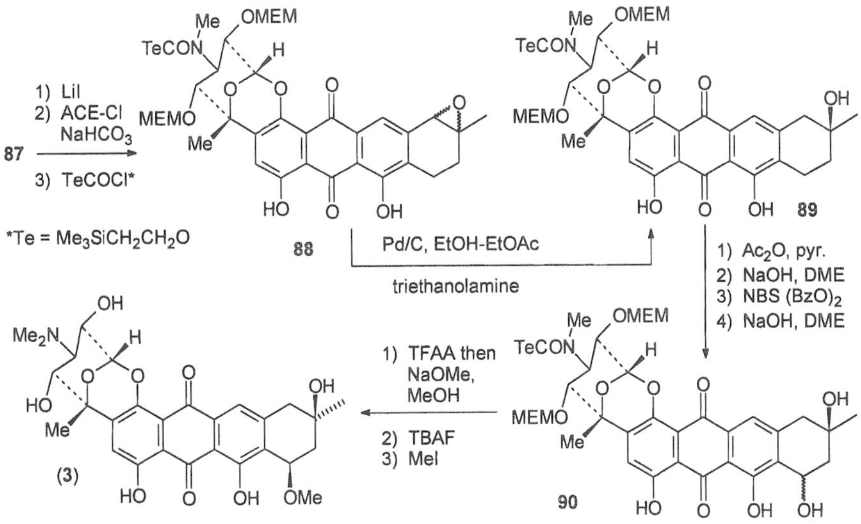

Scheme 18

Weinreb and co-workers[15] used an intramolecular *N*-sulfinyl dienophile Diels-Alder cycloaddition (Scheme 19) to efficiently construct the ketone **97** which can be used to assemble the DEF ring system of nogalamycin by the method of Terashima.[6] Diol **91** was selectively silylated and converted to carbamate **92**. Exposure to thionyl chloride/pyridine produced an intermediate *N*-sulfinyl carbamate that cyclized to a (78:1) mixture of sulfoxides **94** that were oxidized to **95**. Base treatment caused elimination, and carbamate hydrolysis was followed by protection to furnish **96**. Oxidative cleavage then provided the target ketone **97** which contains the proper relative stereochemistry for potential conversion into nogalamycin.

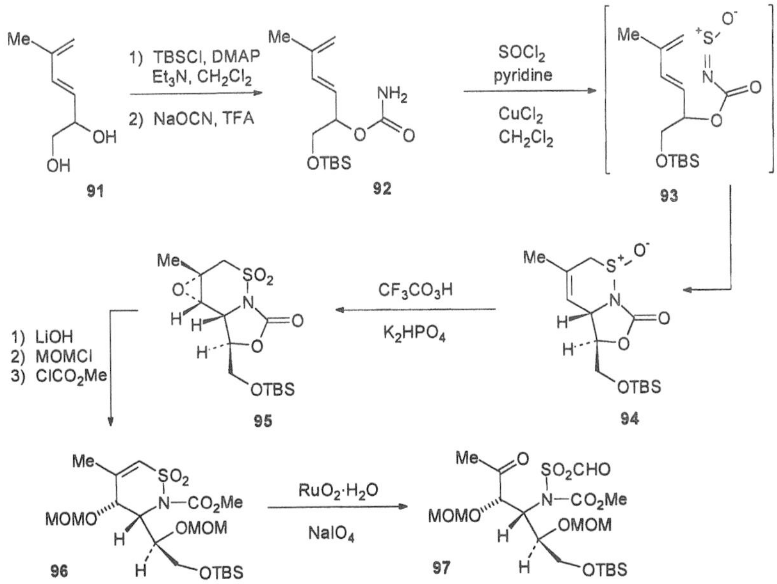

Scheme 19

Semmelhack and Jeong[16] applied arylcarbene-chromium complex alkyl-cycloaddition to construct a suitable aromatic system for potential elaboration to nogalamycin analogs. A key intermediate is the C-glycoside **98** which is accessible from a cycloaddition reaction between **99** and **100**, Scheme 20.

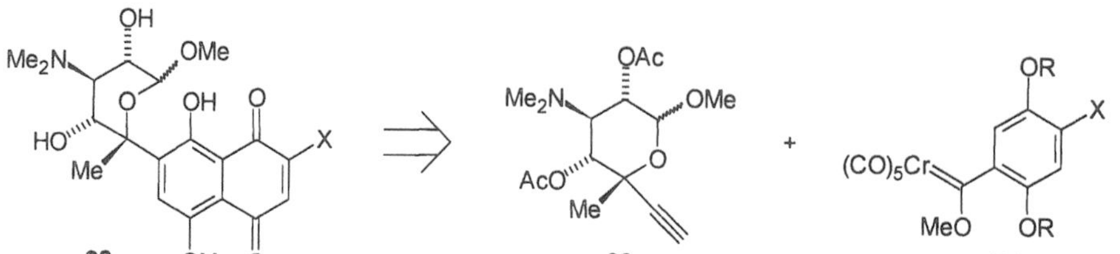

Scheme 20

Oxidation of the furan **101** with m-CPBA gave lactol **102**. Conversion to the hemiacetal was straightforward and epoxidation was found to proceed best with the use of chlorox in dioxane to bring the sequence up to **104**. Reduction with sodium borohydride then provided **105**, after acetylation. Alkyne/carbene-chromium cycloaddition with the chromate ester **106** gave **108** (43%) and **109** (34%). When the carbene-complex **107**, a compound that resembles more closely compounds that would be used in the nogalamycin synthesis, was used no desired adduct was formed! The products formed were furan **111** and the isomeric indenes **112**.[17]

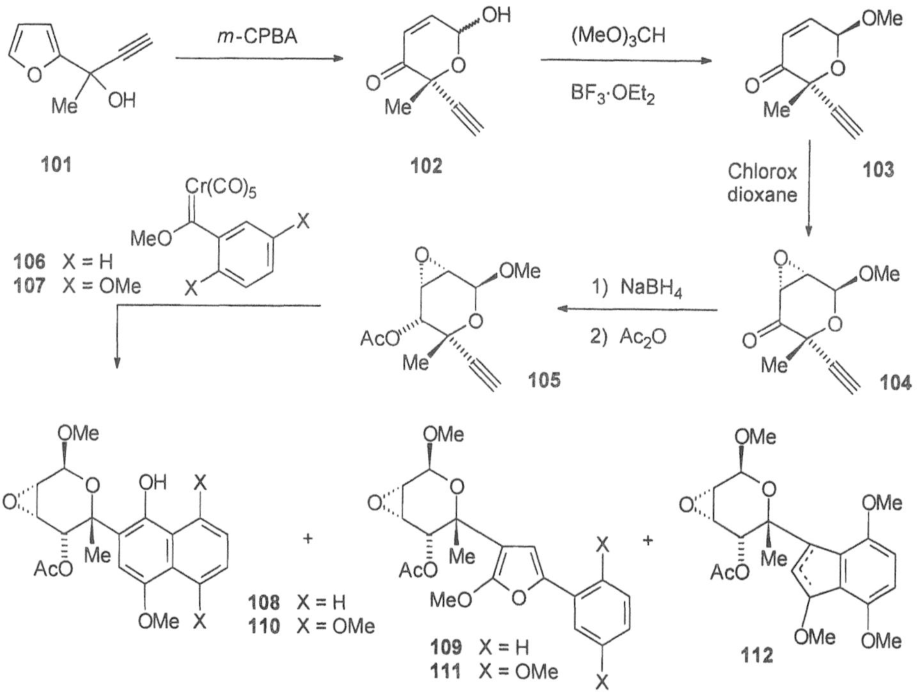

Scheme 21

Smith and Wu[18] also attempted to synthesize the DEF ring system of nogalamycin by a unique approach. They converted D-glucose to the olefin **113** by several steps in order to affect cyclization between the aromatic ring and *C*-5 of the sugar to try and obtain **114**. They observed that external nucleophiles could be trapped by the olefin, but little if any of the desired product could be formed by the use of TFA, NBA, or iodonium dicollidine perchlorate, Scheme 22.

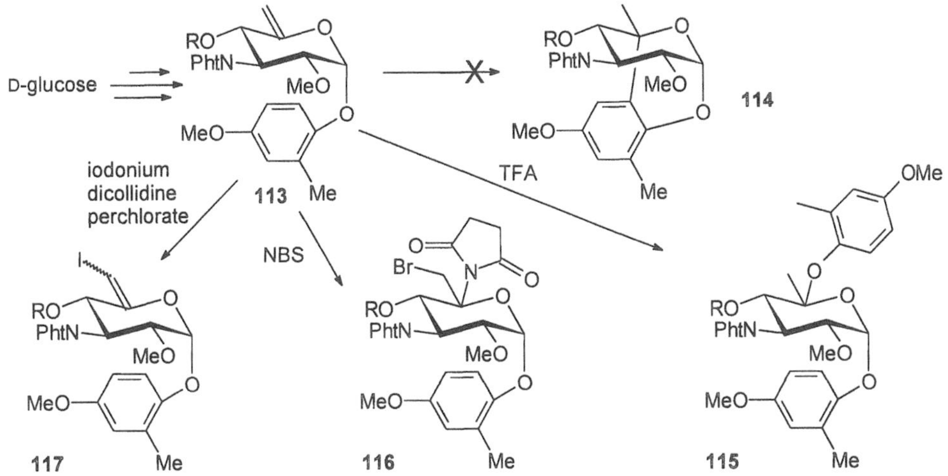

Scheme 22

B.   CHAETIACANDIN AND THE PAPULACANDINS
1.   Structure and Properties

The *C*-aryl glycosides papulacandins A (**118**), B (**119**), C (**120**), D (**121**)[19] and chaetiacandin (**122**)[20] are structurally related antifungal antibiotics that inhibit growth[21] or $(1 \rightarrow 3)$-β-D-glucan synthetase[22] (or both) in several fungi, Fig. 4 and 5.[23]

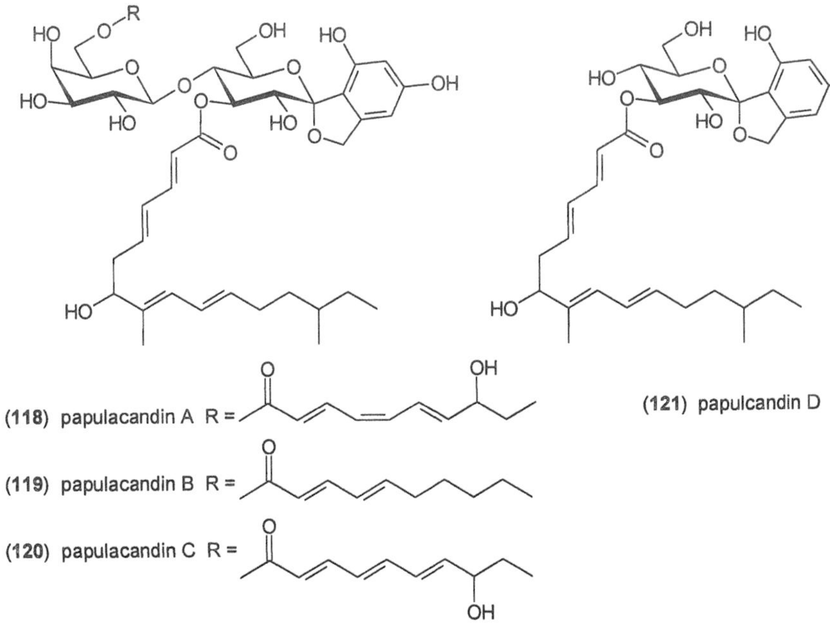

(**118**) papulacandin A R =

(**119**) papulacandin B R =

(**120**) papulacandin C R =

(**121**) papulcandin D

Figure 4

These compounds are characterized by a β-aryl anomeric carbon-carbon bond and in addition the papulacandins also possess a spiroketal function. The fatty acid chain is required for growth inhibition, but not for inhibition of glucan biosynthesis. Traxler *et al.* have carried out several SAR studies in order to determine the functions required for activity.[24]

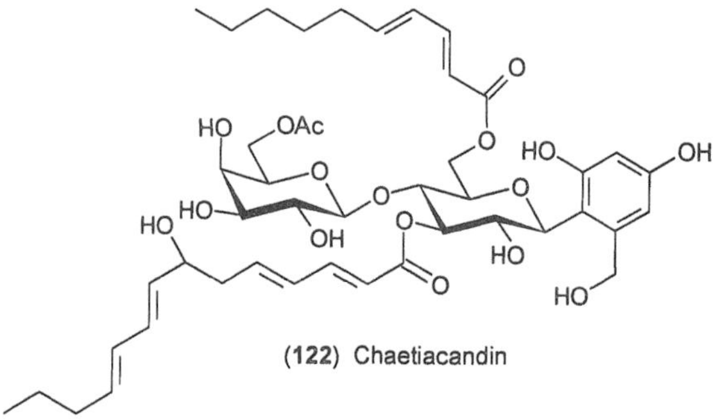

(122) Chaetiacandin

Figure 5

2.   Synthesis of Papulacandins

In order to successfully synthesize the simplest papulacandin one must first address the assembly of the spiroketal function and have a strategy that allows for selective acylation of *O*-3. For the more complex members differentiation between *O*-3 and *O*-4 must be possible in order to not only be able to acylate, but also couple (*O*-4) with an appropriate glycosyl donor.

Work by Schmidt and Frick[25] focused on the construction of the spiroketal moiety of papulacandin D. Their approach employed the condensation of an aryl nucleophile with an open chain sugar. The open chain gluco derivative 123 was oxidized and esterified to 124. This ester was allowed to undergo condensation with the aryllithium 125 to provide 126 that when hydrogenated and acylated gave the required skeleton 127 corresponding to papulacandin D. Alternatively, 126 could be accessed directly from the aldehyde 123 by condensation with the aryllithium followed by oxidation, Scheme 23.

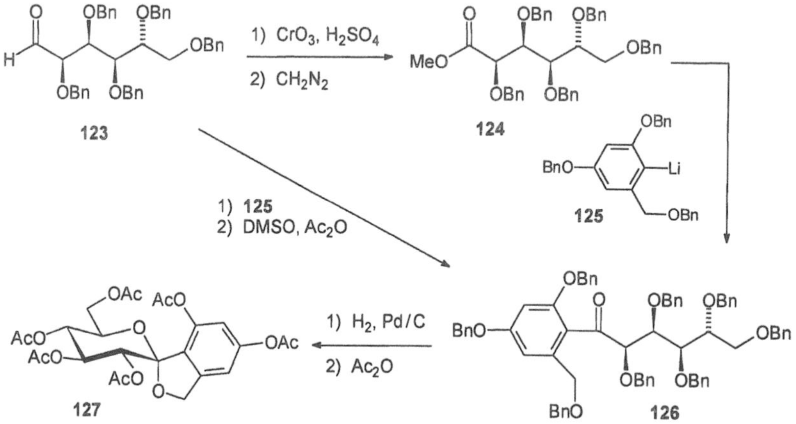

Scheme 23

Both Bihovsky and Czernicki developed similar approaches to the synthesis of the papulacandin spiroketal system. They, like Schmidt, used the condensation of an aryl nucleophile with a sugar carbonyl, but in their cases they carried the condensation reaction on a suitably protected gluconolactone derivative. Bihovsky and Rosenblum[26] condensed dianion **129** with gluconolactone **128** and obtained a mixture of three compounds. The desired product (**130**) was formed in only 14% yield. Protection of the free hydroxyl on the aromatic ring and changing the solvent changed the identity of the main product. In this instance a 42% yield of the open chain ketone **134** was obtained. The silicon protecting group was removed and hydrogenolysis of the benzyl groups caused spontaneous cyclization to give **136** as the sole isomer, after removal of the MOM protecting groups. In order to finish the synthesis, the workers decided to functionalize *O*-3. Accordingly *O*-4 and *O*-6 were protected as a benzylidene acetal and acylation with palmitoyl chloride gave **137**, an analog of papulacandin D.

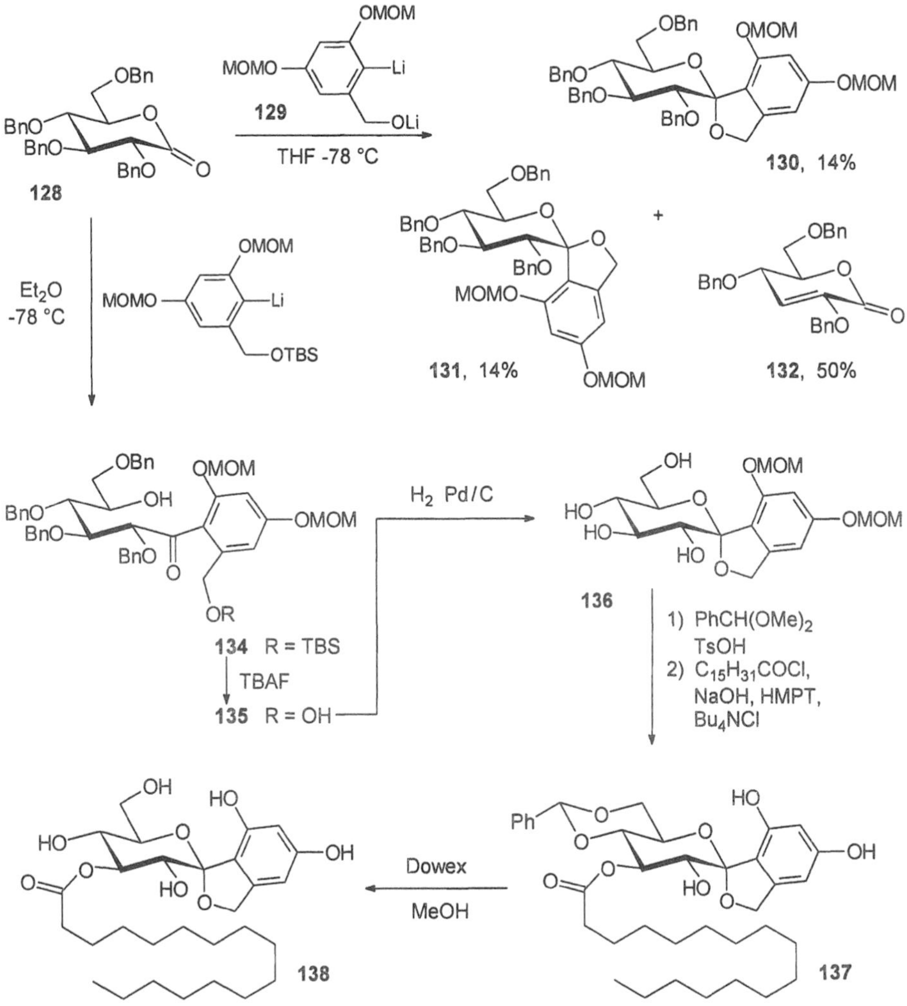

Scheme 24

Czernecki and Perlat's approach[27] at first glance seems very similar to that above, but close scrutiny reveals subtle, but important differences. The workers found that condensation of the anion **139** with the gluconolactone **128** proceeded well in toluene, but poorly in THF. Crude **140** was then treated with boron trifluoride etherate and triethylsilane in acetonitrile at -40°C. Normally this set of conditions will stereoselectively reduce the hemiketal to give a β-*C*-glycoside. Under these reaction conditions the benzylic oxygen participates from the axial direction and the triethylsilane serves to reduce the triphenylmethyl group to triphenylmethane to then provide spriroketal **141**. The sequence was completed by debenzylation to provide the partially protected skeleton **142**.

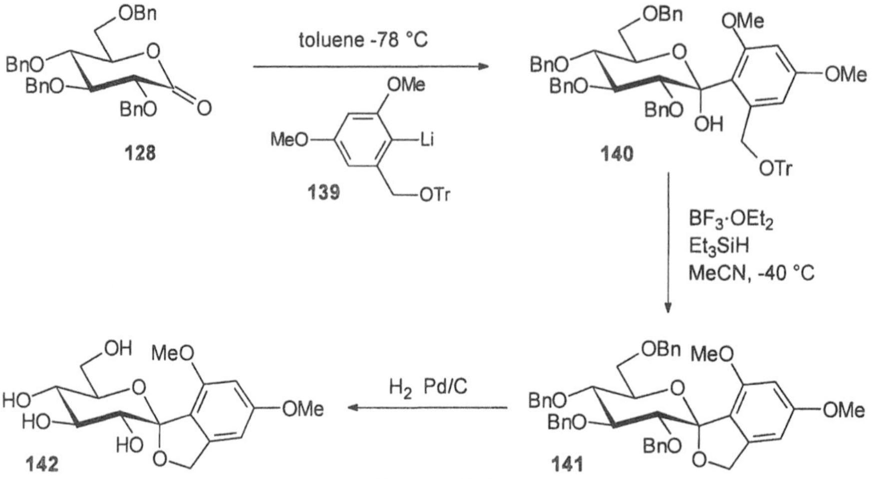

Scheme 25

Both Friesen and Beau simultaneously published palladium mediated routes to the papulacandin core. Beau and Dubois' approach[28] was versatile enough that it allowed them access to both the papulacandin and chaetiacandin cores. The sequence begins with the thioglycoside **143** that was transformed into **144** by protection and oxidation. Elimination then provided a vinyl sulfone which when treated with tributyltin hydride and AIBN furnished the vinyl stannane **145**.

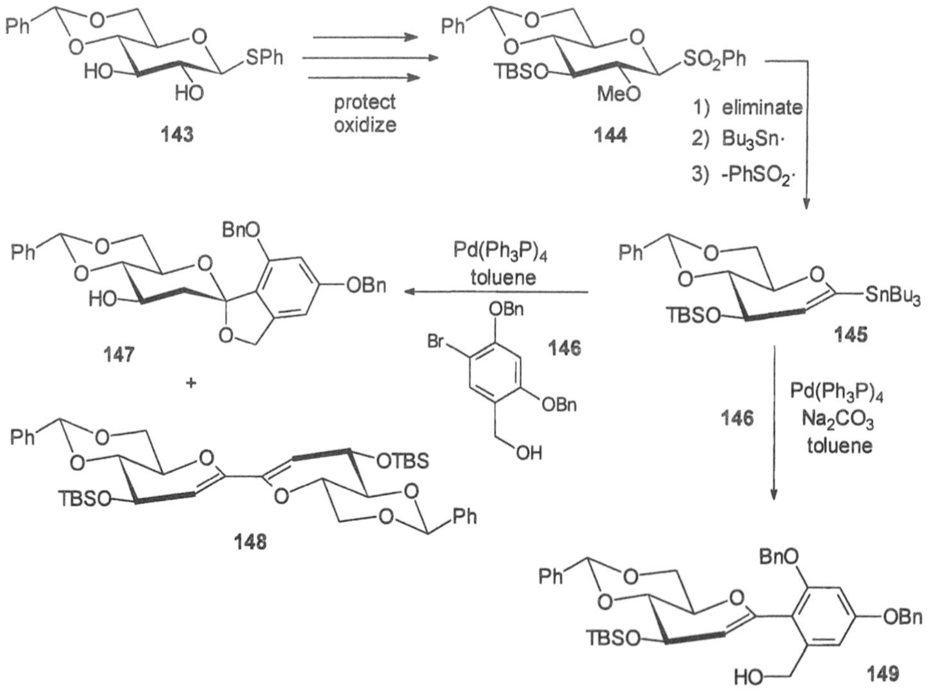

Scheme 26

When the glycal **145** was coupled with the bromide **146** under Pd(Ph₃P)₄ catalysis in refluxing toluene the spiroketal **147** was formed in 72% yield along with some homocoupling product **148**. The workers reasoned that acid produced during the course of the reaction causes spiroketalization leading to the 2-deoxy spiroketal. The reaction was therefore buffered with sodium carbonate and a 78% yield of the desired adduct **149** was obtained. Oxidation of **149** with *m*-CPBA (Scheme 27) gave (8:1) ratio of **150** and **151** in 82% yield. Separate deprotection and acetylation converted both **150** and **151** to the spiroketal **152**.

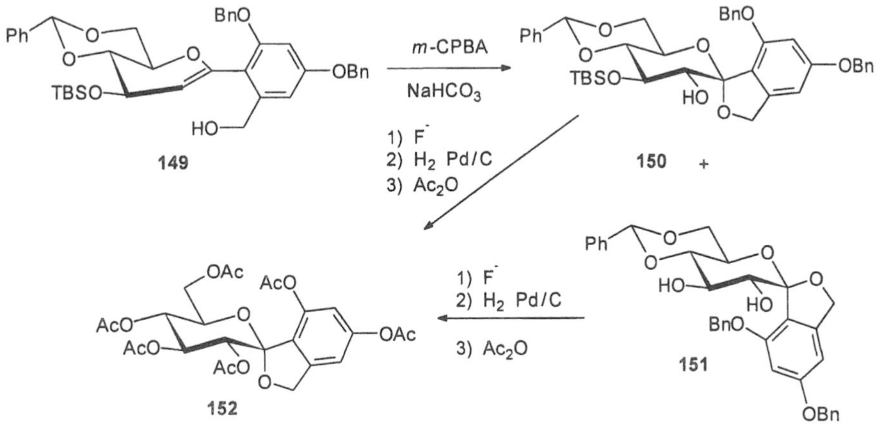

Scheme 27

Conversion of **150** to a 3-*O* ester was accomplished by sodium hydride induced *O*-3→ *O*-2 silyl migration, acylation, and deprotection to give, after acetylation, **154**.

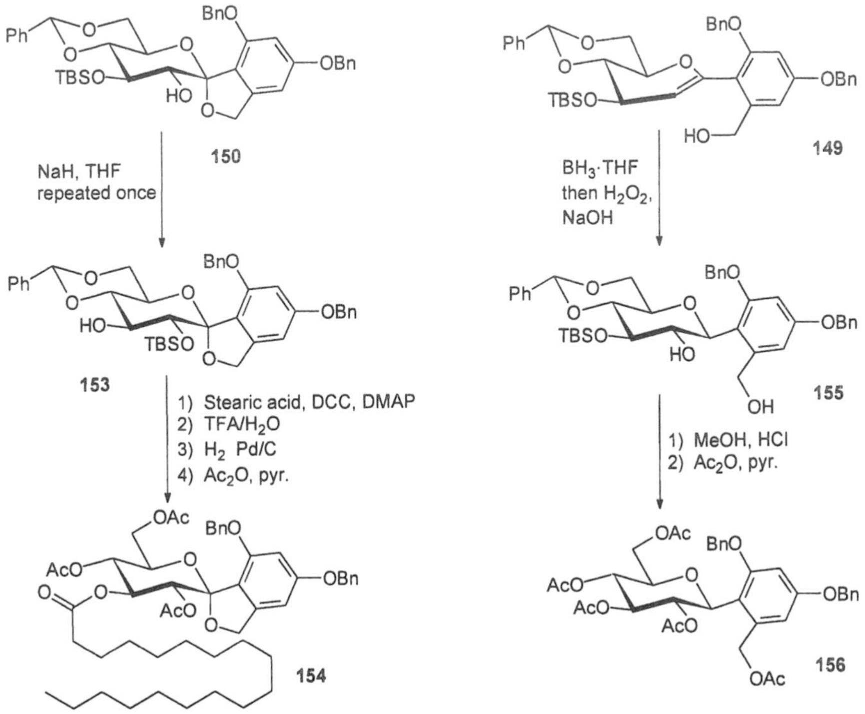

Scheme 28

Access to the chaetiacandin core from **149** was also carried out (Scheme 28). Hydroboration gave a 75% yield of **155** which was subsequently deprotected and acylated to provide **156**, a compound structurally representative of the chaetiacandin core.

Friesen and Sturino[29] used the silylated glycal **157** in a coupling reaction with the bromide **158** under palladium catalysis to give the C-aryl glycoside **159**. They found that optimum conditions entailed the use of 1.1 equivalents of **158** with Pd(Ph$_3$P)$_2$Cl$_2$ as catalyst. The acetate was reductively cleaved and oxidation of the resultant enol gave a 5:1 of **161** and **162**, respectively. A higher ratio (34:1) could be obtained if the reaction was carried out at -78°C. Acid catalyzed epimerization of **162** to **161** was accomplished by the use of PPTS in chloroform to give a net conversion of **160** to **161** in 80% overall yield after epimerization. The sequence was completed by desilylation, hydrogenolysis of the benzyl groups, and characterization as hexaacetate **163**. Access to the Chaetiacandin skeleton[30] was also possible by hydroboration of **164** (R = TBS) and subsequent oxidation to **165**.

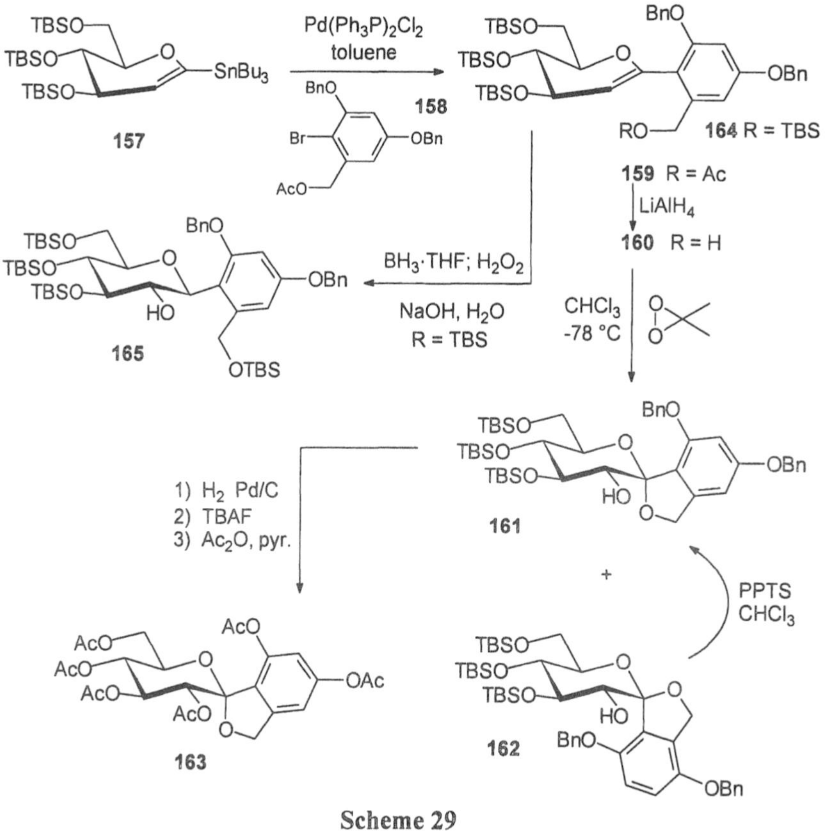

Scheme 29

Danishefsky's racemic approach[31] to the papulacandin core employed hetero Diels-Alder reaction as the key ring forming step (Scheme 30). Aldehyde **167** underwent cycloaddition with diene **166** under Yb(fod)$_3$ catalysis to give pyrone **168** in excellent yield. 1,4-Addition of a vinyl cuprate served to introduce a hydroxy methylene group by oxidative cleavage, reduction, and benzoylation (**168**→**169**). The ketone was deprotonated and converted into what is believed to be a mixture of silyl enol ethers that were oxidized and the desired ketol benzoylated, albeit in modest yield to furnish **170**. Conversion of **170** to **171** was carried out by conversion to the silyl enol ether and exposure to palladium acetate to provide enone **171**. Oxidation of the enone with m-CPBA in methanol then provided the desired methoxyhydrin **172**. The synthesis of the papulacandin core was completed by deacylation, acid treatment to affect spiroketalization, acetylation, debenzylation, and further acylation to provide racemic hexaacetate **174**.

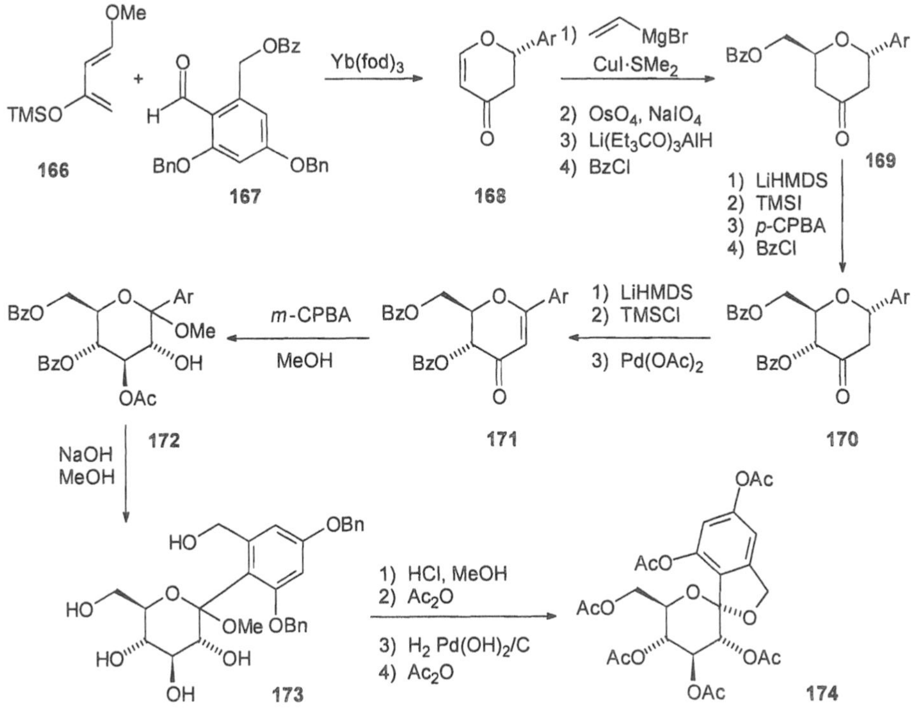

Scheme 30

C. VINEOMYCIN-FRIDAMYCIN ANTIBIOTICS AND RELATED COMPOUNDS
1. Structure and Properties

The vineomycin class of antibiotics (Fig. 6) are a class of aryl C-glycosides that were isolated from *Streptomyces matensis vineus* by Õmura and co-workers.[32] They have been found to be antibacterial agents and also exhibit antitumor activity against solid tumors. These include vineomycin B$_2$ (**175**), vineomycinone B$_2$ (**176**) (fridamycin A) and its corresponding methyl ester (**177**) Vineomycinone B$_2$ (**176**) has attracted considerable synthetic attention over recent years. The core consists of an anthrufin structure with a β-hydroxy acid side chain. In the case of the structurally related compound aquayamycin (**180**)[33] (Fig. 7) the side chain has been elaborated into a fourth ring. The most interesting feature is the presence of an aryl β-C-glycosidic bond. Some workers believe that these antibiotics hold "special promise" since theoretically they should be less vulnerable to deglycosidation than the structurally related O-glycosyl moieties found in the antitumor agents adriamycin and daunomycin.[34] Vineomycin has attracted the most synthetic attention and as a result has sparked considerable interest in the stereoselective preparation of aryl C-glycosides. Almost all the work published thus far has concentrated on synthesizing the derivative of vineomycin, vineomycinone B$_2$ methyl ester (**177**). Several strategies have been applied to the stereoselective construction of the aryl anomeric carbon-carbon bond. Palladium mediated approaches, a cycloaddition route, and Lewis acid catalyzed C-glycosidation reaction have all been successfully employed to synthesize vineomycinone B$_2$ methyl ester.

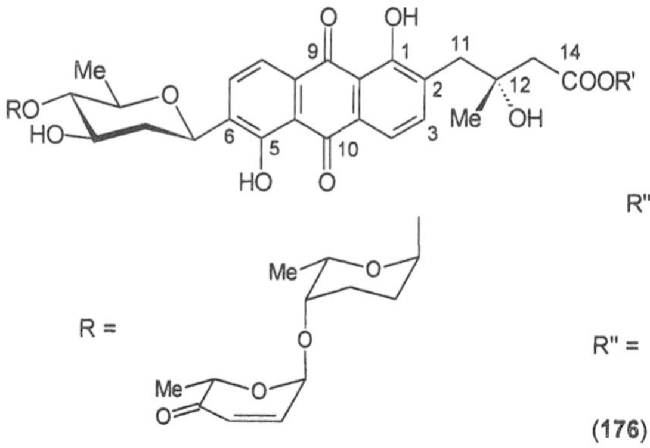

$R =$

$R' = H$
(175) Vineomycin B_2

$R = R' = H$
(176) Vineomycinone B_2

$R = H, R' = Me$
(177) Vineomycinone B_2 methyl ester

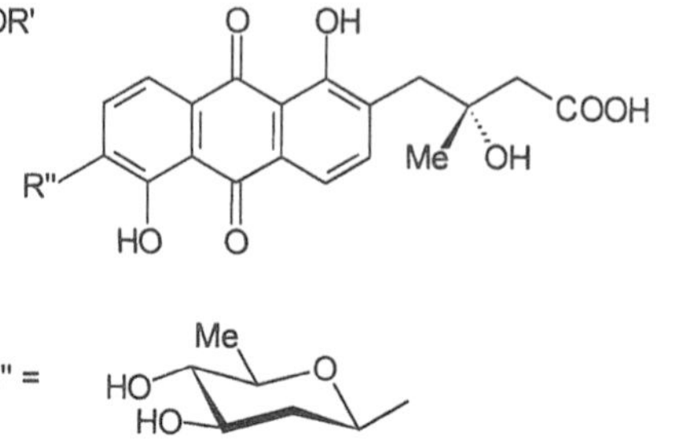

$R'' =$
(176) Fridamycin A

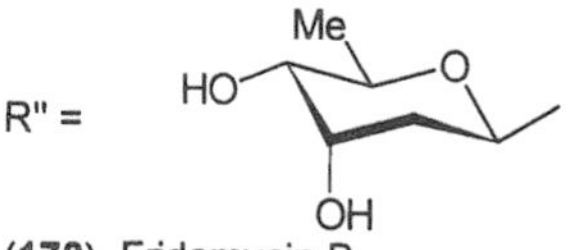

$R'' =$
(178) Fridamycin B

$R'' = H$
(179) Fridamycin E

Figure 6

Medermcyin (181) has also been synthesized as well and from this work it has been concluded that medermycin[35] and lactoquinomycin[36] are indeed the same molecule.

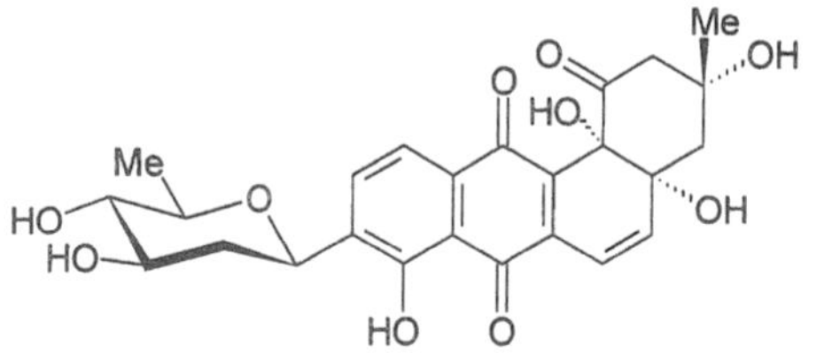

(180) Aquayamycin

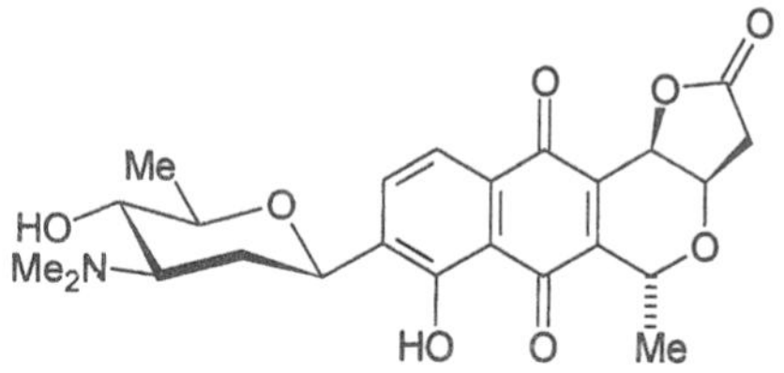

(181) Medermycin or
Lactoquinomycin

Figure 7

2.   Synthesis of Vineomycinone B₂ Methyl Ester

Danishefsky, Uang, and Quallich[37] used two homo Diels-Alder reactions to construct the tricyclic aromatic portion of vineomycinone B_2 methyl ester and a hetero Diels-Alder reaction to assemble the sugar unit. In this sequence, the aryl anomeric carbon-carbon bond is built into the diene 182. The first Diels-Alder reaction between diene 182 and quinone 183 gave 70% yield of 184. Methylation (Ag₂O, MeI) and double isomerization then provided 185. A second Diels-Alder reaction, this time with diene 186, gave, after methylation, a compound containing the requisite tricyclic ring system, 187. The stage was now set for the key hetero Diels-Alder reaction. The dienophile was generated from 187 by ozonolysis and reaction of this keto-aldehyde with diene 189 in the presence of Eu(fod)₃ gave a high yield of 190. Hydroboration then furnished 191 in 49% yield but with a 30% recovery of starting material. It is noteworthy that the hindered ketone survived the reaction conditions and that four contiguous centers are built up in these two steps. The methyl ethers were deblocked at this stage and a two carbon homologation of the ketone tried with carbomethoxy magnesium bromide to give a mixture of isomers **192** that proved to be inseparable and quite difficult to resolve. The workers then reacted the magnesium salt of **193** (derived from *l*-menthone) with **191** and obtained a separable mixture of four compounds from which the desired one was treated with potassium carbonate in methanol to give the target compound, vineomycinone B$_2$, in optically pure form.

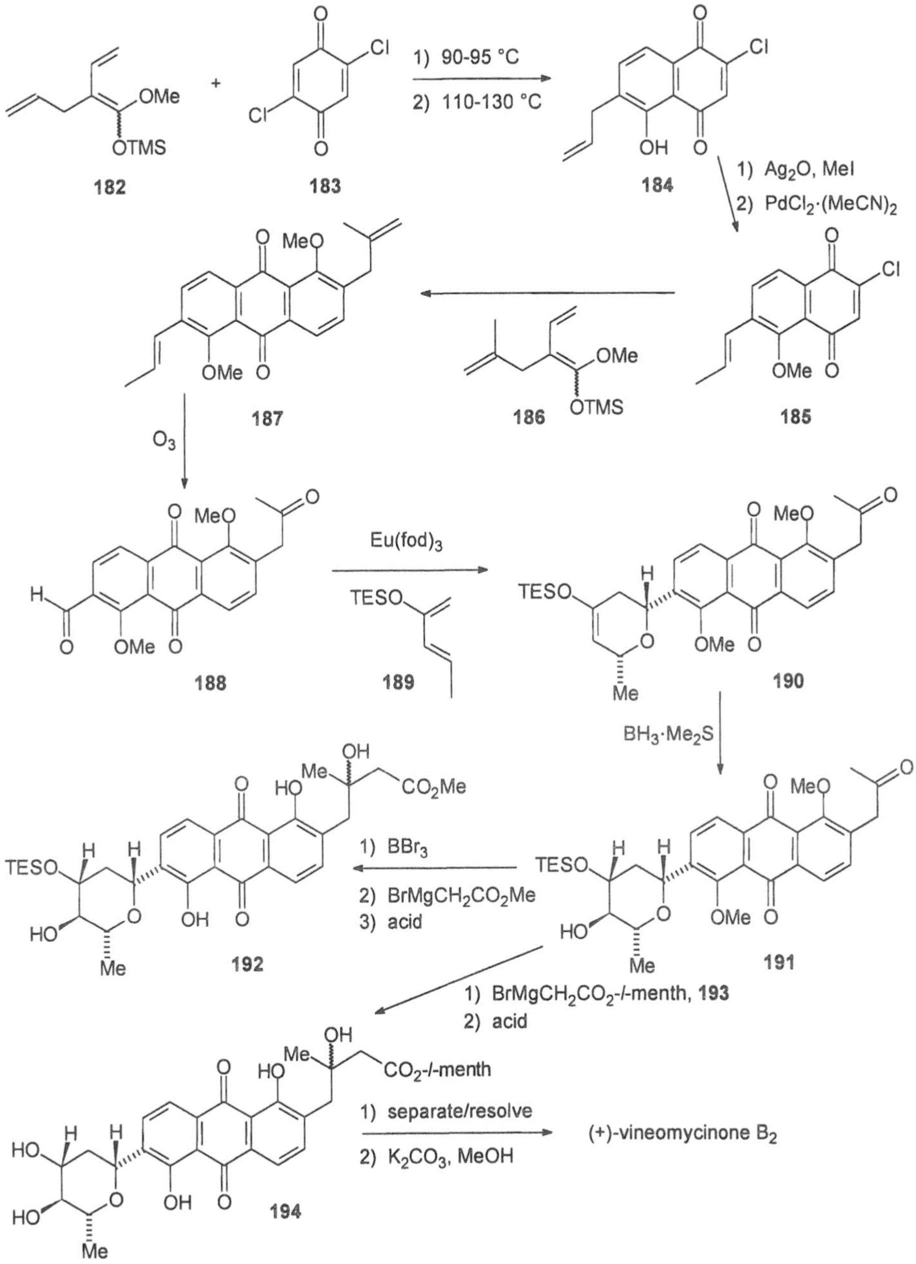

Scheme 31

Rutledge and co-workers[38] published a very short and concise route to the anthraquinone intermediate **200** via the use of two sequential reductive Claisen rearrangements. The symmetrical quinone **195** underwent reductive Claisen rearrangement to give a 98% yield of the mono-rearranged compound **196**. Exposure to potassium hydroxide in ethanol then gave the anthrafuran**197**. A second reductive rearrangement then gave **198**, after methylation. Hydrolysis of the vinyl chloride gave a furan-ketone (along with some diketone that could be recyclized to the furan-ketone) that when exposed to ozone provided the target compound **200** in 80% overall yield from **195**.

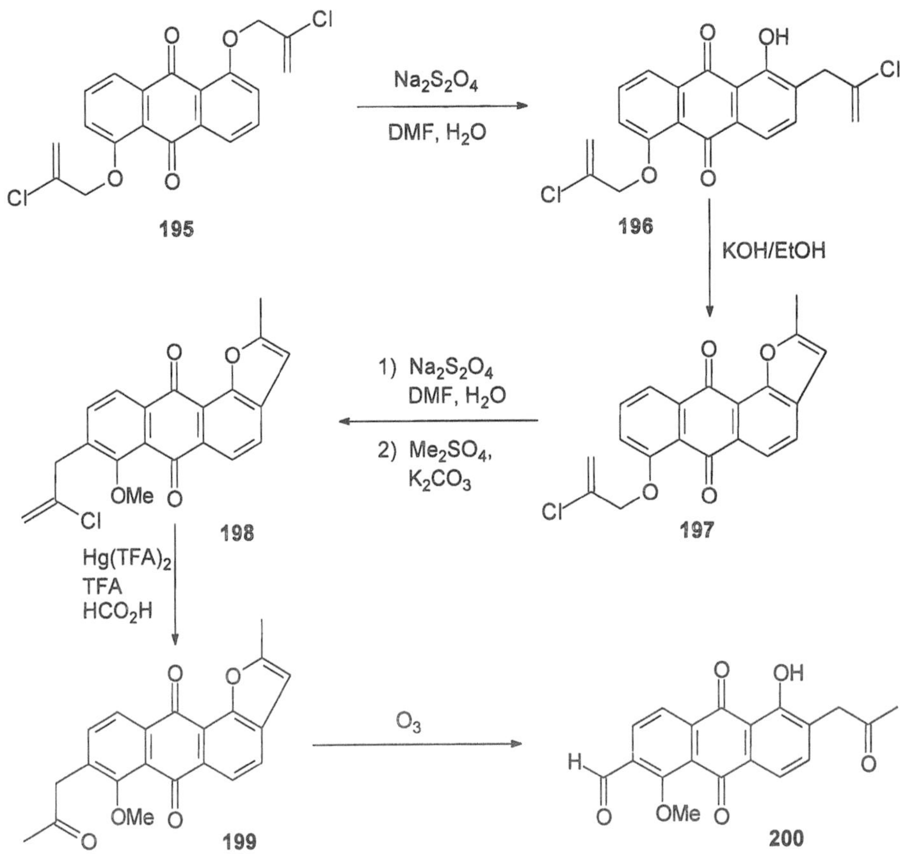

Scheme 32

Mioskowski, Falck, and co-workers[39] used a novel approach in their synthesis of vineomycinone B$_2$ that involved using the inherent symmetry of the molecule and by taking advantage of the Bradsher cycloaddition reaction. The *C*-glycoside **204**, which is used in the second cycloaddition, was made (Scheme 33) by addition of the cerium salt of acetylene **202** to the lactone **201** followed by stereoselective reduction, desilylation, and methoxy-mercuration/demercuration.

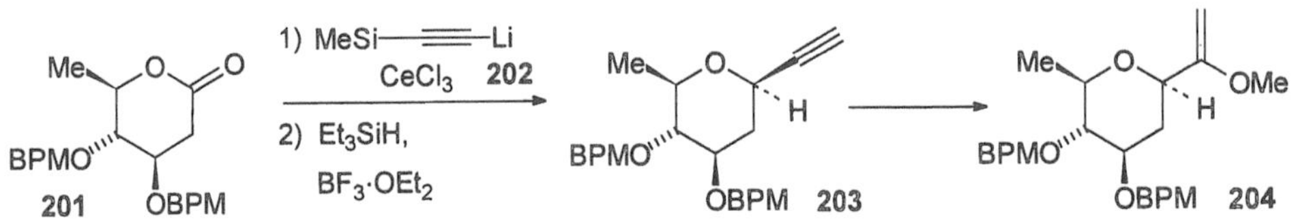

Scheme 33

The first cycloaddition takes place between the salt **205** and the cyclic enol ether **206** to give **207**. Von Braun cleavage of the mixed azaacetal followed by hydrolysis and aromatization then gave aldehyde **208** which contains the side chain with the correct configuration. A second Bradsher cycloaddition (after salt formation), this time on the other end of the molecule with the *C*-glycoside **204**, gave dialdehyde **210**, after hydrolysis and aromatization. The sequence was completed by modified Dakin reaction and addition of singlet oxygen to the central ring followed by oxygen-oxygen bond cleavage and oxidation to the anthraquinone **211**. Oxidation, esterification and deprotection then furnished the product, vineomycinone B$_2$, Scheme 34.

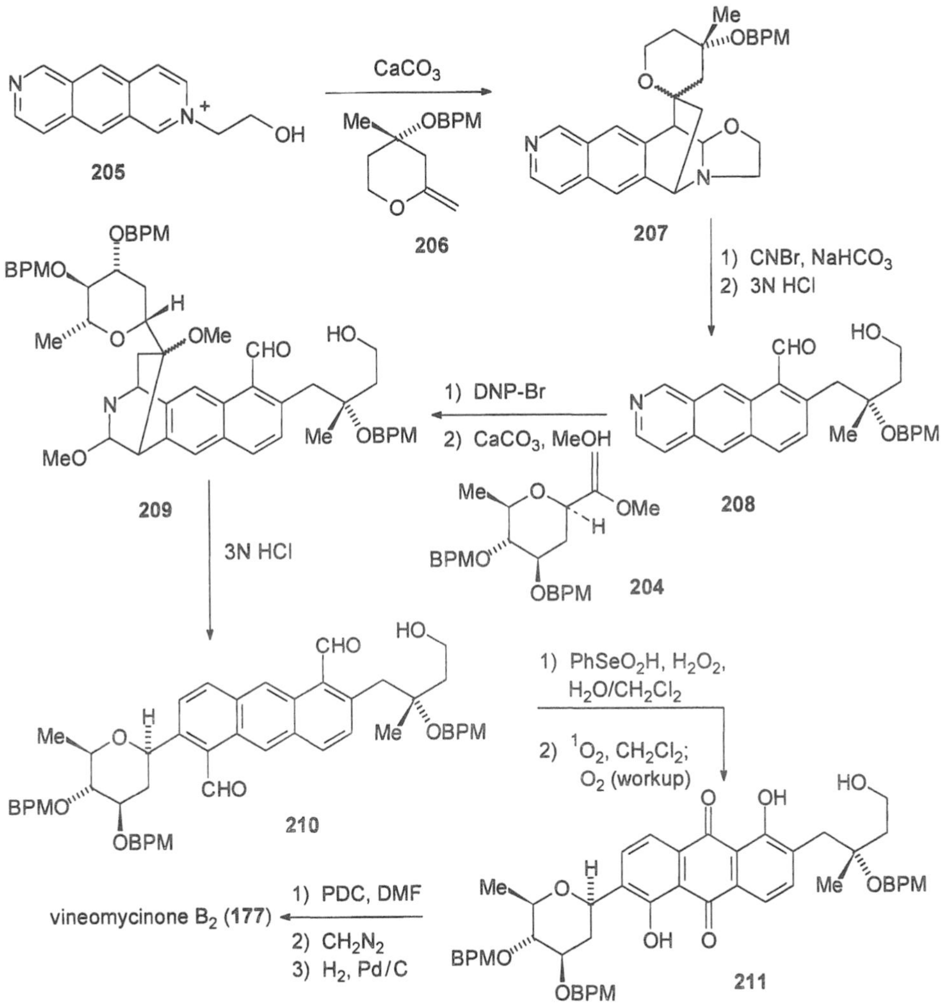

Scheme 34

Tius *et al.*[40] also attached two optically pure pieces to a symmetrical anthrarufin derivative in their synthesis of vineomycinone B$_2$ methyl ester. The sequence (Scheme 35) begins with anthrarufin (**212**) which is protected, reduced, and metallated and the resulting lithio species quenched with tributyltin chloride to furnish **216**. Iodination gave aryl iodide **217** and after considerable experimentation it was found that the coupling reaction proceeded best with the organozinc **218** and with an activated palladium catalyst to provide the desired adduct **219** (78.5%). The glycal was selectively reduced from the axial direction using acidic reductive conditions to provide **220**. A second metallation and tributyltin chloride quench gave **221** and Stille coupling with the optically pure bromide **222** installed the chain to give **223** in fair yield. The required methyl group was introduced by cuprate addition and completion of the synthesis was accomplished by oxidation to the anthraquinone and acidic hydrolysis in methanol to give vineomycinone B_2 methyl ester.

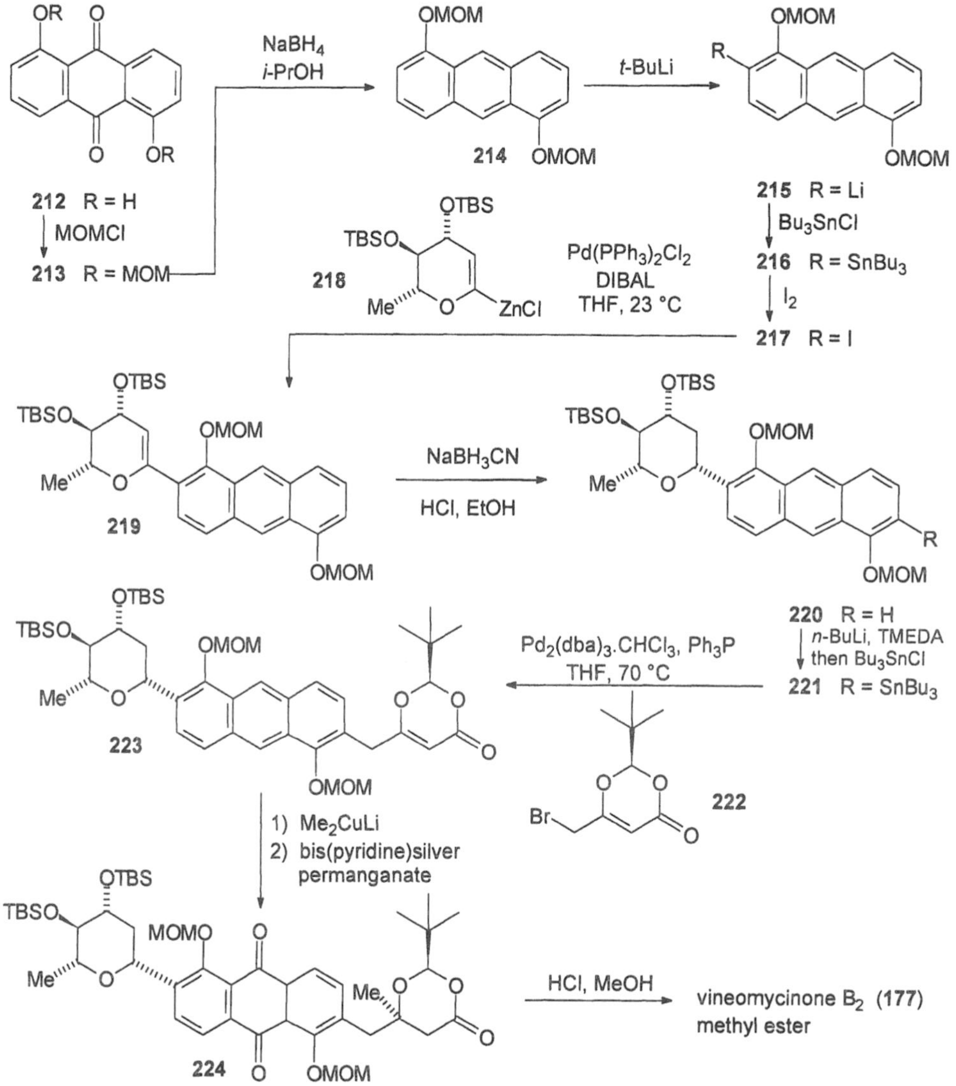

Scheme 35

Suzuki and co-workers[41] used an O→C rearrangement reaction of glycosylated phenols as the key carbon-carbon bond forming step in their total synthesis of vineomycinone B_2 methyl ester (Scheme 36). The fluoride **225** which was coupled with anthrol **226** using the promoter Cp_2HfCl_2-$AgClO_4$ in dichloromethane gave an 86% yield of the desired β-C-glycoside **227**. The benzoate protecting groups were removed and replaced with the base stable silyl protecting TBS group. Introduction of the side chain was now the issue at hand and accordingly anthrol **227** was selectively metallated and the resulting anion trapped with trimethyltin chloride to provide **228**. Lithiation and exposure to optically pure aldehyde **229** then gave a mixture of isomers **230** which were separated and carried on separately. The alcohol was benzoylated and the anthrol converted to the anthraquinone **231** by treatment with CAN. The benzyl group was removed by the action of DDQ and exposure of the resulting alcohol to sodium dithionite caused deoxygenation to furnish **232**. The synthesis was completed by oxidative cleavage of the double bond by a two step protocol to provide protected vineomycinone B_2 that was converted to the target by the action of boron tribromide that caused cleavage of both the two methyl groups and the two silyl protecting groups to deliver natural vineomycinone B_2 methyl ester (**177**).

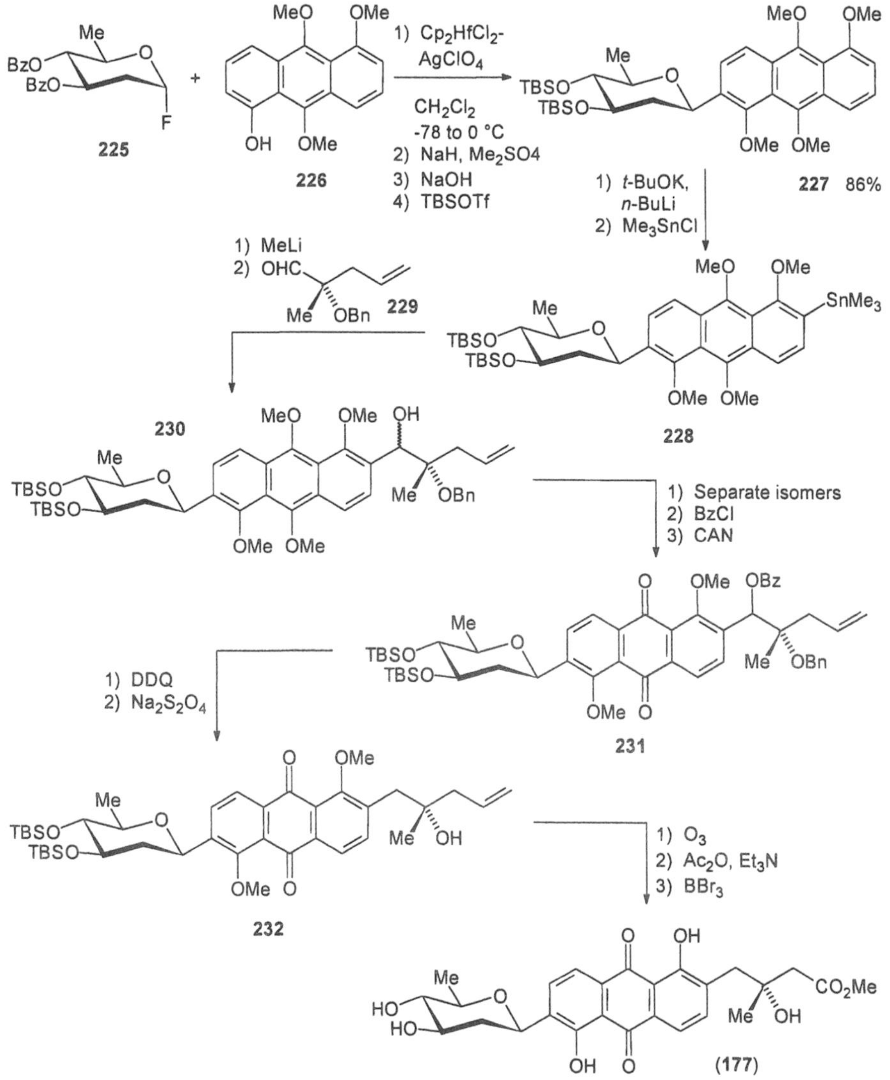

Scheme 36

3. Medermycin

Tasuta and co-workers[42] employed an aryl-organometallic addition to a sugar lactone to create the carbon-carbon aryl anomeric bond in their total synthesis of medermycin (**241**) (lactoquinomycin). Aryl lithium **233** was condensed with lactone **234** to give, after acetal oxidation, a single compound presumed to be the β-anomer **235**. Reduction with triethyl silane and TFA was followed by conversion to the amide and tin based dehalogenation to give the 2-deoxy-β-*C*-glycoside **236**.   The benzyl groups were exchanged for MOM protecting groups and the resulting amide ortho-lithiated and quenched with DMF to give, after hydrolysis, exposure to thiophenol and tosic acid, oxidation, and ketalization **237**. The completion of the synthesis then relied on previous work that culminated with the preparation of kalafungin which is the aglycone of medermycin. Anion formation with **237** and exposure to dienophile **238** gave an anthraquinone which was methylated and reduced to **239**. The sequence was completed by transforming the eastern portion into lactone **240** via Wittig chemistry on the lactol and this was followed by conversion of *O*-3' into the requisite dimethylamine to provide medermycin (**241**).

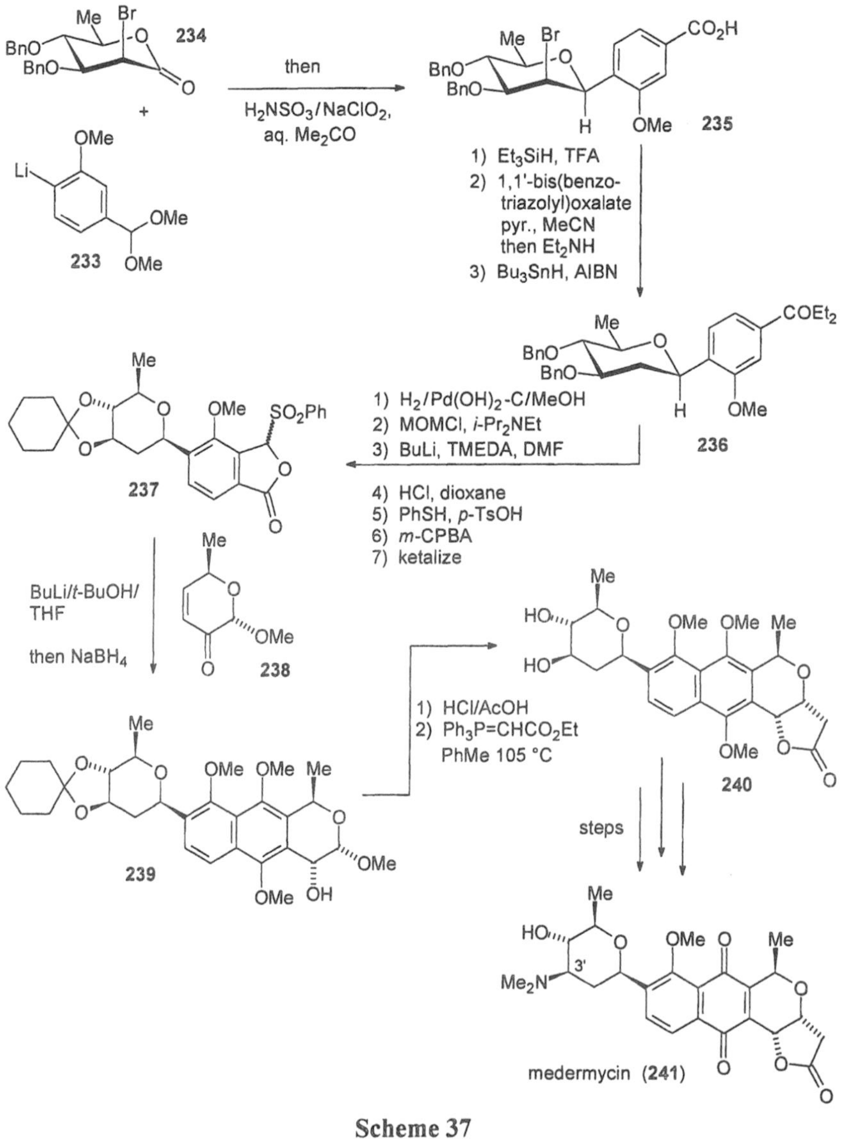

Scheme 37

D.   GILVOCARCIN ANTIBIOTICS
1.   Introduction

The gilvocarcin class of antitumor antibiotics are comprised of a number of polyketide derived natural products that show excellent antitumor activity with low related toxicity.[43]

The class is made up of several members and as of this writing only gilvocarcin M and V have been synthesized.[44] The aglycone is characterized by a δ-lactone and four oxygenated aromatic rings. The sugar is bonded *para* to the hydroxy group and Fig. 8 shows the natural products. When no sugar is present, S = H, one obtains defucogilvocarcin V. Many synthetic approaches have been reported[45] to this natural product, and in doing so have avoided the issue of C-glycoside formation.

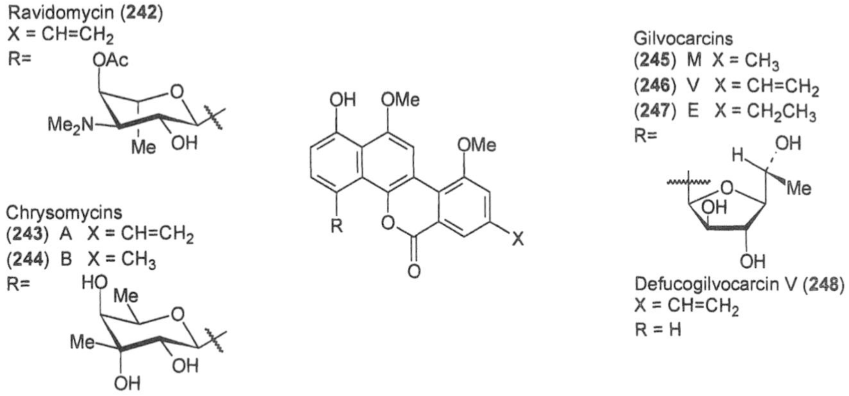

Figure 8

The biological activity of gilvocarcin M is quite interesting.[46] The glycosyl portion is required for efficient binding to DNA and once intercalation takes place strand nicking and covalent modification occurs in the presence of light. McGee and Misra[47] have succeeded in isolating the DNA photoadduct that results when gilvocarcin V is exposed to light in the presence of DNA, eq. 1.

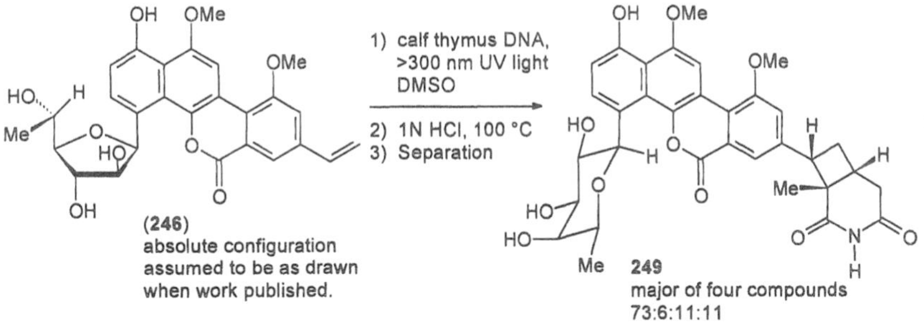

The acid catalyzed cleavage of the purine and pyrimidine residues resulted in an expected acid catalyzed rearrangement of the fucosyl group to its pyranose form. The reaction gave a mixture of four isomers of which the major one, **249**, was fully characterized.

2. Synthetic Work

Suzuki's successful approach[48] to the enantiomer of gilvocarcin M involved the initial formation of an aryl C-glycosidic bond with a substituted benzene ring that would later be modified into the required 6H-benzo[d]naphtho[1,2-b]pyran-6-one unit through the use of cycloaddition chemistry. The first, and probably most difficult, step involved the use of their special glycosidation catalyst to affect bond formation *syn* to the adjacent OBn group with an impressive ratio of 8:1. Phenol **252** was converted to its triflate and benzyne formation followed by trapping with 2-methoxyfuran (**258**) gave, after aromatization, naphthol **254**.

The sequence was completed by ester formation (254→256) and palladium induced cyclization to afford, after cleavage of the benzyl groups, unnatural gilvocarcin M. Synthetic gilvocarcin had the opposite rotation of natural gilvocarcin M. This sequence is short and highly convergent as well as high yielding. The overall yield of the sequence is 38% starting from the acetates 250.

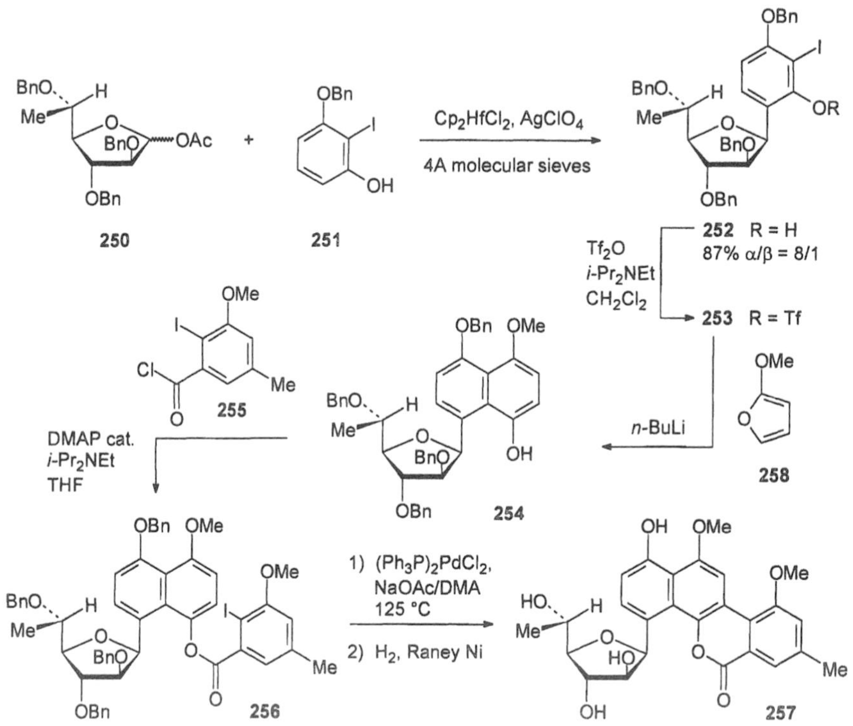

Scheme 38

Further work[44] concentrated on synthesizing the naturally occurring enantiomer of gilvocarcin M (245), Scheme 39.

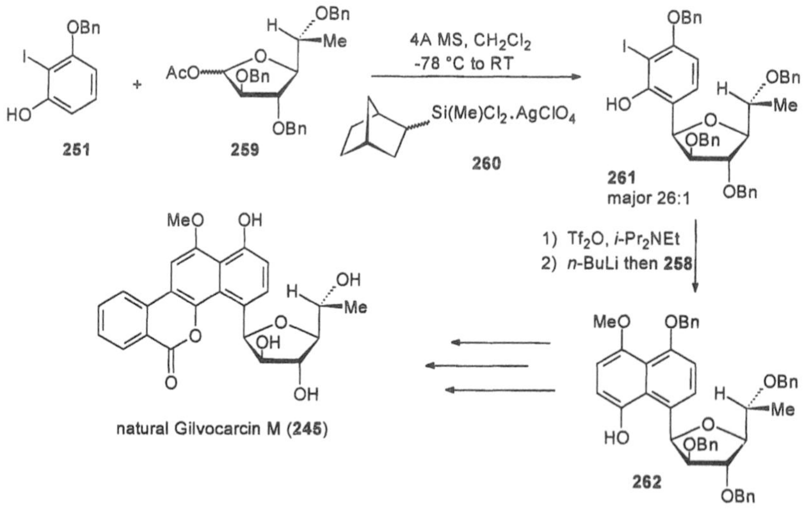

Scheme 39

The route is very similar to that shown in Scheme 38, with one important improvement. When the promoter used in the key C-glycosidation reaction was changed to 260 a 26:1 ratio of α to β isomers was formed. Cycloaddition with 258 gave naphthalene 262 which was carried on to natural gilvocarcin M. Their route was flexible enough to allow for a synthesis of gilvocarcin V as well.[44] Accordingly, acid 263 was coupled with the naphthol 262 under dehydrating conditions to provide ester 264. Palladium mediated coupling followed by protective group manipulations then gave 265. The hydroxyl group was converted to the o-nitrophenyl selenide and oxidative fragmentation gave, after deacetylation, natural gilvocarcin V (246), Scheme 40.

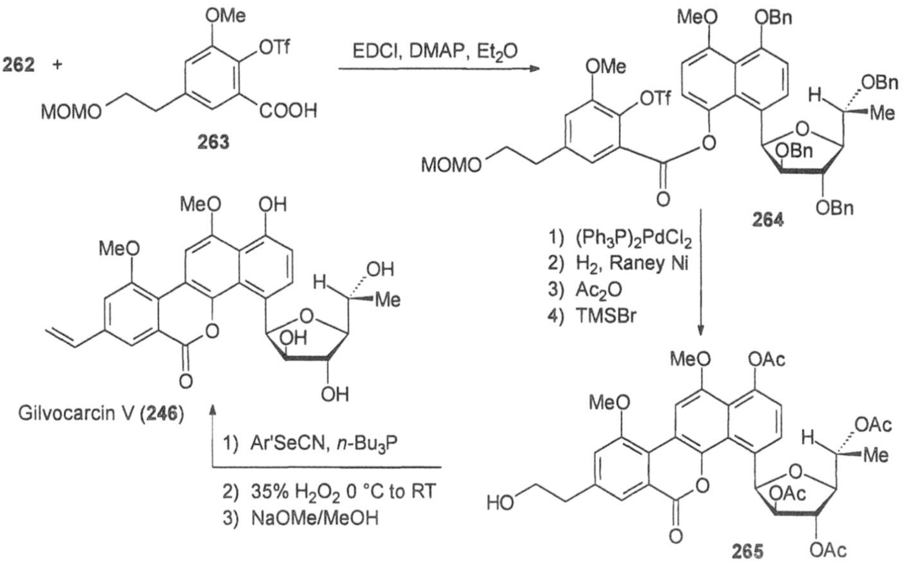

Scheme 40

Daves and co-workers have published several papers on synthetic studies leading to the gilvocarcin class of antibiotics. One of their approaches[49] to a potential analog (that possesses the requisite vinyl group and carbohydrate portion) is based on a Lewis acid catalyzed glycosidation reaction as depicted in Scheme 41.

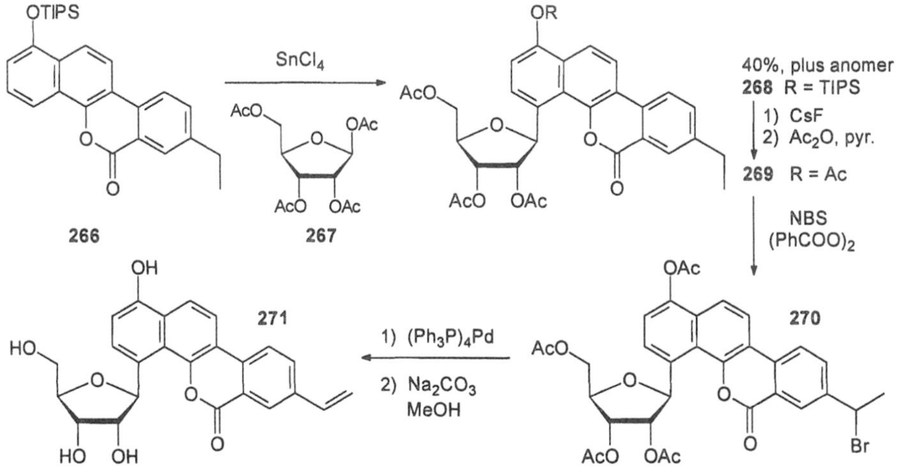

Scheme 41

The acetylated furanose was coupled with **266** under stannic chloride catalysis to give an 80% yield of a 1:1 anomeric mixture. The silyl group was exchanged for an acetate and benzylic bromination followed by dehydrobromination then provided the analog **271**, after deacetylation.

Further work from the same group utilized a palladium mediated coupling of an appropriate aromatic stannane with a furanoid glycal (Scheme 42).[50] Stannane **272** was coupled with glycal **273** in the presence of palladium acetate to give a 66% yield of *C*-glycoside **275**. When the protecting group on the hydroxymethyl group was changed to TIPS a modest 28% yield of adduct was isolated. This is probably a result of the steric bulk of the TIPS group shielding the β-face from attack. The sequence was taken as far as **279** and **280** by exposure of the silyl enol ether to TBAF followed by reduction of the resulting ketone with sodium borohydride to give the alcohols **279** and **280** in a 7:5 ratio, respectively. Reduction in the TIPS series led only to **281**.

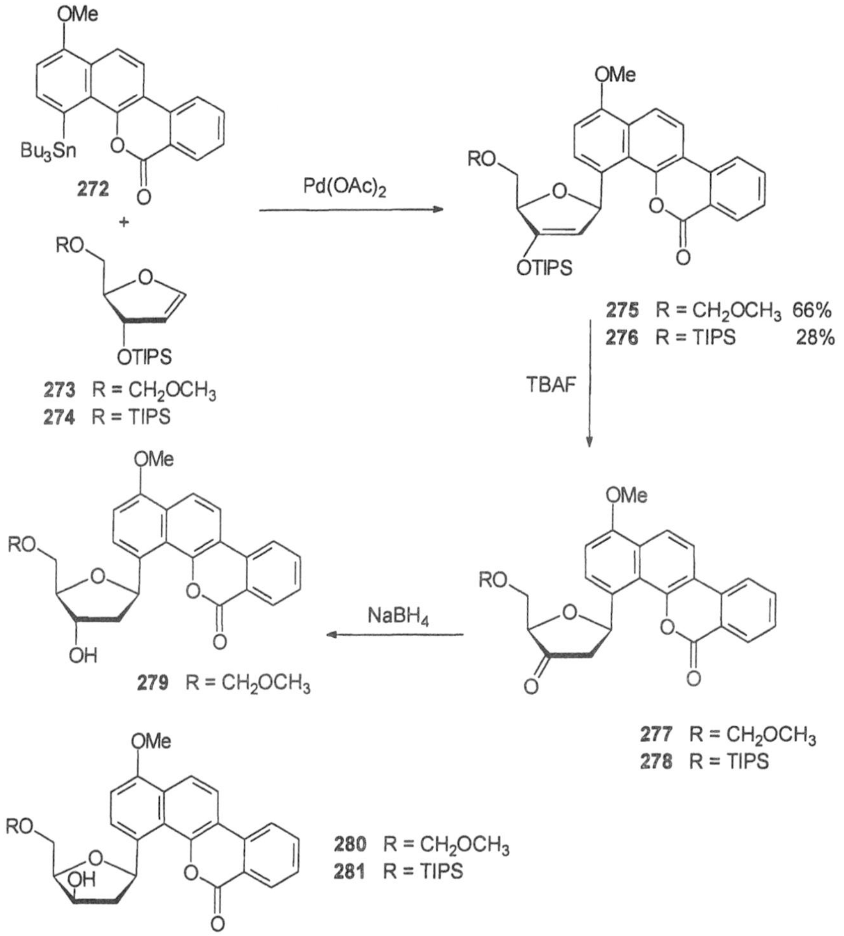

Scheme 42

E.   MISCELLANEOUS ARYL *C*-GLYCOSIDES
1.   (-)-Urdamycinone B

The polyketide natural product (+)-urdamycin B was isolated from *Streptomyces fradie* in 1986 and careful anomeric hydrolysis gave (+)-urdamycinone B (**282**).[51] Yamaguchi and co-workers employed an original approach to the synthesis of this aryl *C*-

glycoside. They chose to begin with an alkyl *C*-glycoside and convert it, through the use of polyketide condensation reactions, to the target (**282**). Their approach is unique since many approaches to aryl *C*-glycoside synthesis involve the coupling of an aromatic fragment with a suitable sugar derivative to obtain the requisite aryl *C*-glycoside. The synthesis is shown in Scheme 43.[52] *C*-Glycoside **283** underwent condensation with acetoacetate dianion, to give after aromatization, the β-*C*-glycoside **284**. Condensation, palladium mediated dealkoxy-carbonylation, and protection then provided the enol lactone **285**. The lactone was reduced and the resulting aldehyde condensed with the anion **286** to give, after aromatization, anthracene **287**. Conversion to the anthroquinone was accomplished by sequential protecting group removal, oxidation, and dethioketalization to bring the sequence up to **288**. The route was completed by aldol condensation to provide (**282**) in 34% yield, Scheme 43.

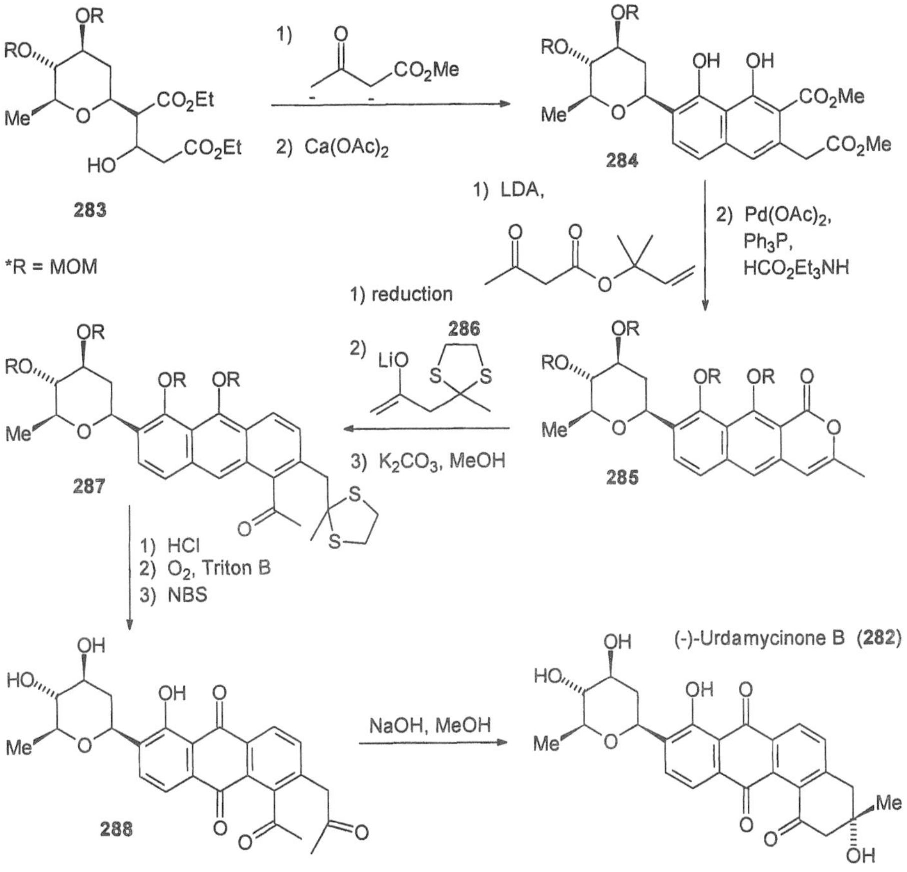

Scheme 43

2. Carminic Acid

Carminic acid (**289**) is obtained from the dried female bodies of the insect species *Dactylopius coccus*.[53] It is characterized by the presence of a β-*C*-glucoside bond[54] and an anthraquinone aromatic portion. The Allevi group[55] has recently completed the first total synthesis of (**289**), Scheme 44. Their approach involves the Lewis acid catalyzed attachment of a tricyclic aromatic molecule with a suitably derivatized sugar. Diels-Alder reaction of diene **291** with the naphthazarin **290** gave **292**, after desilylation and methylation.   Reductive methylation and coupling with the gluco derivative **294** under boron trifluoride catalysis in acetonitrile gave the aryl β-*C*-glucoside **295** in 40% yield. Oxidation and protective group manipulations then provided **296**. Oxidation with lead tetraacetate and Thiele acetylation gave **297** which was deacetylated to furnish carminic acid (**289**).

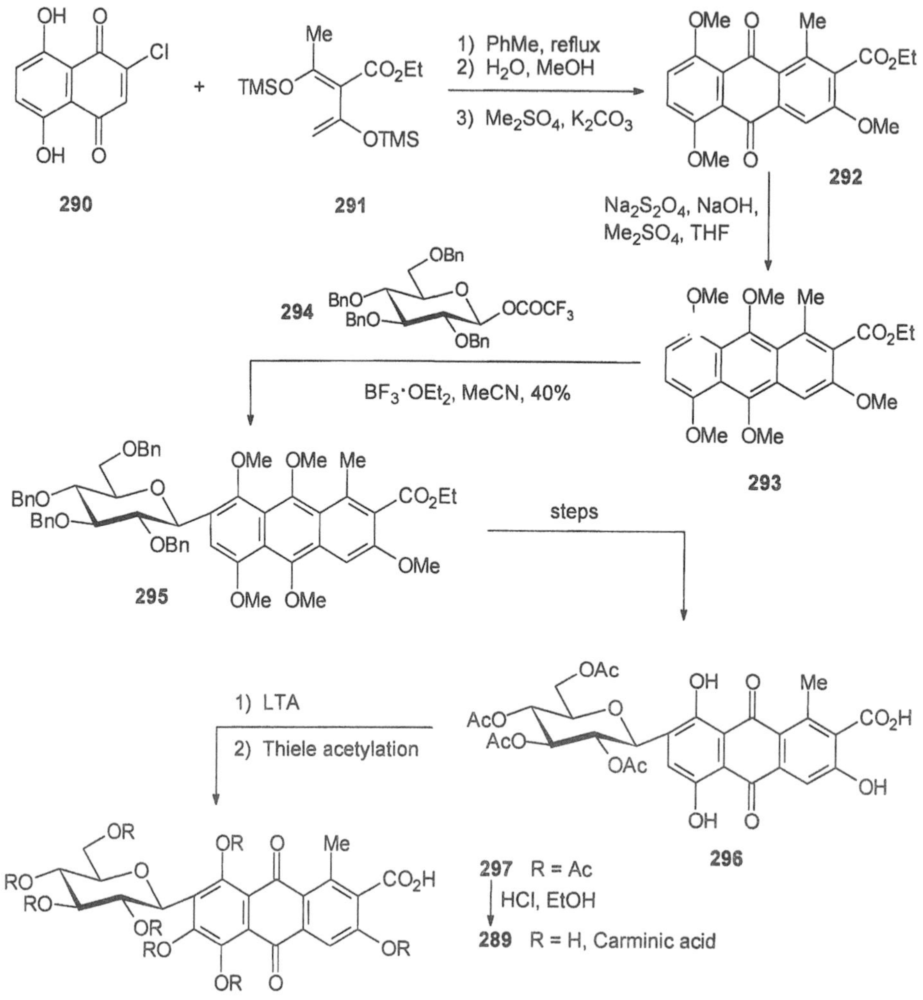

Scheme 44

3.   Bergenin

Bergenin (**298**) is an aryl β-*C*-glucoside that also possesses a δ-lactone ring. It has been isolated from various sources and has had several medical applications.[56]   A second compound isolated along bergenin, 8,10-di-*O*-methyl bergenin (**299**), has been synthesized by Schmidt and Frick[57] through the use of a Lewis acid catalyzed carbon-carbon bond forming process. The workers chose to attach the aromatic ring first and then elaborate it into the required *C*-6 ester followed by lactonization. Accordingly, condensation of **294** with 1,2,3-trimethoxy benzene (**300**) under boron trifluoride etherate catalysis gave a 59% yield of the β-*C*-glucoside **301**. Protective group manipulations provided **302** which was converted to the lithio species **304** (via the intermediacy of the corresponding bromide) that was quenched with diphenyl disulfide to **305**. Oxidation gave a 1:2 mixture of diastereomeric sulfoxides that were separable and the major was lithiated and quenched with methyl chloroformate to furnish **306**. Desulfurization and deacetylation then gave authentic (**299**).

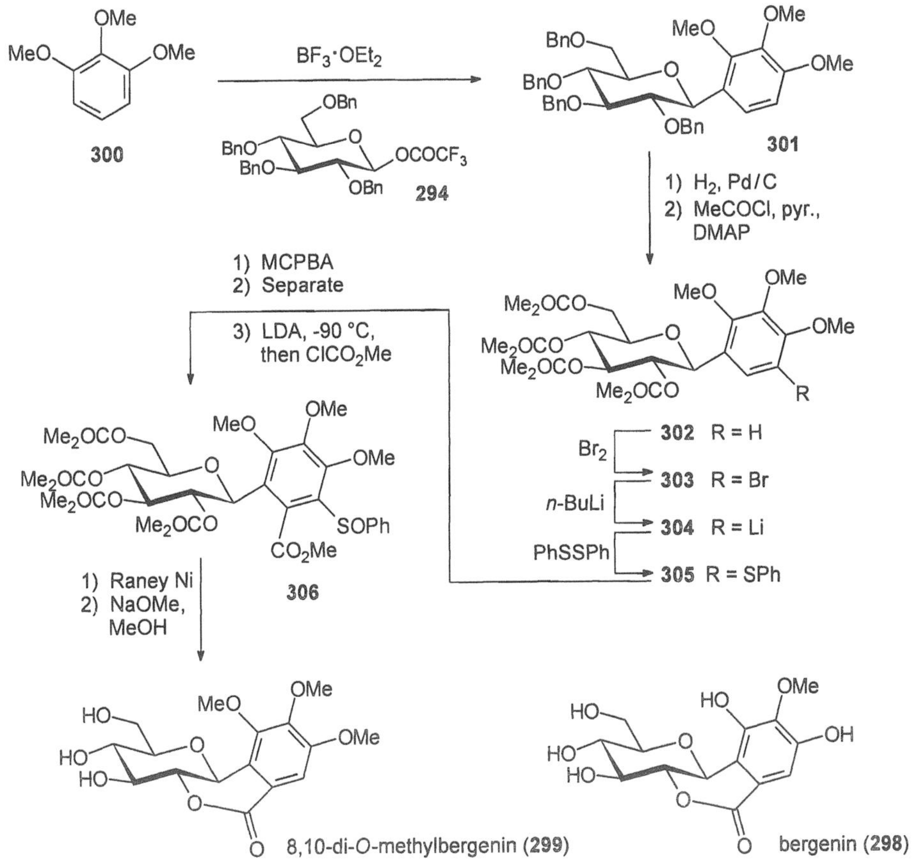

Scheme 45

4. *C*-Glycosyl Flavanoids

a. *5,7,4'-Tri-O-Methylvitexin*

A *C*-glycosyl flavanoid occurs when the anomeric carbon of a sugar is bonded to the aromatic carbon of a suitable flavanoid. Frick and Schmidt[58] have utilized an open electrophilic sugar aldehyde in the condensation with a flavanoid nucleophile in their synthesis of 5,7,4'-tri-*O*-methylvitexin (**313**). The lithio species **307** was condensed with aldehyde **123** to give a mixture of alcohols **308**. Depending on the reaction conditions used either the *C*-furanosyl flavanoids **309** or the desired *C*-glucosyl flavanoid **311** could be obtained as a 1:1 mixture of isomers at *C*-2. The furanosyl compounds **309** were convertible to the glucosyl derivative **311** by treatment with hydrochloric acid in dioxane. Acetylation followed by oxidation with selenium dioxide gave the *C*-furanosyl flavanoid **310**. Similarly, acetylation followed by oxidation provided the *C*-glucosyl flavanoid, 5,7,4'-tri-*O*-methylvitexin **313**.

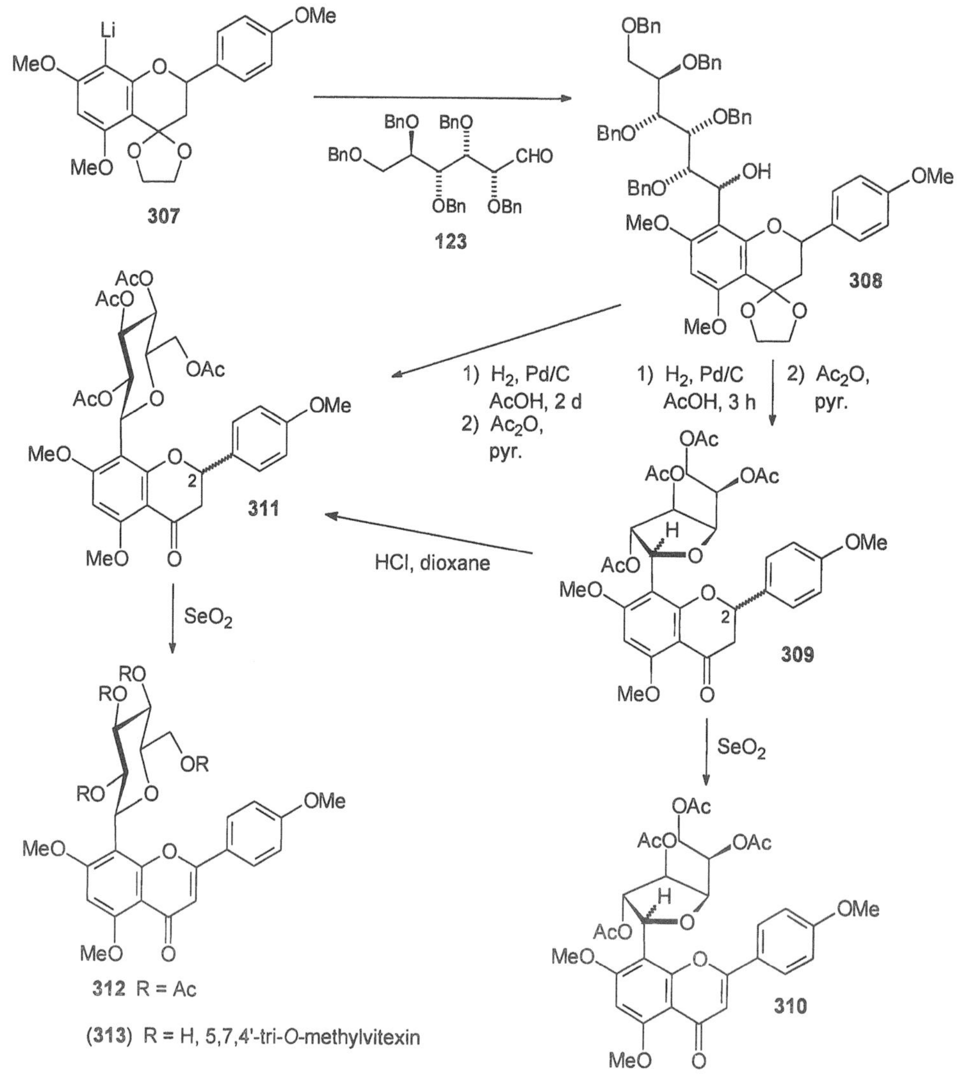

Scheme 46

b. Bayin Flavanoids

Work by Tschesche and Widera[59] used the condensation of an aryl Grignard with a perbenzylated glucopyranose to produce a β-*C*-glycoside, Scheme 47. Thus reaction of **314** with 2,4-dimethoxyphenylmagnesium bromide (**315**) gave a 51% yield of *C*-glycoside **316**. Debenzylation was followed by acetylation to provide **318**.

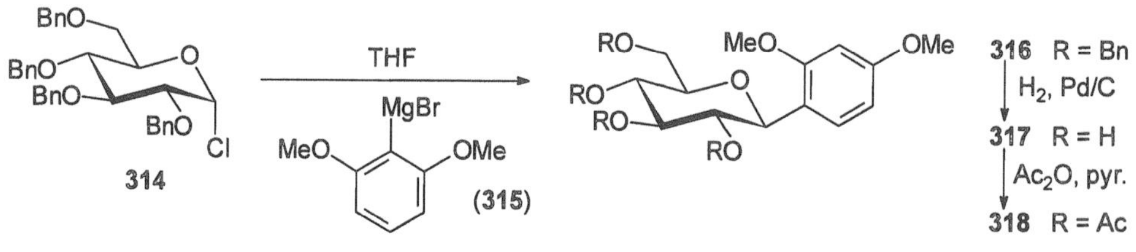

Scheme 47

Transformation into the flavanoid was accomplished along the lines shown in Scheme 48. Friedel-Crafts acylation of **318** was selective and gave **319**, a compound in which the *o*-methoxy group has been lost. Base catalyzed condensation with anisaldehyde gave olefin **321** which was cyclized with selenium dioxide to furnish 4',7-di-*O*-methylisobayin (**322**).

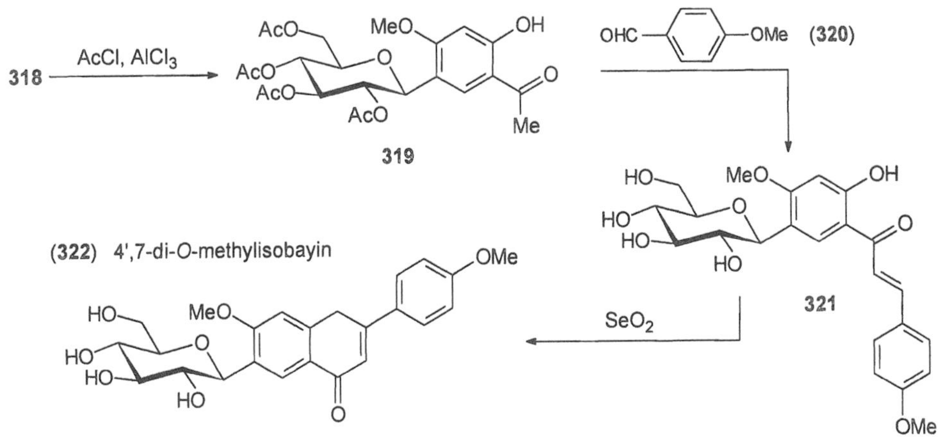

Scheme 48

In earlier work, Eade, McDonald, and Simes[60] condensed aryl Grignard **315** with the peracetylated glucopyranosyl chloride **323** to obtain a mixture of anomeric *C*-glycosides, Scheme 49.

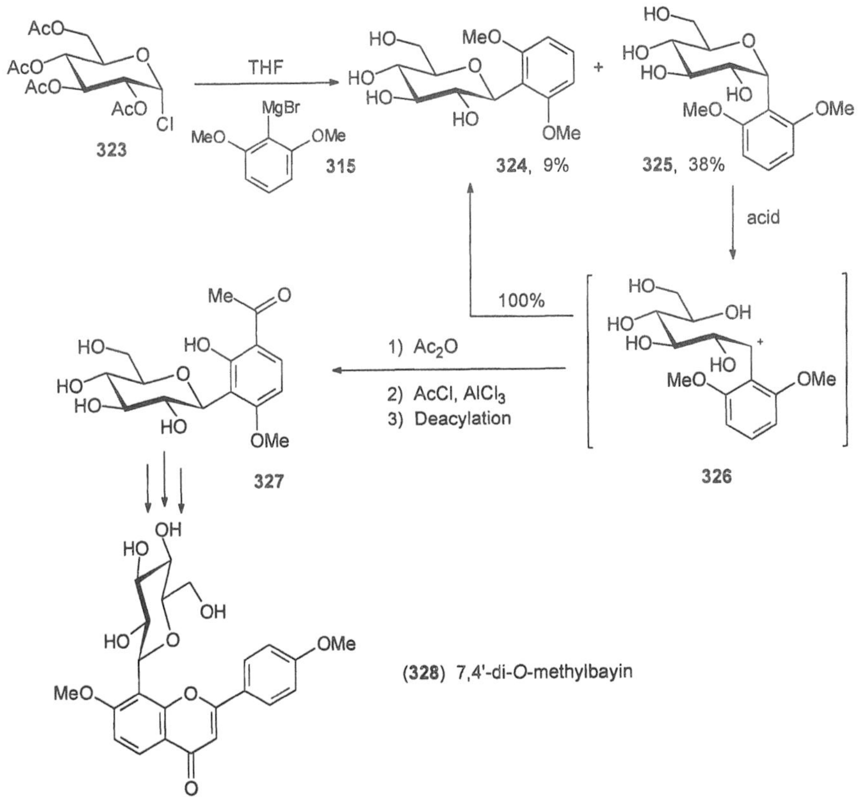

Scheme 49

They found that the major isomer **325** could be converted to the more thermodynamically stable β-*C*-glucoside **324** by treatment with acid, presumably via the intermediacy of the ring open form **326**. Acylation, chalcone formation, and cyclization then gave 7,4'-di-*O*-methylbayin (**328**).

III. CONCLUSION

The aryl *C*-glycoside natural products have attracted considerable synthetic attention. As a result several methods have been developed in order to be able to synthesize these molecules. One approach involves joining a functionalized aromatic ring to a suitable sugar acceptor. An alternative strategy is to first form the anomeric carbon-carbon bond and then elaborate the aromatic portion accordingly. Lewis acid and transition metal catalyzed carbon-carbon bond forming processes have been employed with high success in the preparation of these natural products. More traditional methods such as condensations with aryl Grignards or aryl lithiums have also found their place in aryl *C*-glycoside natural product synthesis. In one case, an existing alkyl *C*-glycoside was further transformed into an aryl *C*-glycoside.

The synthesis of these compounds is often a challenge and in many cases exciting new chemistry is developed. For the gilvocarcins, total synthesis has shown that the absolute configuration, as assigned initially, was incorrect. In other cases, the assignment of configuration was confirmed by total synthesis. With the chemical behavior of some of these molecules having been mapped out (at least partially), workers can, with some degree of confidence, focus on the preparation of analogs that may prove to possess more therapeutic value than the natural products themselves. Although the described syntheses are impressive, there still exist some natural products that have yet to be synthesized in the laboratory. For example hedamycin[61] is a bis aryl *C*-glycoside that contains two sugar units bonded to the same aromatic ring.

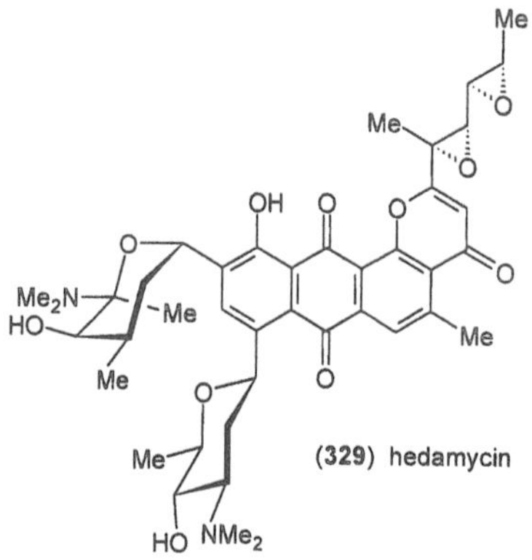

Figure 9

It is certain that more complex and exciting aryl *C*-glycoside natural products will be isolated and eventually synthesized. One can also expect that methodology will be developed so that it will allow for selective anomeric carbon-carbon bond formation under very mild conditions. The key to realizing this goal is to develop new promoters and catalysts that will permit these transformations to be carried out. Suzuki's work in the vineomycin and gilvocarcin classes of compounds led to the development of a new and selective catalyst that allows for selective aryl β-*C*-glycoside formation.

IV. REFERENCES

1. Wiley, P.F.; MacKellar, E.L.; Carton, E.L.; Kelly, R.B. *Tetrahedron Lett.* **1968**, 663.
2. Wiley, P.F. in *Anthracycline Antibiotics*, El Khadim, H.S., Ed., Academic Press, New York , 1982, 97-117.
3. For a recent report that addresses biological activity see: Boldt, M.; Gaudiano, G.; Haddadin, M.; Koch, T.H. *J. Am. Chem. Soc.* **1989**, *111*, 2283.
4. Most of the synthetic attention has been focused on menogaril and not on nogarol.
5. Bates, M.A.; Sammes, P.G.; Thomas, G.A. *J. Chem. Soc., Perkin Trans I* **1988**, 303.
6. Kawasaki, M.; Matsuda, F.; Terashima, S. *Tetrahedron* **1988**, *44*, 5695.
7. Kawasaki, M.; Matsuda, F.; Terashima, S. *Tetrahedron* **1988**, *44*, 5713.
8. Kawasaki, M.; Matsuda, F.; Terashima, S. *Tetrahedron* **1988**, *44*, 5727.
9. Kawasaki, M.; Matsuda, F.; Ohsaki, M.; Yamada, K.; Terashima, S. *Tetrahedron* **1988**, 44, 5745.
10. DeShong, P.; Li, W.; Kennington, J.W. Jr.; Ammon, H.L.; Leginus, J.M. *J. Org. Chem.* **1991**, *56*, 1364.
11. Hauser, F.M.; Ellenberger, W.P.; Adams, T.C.Jr., *J. Org. Chem.* **1984**, *49*, 1169.
12. Hauser, F.M.; Ellenberger, W.P. *J. Org. Chem.* **1988**, *53*, 1118.
13. Hauser, F.M.; Adams, T.C. Jr. *J. Org. Chem.* **1984**, *49*, 2296.
14. Hauser, F.M.; Chakrapani, S.; Ellenberger, W.P. *J. Org. Chem.* **1991**, *56*, 5248.
15. Joyce, R.P.; Parvez, M.; Weinreb, S.M. *Tetrahedron Lett.* **1986**, *27*, 4885.
16. Semmelhack, M.F.; Jeong, N. *Tetrahedron Lett.* **1990**, *31*, 605.
17. Semmelhack, M.F.; Jeong, N.; Lee, G.R. *Tetrahedron Lett.* **1990**, *31*, 609.
18. Smith, T.H.; Wu, H.Y. *J. Org. Chem.* **1987**, *52*, 3566.
19. (a) Gruner, J.; Traxler, P. *Experentia* **1977**, *33*, 137. (b) Traxler, P.; Gruner, J.; Auden, J.A.L. *J. Antibiot.* **1977**, *30*, 289. (c) Traxler, P.; Fritz, H.; Fuhrer, H.; Richter, W.J. *J. Antibiot.* **1980**, *33*, 967.
20. Komori, T.; Yamashita, M.; Tsurumi, Y.; Kohsaka, M. *J. Antibiot.* **1985**, *38*, 455 and Komori, T.; Itoh, Y. *J. Antibiot.* **1985**, *38*, 544.
21. Pérez, P.; García-Acha, I.; Durán, A. *J. Gen. Microbiol.* **1983**, *129*, 245.
22. (a) Baguley, B.C.; Roemmele, G.; Gruner, J.; Wehrli, W. *Eur. J. Biochem.* **1979**, *97*, 345. (b) Pérez, P.; Varona, R.; García-Acha, I.; Durán, A. *FEBS Lett.* **1981**, *129*, 249. (c) Varona, R.; Pérez, P.; Durán, A. *FEMS Microbiol. Lett.* **1983**, *20*, 243.
23. Kang, M.S.; Szaniszlo, P.J.; Notario, V.; Cabib, E. *Carbohydr. Res.* **1986**, *149*, 13.
24. Traxler, P.; Tosch, W.; Zak, O. *J. Antibiot.* **1987**, *40*, 1146.
25. Schmidt, R.R.; Frick, W. *Tetrahedron* **1988**, *44*, 7163.
26. Rosenblum, S.B.; Bihovsky, R. *J. Am. Chem. Soc.* **1990**, *112*, 2746.
27. Czernecki, S.; Perlat, M.-C. *J. Org. Chem.* **1991**, *56*, 6289.
28. Dubois, E.; Beau, J.-M. *Carbohydr. Res.* **1992**, *223*, 157.
29. Friesen, R.W.; Sturino, C.F. *J. Org. Chem.* **1990**, *55*, 5808.
30. Friesen, R.W.; Daljeet, A.K. *Tetrahedron Lett.* **1990**, *31*, 6133.
31. Danishefsky, S.; Phillips, G.; Ciufolini, M. *Carbohydr. Res.* **1987**, *171*, 317.
32. (a) Ōmura, S.; Tanaka, H.; Ōiwa, R.; Awaya, J.; Masuma, R.; Tanaka, K. *J. Antibiot.* **1977**, *30*, 908. (b) Imamura, N.; Nakinuma, K.; Ikekawa, N.; Tanaka, H.; Ōmura, S. *J. Antibiot.* **1981**, *34*, 1517. (c) Imamura, N.; Nakinuma, K.; Ikekawa, N.; Tanaka, H.; Ōmura, S. *Chem. Pharm. Bull.* **1982**, *35*, 602.
33. Sezaki, M.; Kondo, S.; Maeda, K.; Umezawa, H.; Ohno, M. *Tetrahedron* **1970**, *26*, 5171.
34. For example see: *The Chemistry of Anti-Tumor Antibiotics*, Remers, W.A. Ed., Wiley, New York, **1978**.

35. Takano, S.; Hasuda, K.; Ito, A.; Koide, Y.; Ishii, F.; Haneda, I.; Chihara, S.; Koyama, Y. *J. Antibiot.* **1976**, *29*, 765 and Omura, S.; Ikeda, H.; Malpartida, F.; Kieser, H.M.; Hopwood, D.A. *Antimicrob. Agents Chemother.* **1986**, *29*, 13.

36. Okabe, T.; Nomoto, K.; Funabashi, H.; Okuda, S.; Suzuki, H.; Tanaka, N. *J. Antibiot.* **1985**, *38*, 1333.

37. Danishefsky, S.J.; Uang, B.J.; Quallich, G. *J. Am. Chem. Soc.* **1985**, *107*, 1285.

38. Cambie, R.C.; Pausler, M.G.; Rutledge, P.S.; Woodgate, P.D. *Tetrahedron Lett.* **1985**, *26*, 5341.

39. Bolitt, V.; Mioskowski, C.; Kollah, R.O.; Manna, S.; Rajapaksa, D.; Falck, J.R. *J. Am. Chem. Soc.* **1991**, *113*, 6320.

40. Tius, M.A.; Gomez-Galeno, J.; Gu, X.-q.; Zaidi, J.H. *J. Am. Chem. Soc.* **1991**, *113*, 5775.

41. Matsumoto, T.; Katsuki, M.; Jona, H.; Suzuki, K. *J. Am. Chem. Soc.* **1991**, *113*, 6982.

42. Tatsuta, K.; Ozeki, H.; Yamaguchi, M.; Tanaka, M.; Okui, T. *Tetrahedron Lett.* **1990**, *31*, 5495.

43. Hacksell, U.; Daves, G.D., Jr. *Prog. Med. Chem.* **1985**, *22*, 1-65.

44. Hosoya, T.; Takashiro, E.; Matsumoto, T.; Suzuki, K. *J. Am. Chem. Soc.* **1994**, *116*, 1004.

45. See footnote #7 in reference #44.

46. See footnote #6 in reference #44.

47. McGee, L.R.; Misra, R. *J. Am. Chem. Soc.* **1990**, *112*, 2386.

48. Matsumoto, T.; Hosoya, T.; Suzuki, K. *J. Am. Chem. Soc.* **1992**, *114*, 3568.

49. Kwok, D.-I.; Daves, G.D., Jr. *J. Org. Chem.* **1989**, *54*, 4496.

50. Outten, R.A. ; Daves, G.D., Jr. *J. Org. Chem.* **1987**, *52*, 5064.

51. Drautz, H.; Zähner, H.; Rohr, J.; Zeeck, A. *J. Antibiot.* **1986**, *39*, 1657. Rohr, J.; Zeeck, A. *J. Antibiot.* **1987**, *40*, 459. Henkel, T.; Ciesiolka, T.; Rohr, J. and Zeeck, A. *J. Antibiot.* **1989**, *42*, 299. Rohr, J.; Beale, J.M.; Floss, H.G. *J. Antibiot.* **1989**, *42*, 1151. Zeeck, A.; Rohr, J.; Sheldrick, G.M.; Jones, P.G.; Paulus, E.F. *J. Chem. Res., Synop.* **1986**, 104.

52. Yamaguchi, M.; Okuma, T.; Horiguchi, A.; Ikeura, C.; Minami, T. *J. Org. Chem.* **1992**, *57*, 1647.

53. Haynes, L.J. *Adv. Carbohydr. Chem.* **1965**, *30*, 357.

54. Structure: Fiecchi, A.; Anastasia, M.; Galli, C.; Gariboldi, P. *J. Org.Chem.* **1981**, *46*, 1511 and references cited therein. See also: Schmitt, P.; Günther, H.; Hägele, G.; Stilke, R. *Org. Magn. Reson.* **1984**, *22*, 446.

55. Allevi, P.; Anastasia, M.; Ciuffreda, P.; Fiecchi, A.; Scala, A.; Bingham, S. Muir, M.; Tyman, J. *J. Chem. Soc., Chem. Commun.* **1991**, 1319.

56. See references cited in: Frick, W.; Hofmann, J.; Fischer, H.; Schmidt, R.R. *Carbohydr. Res.* **1991**, *210*, 71.

57. Frick, W.; Schmidt, R.R. *Carbohydr. Res.* **1991**, *209*, 101.

58. Frick, W.; Schmidt, R.R. *Liebigs Ann. Chem.* **1989**, 565.

59. Tschesche, R.; Widera, W. *Liebigs Ann. Chem.* **1982**, 902.

60. Eade, R.A.; McDonald, F.J.; Simes, J.H. *Aust. J. Chem.* **1975**, *28*, 2011.

61. Zehnder, M.; Sequin, U.; Nadig, H. *Helv. Chim. Acta.* **1979**, *62*, 2525.

Chapter 11

C-NUCLEOSIDE SYNTHESIS

I. SOME BIOLOGICALLY RELEVANT *C*-NUCLEOSIDES

A *C*-nucleoside occurs when the anomeric nucleosidic bond is comprised of two carbon atoms and not a carbon and a nitrogen atom. There exist several naturally occurring *C*-nucleosides that possess interesting biological activities both *in vitro* and *in vivo* tests.[1] The *C*-nucleosides are structurally similar to nucleosides except that they possess a hydrolytically and enzymatically more stable sugar-base bond. The naturally occurring *C*-nucleosides all have the common feature of the D-ribose sugar and most (pyrazomycin B is an exception)[2] have the β-configuration. Figure 1 illustrates some of the common *C*-nucleosides. Pseudouridine (**1**),[3] oxazinomycin (**2**),[4] and showdomycin (**3**)[5] all have the common feature of having the sugar portion bonded to a mono-heterocycle.

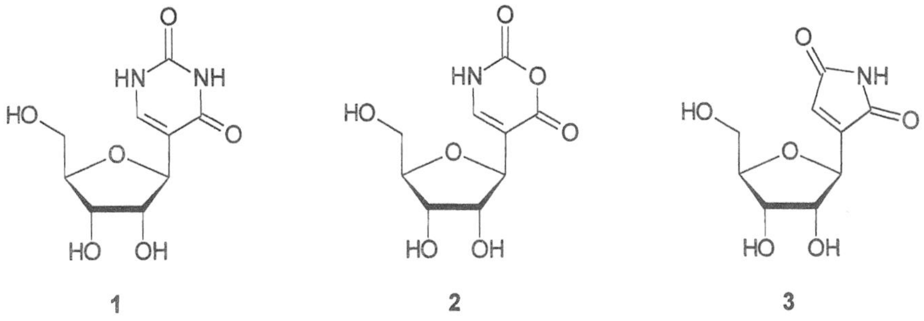

Figure 1

As in the regular nucleosides the similarity to the bases that make up DNA or RNA are quite striking. Both have the common feature of a D-ribo sugar, excepting that DNA is made up of 2-deoxy ribo units. The sugar heterocycle bond is β in both cases and for pseudouridine the heterocycle closely resembles thymine. Pseudouridine was the first *C*-nucleoside isolated. A closely related *C*-nucleoside natural product was also isolated and was termed oxazinomycin. It differs from pseudouridine in the sense that it contains an oxygen atom in the ring in lieu of a nitrogen atom. Of the three mono-heterocycle antibiotics pictured above, showdomycin has received the most synthetic attention. It was isolated in 1964 from Streptomyces Z-452 and was immediately shown to possess moderate anti-bacterial against both Gram positive and negative bacteria and strong activity against *Streptococcus hemolyticus*. In preliminary tests, it showed promising antitumor activity against ascites tumor in Erhlich mouse. The biological activity of this compound has since been studied and reviewed.[1]

Formycin (4),[6] formycin B (5),[7] and oxoformycin (6)[8] all closely resemble the purine group of nucleotides, the main difference being that the heterocycle is no longer a purine based structure, but of the structural type shown in Figure 2 with the anomeric-base bond in the same position as the purine nucleotides; the result is a *C*-nucleoside.

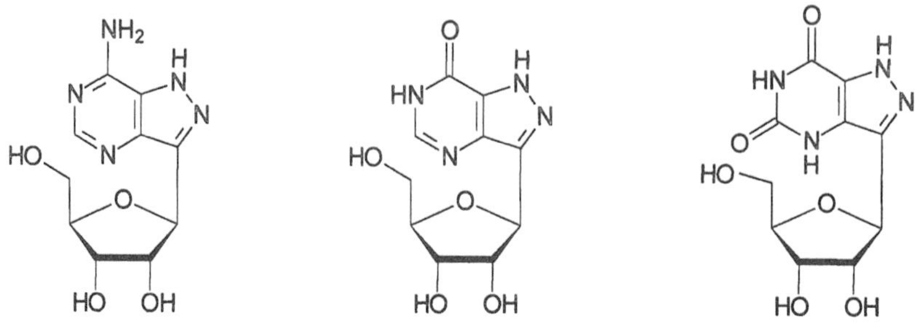

Figure 2

Perhaps one of the most impacting of the *C*-nucleoside antibiotics has been formycin. It is an isomer of adenosine and it is believed that its observed antibacterial, antiviral and antitumor effects are brought about by its ability to replace adenosine and thereby affect the metabolism of various nucleic acids. Its biological activity and synthetic approaches have been reviewed.[1] In this work, recent selected synthesis of analogs of formycin that are deemed interesting will be examined. Formycin is metabolized to formycin B in the body and formycin B shows less medicinal application. As a result, work has progressed towards synthesizing more metabolically stable analogs of formycin in the hope of increasing the serum lifetime of the molecule.

Tiazofurin (7) is a synthetic *C*-nucleoside thiazole that has demonstrated antitumor activity in several model tumor systems, Figure 3.[9] Tiazofurin has been found to be curative for Lewis lung carcinoma in mice and exhibited both *in vivo* and *in vitro* activity against several cancers in humans. Tiazofurin is an analog of the potent antiviral agent ribavirin (8)[10] which itself has shown to be only weakly active against tumors.

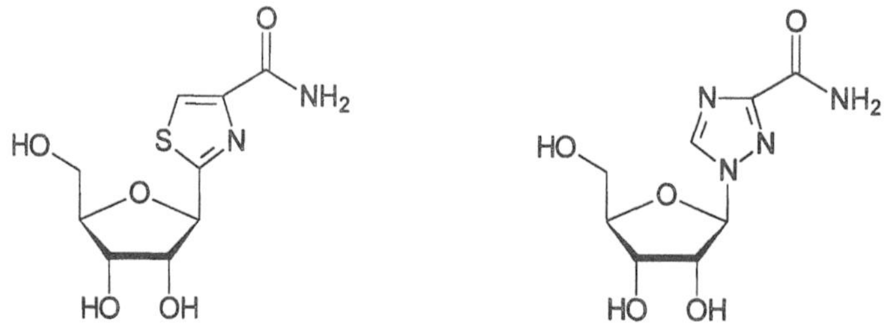

7   Tiazofurin   8   Ribarvirin

Figure 3

C-Nucleoside analogs of biologically interesting *N*-nucleoside medicinal agents will also be discussed where deemed appropriate. It should be mentioned that the topic of *C*-nucleoside synthesis is an enormous one. A comprehensive review would be well out of the scope of this book. The aim of this chapter is to give the reader an overview of the common and more interesting strategic and chemical approaches toward *C*-nucleoside assembly.

II. SYNTHETIC APPROACHES TO C-NUCLEOSIDES

There have been several types of strategic approaches to the assembly of C-nucleosides. Generally speaking, these can be divided into two main classes. The first (Fig. 4) involves direct attachment of the base heterocycle to the C-1 carbon atom of the D-ribo sugar. Typical examples would include the reaction of heterocyclic organometallics with appropriate electrophilic sugar acceptors, reaction of electron rich heterocycles with the same acceptors, and heterocyclic Wittig reagent in reaction with reducing sugars. The challenge here is to stereoselectively introduce the heterocycle onto the β-side of the ribo sugar. Several novel methods have been developed to achieve this goal.

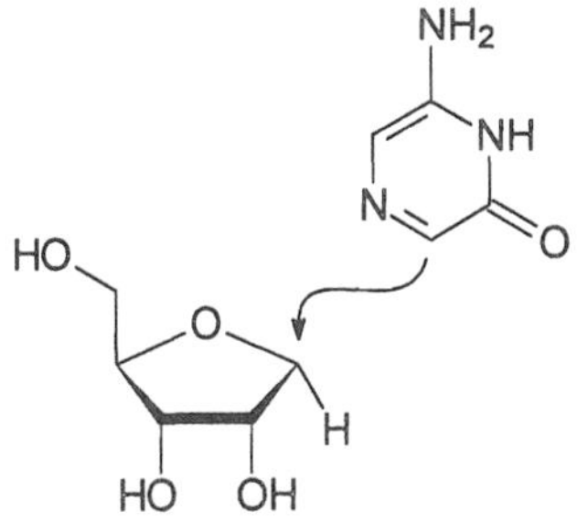

Figure 4

The second classification involves elaboration of the heterocycle onto an existing carbon atom(s) that makes up the pre-existing C-glycoside (Fig. 5). The carbon fragment on the sugar can be an acid, an ester, a nitrile or an acetylene function. Here the stereochemistry of the C-1 carbon group usually dictates the configuration of the C-nucleoside. So in effect, the goal here is to stereospecifically make β-C-ribosides. Methods to construct the α-C-nucleosides are not abundant since this would lead to the opposite configuration normally found in the biologically active naturally occurring C-nucleosides and most workers have concentrated on developing methods for β-carbon-carbon bond formation. A typical example would be the condensation of a C-1 acid with an appropriate segment to give a C-nucleoside.

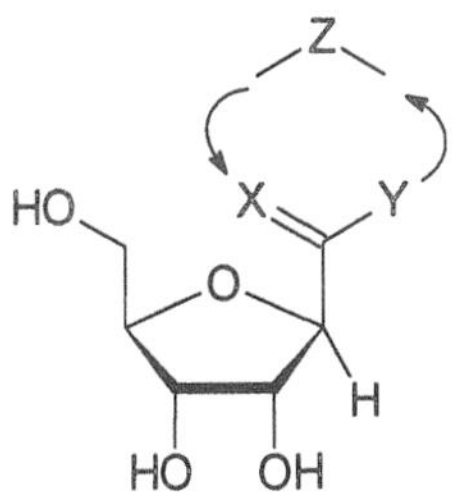

Figure 5

The organization of the following section will be in line with the rest of the text, that is the strategic approach will be used as the basis for classification. Other information such as biological activity or synthesis of analogs may also be discussed. It should be noted that the field of C-nucleoside chemistry is rather mature and both the synthetic and biological aspects have been reviewed previously.[11] This chapter will outline the synthetic approaches to C-nucleoside construction using strategy as the key points of difference. Examples from the recent literature (~last five years) will be used as much as possible, but some classical work may also be cited for completeness. The organization is by reaction type and free radical approaches will start things off, and this will be followed by Wittig approaches and then by Friedel-Crafts approaches. Condensation of organometallic heterocycles with appropriate acceptors will comprise one section and approaches based on *C*-glycosides (and their preparation) will be followed by the use of bicyclo compounds to wrap up the chapter.

A. FREE RADICAL APPROACHES
1. Multiple Bond Addition
a. *Addition to a Base Precursor*

Araki[12] has used intermolecular free radical addition of the D-ribo anomeric radical derived from the xanthate **9** to dimethyl maleate to provide the β-C-glycoside **11** in 62% yield as a mixture of isomers at *C*-1'.

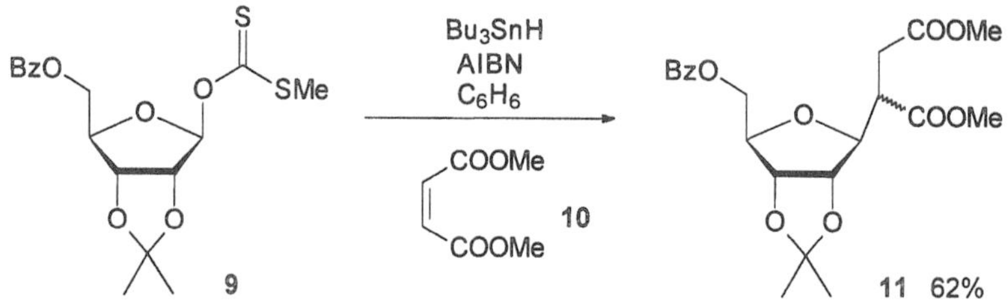

Scheme 1

Saponification of the diester gave the diacid which was dehydrated to give the anhydrides which were then exposed to ammonia and acetyl chloride to furnish the imides **13**. Deacetylation of the primary acetate was followed by silylation to give **14**, a compound which has been previously transformed into the product showdomycin.

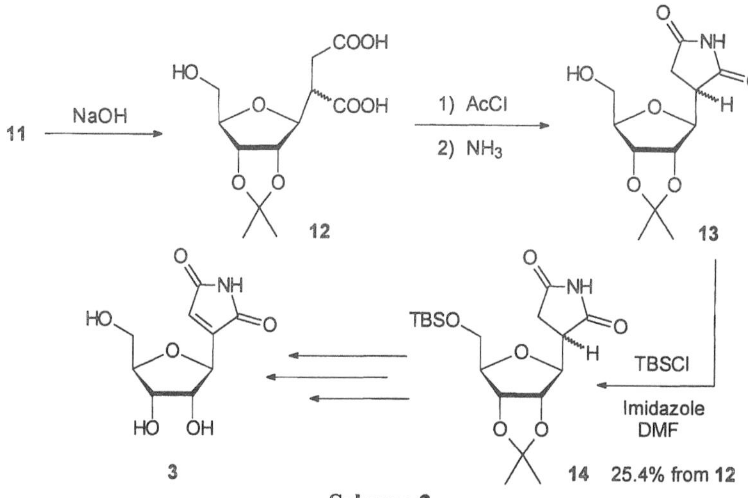

Scheme 2

b. *Addition to a Heterocycle*

Barton and Ramesh[13] have also synthesized showdomycin, but via a tandem nucleophilic-radical method as shown in Scheme 3. The telluride **16** was easily accessed from the protected D-ribo derivative **15** by mesylation and displacement with telluride anion. Exposure of **16** to light (150-W lamp) in the presence of maleimide and the acetyl derivative **17** gave the β-adduct **18** which was then oxidatively eliminated to provide protected showdomycin. The sequence was completed by acid hydrolysis of the trityl and ispropylidene groups to give showdomycin (**3**).

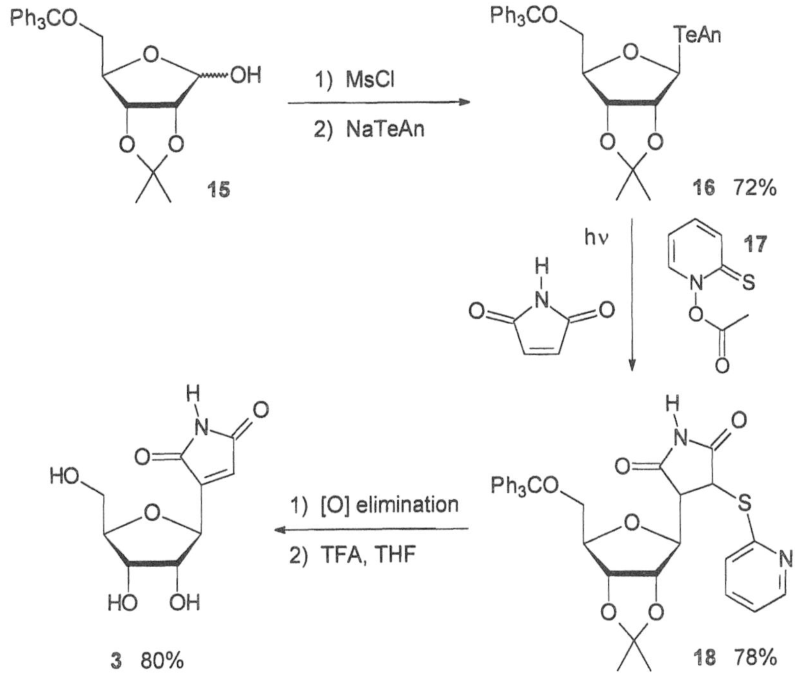

Scheme 3

Interestingly enough, a recent radical approach to *C*-nucleoside assembly employed Barton type chemistry. Another approach[14] relied on the addition of an anomeric radical to a heterocycle to give a *C*-nucleoside directly. They chose to generate the radical from the corresponding acid **21** which itself is available from 2-deoxy-D-ribose (**19**) in three steps. Radical generation via the Barton protocol then gave the anomeric radical which added to lepidine (**24**) to give a mixture of anomers **25** (α:β ratio of 14 to 86).

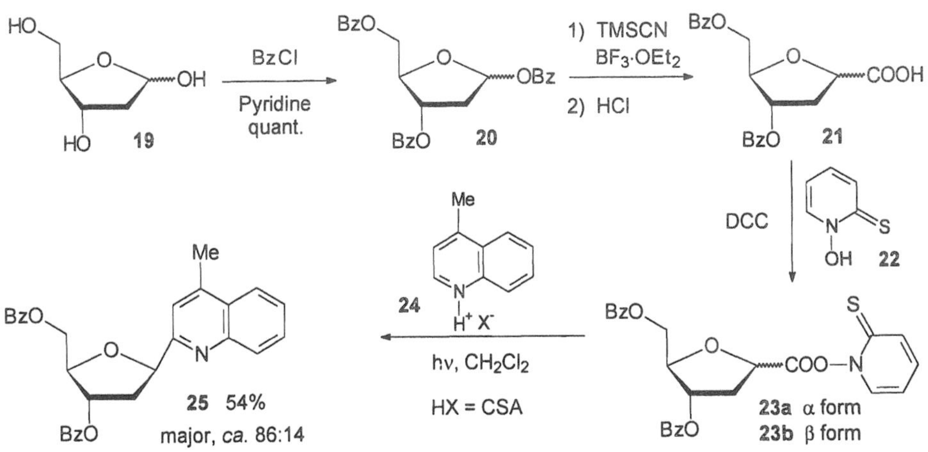

Scheme 4

Other bases were examined and these gave the *C*-nucleosides **26** and **27** in 26 and 30% yield, respectively. The α-anomers were the only products isolated in both cases (Fig. 6).

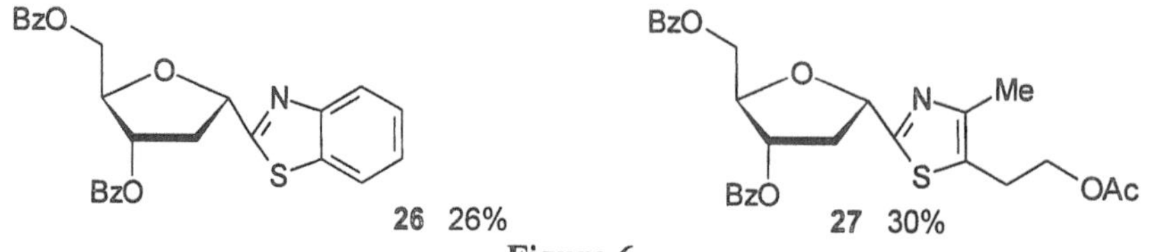

Figure 6

B.   WITTIG APPROACHES
1.   Attachment of a Carbon Fragment

Kato *et al.* used a Wittig olefination to attach a keto-ester function onto the anomeric carbon of a D-ribo sugar to give a *C*-nucleoside precursor.[15] Reaction of ylide **28** with the protected D-ribo sugar **15** gave **29** as an anomeric mixture (α:β of 1:2). The reaction must proceed via the open chain form which then recyclizes with the oxygen atom cyclizing in a Michael fashion onto the α,β-unsaturated ketone to give the *C*-glycosides **29**. The synthesis was completed by elaboration of the keto-ester function into the pyrazole. This was accomplished by introduction of a diazo function at *C*-3' followed by base induced cyclization to furnish **31**. The sequence was completed by treatment of **31** with ammonia and deprotection with TFA to give pyrazofurin (**32**) and its epimer pyrazofurin B (**33**).

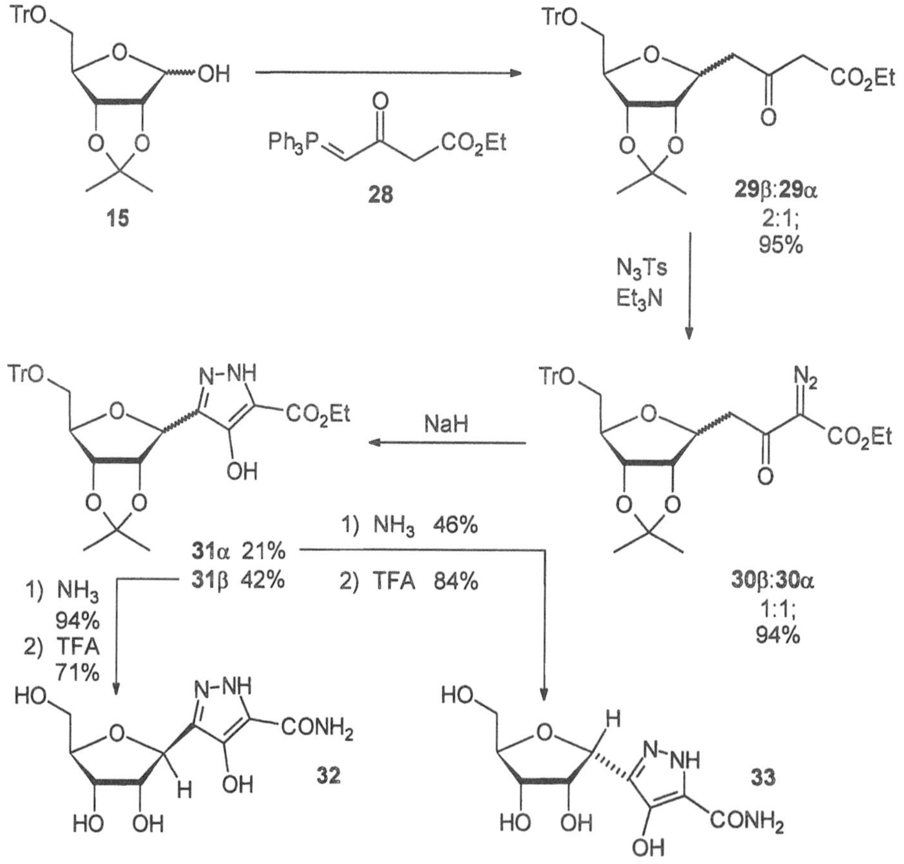

Scheme 5

Synthetic endeavors towards tiazofurin type compounds have also attracted the use of Wittig olefination chemistry. Secrist III and Clingerman[16] also used a relatively highly functionalized ylide to make *C*-nucleosides. In their case, a halogen on *C*-1' increased the stereoselectivity of the Wittig/cyclization sequence. For example, reaction of the D-ribo compound 15 with the ylide 34 in the presence of a catalytic amount of DBU gave a 85% yield of diastereomeric β-isomers 36. The addition of DBU aids in cyclizing the small amount of open form compound that is formed during the course of the reaction.

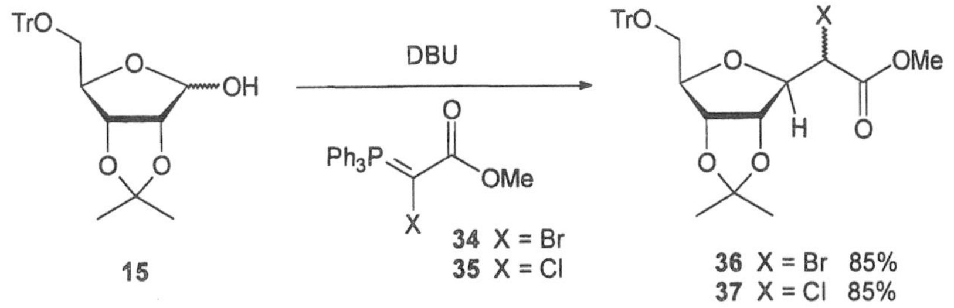

The corresponding chloro ylide 35 also gave good yields of β-products 37. It would seem that the bulkiness of the halogen substituent undergoes steric interference when in the α-configuration thereby favoring the β-form. This is supported by the fact that attempted epimerization led only to decomposition and no trace of the α-isomer. In order to apply this methodology to the synthesis of a *C*-nucleoside the ylide 38 was condensed with 15 to give 39 in ~45% yield. Condensation of 39 with thioacetamide (40) in HMPA then gave the protected thiazole *C*-nucleoside 42.

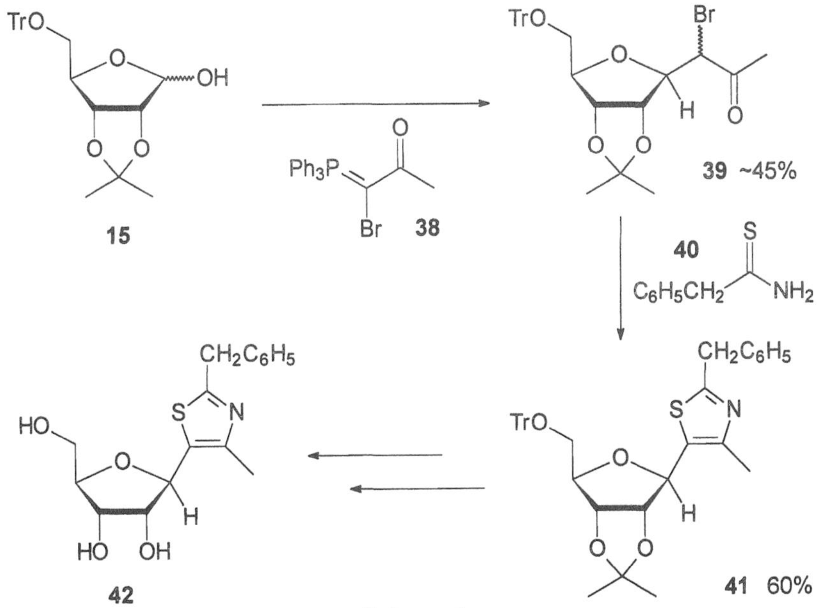

Scheme 6

Gonzalez[17] has used a similar ylide in a synthesis of showdomycin. Compound 36 (derived as above) was treated with silver acetate in DMSO to furnish the epimeric acetates 43 (53%). Methanolysis gave the corresponding hydroxy esters which were oxidized with

PCC to yield the ketone **45**. Wittig reaction with the ylide **46** in dry chloroform gave the cyclized precursor **47**. Hydrolysis then furnished showdomycin (**3**).

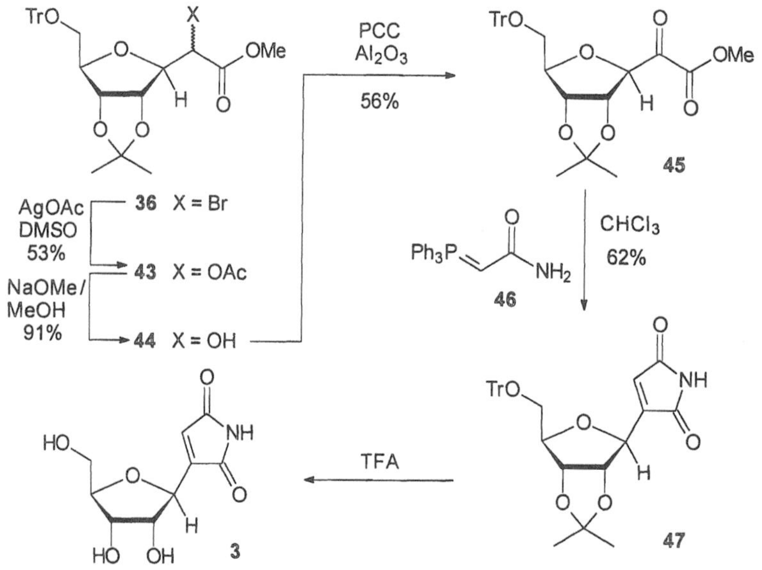

Scheme 7

It would seem that in order to gain access to the β-isomers via the use of Wittig chemistry one would only need a (removable!?) substituent on *C*-1 of the ylide. If desired the group can be retained or by judicious choice removed to give the corresponding β-*C*-furanoside.

Weigele and De Bernardo[18] have relied, initially, on Horner Emmons type chemistry and subsequently formed an aminoacrylonitrile species to develop a total synthesis of minimycin. Horner modification using the D-ribo sugar and the ylide **48** in DME gave a 2:1 mixture of β to α isomers. Formylation of either pure **49** or the mixture of anomers with bis(dimethylamino)-*tert*-butyloxymethane then gave a mixture of **51** and **52** (2:1), respectively, Scheme 8.

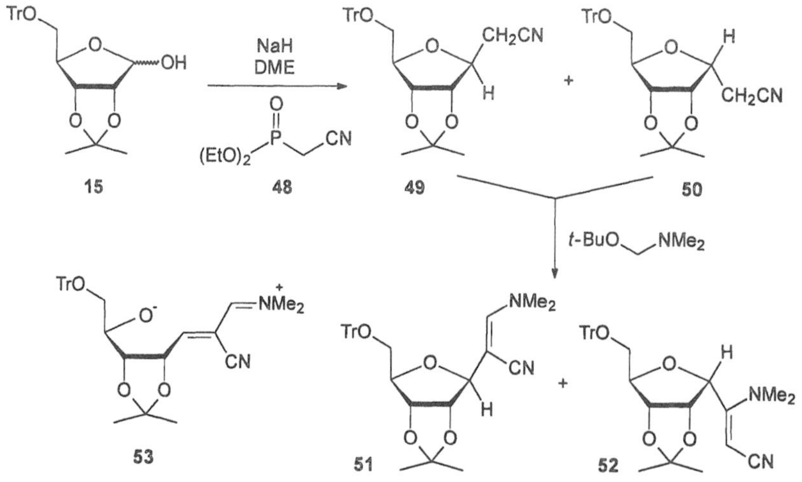

Scheme 8

Presumably the epimerization occurring between **51** and **52** is partially due to the stabilization of the open chain ring **53** which can then close to give a mixture of α and β compounds. Condensation of either **51** or the mixture (**51** and **52**) with hydroxylamine then gave the isoxazole **54** and its anomer. Hydrogenation afforded **55**. This was followed by hydrolysis and condensation of *N,N'*-carbonyldiimidazole to give **56**. Deprotection then furnished minimycin **57**.

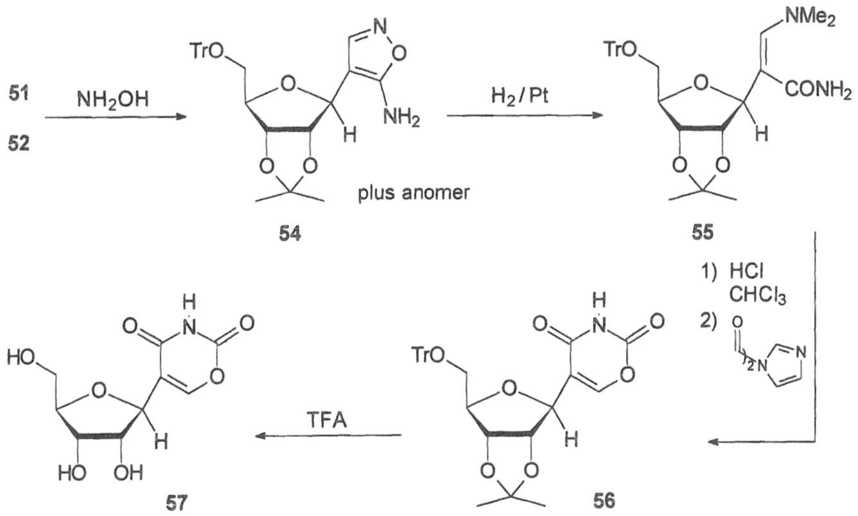

Scheme 9

A more recent example further illustrates the utility of the dimethylaminonitrile functionality in *C*-nucleoside assembly.[19] In this case an inosine analog was required and a bicyclic heterocyclic system was constructed. Compound **58** was hydrolyzed with TFA in dichloromethane to give **59** and *O*-alkylation with chloroacetonitrile then gave the bis nitrile which underwent Thorpe Ziegler reaction to give the furan **61**.

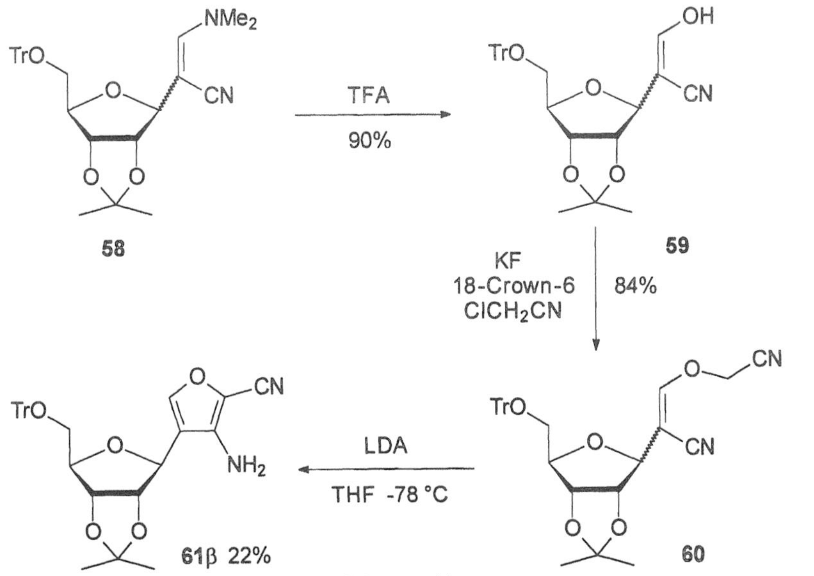

Scheme 10

The two isomers could be separated by chromatography and transformed into a variety of heterocycles as shown below in Scheme 11. For example, treatment of **61** with formamidine acetate gave the *C*-nucleoside **62** after deprotection. Alternatively, **61**, when exposed to carbon disulfide in pyridine, gave the dithiones **63**. In a three step sequence, **61** could be fashioned into the methylthio derivative **64** by sequential treatment with hydrogen sulfide in pyridine, triethyl orthoformate, and finally methyl iodide to give after acidic deprotection the requisite *C*-nucleoside.

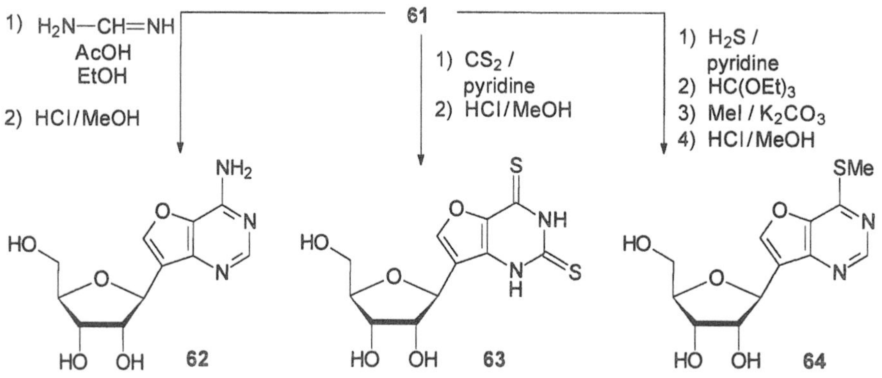

Scheme 11

2.   Attachment of the Heterocycle

Barrett and co-workers'[20] approach to showdomycin also involved the use of Wittig chemistry. Their ylide **66** was heterocycle based and this was reacted with an unprotected sugar **65** to obtain a good yield of the condensation product **67**.

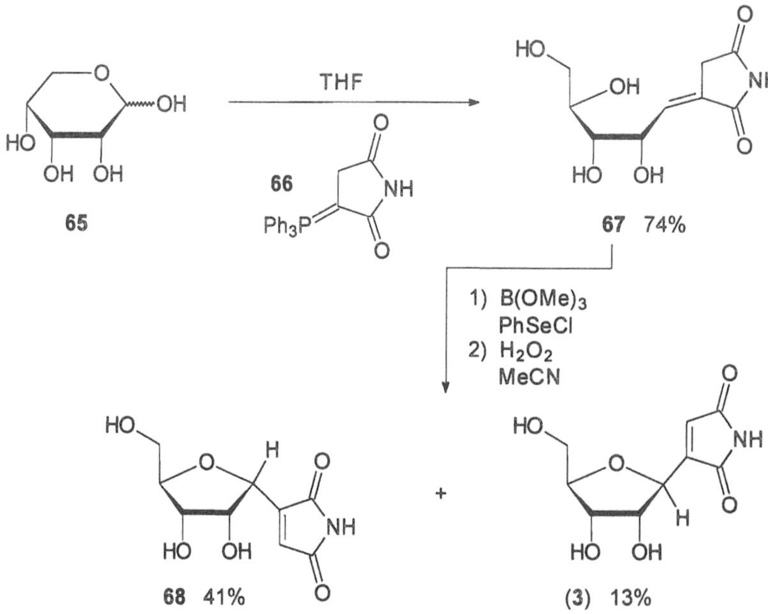

Scheme 12

This is extremely attractive since no need for elaborate protection-deprotection schemes are required. Next they needed to close the ring and introduce endocyclic double bond.

Selenoetherification proceeded to give an unstable furanose adduct which was not isolated, but directly oxidized and fragmented to give a mixture of two products. Showdomycin was obtained in 13% yield while the major compound isolated was the α-isomer epishowdomycin **68** in 41% yield. Although the yield in the final step is low, the sequence has the merits of being short and easy to carry out.

Koomen and Wanner[21] also used a similar strategy as Barrett for assembling a *C*-nucleoside via Wittig methodology. In their case they were after a glutarimide type *C*-nucleoside. Condensation of the ylide **69** gave a 46% yield of a mixture (*E:Z* of 62:38) of adducts. The *E*-isomer was then treated with DBU and a 89% yield of products (epimeric at *C*-1') was obtained. The sequence was completed by standard deprotection to give the *C*-nucleosides **72**, Scheme 13.

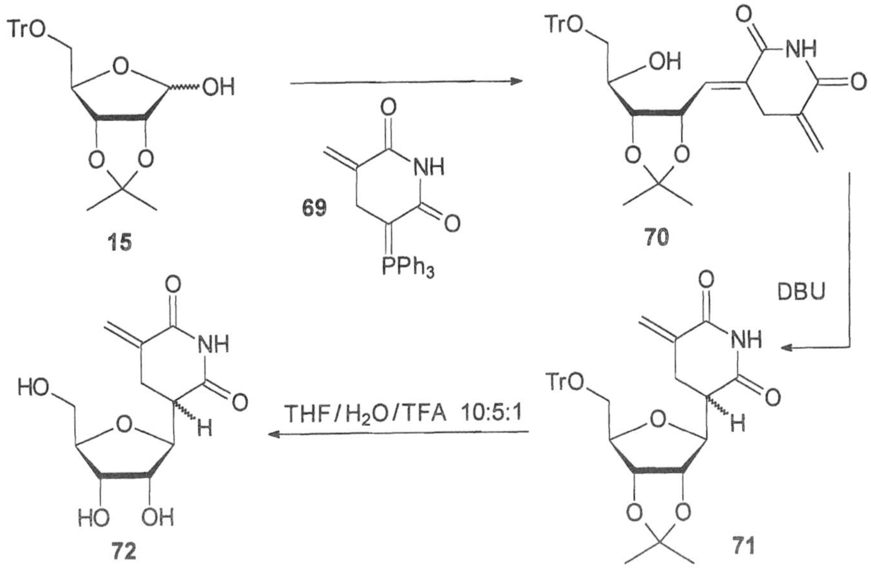

Scheme 13

When the sugar and heterocycle are separated by only one carbon atom, this type of *C*-nucleoside is termed a homo-*C*-nucleoside. Kato[22] also employed a heterocycle based ylide in a condensation reaction with the protected D-ribo sugar **15** to assemble such a compound. They examined several ylides in their study as shown in Schemes 14-16.

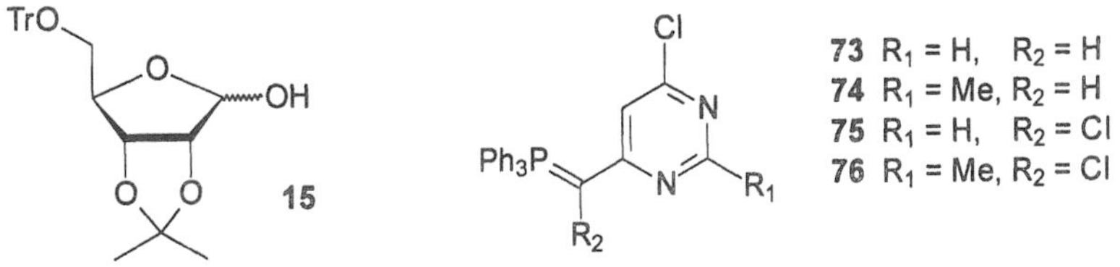

Scheme 14

When the ylides **73** and **74** were used, the β-isomer was found to be the major product formed. Some α-anomer and open chain products were also isolated. When a chlorine atom was placed on the ylide as in **75** and **76**, it was found that α-isomer was the major product (Scheme 15).

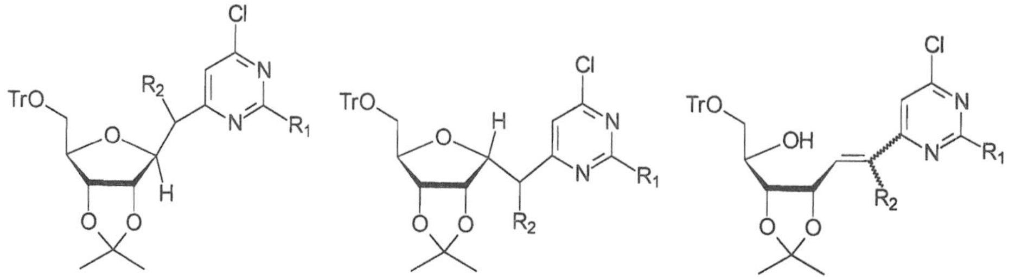

77β $R_1 = H$, $R_2 = H$ 64%
78β $R_1 = Me$, $R_2 = H$ 63%
79β $R_1 = H$, $R_2 = Cl$ 29%
80β $R_1 = Me$, $R_2 = Cl$ 29%

77α $R_1 = H$, $R_2 = H$ 14%
78α $R_1 = Me$, $R_2 = H$ trace
79α $R_1 = H$, $R_2 = Cl$ 43%
80α $R_1 = Me$, $R_2 = Cl$ 43%

E-81 $R_1 = H$, $R_2 = H$ 6%
Z-81 $R_1 = H$, $R_2 = H$ 7%
Z-82 $R_1 = Me$, $R_2 = H$ 24%
83 $R_1 = H$, $R_2 = Cl$ 11%
84 $R_1 = Me$, $R_2 = Cl$ 17%

Scheme 15

The workers also examined the base induced cyclization of some of the open chain products. Compound *Z*-**81** in the presence of triethylamine gave the β-isomer **77β** exclusively, whereas the *E*-**81** isomer cyclized to give a mixture of compounds α:β ratio of 82:18.

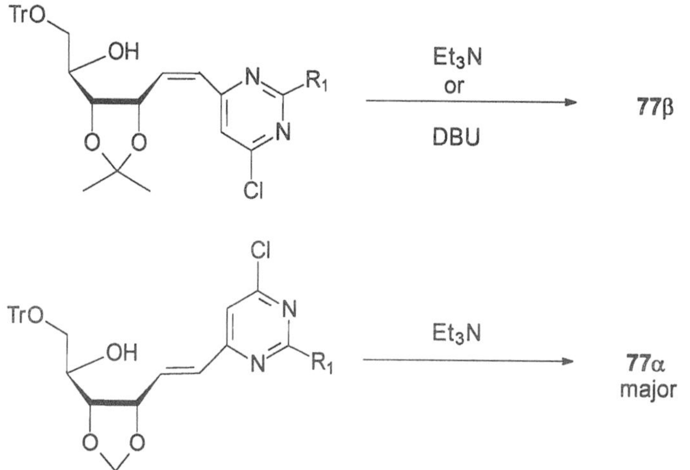

Scheme 16

Wise and co-workers[23] have made homo-*C*-nucleosides in a similar fashion. The ylide **85** was condensed with the familiar D-ribo sugar **15** to give three compounds **86-88**.

Figure 7

Hydrolysis of the protecting groups on **86** with hydrochloric acid in dioxane led, as expected, to the β-anomer **89**. Similar treatment of the α-isomer **87** gave an 87% yield of the β-isomer **89**. Therefore, access to the β-isomer was available from both protected anomers. The small amount of uncyclized material **88** was easily cyclized to **86** and **87** in the presence of a catalytic amount of sodium methoxide in methanol.

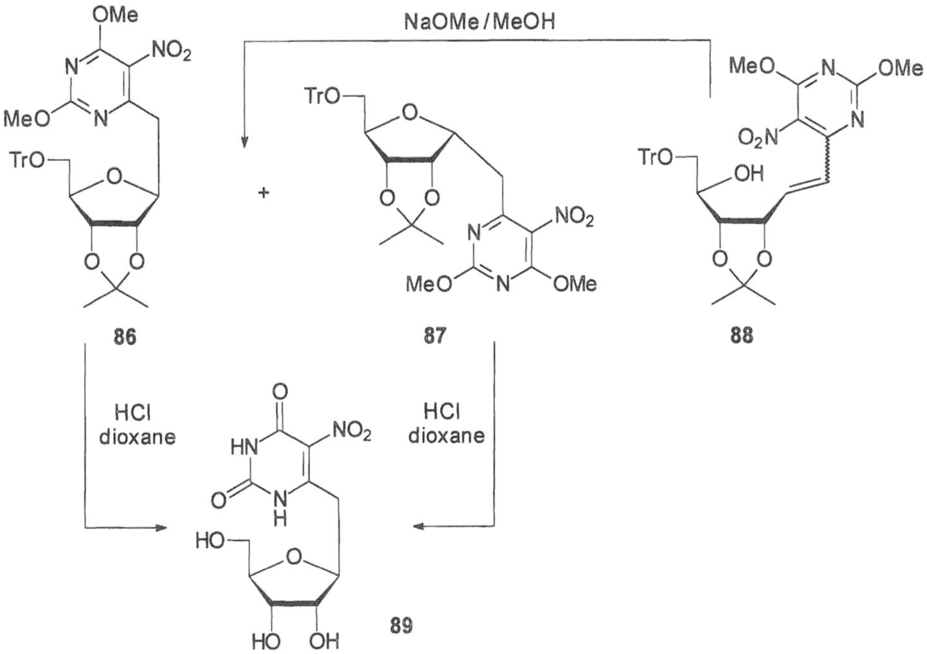

Scheme 17

C. FRIEDEL-CRAFTS APPROACHES

1. Glycosidation with a Heterocycle

Klein *et al.*[24] used a Lewis acid catalyzed glycosidation to affect carbon-carbon bond formation between the thiophene **91** and the D-ribo sugar **90** to give rise to a mixture of anomers **92** and **93** in 50 and 10% yield, respectively, Scheme 18.

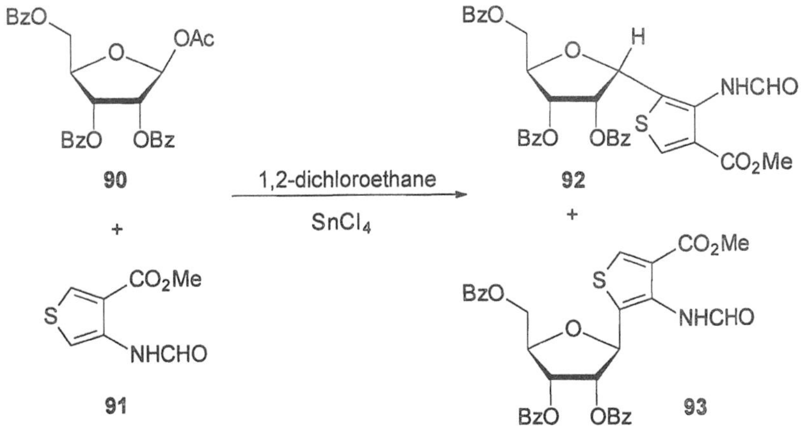

Scheme 18

Deformylation of either **92** or **93** gave rise to a mixture of anomers **94**. This mixture was treated with formamidine acetate to then give a separable mixture of *C*-nucleosides **95**.

Alternatively, **93** could be treated with ammonia in methanol to also give the product **95β**. Deprotection of **95β** with sodium methoxide in methanol then furnished the free *C*-nucleoside **96**.

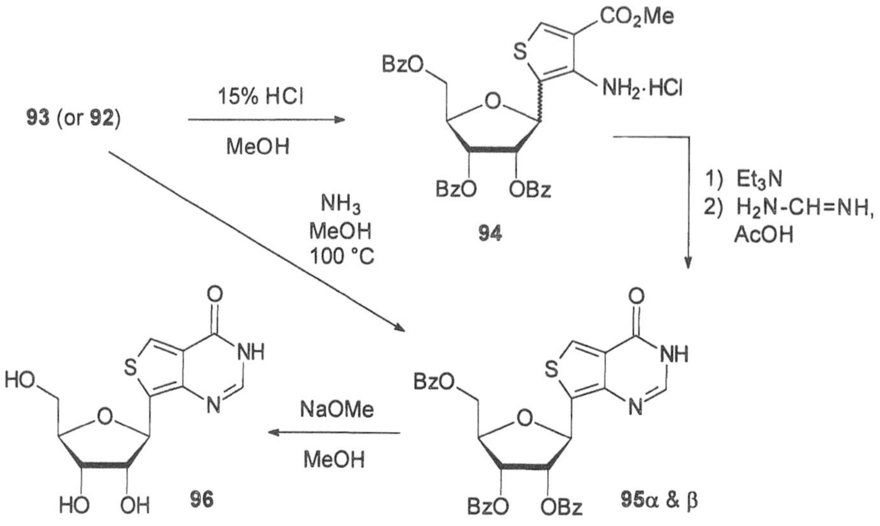

Scheme 19

The mechanism of the anomerization is thought to be as depicted below in Scheme 20. Protonation of the ring oxygen is followed by an electron push starting from the amino group to give the open chain compound **97c** which can then close from the opposite side.

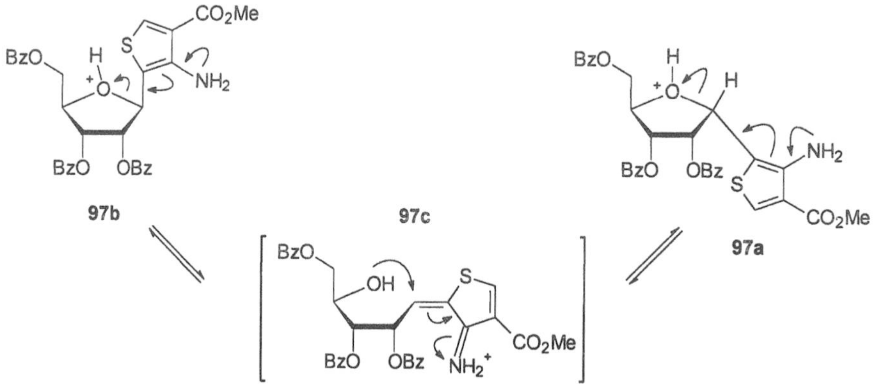

Scheme 20

2.   Glycosidation with a Carbon Fragment Followed by Heterocycle Formation

The cationic approach to carbon-carbon bond formation at the anomeric center is dealt with in greater detail in Chapters 1 and 2. This section will illustrate a few examples that have shown utility in the forum of *C*-nucleoside preparation.

Kalvoda and co-workers[25] have used the Lewis acid catalyzed condensation of an electron rich aromatic with a furanosyl bromide in their synthesis of the natural product showdomycin. 2,3,5-Trimethoxy benzene (**99**) condensed with the furanosyl bromide **98** in the presence of zinc oxide to give the β-isomer. This glycosidation is presumably governed by neighboring group participation from the *O*-2 benzoyl group to give an intermediate which then predominantly undergoes attack from the top face. The *C*-glycoside was debenzolyated, characterized as its 2,3-phenylboronate, and then reacetylated to give **100**. Exposure of **100** to ozone then gave **101** and Wittig reaction then provided the olefin **102** as a 10:1 mixture of cis and trans isomers which were separated as their acids. Treatment with ammonia then furnished **102** which was cyclized to the maleimide by exposure to ethyl polyphosphate in DMF to give **103**. Transformation into showdomycin was realized by a mild acidic methanolysis.

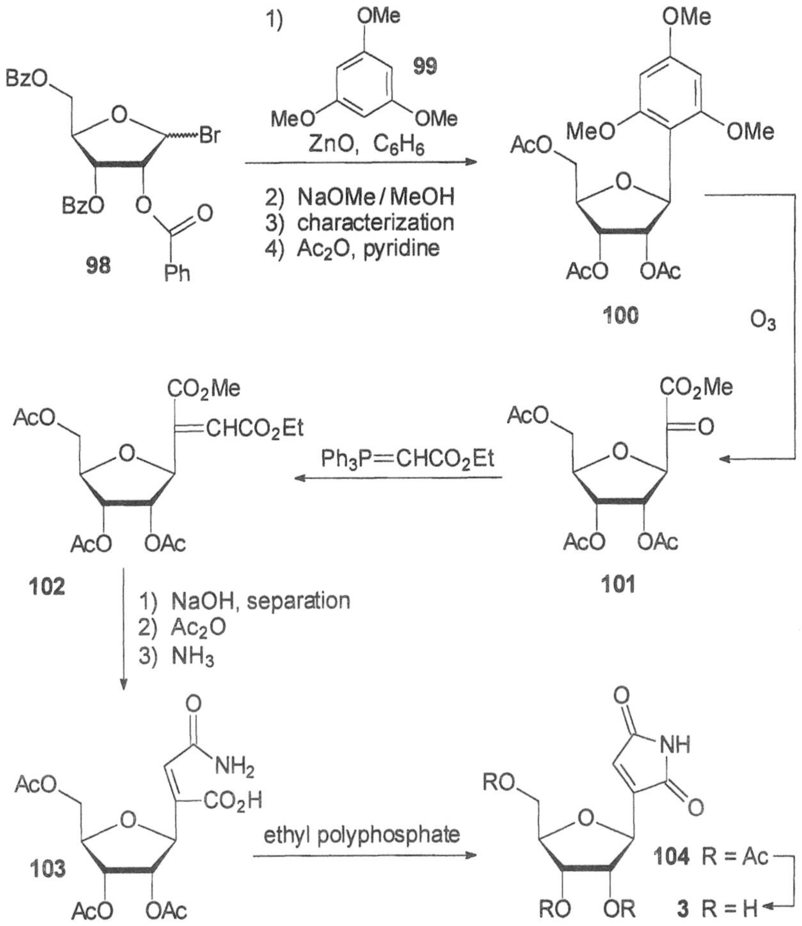

Scheme 21

Williams and Stewart[26] have used pyridyl thioglycosides as anomeric activators to effect carbon-carbon bond formation. Their work (Scheme 22) was also applied to the synthesis of showdomycin. Their route is similar to that described in Scheme 21. The requisite starting material was easily made in three steps from D-ribose. Ketalization of the 2,3 *syn*-diol was followed by silylation of the primary hydroxyl with *tert*-butyl dimethyl silyl chloride and reaction of the resulting hemiacetal with tributyl phosphine and dipyridyl disulfide then gave **105** as a mixture of anomers (α:β of 3:1). Activation with silver(I) triflate in the presence of trimethoxy benzene then gave the aryl β-*C*-glycoside **106** in 61% yield. The resulting β configuration was proven by eventual transformation into showdomycin. Ozonolysis of **106** was followed by Wittig reaction and hydrolysis of the protecting groups with aqueous TFA then gave synthetic showdomycin (**3**).

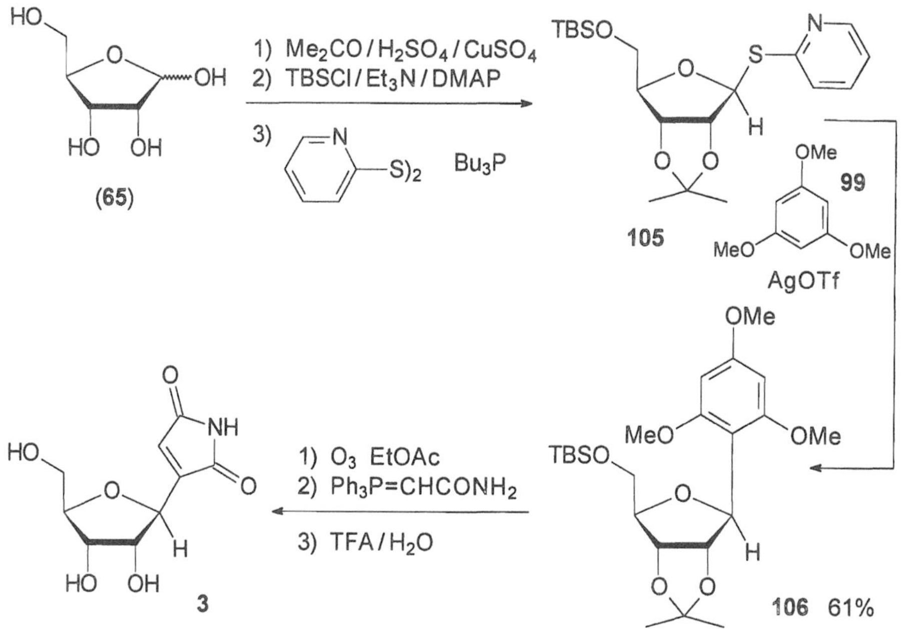

Scheme 22

Kuwajima and Inoue[27] have also attached a suitable maleimide precursor on the ribofuranose via a cationic carbon glycosidation mediated reaction. The fully acylated furanose **90** was treated with tin(IV) chloride in the presence of 1,2-bis(trimethylsiloxy)cyclobut-1-ene (**107**) and this gave rise to the formation of a single product **108** in 92% yield, Scheme 23.

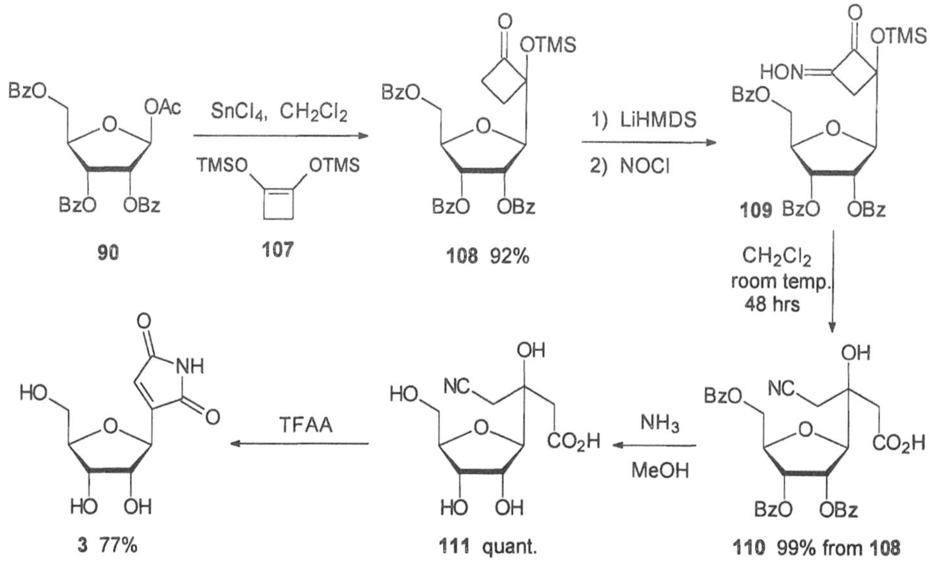

Scheme 23

Once again we see the influencing effect of neighboring group participation at work in helping to determine the stereochemical course of the reaction. Enolate formation of **108** was followed by exposure to nitrosyl chloride which then gave the oximino compound 109. Ring cleavage was effected by letting 109 stand in dichloromethane for two days at ambient temperature to give the cyano-acid 110. Deprotection then afforded 111 which was cyclized and dehydrated (-H$_2$O) to showdomycin (3) by treatment with trifluoracetic anhydride (TFAA) in benzene (77%). The overall yield of showdomycin from the protected ribose 90 was 70%. The above route is very short, highly stereoselective, as well as high yielding.

D.   ONE CARBON INTRODUCTION IN C-NUCLEOSIDE SYNTHESIS

Furanoses possessing a one carbon substituent on C-1 have been used fairly often in C-nucleoside synthesis. Hence methods to make these latter types of compounds are rather numerous. Since some of this chemistry has been discussed by Hanessian,[11] only recent developments and applications will be considered.

1.   Esters, Nitriles and Acids

Nitriles and anomeric acids have been widely used in the preparation of C-nucleosides. The nitriles can be easily transformed into thioamides and utilized or employed as such. The reaction of cyanide ion with a halo sugar under catalysis is a very old reaction[28] and this method of C-glycosidation is still sometimes used today.

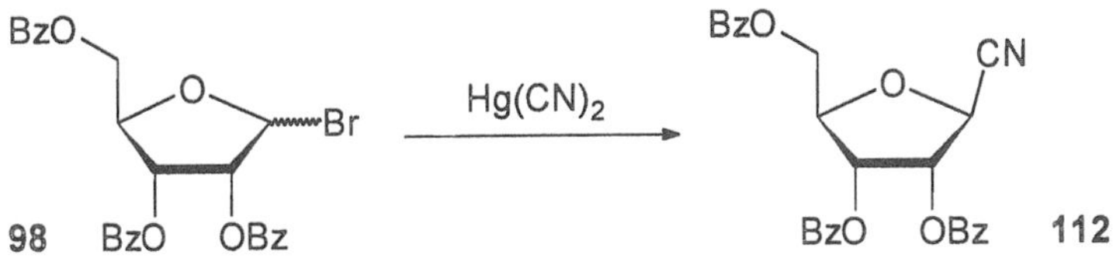

eq. 2

The reaction is controlled by participation of the benzoyl group on O-2. The formed nitriles are useful starting materials for C-nucleoside preparation.

Kalvoda used 112 as a starting point for a synthesis of showdomycin.[29] The nitrile was hydrolyzed to the acid and converted to its methyl ester 113. The benzoyl groups were replaced with acetyl groups and the formed acid converted to an acid chloride which was reacted with cyanide ion to give the intermediate acyl cyanide 114. Wittig reaction then gave a mixture of cis and trans isomers. The cis isomer 115 was then treated with acid and acetic anhydride to give the fully acetylated showdomycin 116 which was deprotected with acidic methanol to give showdomycin.

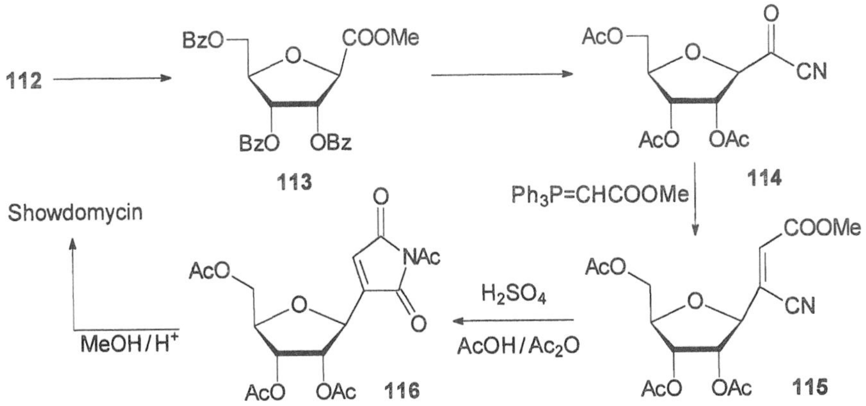

Scheme 24

El Khadem and El Ashry[30] have also employed a cyano sugar as a starting point for the synthesis of a *C*-nucleoside. Their target was an antibiotic named cordycepin C (**117**) which is an analog of cordycepin (**118**).[31] Cordycepin is a cytotoxic agent which has been found to inhibit the growth of *Bacillus subtillis*, *avaim tubercle bacillus* and Ehrlich ascites tumor cells.

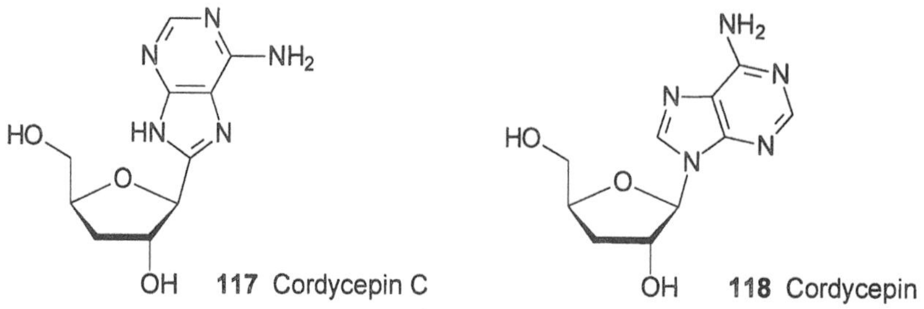

117 Cordycepin C   **118** Cordycepin

Figure 8

The nitrile **120** was readily available from the bromo sugar **119** by standard methods and hydrolysis then provided acid **121**. Removal of the protecting groups was followed by condensation with the diamine **123** to give after purification **124** which cyclized under thermal conditions to cordycepin C.

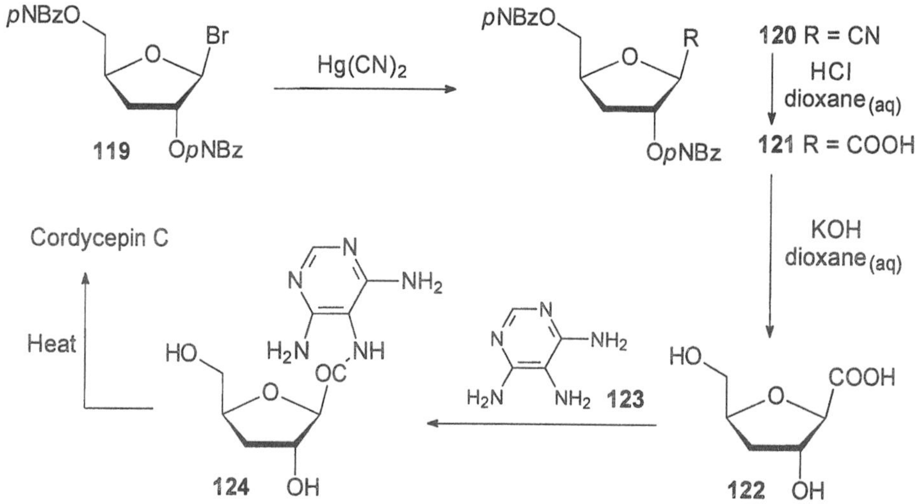

Scheme 25

The nitrile functionality can also serve as a precursor to the thioamide group as shown below in eq. 3. The nitrile **125** was exposed to a stream of hydrogen sulfide in ethanol in the presence of DMAP to give the thioamide **126** in ~80% yield.[32]

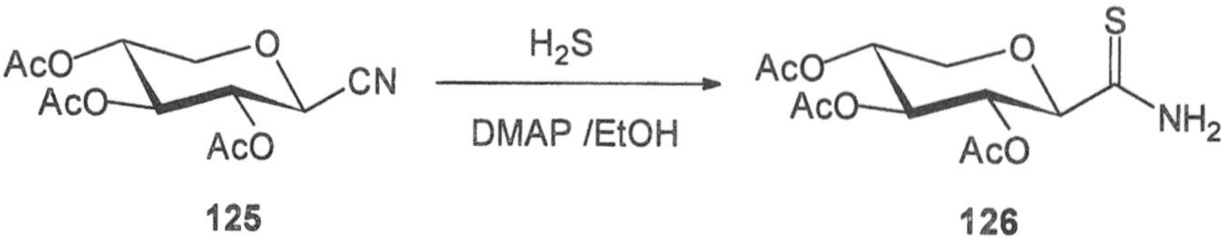

eq. 3

125   **126**

Condensation of **126** with ethyl bromopyruvate then gave the requisite *C*-nucleoside in, albeit, low yield. The reaction mixture was contaminated by another product, the furan derivative **128**, which comes about as a result of ring contraction and elimination. Further reaction of **127** with ammonia in methanol then provided amide **129**. Improvement in the sequence was realized when the cyclization reaction was conducted in the presence of barium carbonate to give the product **130**, which could then be dehydrated under partially controlled conditions, in this instance, to give the glycal *C*-nucleoside **131**. Deprotection and amide formation were realized in a single operation to give the pyranose *C*-nucleoside, **132**, a pyranoside analog of tiazofurin.

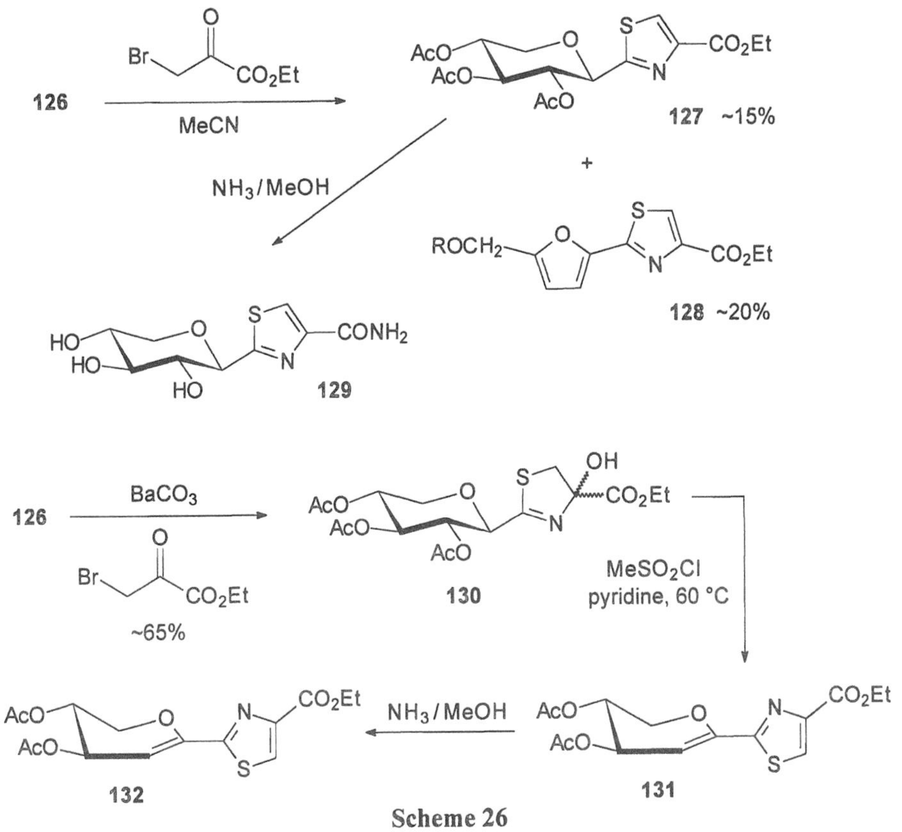

Scheme 26

Another analog of tiazofurin was made using related methodology.[33] Treatment of the anomeric *p*-nitrobenzoate ester **133** with trimethylsilyl cyanide gave a **134β** and **134α** in a 3:7 ratio. It was hoped that the benzyl protecting groups would preclude participation and that the *C*-glycosidation reaction would proceed predominantly from the top face. Obviously steric effects forced the reaction to occur from the more accessible α-face.

The β-isomer was then transformed into the thioamide in the usual way and condensation with ethyl bromopyruvate in cold acetonitrile gave a mixture of three compounds **136α**, **136β** and **137**. The α and β-isomers were obtained as a 1:1 mixture with a side product, the furan, formed in about 11% yield. Similar ratios of products were obtained when the α-isomer of **135** was used. The sequence was completed by formation of the amide and boron tribromide induced debenzylation to give the *C*-2 analog of tiazofurin **138**.

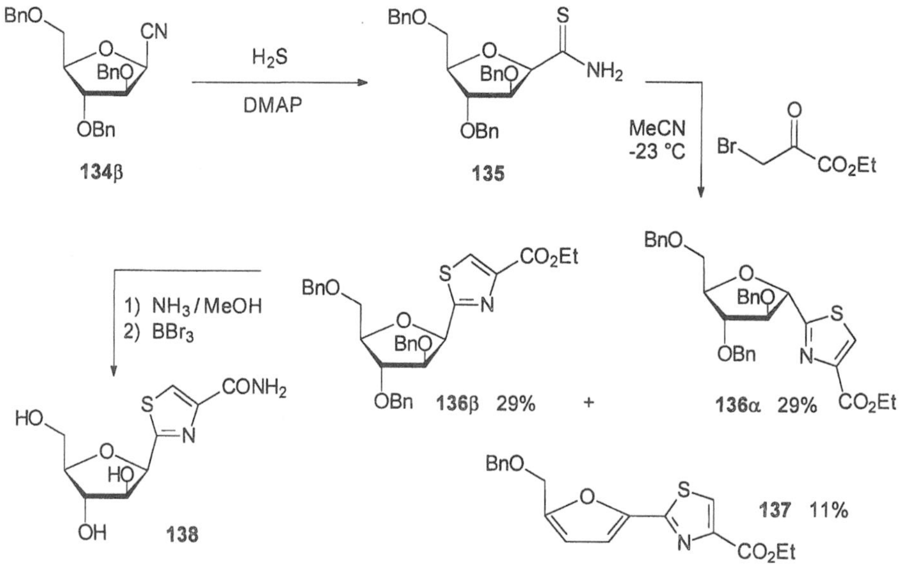

Scheme 27

The synthetic approach to tiazofurin itself also relied on similar chemistry using an acylated β-cyano-D-ribo sugar as the starting material.[34]

Due to the considerable biological activity of tiazofurin, several analogs have been synthesized and evaluated.[35] The preparation oxazofurin began with the acid chloride **139** which was treated with ethyl 2-amino-2-cyanoacetate in pyridine to give two compounds **140** and **141** as mixtures of diastereomers.[36]

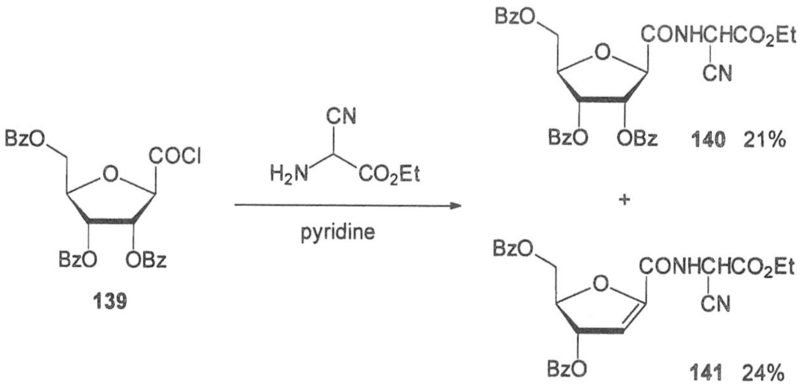

Scheme 28

Treatment of **140** (Scheme 29) with anhydrous hydrogen chloride affected cyclization to afford **142**. The amino group was reductively cleaved via a diazonium species and de-esterification was accompanied by ester exchange to give the triol methyl ester **144**. Ammoniolysis then gave the oxazofurin (**145**), the oxygen analog of tiazofurin.

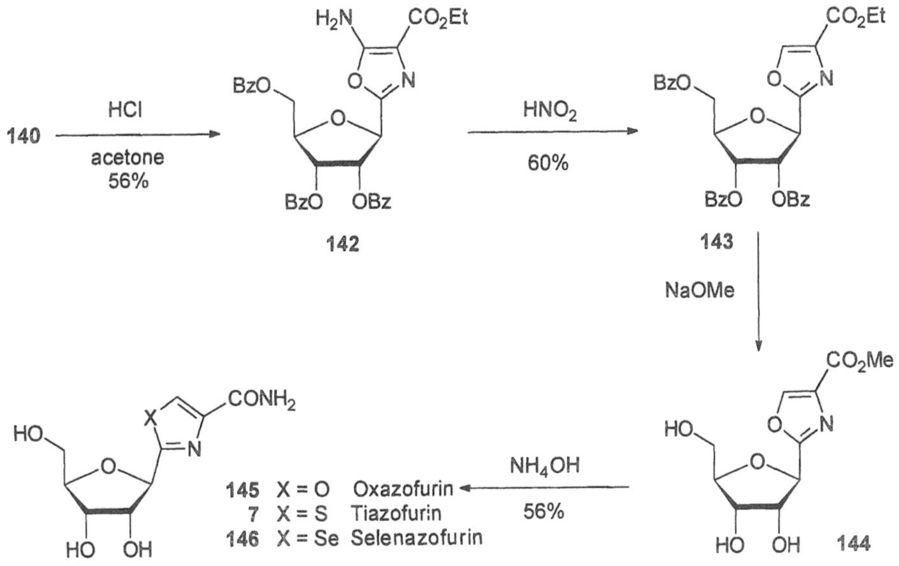

Scheme 29

The isosteric replacement of the sulfur atom with oxygen resulted in a compound that demonstrated poor biological activity. It was found to be only weakly active against B16 murine melanoma. This is compared to the activity of the selenide analog **146**[37] that exhibits ten times the activity of tiazofurin.

C-1 Acids have also found application in *C*-nucleoside synthesis. For example, DCC mediated coupling of the hydrazine **148** with acid **147** gave compound **149**. Cyclization was affected in ethylene glycol to eventually give the formycin analog **151**.[38]

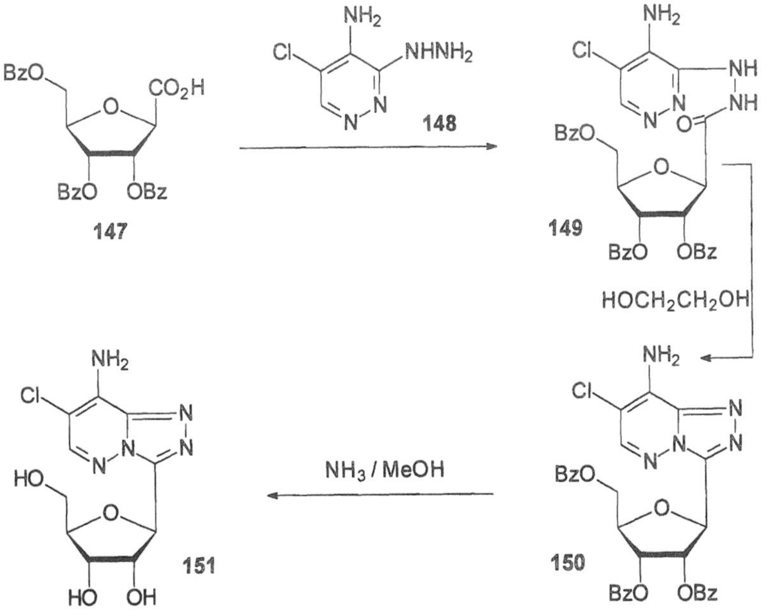

Scheme 30

2.   Aldehydes and Other Groups

C-1 Aldehydes have also found use as precursors in *C*-nucleoside synthesis. This section will illustrate a few syntheses that have utilized *C*-1 aldehydo sugars as well as a few of the methods used to generate these starting materials.

Moffatt and Trummlitz[39] employed a *C*-1 aldehydo sugar as their starting point for the preparation of showdomycin. The aldehyde 153 was liberated from the imidazolidine 152 by acid hydrolysis. The latter compound was treated with sodium cyanide and this was followed by oxidation with hydrogen peroxide to give a mixture of hydroxy amides 154 (93%) that could be separated, but were used as such. Attempted oxidation to the amido ketone proved troublesome so the amide was transformed into the ester and oxidation gave the keto-ester 155. Wittig reaction with the ylide 46 provided the tribenzylated compound 43. Deprotection with boron trichloride afforded showdomycin (3).

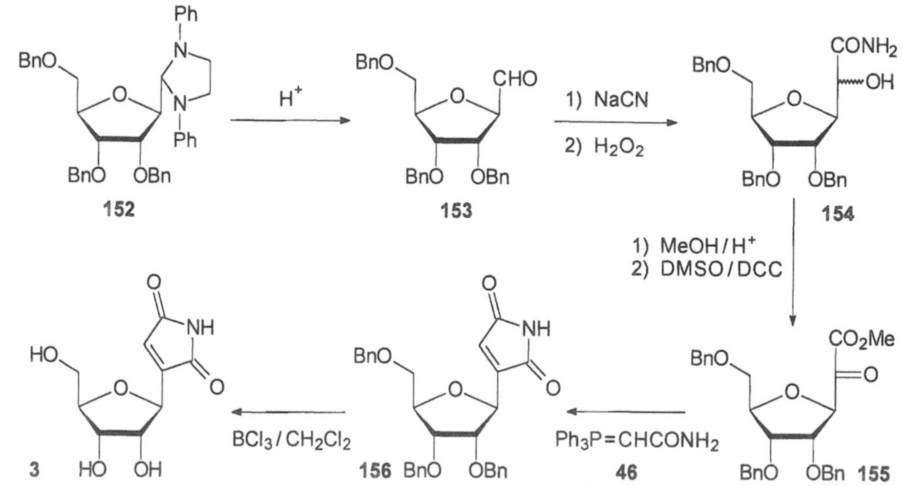

Scheme 31

Moffatt and co-workers[40] have also used Wittig chemistry on a *C*-1 aldehydo sugar to synthesize a *C*-nucleoside as shown below in Scheme 32.

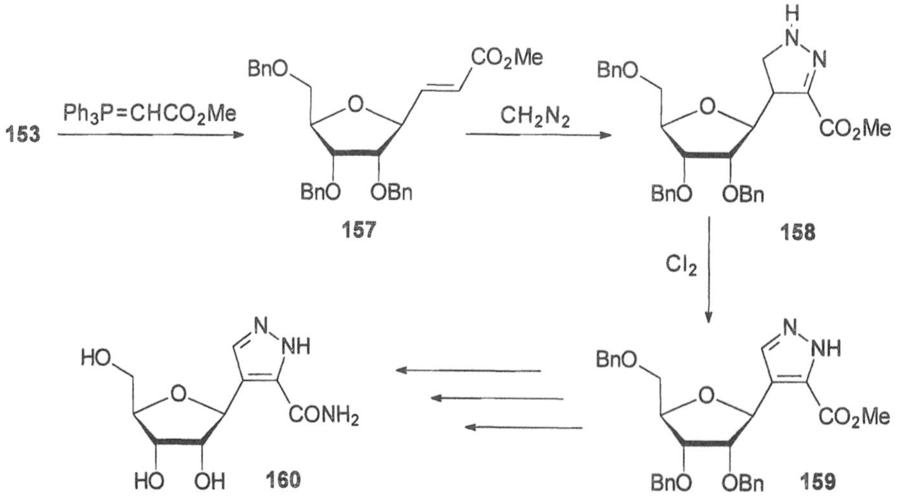

Scheme 32

Wittig reaction on **153** with the ylide gave a 94% yield of isomerically pure **157**. 1,3-Dipolar addition of the α,β-unsaturated ester with diazomethane gave **158**. Oxidation with molecular chlorine furnished **159** and this followed by deprotection and conversion of the ester to the amide to provide the product **160**.

Yamamura *et al.*[41] utilized the aldehyde **163** to synthesize the *C*-nucleoside **162** which has a blend of the structural features from both showdomycin and oxetanocin.[42]

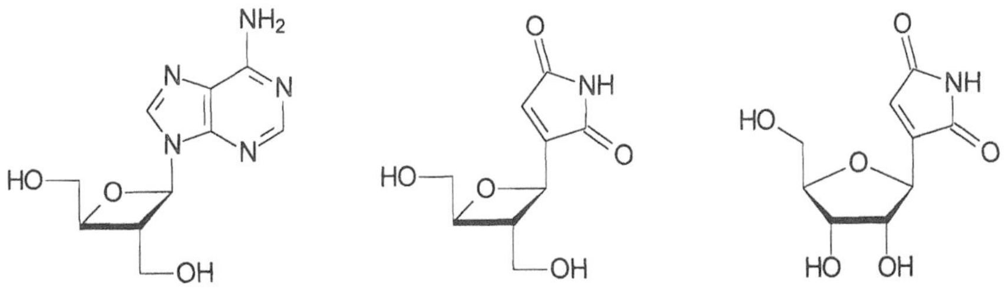

161 Oxetanocin **162** **3** Showdomycin

Figure 9

The aldehyde **163** was condensed with vinyl magnesium bromide to give the allyl alcohols **164** as a 2:1 mixture of isomers. Silylation was followed by oxidative cleavage of the double bond to give the desilylated esters **166** which were then oxidized to the keto-ester **167**. Wittig reaction followed by deprotection then led to **162**.

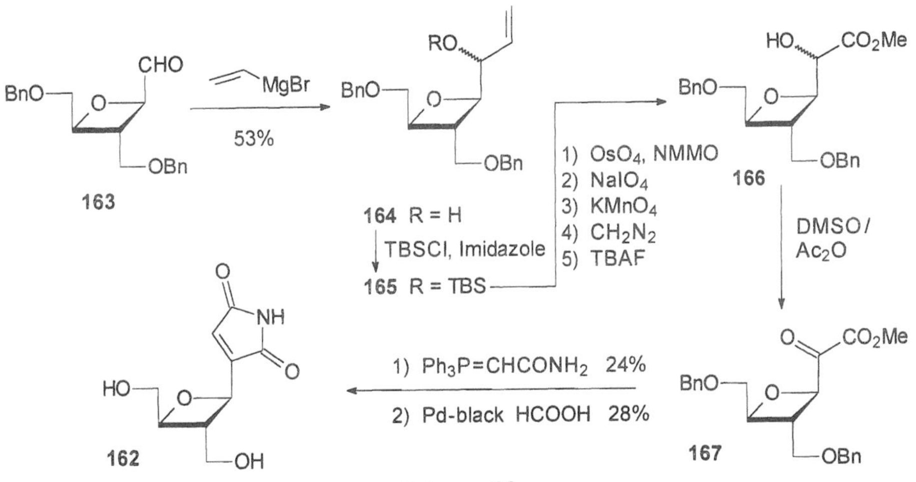

Scheme 33

The aldehydes themselves are available by several methods. Partial reduction of the corresponding nitrile is a common technique while ring contraction of aminohexoses also leads to the corresponding aldehydo furanoses.

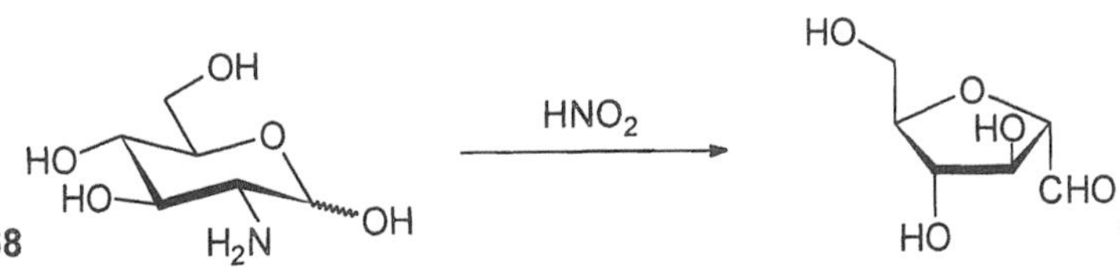

168 **169** eq. 5

For example, **168** gives **169** upon treatment with nitrous acid.[43]   The subject has been reviewed[11] and will therefore not be detailed here.

Reese and Kaye[44] have used a thallium (III) based ring contraction to gain access to the aldehydo furanose **172**. The glycal **170** was treated with thallium (III) nitrate trihydrate to give the aldehyde **172**.   This compound was not isolated but transformed into the showdomycin analog **173** in six steps in an 18% overall yield. The reaction is believed to proceed through the intermediate **171** (Scheme 34).

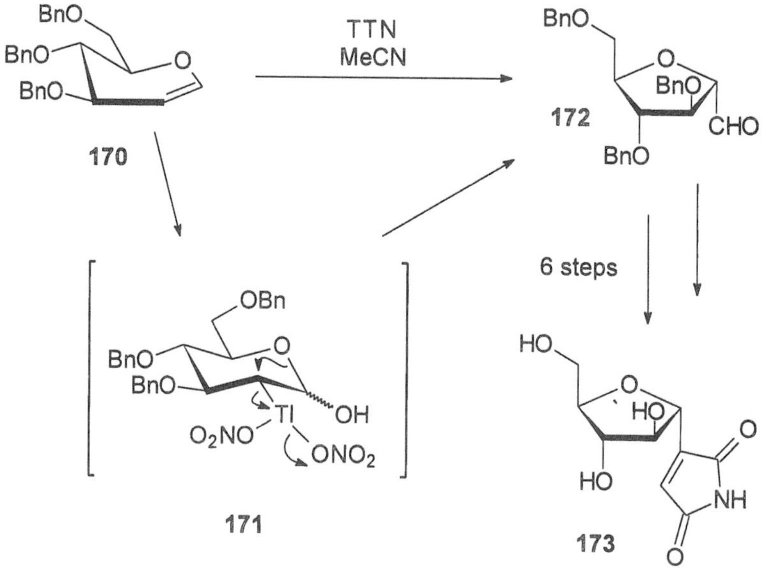

Scheme 34

Ring contraction has also been applied to furanose sugars to yield oxetanes[45] (Scheme 35). Reaction of the triflates **174** and **175** gave the ring contracted esters **176** and **177** in 51 and 57% yield, respectively.

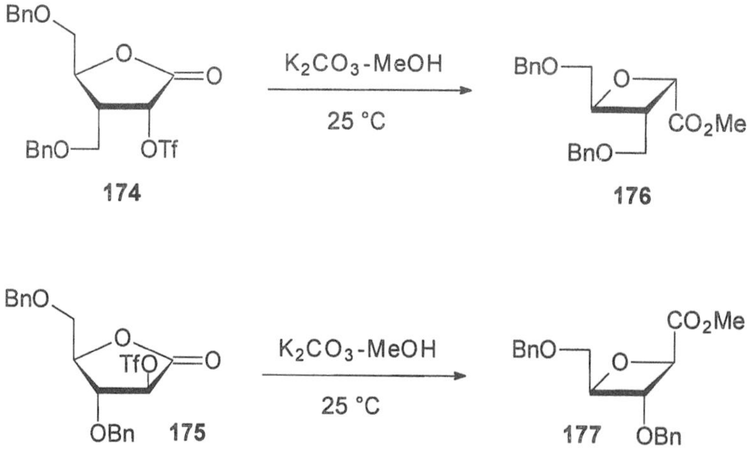

Scheme 35

Jung[46] has recently applied ring contraction strategy (Scheme 36) to the formal construction of several anti-viral agents. His strategy is versatile and should be amenable to scale up. D-Glucoseamine was diazotized and the ring contracted aldehyde was converted to its oxime **178**. Dehydration was followed by elimination to give the olefin **179**. Hydrogenation gave a 2:1 mixture of 2-deoxy sugars with **180** predominating. Since **180** had been previously converted to **181**, **182**, and **183** (via the acid), the above constitutes three formal total syntheses.

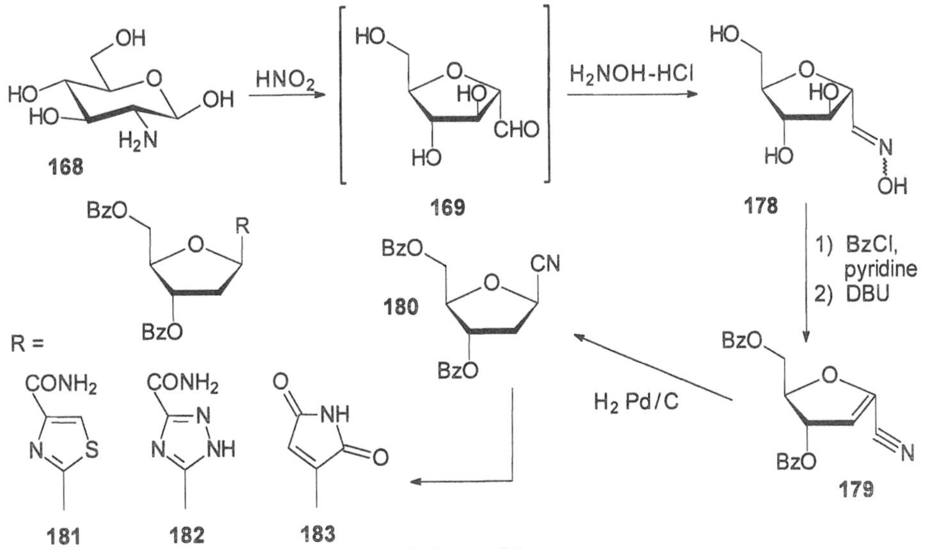

Scheme 36

Scheme 37 shows how **184** was transformed into **189**.[47] Osmylation of **185** was followed by selective protection to give **186** and treatment with TBAF then gave **187**. Exposure to an excess of sodium hydride then provided the *C*-nucleoside precursor **188**.

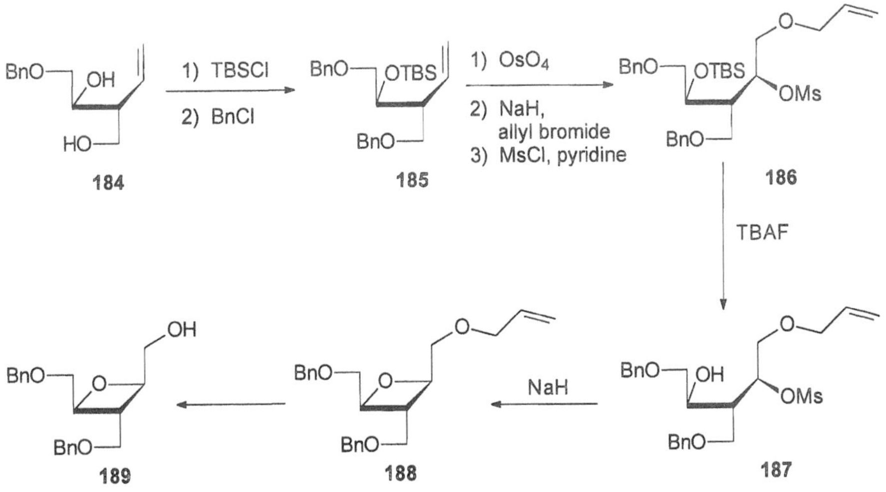

Scheme 37

Maeba[48] has developed chemistry based on the furan furanose **190**[49] which itself is available by Lewis acid based chemistry. Oxidation of **190** provides the **191** which is the pivotal intermediate. Elimination then provides the keto-acid **192**. Hydrogenation gave a 1:1 ratio (43%) of anomers and esterification of the major was followed by treatment with hydrazine to give **194**.   Oxidation with DDQ and deprotection formed the 2-deoxy-*C*-nucleoside **195**.

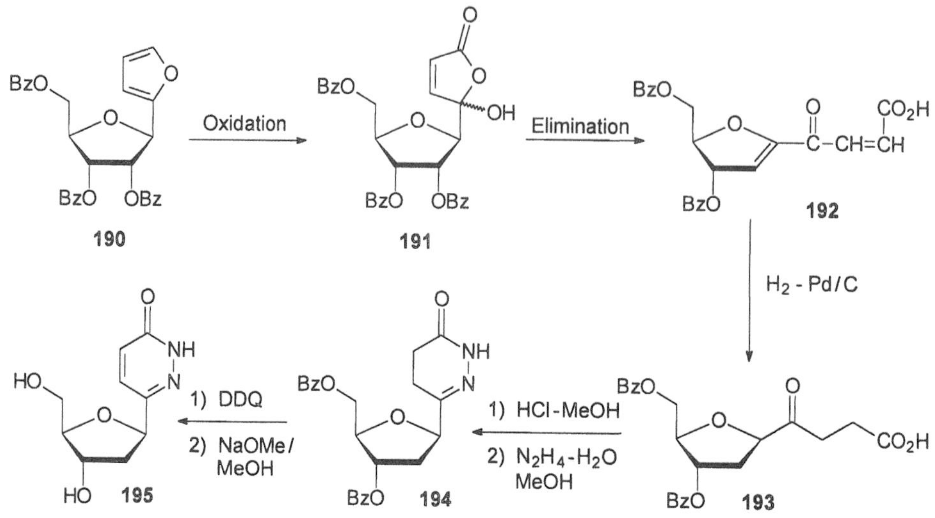

Scheme 38

E.   NUCLEOPHILIC ADDITION
1.   Lithiated Heterocycles

Lithiated pyridines have been used extensively as the heterocycle of choice since the corresponding halo compounds are readily available.   2-Lithiopyridine was condensed with the lactone **196** to furnish the *C*-nucleoside **197**.[50]

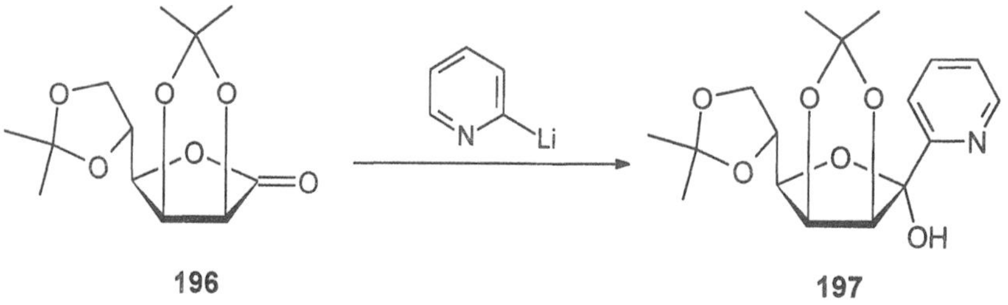

The above approach is slightly limited since the product still contains an anomeric hydroxyl group.   Benner and co-workers[51] have found that condensation of the lithiated pyridine **199** with the lactone **198** gave an anomeric mixture **200**.

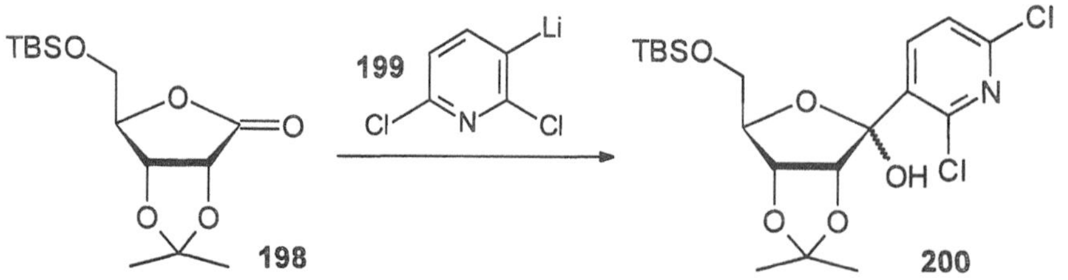

Reduction with triethyl silane with a variety of Lewis acids resulted in varying α:β ratios. The best results were found with boron trifluoride etherate in toluene. Desilylation and acetonide hydrolysis gave **202**. Compound **202** was then selectively protected, derivatized (*O*-2), and radically deoxygenated to provide the 2-deoxy sugar **203**. Deprotection then furnished the target 2-deoxy-*C*-nucleoside **204**.

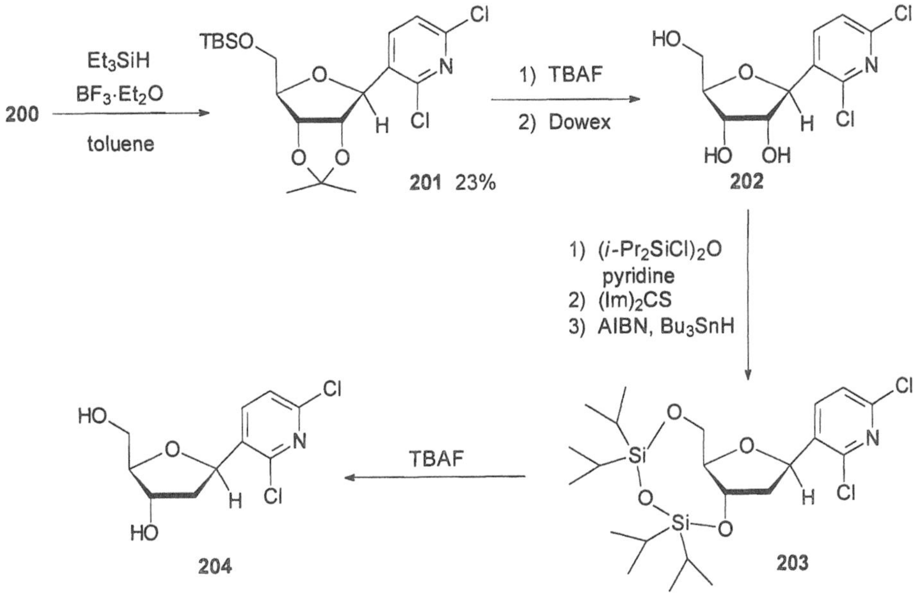

Scheme 39

A lithiated pyridine can also be added to an open chain sugar that needs to be cyclized to assemble a *C*-nucleoside.[52] The lithiated pyridine **206** was condensed with the protected D-ribo sugar **205** to give a mixture of allo and altro products which were not separated but mesylated to give **208**. Cyclization was accompanied by some hydrolysis to give a mixture of **209** and **210**. Presumably some of the isomer also cyclized, however no effort was made to isolate or characterize it.

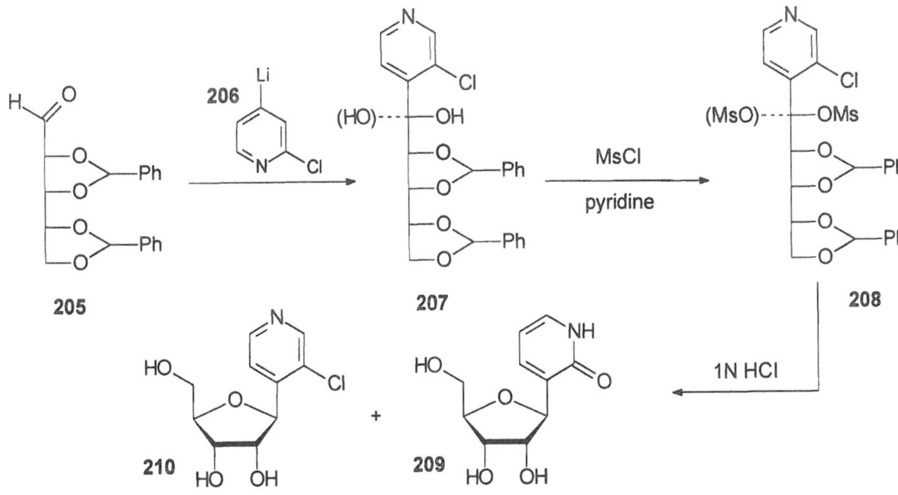

Scheme 40

The nucleophile need not be a pyridine; indoles can also be used. Condensation of the indole bromomagnesium salt **212** with **211** in dichloromethane furnished the open chain compound **214** and the cyclized form **213** in a 86:14 ratio in 75% combined yield. Dehydration of **214** gave a mixture of anomers with the β-anomer **215** predominating.[53]

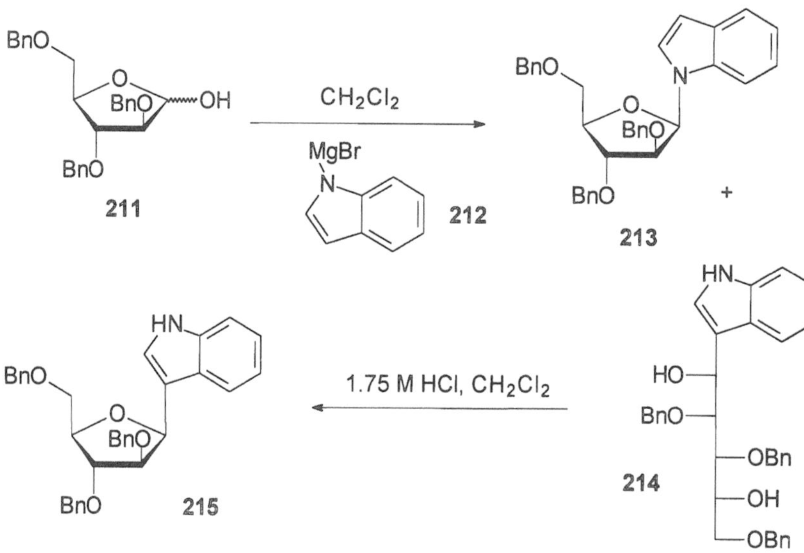

Scheme 41

A similar approach was taken by Eaton and co-workers.[54] Their research was driven by a search for nucleoside analogs that can be incorporated into synthetic DNA strands in order to study the factors affecting helix stability. They condensed 3-lithiopyridine (**217**) with the differentially protected aldehyde **216**. Mesylation was followed by cyclization and concomitant deprotection to yield an anomeric mixture of *C*-nucleosides **219**. Removal of the dimethoxytrityl group was achieved using acetic acid-water and at this point the anomers were separated and further deblocked to furnish the α and β-*C*-nucleoside **220**. Compound **220** exhibited no anti-tumor activity against mouse leukemia L1210 or human lymphoblast Raji cells.

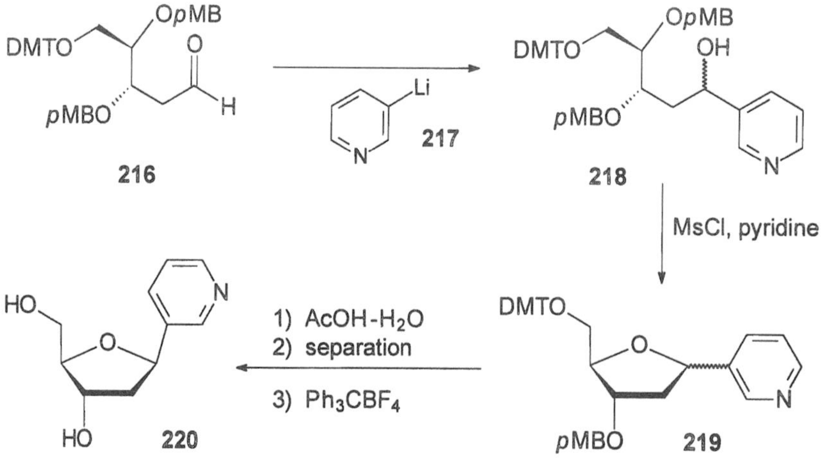

Scheme 42

Watanabe *et al.*[55] have condensed the lithiated pyridine **222** with the aldehyde **221** to give a 45:55 mixture of isomers. Lithiation and treatment with carbon dioxide provided the acid which was esterified to give the ester and then treated with ammonia in methanol to furnish the amide **226**. Mesylation was followed by acid hydrolysis to provide a mixture of anomers which were separated and the β-isomer was methylated to give **228**. Compound **228** was found to be 1-2 log orders less active than tiazofurin in anti-tumor tests.

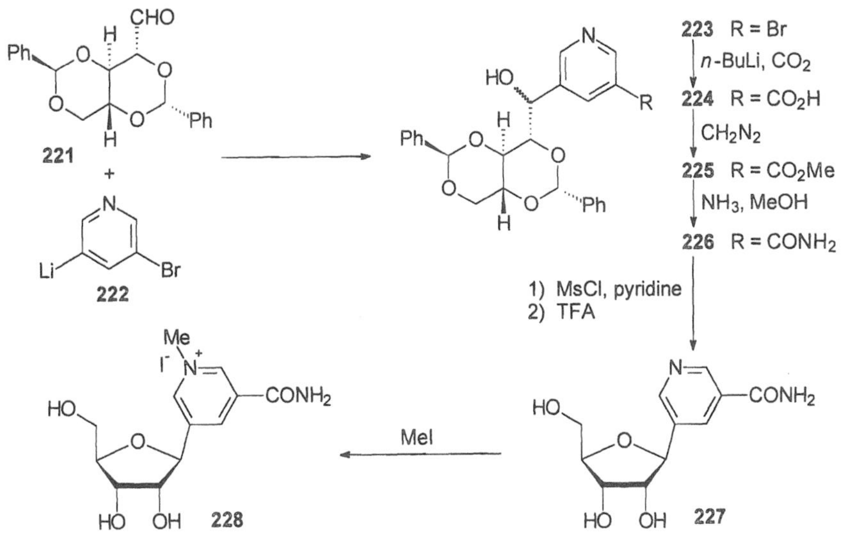

Scheme 43

Daves and Hacksell[56] have utilized an organomercurial in a palladium mediated coupling between a heterocycle and a furanose glycal. Scheme 44 illustrates the types of compounds that can be assembled using this technology. Chapter 5 deals with palladium based approaches to *C*-nucleosides.

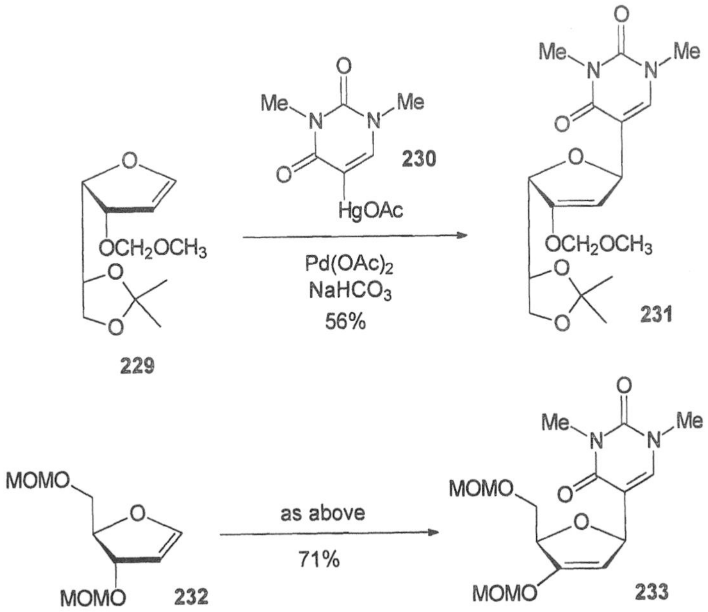

Scheme 44

2.   Carbon Based Nucleophiles

Buchanan has made considerable contributions in the area of *C*-nucleoside synthesis.[57] Not only has he prepared several *C*-nucleosides, as well as acyclic analogs, but has also reviewed the subject.[58] This section will only briefly highlight his chemistry since it has been reviewed previously. His methodology begins with the ethynyl *C*-glycoside **235**.[59] This compound is prepared by condensation of an appropriate acetylene with the D-ribo sugar **234** to produce a mixture of isomers which are cyclized via the *O*-3 sulfonate to give the desired β-isomer **235** in about 50% yield. The reaction also works well to give the protected aldehydo alkyne **236**. These compounds are pivotal intermediates for the preparation of various *C*-nucleosides.

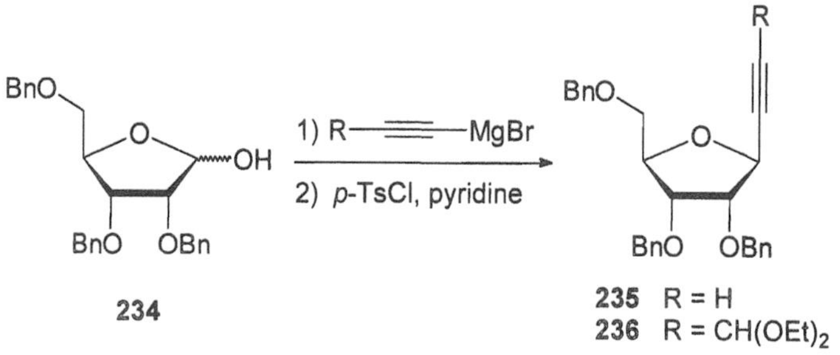

eq. 8

234

235 R = H
236 R = CH(OEt)$_2$

The acetylene **235** could be made to undergo cycloaddition with benzyl azide to give a mixture of regioisomers **237** and **238**.

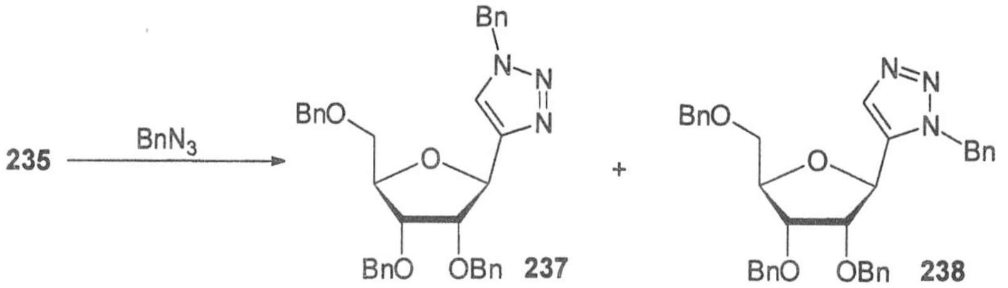

eq. 9

Compound **235** could also be carbonylated to provide **239** which was then easily transformed into showdomycin.[60]

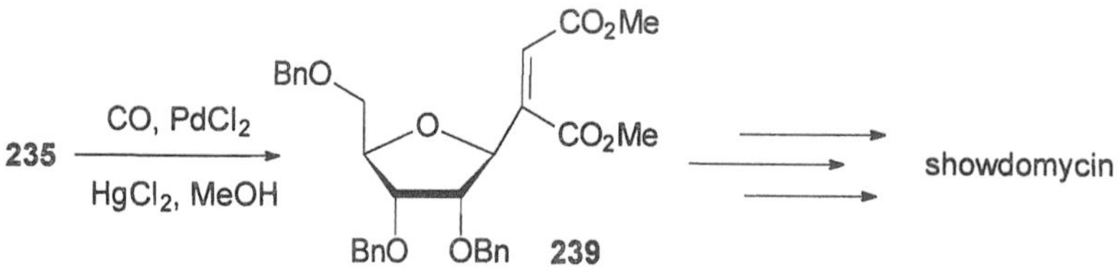

showdomycin

Scheme 45

The ethyne **236** was also used, in this case, to construct both formycin and pyrazofurin. Compound **236** was condensed with hydrazine to afford **240**. Conversion to **241** was accomplished via four steps and several more steps afforded the amino pyrazine **242** which was converted into both title compounds.[61]

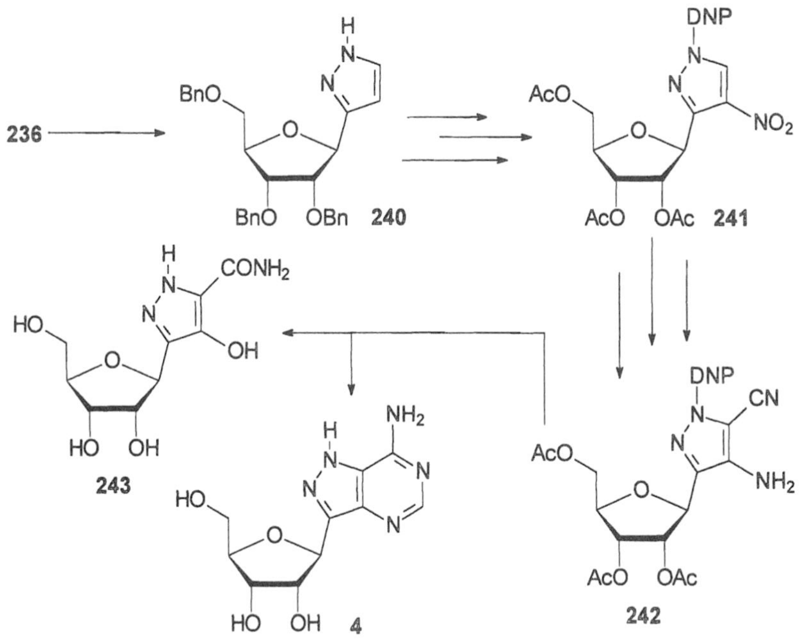

Scheme 46

More recently, Buchanan[62] has utilized the above methodology to synthesize a potentially HIV potent *C*-nucleoside that incorporates the structural characteristics of both AZT and formycin. The arabinose derivative **244** was condensed with ethynyl magnesium bromide to afford the expected product. Selective pivaloylation of the primary alcohol gave **246** which was tosylated and concomitantly cyclized to give **246**. Treatment with hydrazine gave **248** and deprotection then afforded the target compound **249**.

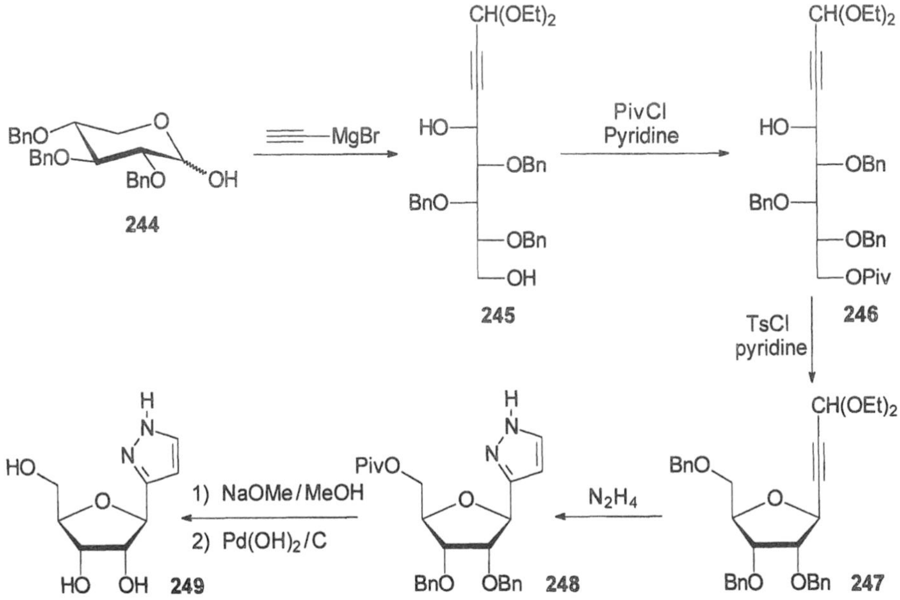

Scheme 47

Carbene insertions have also been used, the nucleophile here being of the malonate type.[63] The β-thioglycoside was treated with dimethyl diazomalonate in the presence of rhodium (II) acetate to give **251** in 71% yield. Decarboxymethylation was followed by oxidation to afford **252**. Pummerer rearrangement and desulfurization then provided **101** which underwent Wittig reaction to furnish **104** which has been previously converted into showdomycin.

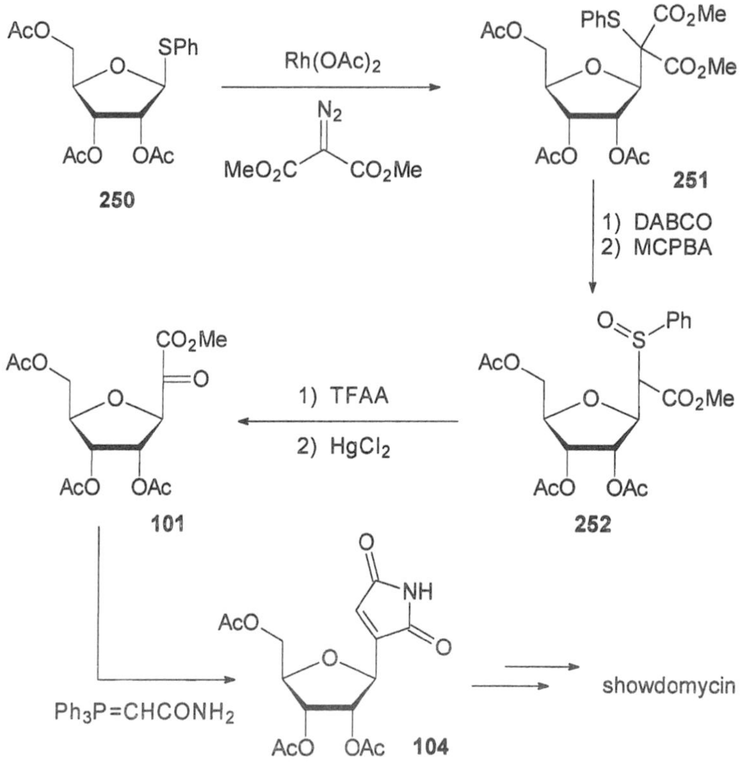

Scheme 48

F.   THE BICYCLO APPROACH

A not so obvious disconnection for the preparation of *C*-nucleosides involves a cleavage of the double bond in the bicyclo compound **253** to provide a suitable *C*-nucleoside precursor. There have been several variations published on this theme.

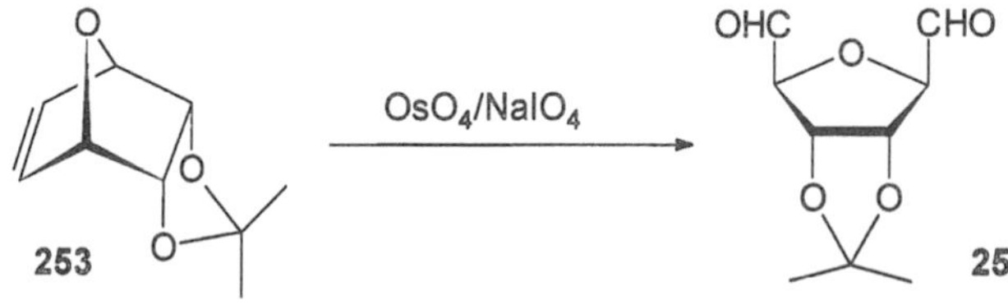

Noyori[64] has developed an iron catalyzed cyclocoupling of dibromo ketones with dienes to assemble cleavage precursors for the preparation of novel *C*-nucleoside analogues. The reaction utilizes the symmetrical tetrabromo acetone compound **255** with furan in the presence of the catalyst $Fe_2(CO)_9$. Once the cyclocoupling is complete, the remaining bromine atoms are removed by the action of Zn/Cu couple and hydroxylation of the double bond is followed by protection to produce compound **259** in 65% yield as a single acetonide.

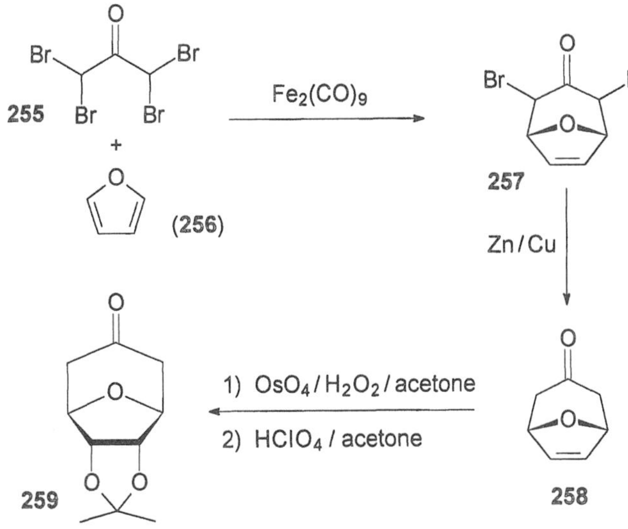

Scheme 49

Oxidative cleavage of the ketone **259** with trifluoroperacetic acid then gave the lactone **260**. Hydrolysis gave an acid which could be resolved via its cinchodine salt to furnish optically pure β-*C*-nucleoside precursor **261**. The optically pure lactone could be reacted with **264** to afford the dimethylamino methylene lactone **263**. This compound, when treated with urea, gave compound **265** which was deprotected to afford pseudouridine (**1**).[65]

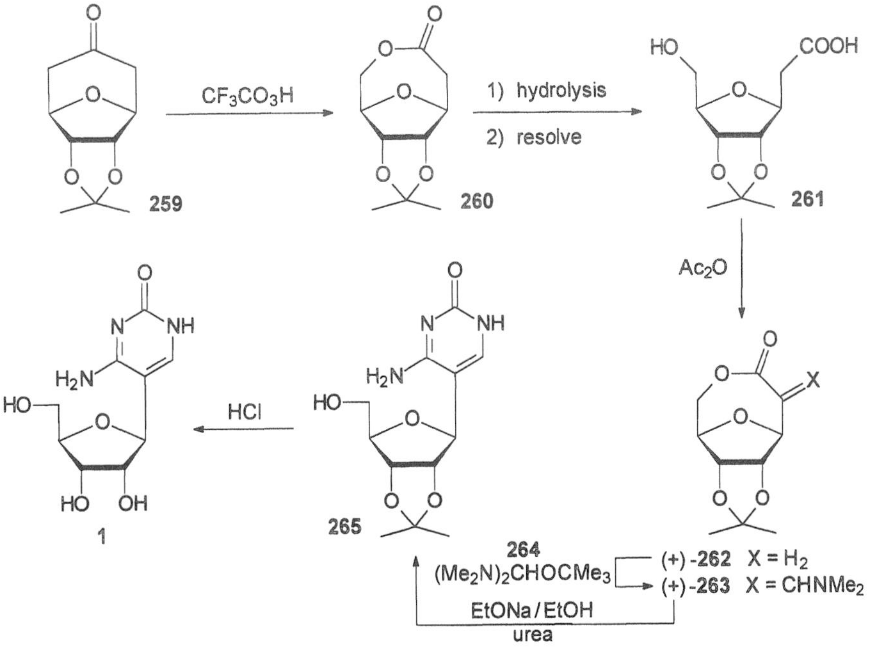

Scheme 50

Alternatively, **263** also served as an intermediate in the synthesis of showdomycin. Ozonolytic cleavage of the double bond gave compound **266** which underwent a standard Wittig olefination to provide compound **267**. Deprotection (50% aqueous TFA) then afforded showdomycin. The above route to *C*-nucleosides (and β-*C*-furanosides) is general and fairly straightforward, the only drawback being that the products are racemic, thus necessitating an intermediate resolution step.

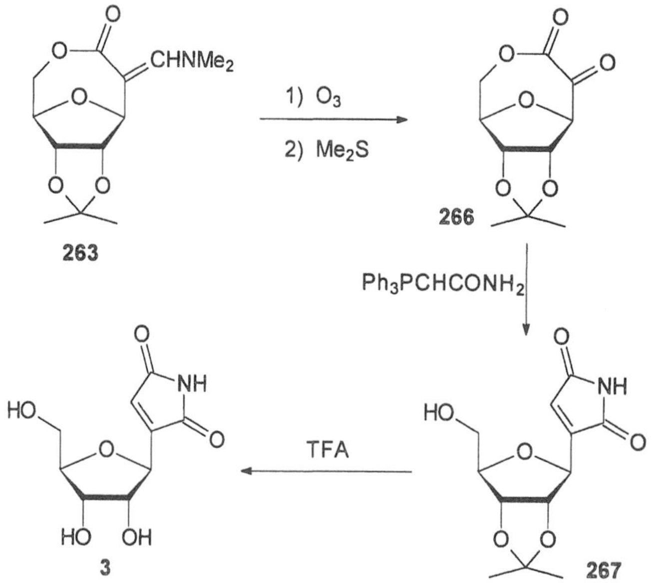

Scheme 51

Ohno[66] has used the Diels-Alder adduct **268** in a similar fashion. Compound **268** was hydroxylated and protected to give **269**. The diester was enantioselectively hydrolyzed by pig liver esterase to give optically pure monoacid **270**. The free acid was converted into its *t*-butyl ester (without affecting the methyl ester) and the methyl ester was saponified to furnish the corresponding monoacid **271**.

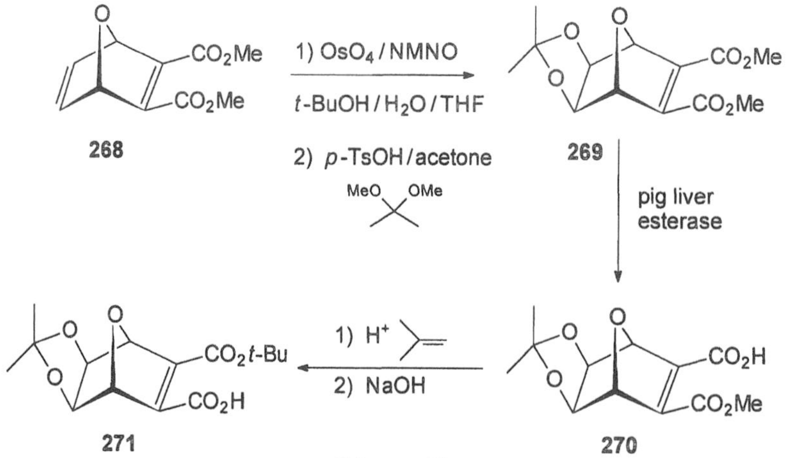

Scheme 52

Ozonolysis of **271** then gave **272**. Wittig reaction provided **273** and the acid was reduced down to the alcohol and acetylated. Acid hydrolysis then caused concomitant deprotection and cyclization to give showdomycin, Scheme 53.

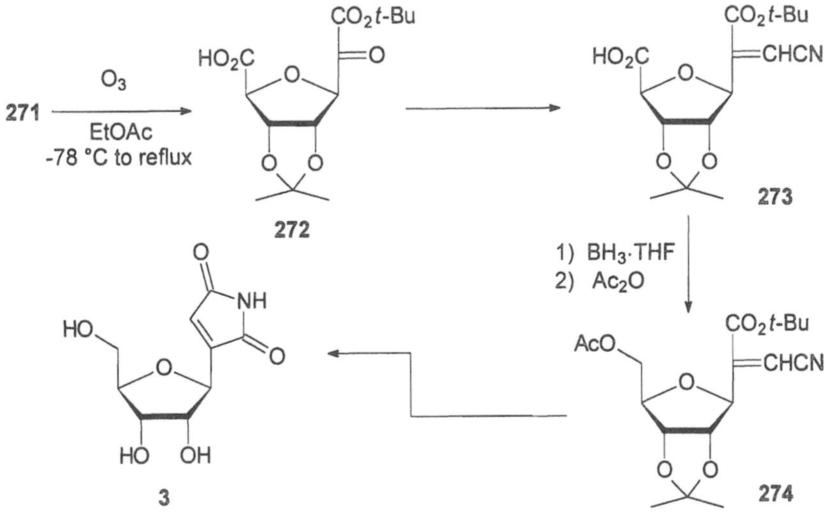

Scheme 53

Just's approach to showdomycin was along similar lines.[67] Ozonolysis of compound **275** gave the dicarbonyl derivative which was reduced with a bulky aluminum reagent to give **276**. Protective group manipulation and oxidation then afforded **277**. This intermediate was then transformed into showdomycin in the usual way.

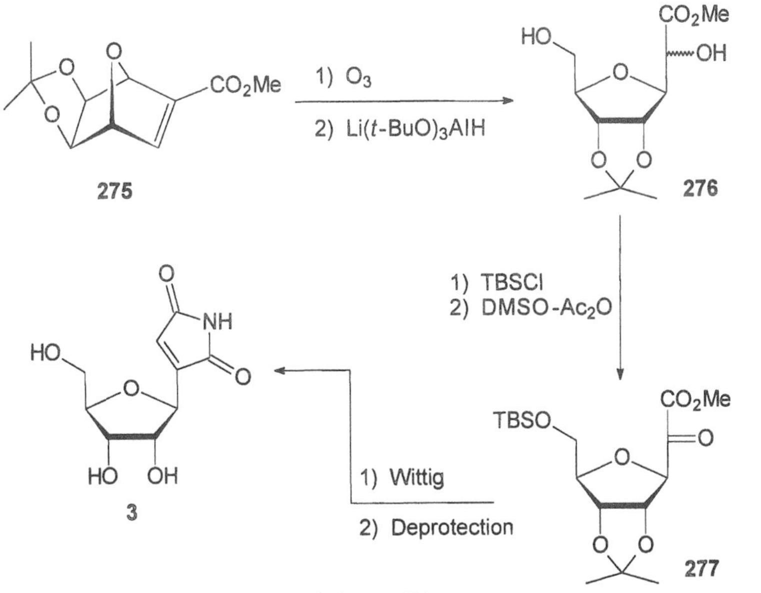

Scheme 54

Vogel's approach to the synthesis of C-nucleosides[68] is in line with his work on "naked sugars."[69] Optically pure **278** was transformed into the azide **279** in three steps. Ozonolysis, acetal formation, and oxidative work-up then furnished the acetal acid **280**. DCC mediated coupling with the triamine **281** provided the amide **282** which was cyclized by the action of CsF in DMF. Acid hydrolysis and reduction of liberated aldehyde then provided the *C*-nucleoside analog of AZT **283**, Scheme 55.

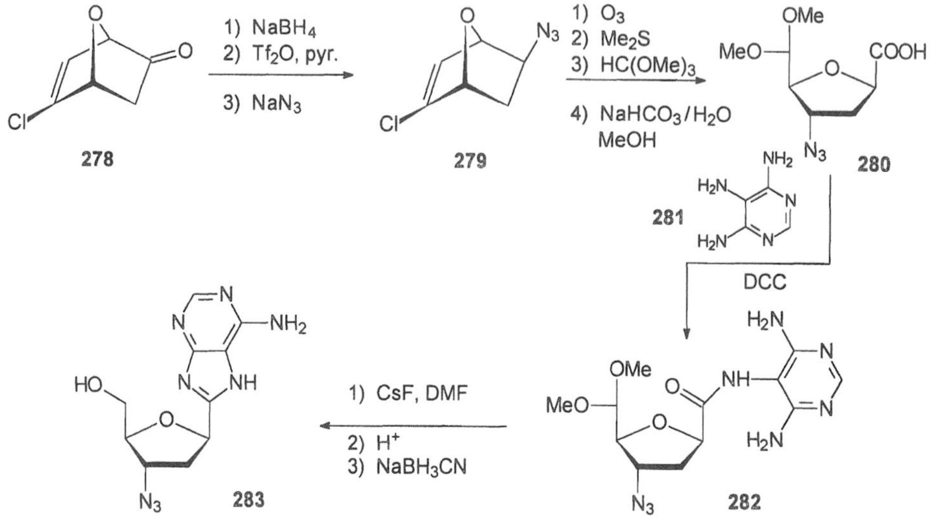

Scheme 55

Kozikowski and Ames[70] utilized a cycloaddition approach based on the Diels-Alder reaction of furan with the dienophile **284**. Hydroxylation and protection then furnished **285**. Alkylation occurred without elimination and this was followed by fragmentation to provide **286**. Selective reduction and silylation was followed by conversion of the ester to the amide and cyclization gave **287**. The double bond was introduced in the regular way and deprotection then gave racemic showdomycin.

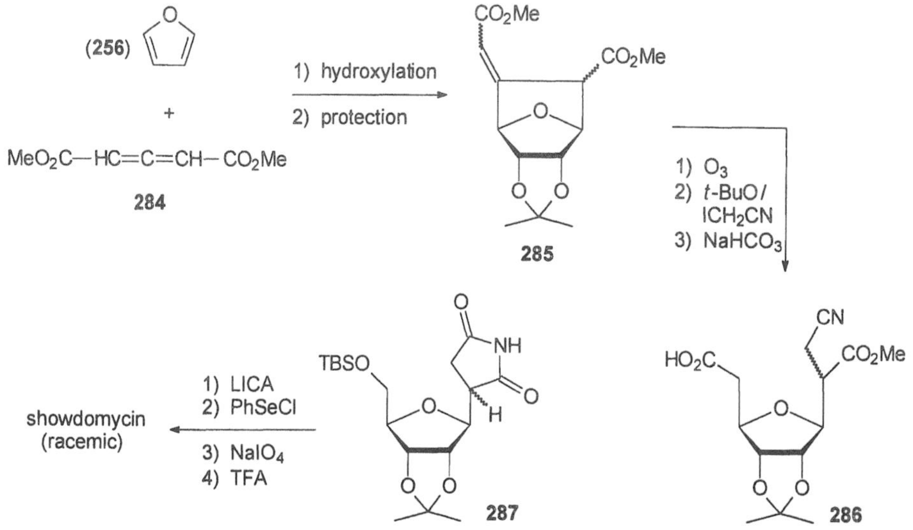

Scheme 56

Simpkins[71] has used chiral base technology to develop a preparation of a chiral intermediate suitable for *C*-nucleoside synthesis. Cyclocondensation by the method of

Noyori was followed by *syn*-hydroxylation and protection to give the ketone **288**. Deprotonation with the homochiral lithium base (HCLA) **289** followed by trapping of the enolate with trimethylsilyl chloride gave **290** in 98% yield in 85% enantiomeric excess. Treatment with PhIO gave the hydroxy ketone **291** which when cleaved with lead tetraacetate provided the target compound **292**. It is noteworthy that ozonolysis of the enol ether led only to complicated reaction mixtures.

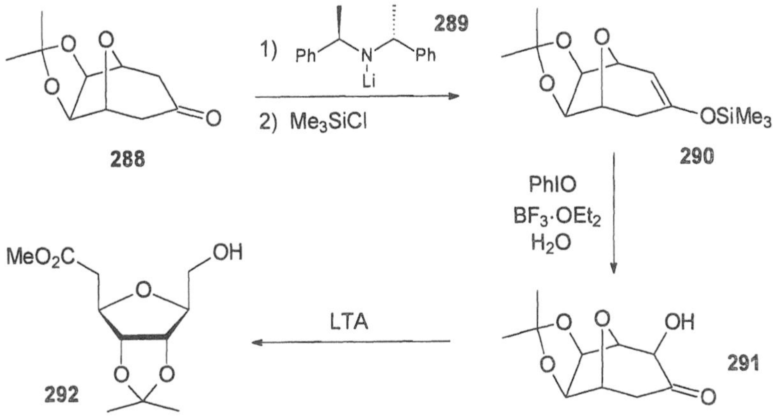

Scheme 57

The final example in this section is a carbocyclic analog of a *C*-nucleoside.[72] The hydroxylated adduct was enantioselectively saponified and then the double bond cleaved to furnish **294**. The ketone was reduced and oxidative cleavage followed by further reduction by sodium borohydride and lactonization gave **295**. The lactone was treated with ammonia in methanol to give an amide which was dehydrated to the nitrile **296**. Azide cycloaddition then furnished **297**. Deprotection afforded the *C*-nucleoside analog **298**.

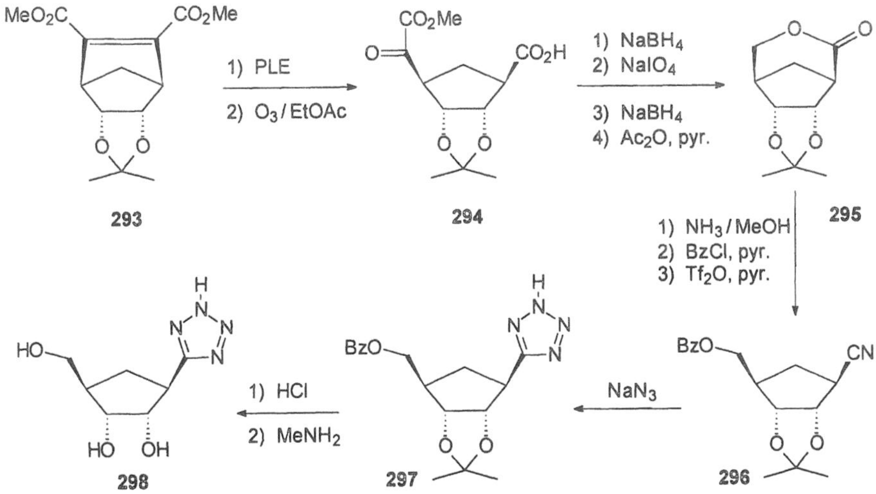

Scheme 58

G. CYCLIZATION OF ALDITOLS

Acid catalyzed dehydration of alditols is one method to cyclize open chain sugars to suitable precursors of *C*-nucleosides.[11] When the protected D-mannitol derivative **299** is

treated with hot acid three products are formed. The one of interest,**300**, is shown in eq. 11.
The reaction is believed to proceed by protonation of one of the secondary hydroxyls which
is then displaced by a second hydroxyl group to close the ring.[73]

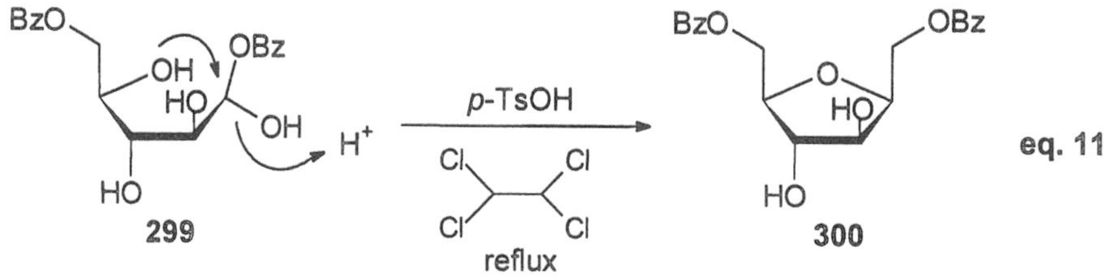

eq. 11

The reader should consult Hanessian's review[11] since the section dealing with this type
of reaction is well treated. A few more recent examples from the modern literature will be
presented to finish the chapter. Scheme 59 shows that the acid catalyzed dehydration of the
pentitol **301** with trifluoroacetic acid gave the *C*-nucleoside **302**. Dehydration of the epimer
303 also gave the same product, Scheme 59.[74]

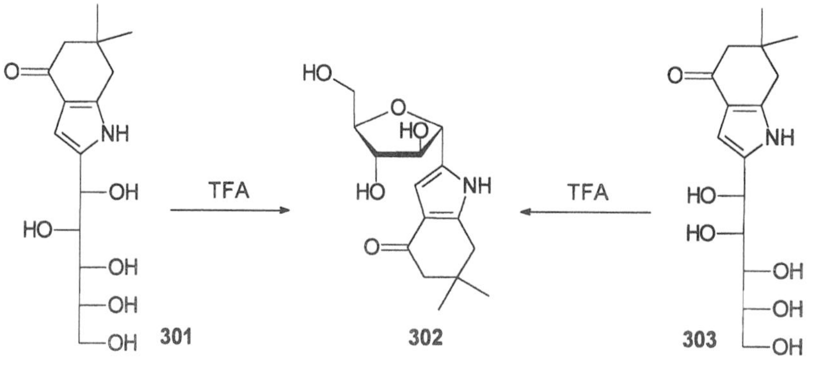

Scheme 59

Cyclization of compound **304** illustrates yet another example of this useful reaction.
The yield is high (82%) and formation of the β-product **305** is favored.[75]

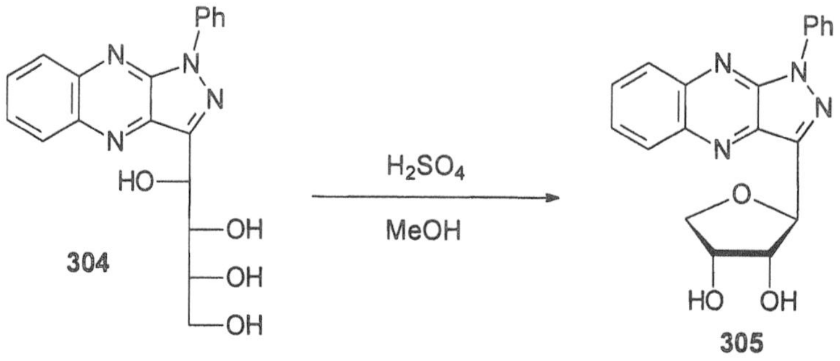

Scheme 60

The following example shows how easily the precursor for cyclization can be assembled. The three components are mixed together and one obtains **309**. Exposure of this compound to sulfuric acid in boiling methanol then leads to the *C*-nucleoside **310**.[76]

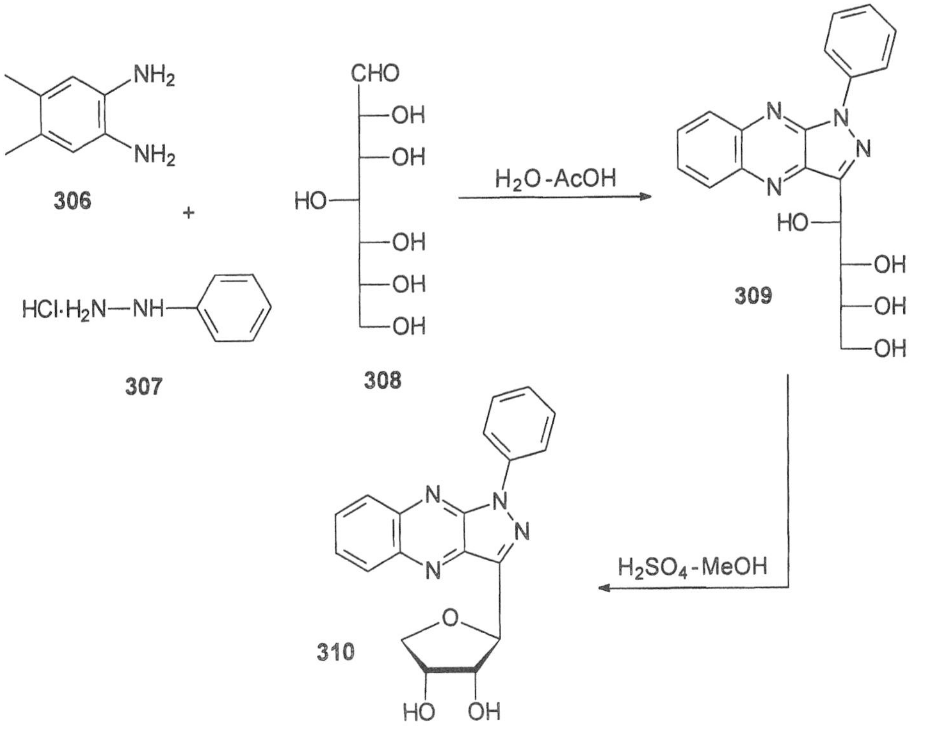

Scheme 61

III. CONCLUSION

This chapter has illustrated some of the more commonly employed methods for *C*-nucleoside synthesis. The reactions have been presented with strategy in mind. Wittig reactions have played an important role in *C*-nucleoside assembly. The condensation of heterocyclic nucleophiles as well as Lewis acid based methods have also been utilized. Radical methods have also seen some utility. *C*-1 Acids, nitriles, aldehydes, and thioamides have also occupied a central place in the preparation of these types of compounds. The most useful methods are those that allow a highly stereoselective method for generation of the β-*C*-nucleoside. The use of bridged compounds nicely addresses this problem since the *syn* disposition of the hydroxymethyl group and the *C*-1 moiety are a direct consequence of the existence of the bridge. Another consideration is the ease in which a sequence can be carried out. If a synthesis is short, but suffers from low stereoselectivity (provided that separation of anomers is fairly facile), it may still be well suited for the preparation of many analogs in a relatively short time since the brevity and simplicity of the sequence make it a worthwhile endeavor. Intramolecular delivery of heterocyclic groups will probably be exploited in the future and as methods for the synthesis of *C*-furanosides improve, it is certain that their application will be exploited for use in *C*-nucleoside assembly.

IV. REFERENCES

1. Suhadolnik, R.J., *Nucleosides as Biological Probes*, John Wiley & Sons, Inc. New York, 1979 and Suhadolnik, R.J., *Nucleoside Antibiotics*, Wiley-Interscience, New York, 1979.

2. Gutowsky, G.E.; Chaney, M.; Jones, H.D.; Hamill, R.L.; Davis, F.A.; Miller, R.D. *Biochem. Biophys. Res. Commun.* **1973**, *51*, 312.

3. Cohn, W.E. *J. Biol. Chem.* **1960**, *235*, 1488.

4. Sasaki, K.; Kasakabe, Y.; Ezumi, S. *J. Antibiot., Ser. A* **1972**, *25*, 151.

5. Nishimura, N.; Mayama, M.; Komatsu, Y.; Katõ, F.; Shimaoka, N.; Tanaka, Y. *J. Antibiot., Ser. A* **1964**, *17*, 148.

6. Koyama, G.; Maeda, K.; Umezawa, H.; Iitaka, Y. *Tetrahedron Lett.* **1966**, 597.

7. Koyama, G.; Umezawa, H. *J. Antibiot. Ser., A* **1965**, *18*, 175.

8. Sheen, M.R.; Martin, H.F.; Parks, H.F. Jr., *Mol. Pharmacol.* **1970**, *6*, 255.

9. Robins, R.K.; Srivastava, P.C.; Narayanan, V.L.; Plowman, J.; Paull, K.D. *J. Med. Chem.* **1982**, *25*, 107.

10. Sidwell, R.W.; Huffman, J.H.; Khare, G.P.; Allen, L.B. Witkowski, J.T.; Robins, R.K. *Science* **1972**, *177*, 705.

11. Hanessian, S.; Pernet, A.G. *Adv. Carbohydr. Chem. Biochem.* **1976**, *33*, 111. See also: Hacksell, U.; Daves, Jr., G.D. *Prog. Med. Chem.* **1985**, *22*, 1.

12. Araki, Y.; Endo, T.; Tanji, M.; Nagasawa, J.; Ishido, Y. *Tetrahedron Lett.* **1988**, *29*, 351.

13. Barton, D.H.R.; Ramesh, M. *J. Am. Chem. Soc.* **1990**, *112*, 891.

14. Togo, H.; Ishigami, S.; Yokoyama, M. *Chem. Lett.* **1992**, 1673 and Togo, H.; Fujii, M.; Ikuma, T.; Yokoyama, M. *Tetrahedron Lett.* **1991**, *32*, 3377.

15. Katagari, N.; Takashima, K.; Kato, T. *J. Chem. Soc., Chem. Commun.* **1982**, 664.

16. Clingerman, M.C.; Secrist III, J.A. *J. Org. Chem.* **1983**, *48*, 3141.

17. Gonzalez, P.M.S.; Aciego, D.R.M.; Herrera, L.F.J. *Tetrahedron* **1988**, *44*, 3715.

18. De Bernardo, S.; Weigele, M. *J. Org. Chem.* **1977**, *42*, 109.

19. Bhattacharya, B.K.; Otter, B.A.; Berens, R.L.; Klein, R.S. *Nucleosides Nucleotides* **1990**, *9*, 1021.

20. Barrett, A.G.M.; Broughton, H.B.; Attwood, S.V.; Gunatilaka, L.A.A. *J. Org. Chem.* **1986**, *51*, 495.

21. Wanner, M.J.; Koomen, G.J. *Tetrahedron Lett.* **1990**, *31*, 907.

22. Katagari, N.; Takashima, K.; Kato, T.; Sato, S.; Tamura, C. *J. Chem. Soc., Perkin Trans. 1* **1983**, 201.

23. Wise, D.S.; Cupps, T.L.; Krauss, J.C.; Townsend, L.B. *Nucleosides, Nucleotides and their Biological Applications*; Academic Press, New York, 1983, 297.

24. Patil, S.A.; Otter, B.A.; Klein, R.S. *Nucleosides Nucleotides* **1990**, *9*, 937.

25. Kalvoda, L.; Farkaš, J.; Šorm, F. *Tetrahedron Lett.* **1970**, 2297.

26. Stewart, A.O.; Williams, R.M. *J. Am. Chem. Soc.* **1985**, *107*, 4289.

27. Inoue, T.; Kuwajima, I. *J. Chem. Soc., Chem. Commun.* **1980**, 251.

28. Helferich, B.; Wedemeyer, K.F. *Ann.* **1949**, *563*, 139.

29. Kalvoda, L. *J. Carbohydr. Nucleosides Nucleotides* **1976**, *3*, 47.

30. El Khadem, H.S.; El Ashry, E.S.H. *Carbohydr. Res.* **1973**, *29*, 525.

31. See for example reference #1 and Jagger, D.V.; Kredich, N.M.; Guarino, A.J. *Cancer Res.* **1961**, *21*, 216.

32. Kovács, L.; Herczegh, P.; Batta, G.; Farkas, I. *Tetrahedron* **1991**, *47*, 5539.

33. Jiang, C.; Baur, R.H.; Dechter, J.J.; Baker, D.C. *Nucleosides Nucleotides* **1984**, *3*, 123.

34. Srivastava, P.C.; Pickering, M.V.; Allen, L.B.; Streeter, D.G.; Campbell, M.T.; Witkowski, J.T.; Sidwell, R.W.; Robins, R.K. *J. Med. Chem.* **1977**, *20*, 256.

35. Avery, T.L.; Hermen, W.T.; Revankar, G.R.; Robins, R.K. in *New Avenues in Development Cancer Chemotherapy*; Harrap, K.R.; Connors, T.A. Eds.; Academic Press Inc., Orlando, FL, 1987, 367.

36. Franchetti, P.; Cristalli, G.; Grifantini, M.; Cappellacci, L.; Vittori, S.; Nocentini, G. *J. Med. Chem.* **1990**, *33*, 2849.

37. Srivastava, P.C.; Robins, R.K. *J. Med. Chem.* **1983**, *26*, 445.

38. Kang, Y.; Larson, S.B.; Robins, R.K.; Revankar, G.R. *J. Med. Chem.* **1989**, *32*, 1547.

39. Trummlitz, G.; Moffatt, J.G. *J. Org. Chem.* **1973**, *38*, 1841.

40. Albrecht, H.P.; Repke, D.B.; Moffatt, J.G. *J. Org. Chem.* **1974**, *39*, 2176.

41. Watanabe, T.; Nishiyama, S.; Yamamura, S.; Kato, K.; Nagai, M.; Takita, T. *Tetrahedron Lett.* **1991**, *32*, 2399.

42. Shimida, N.; Hasegawa, S.; Harada, T.; Tomisawa, T.; Fujii, A.; Takita, T. *J. Antibiot.* **1986**, *39*, 1623.

43. For example see: Horton, D.; Philips, K.D. *Carbohydr. Res.* **1973**, *30*, 367.

44. Kaye, A.; Reese, C.B.; Neidle, S. *Nucleosides Nucleotides* **1988**, *7*, 609 and Kaye, A.; Neidle, S.; Reese, C.B. *Tetrahedron Lett.* **1988**, *29*, 1841.

45. Izawa, T.; Nakayama, K.; Nishiyama, S.; Yamamura, S.; Kato, K.; Takita, T. *J. Chem. Soc., Perkin Trans. I* **1992**, 3003.

46. Jung, M.E.; Trifunovich, I.D.; Gardiner, J.M.; Clevenger, G.L. *J. Chem. Soc., Chem. Commun.* **1990**, 84.

47. Nishiyama, S.; Yamamura, S.; Kato, K.; Takita, T. *Tetrahedron Lett.* **1988**, *29*, 4739.

48. Maeba, I.; Iijima, T.; Matsuda, Y.; Ito, C. *J. Chem. Soc., Perkin Trans. I* **1990**, 73.

49. Maeba, I.; Ito, Y.; Wakimura, M.; Ito, C. *Heterocycles* **1993**, *36*, 1617 and references cited therein.

50. Takahashi, H.; Ogura, H. *J. Org. Chem.* **1974**, *39*, 1374.

51. Piccirilli, J.A.; Krauch, T.; MacPherson, L.J.; Benner, S.A. *Helv. Chim. Acta.* **1991**, *74*, 397.

52. Belmans, M.; Vrijens, I.; Esmans, E.L.; Lepoivre, J.A.; Alderweireldt, F.C. *Nucleosides Nucleotides* **1987**, *6*, 245.

53. Cornia, M.; Casiraghi, G.; Zetta, L. *J. Org. Chem.* **1991**, *56*, 5466.

54. Eaton, M.A.W.; Millican, T.A.; Mann, J. *J. Chem. Soc., Perkin Trans. I* **1988**, 545.

55. Kabat, M.M.; Pankiewicz, K.W.; Watanabe, K.A. *J. Med. Chem.* **1987**, *30*, 924.

56. Hacksell, U.; Daves, G.D. Jr., *J. Org. Chem.* **1983**, *48*, 2870.

57. Buchanan, J.G. *Nucleosides Nucleotides* **1985**, *4*, 13.

58. Buchanan, J.G. *Prog. Chem. Org. Nat. Prod.* **1983**, *44*, 243.

59. Buchanan, J.G.; Edgar, A.R.; Power, M.J. *J. Chem. Soc., Perkin Trans. I* **1974**, 1943.

60. Buchanan, J.G.; Edgar, A.R.; Power, M.J.; Shanks, C.T. *J. Chem. Soc., Perkin Trans. I* **1979**, 225.

61. Buchanan, J.G.; Stobie, A.; Wightman, R.H. *Can. J. Chem.* **1980**, *58*, 2624 and *ibid. J. Chem. Soc., Perkin Trans. I* **1981**, 2374.

62. Buchanan, J.G. Quijano, M.L.; Wightman, R.H. *J. Chem. Soc., Perkin Trans. I* **1992**, 1573.

63. Kametani, T.; Kawamura, K.; Honda, T. *J. Am. Chem. Soc.* **1987**, *109*, 3010.

64. For a review on this topic see: Noyori, R. *Acc. Chem. Res.* **1979**, *12*, 61.

65. Noyori, R.; Sato, T.; Hayakawa, Y. *J. Am. Chem. Soc.* **1978**, *100*, 2561.

66. Ohno, M.; Ito, Y.; Arita, M.; Shibata, T.; Adachi, K.; Sawai, H. *Tetrahedron* **1984**, *40*, 145.

67. Just, G.; Liak, T.J.; Lim, M.; Potvin, P.; Tsantrizos, Y.S. *Can. J. Chem.* **1980**, *58*, 2024.

68. For a recent account of this topic see: Vogel, P.; Fattori, D.; Gasparini, F.; Le Drian, C. *Synlett* **1990**, 173.

69. Jeanneret, V.; Gasparini, F.; Péchy, P.; Vogel, P. *Tetrahedron* **1992**, *48*, 10637.

70. Kozikowski, A.P.; Ames, A. *J. Am. Chem. Soc.* **1981**, *103*, 3923.

71. Cox, P.J.; Simpkins, N.S. *Synlett*, **1991**, 321.

72. Mohar, B.; Štimac, A.; Kobe, J. *Nucleosides Nucleotides* **1993**, *12*, 793.

73. See reference Hanessian in reference #11 p. 120 and Hockett, R.C.; Zief, M.; Goepp, R.M. *J. Am. Chem. Soc.* **1946**, *68*, 935.

74. Gonzalez, F.G.; Guillen, M.G.; Perez, J.A.G.; Galán, E.R. *Carbohydr. Res.* **1980**, *80*, 37.

75. Sallam, M.A.E. *Carbohydr. Res.* **1978**, *67*, 79.

76. Sallam, M.A.E.; Abdel Megid, S.M.E. *Carbohydr. Res.* **1984**, *125*, 85.

Chapter 12

BIOLOGICAL AND PRACTICAL APPLICATIONS OF C-GLYCOSIDES

I. INTRODUCTION

The majority of this book has dealt with the preparation of *C*-glycosides and some of their potential applications. *C*-Glycosides possess a rich source of chirality with more carbon functionality than *O*-glycosides, and have been used primarily as building blocks in natural product synthesis; however, there has also been a need to develop mild and stereoselective methods for carbon-carbon bond formation at the anomeric center in order to efficiently synthesize naturally occurring alkyl and aryl *C*-glycosides. While many introductions to papers dealing with *C*-glycoside preparation mention the fact that *C*-glycosides could be useful compounds as enzyme inhibitors, it has been only recently that studies dealing with enzyme inhibition and biological testing of *C*-glycosidic compounds have been published. The focus of this chapter will be to bring this emerging aspect of the field into perspective and outline the results presented to date in the literature. *C*-Disaccharides have also been a class of compounds that are associated with important biological processes and a few examples of their potential uses will be discussed here. The potential use of carbon glycosides as anti-viral and chemotherapeutic agents will also be presented in this chapter. Other practical applications, such as their use in molecular scaffolding and surfactants will also be briefly discussed.

II. BIOLOGICAL APPLICATIONS

The general mechanism for enzymatic hydrolysis of glycosides has been studied for a number of glycosidases and the topic has been reviewed.[1] It is thought that a carboxylate function at the enzymatic active site is responsible for charge stabilization of the transition state, Figure 1.

Transition State

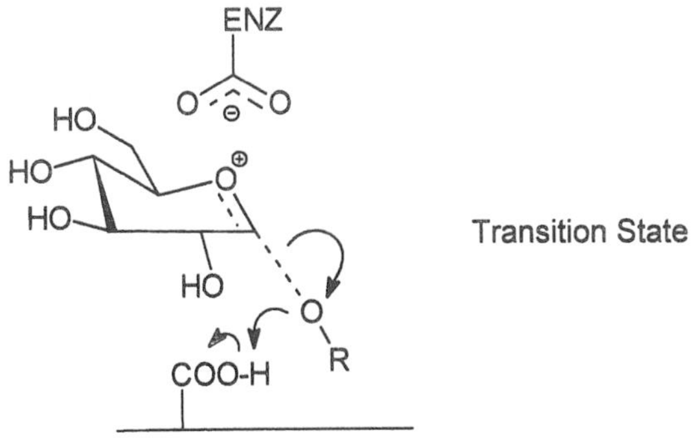

Figure 1

345

The relative ease of glycoside hydrolysis is partly due to the ability of the oxygen to act as a leaving group when protonated. This is consistent with what is understood for the acid hydrolysis of hemiacetals. If the aglycone oxygen is replaced carbon hydrolysis via a similar mechanism now becomes difficult just as would the hydrolysis of any alkyl ether. Accordingly, several workers have postulated that replacement of the oxygen group with a carbon group *e.g.* a *C*-glycoside or *C*-disaccharide would prevent this simple enzymatic hydrolysis.

A.   ENZYME INHIBITORS

Work by Liu[2] focused on the preparation of a compound that combines the inhibitory properties of norjirimycin (1) with the hydrolytic stability of a *C*-glycoside. The general preparation of 6 is as outlined in Scheme 1. Coupling of 3 with bromide 4 gave the target compound 6 after deprotection. Compound 6 was found to be a potent competitive inhibitor for intestinal sucrase $K_i = 2$ μM while K_m for sucrose is 2000 μM. The compound was also found to inhibit maltase, trehalase, glucoamylase, and α-amylase.

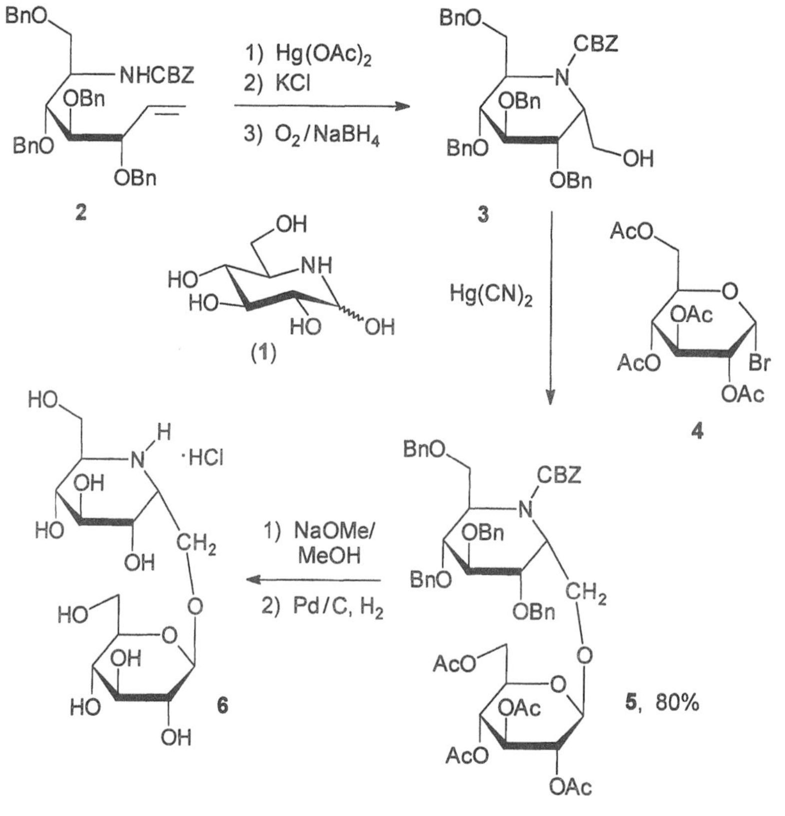

Scheme 1

The biological impetus for some of this chemistry was to determine if 6 was a useful inhibitor of intestinal α-glucosidases, since it has been shown that inhibition of these enzymes is a useful adjunctive therapy for the treatment of diabetes mellitus. When 6 was given to mice simultaneously with sucrose or starch a significant suppression of postprandial hyperglycemia was observed. This result indicates that inhibitors of the type 6 hold promise for use in diabetes therapy.

Due to the inhibitory activity of the class of compounds known as the Nojirimycins[3] several workers have become interested in amino C-glycosides as enzyme inhibitors. Schmidt and Dietrich[4] have studied the inhibition of β-glucosidase (from sweet almonds) with the C-glycosides **7** and **8** (preparation is outlined in Chapter 3). The K_i values for **9a** and **9b** were very different. Compound **9a** had a K_i = 70 μM which is similar to the value for 1-deoxynojirimycin (**10**) (K_i = 18 μM) while K_i = 7600 μM for **9b**. This gives some indication on the relative position of the acid groups at the enzyme active site. The K_i values for **7** and **8** fell into the same range as for **9b**.

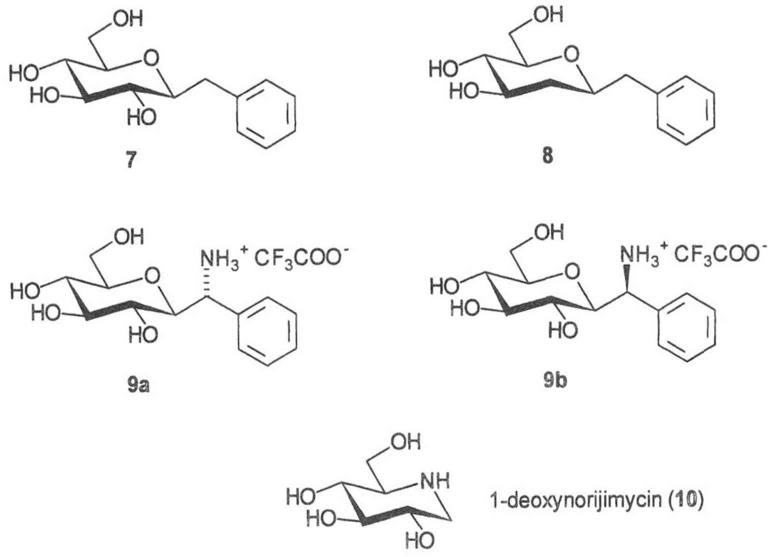

Figure 2

The corresponding α-C-glycosides[5] **12a** and **12b** were made via a mercuriocyclization-oxymercuration route as shown in Scheme 2. Aldehyde **13** was condensed with phenylmagnesium bromide to give the (1S)-diastereomer **14**, exclusively.

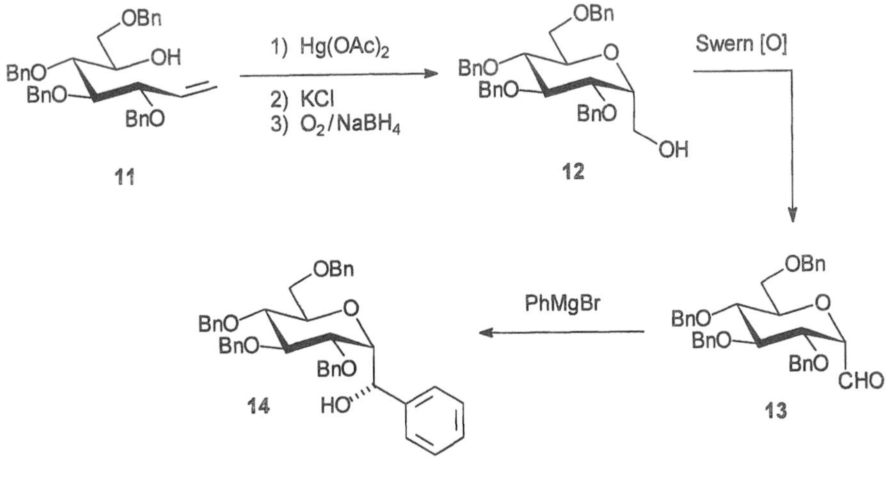

Scheme 2

Replacement of the alcohol function by an amino group, via azide displacement of the corresponding mesylate, proved problematic. So the workers turned to an oximation-reduction route. When oxime **15** (available by oximation of the ketone obtained from Swern oxidation of **14**) was treated with lithium aluminum hydride in THF three compounds **16a**, **16b**, and **17** were isolated in a 1:1:1 ratio in good yield, after deprotection Compound **17** arises from Beckman rearrangement and reduction.

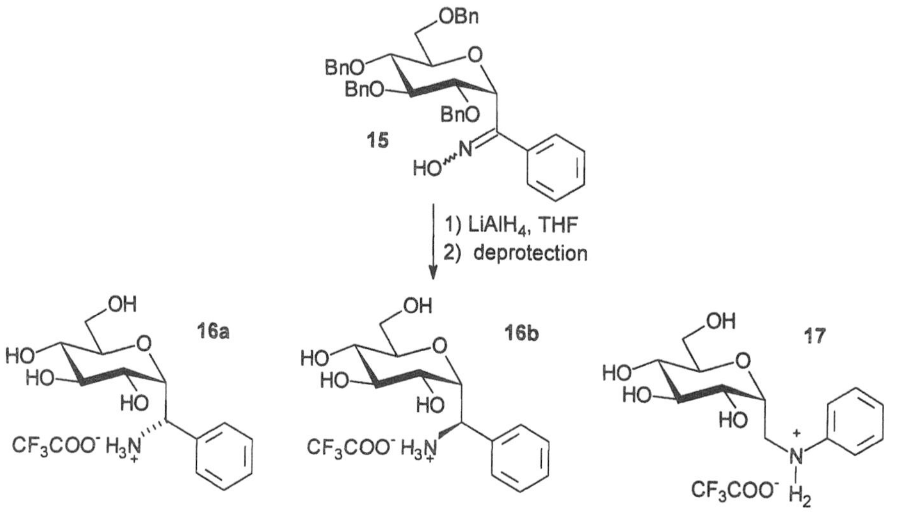

Scheme 3

The compounds were assayed by measuring the release of *p*-nitrophenol from *p*-nitrophenyl α-D-glucopyranoside by the action of yeast α-D-glucosidase. Benzyl α-*C*-glycoside (not shown) showed only a low inhibition (K_i = 1300μM). Compounds **16a** and **16b** showed K_i values of 38 μM and 1100 μM. The aniline **17** showed a K_i value of 11 μM which is similar to that for deoxynorijimycin (K_i = 18 μM).

Martin and Lai[6] have studied the inhibitory activity of several amino-*C*-glycosides. Their preparation was achieved by reduction of either the appropriate glucopyranosyl(nitromethanes) or nitriles. Compounds **18**, **19**, and **20** all proved to be fairly weak inhibitors of the sweet almond β-glucosidase with K_i values of 5500, 17000, and 2800 μM respectively.

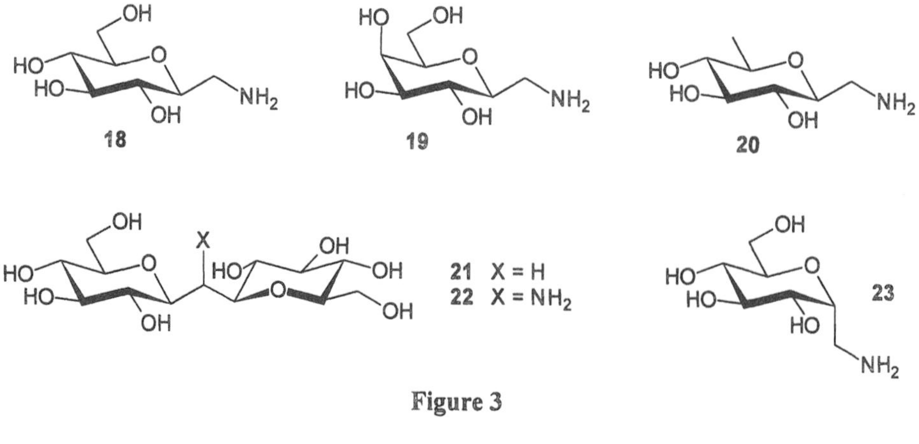

Figure 3

Compounds **21** and **22** proved to have no activity when tested with the same enzyme. This is probably due to the lack of hydrophobic functionality on the aglycone, which seems to be a necessary requirement for recognition by this enzyme (*vide supra*). The α-*C*-glycoside **23** was tested against yeast α-glucosidase and was found to inhibit the hydrolysis of *p*-nitrophenyl α-D-glucopyranoside with a K_i value in the mM range.

Diazocompounds, known for their ability to form carbenes, have begun to emerge as useful species for enzyme labelling or inactivation by covalent modification. *C*-Glycosides are well suited to serve as delivery mechanisms for these types of intermediates since the sugars can bind in certain enzyme active sites and the carbon chain is robust enough to tolerate the required chemistry to install the carbene precursor. Diazoketones are suitable carbene precursors and both Schmidt and BeMiller have used this concept to synthesize irreversible *C*-glycoside based enzyme inhibitors. The available amine **24** was converted to the urethane **25**.[7] Protective group manipulations then gave **26** which was exposed to gaseous N_2O_4 to provide the nitroso-urethane **27**. When exposed to base, **27** collapsed to the diazoketone **28** which proved to be fairly stable in buffer (pH = 8) for a short time. The identity of the diazo ketone was confirmed by exposure to an acidic aqueous solution and per *O*-acetylation to provide **29** as a 1:1 mixture of isomers (Scheme 4).

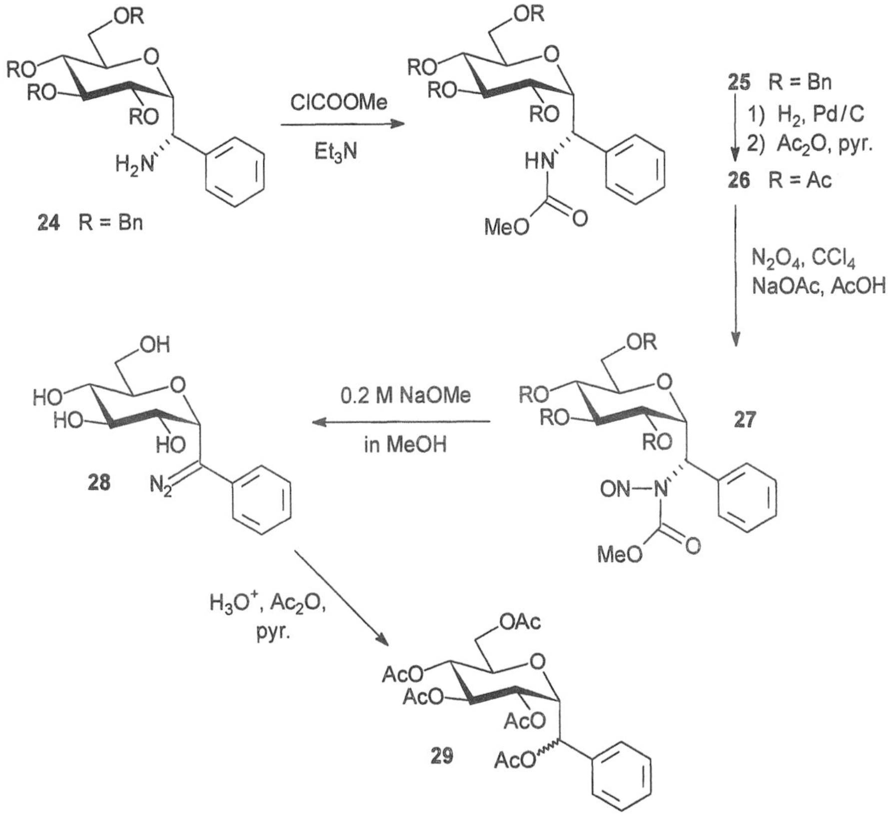

Scheme 4

Photolysis of **28** in the presence of yeast α-glucosidase caused irreversibly inhibition of the enzyme. The activity dropped to almost zero after 14 min, and could not be regained by dialysis. Competition experiments with native substrate showed that the inhibitor does indeed block the active sight of the enzyme. No inhibition was observed when **28** was assayed with β-glucosidase from sweet almonds.

The work of BeMiller and co-workers[8] focused on the diazomethyl β-D-galactopyransoyl ketone. One of the possible mechanisms of inactivation is given in Scheme 5. Formation of a carbene is followed by insertion to either an acidic or basic group. The result is a covalent modification of the enzyme active site.

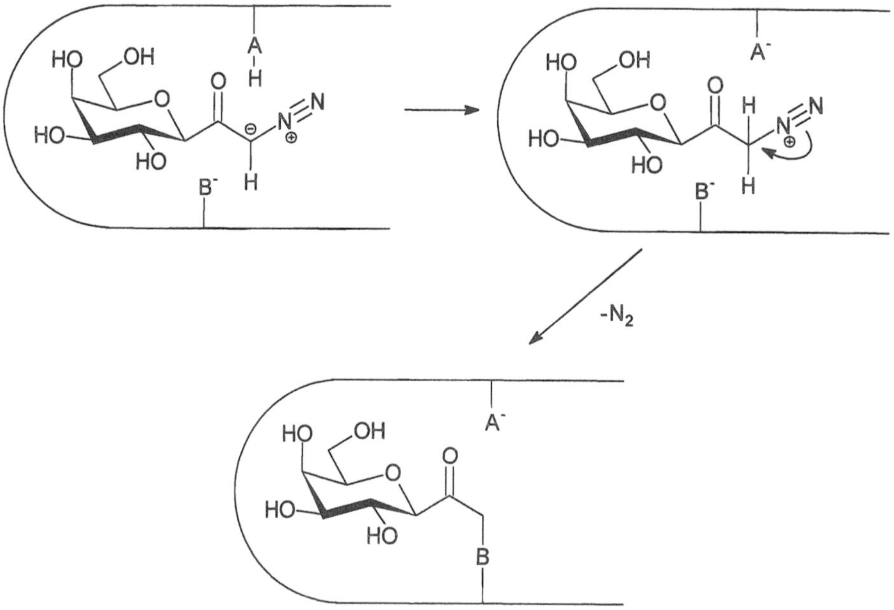

Scheme 5
Reprinted from ref. 8, pg. 103, with kind permission from
Elsevier Science Ltd, The Boulevard, Langford Lane, Kidlington, 0X5 1GB, UK

The galactopyranosyl diazoketone was found to be a mechanism-based irreversible (suicide-substrate) inactivator of *Aspergillus orzyae* β-D-galactosidase. It was found that the inhibitor was covalently bound to the enzyme and it was proposed that the mechanism of inactivation is that proposed in Scheme 5.

Baasov and collaborators[9] used a *C*-glycosidic analog of KDO to inhibit the enzyme KDO8P synthase which is responsible for condensation between phosphoenolpyruvate with D-arabinose-5-phosphate to give KDO8P, Scheme 6.

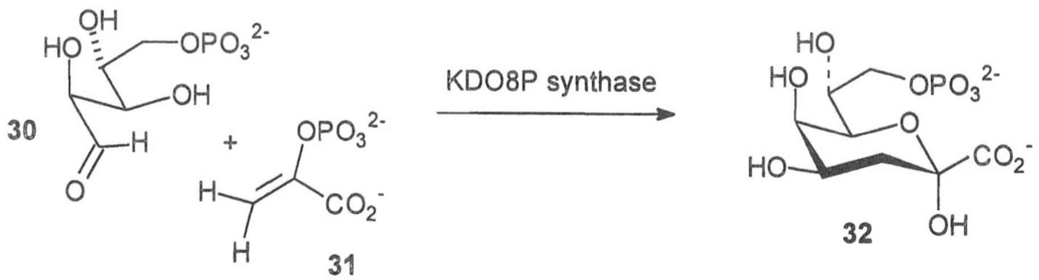

Scheme 6

Analog **39** was made from **34** by alkylation with triflate **35**. Protecting group manipulations gave **37** which was deprotected to **39**. Analog **39** was found to be a potent competitive inhibitor of KDO8P from *Escherichia coli* with a K_i of 5 μM.

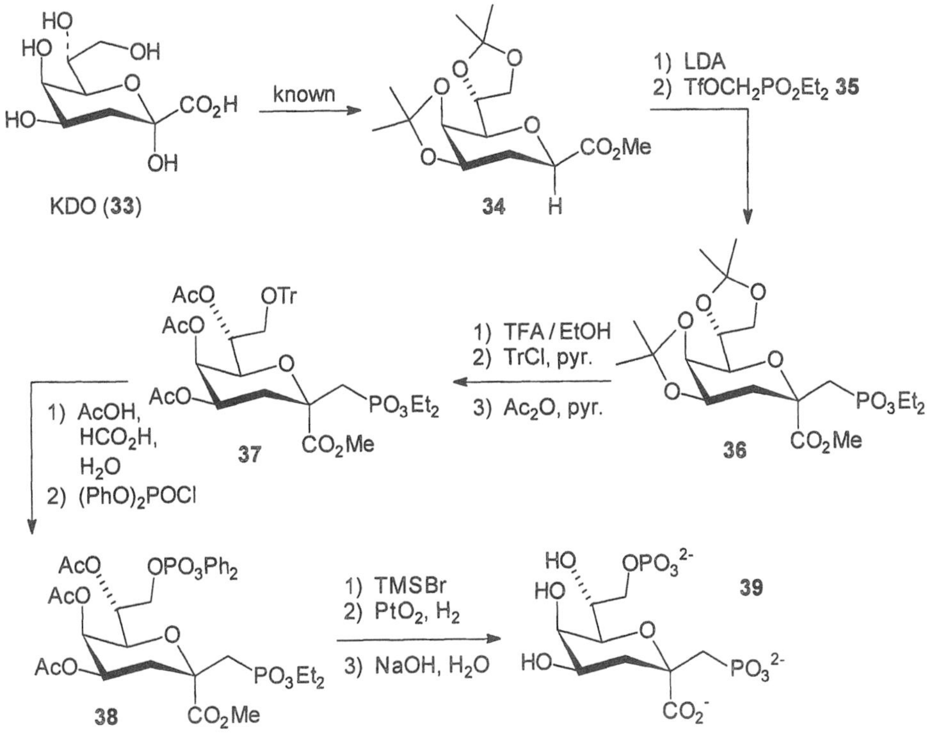

Scheme 7

Schmidt and Frische[10] used a *C*-glycosidic analog of UDP-galactose as a transition state analog that would hopefully inhibit the activity of galactosyltransferase. The *C*-glycoside serves as a stable analog that cannot undergo enzymatic cleavage and the required conformational change required for glycosyl transfer. Analogs **40** and **41** have been made and show good activity for inhibition against β-galactosyl transferase.[11]

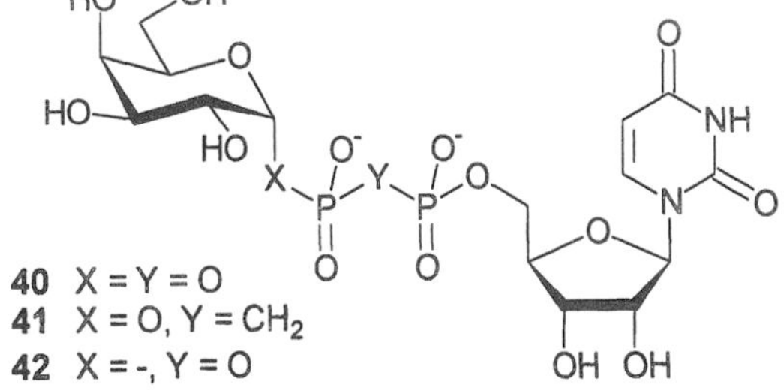

40 X = Y = O
41 X = O, Y = CH$_2$
42 X = -, Y = O

Figure 4

Schmidt and Frische proposed to replace "X" (Fig. 4) with a methylene group and use a glycal instead of the pyranosyl ring, since during glycosylation the pyranosyl ring must flatten to a structure best mimicked by a glycal. The β-galactosyl cyanide **43** was reduced to the aldehyde and dehydrated to glycal **44**. Standard manipulations then gave the protected phosphonate **46**. Saponification gave analog **47**. Alternatively, **48** could be coupled with **49** to give the analog **50**. The K_i values for **47** and **50** against β-galactosyltransferase from bovine milk were found to be 1430 μM and 62 μM respectively.

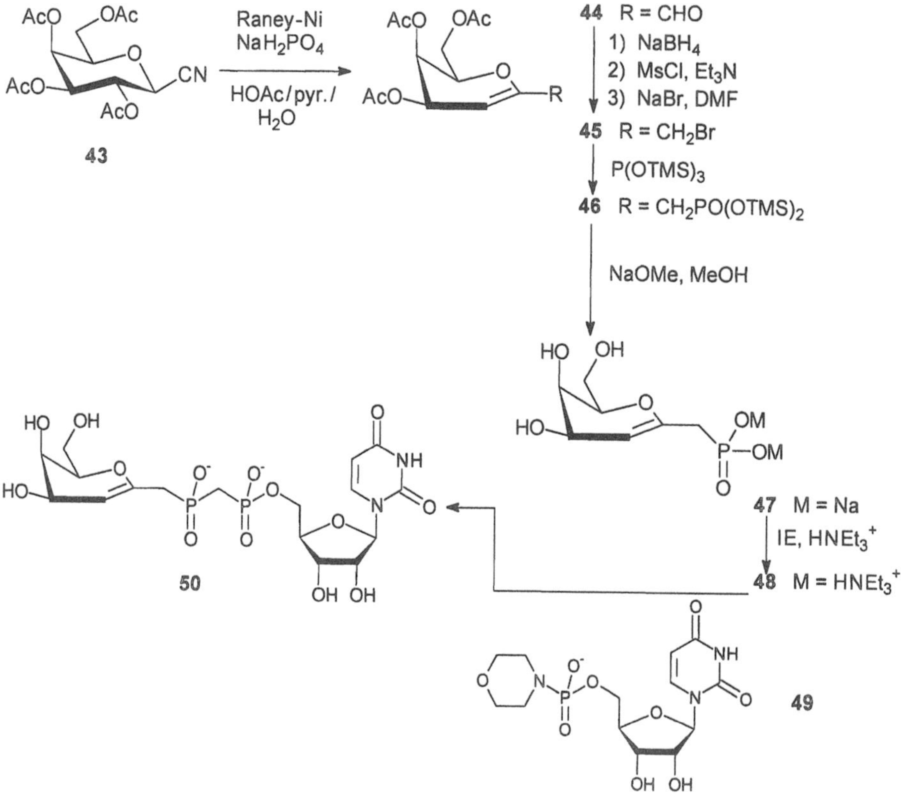

Scheme 8

Shulman *et al.*[12] synthesized epoxides **51** and **52** via epoxidation of the corresponding olefins which were in turn prepared by displacement of an anomeric bromide with a suitable olefinic organometallic reagent. These epoxides were found to irreversibly deactivate sweet almond β-D-glucosidase.

Figure 5

Luengo and Gleason[13] prepared *C*-glycoside analogs of GDP-fucose (**63**, where X = O) for use as competitive inhibitors of fucosyltransferases, since the pyranose ring is now inactivated towards glycosyl transfer. Compound **53** underwent siloxymethylation to give, after desilylation, **54**. The hydroxyl group was converted to its phosphate **55** by bromination, Arbuzov reaction, deprotection, and salt formation. Coupling then gave the analog **56**. The corresponding *O*-analog was prepared by phosphorylation of **57** and conversion to **58**. Coupling then gave **59**. The homologue of **59** was available by a non-selective *C*-allylation by treating **53** with zinc bromide in neat allyltrimethylsilane at 60°C to give the isoemric ally *C*-glycosides (not shown) in roughly equal amounts. Compound **60** was then converted into **62** in the same way as for **55→56**. The axial anomer of **62** was available by analogous chemistry (Scheme 9).

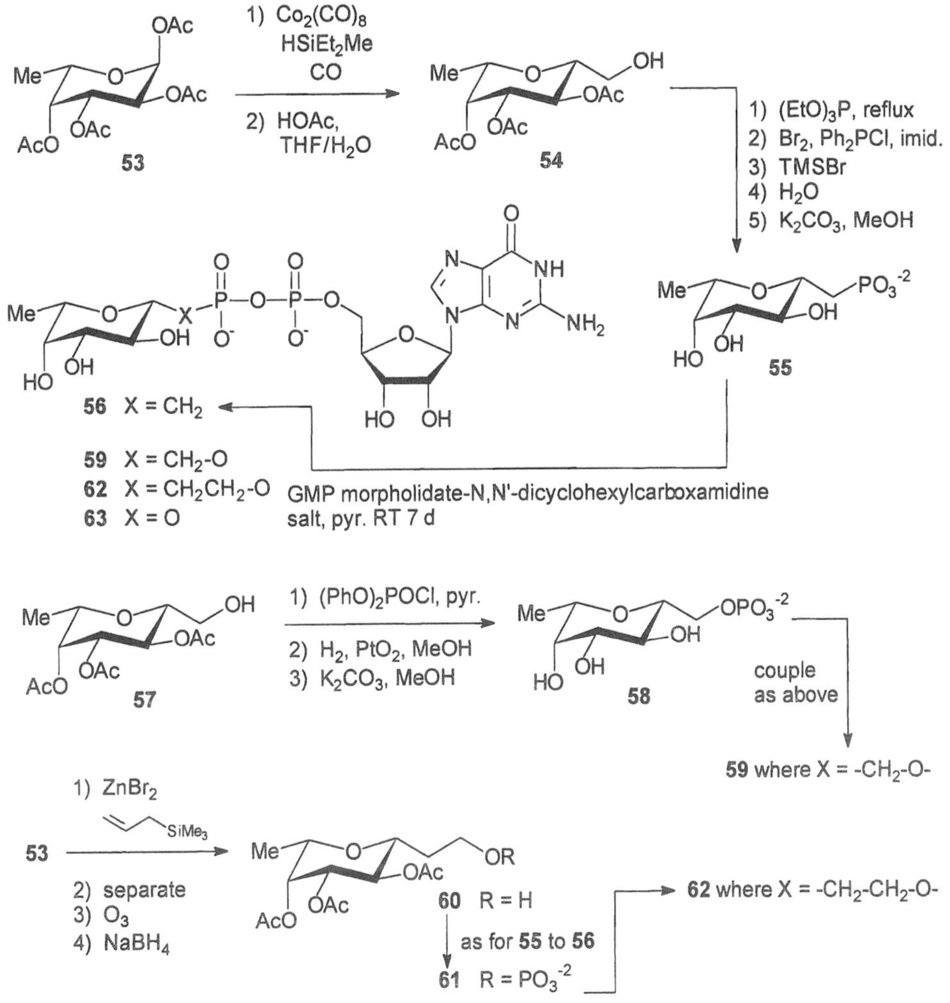

Scheme 9

Brockhaus and Lehmann[14] have used the *exo*-methylenic sugar **66** to probe the activity of β-D-galactosidase. Compound **66** was prepared from ester **64** by reduction and Hoffman type chemistry to install the olefin. Compound **66** was incubated with β-D-galactosidase and it was found that hydration to **67** occurred quite rapidly as compared to non-enzymatic hydration. It was also found that incubation of **66** with glycerol gave the β-anomer **68** exclusively, and that the reverse reaction (**68→67**) was catalyzed by the same enzyme.

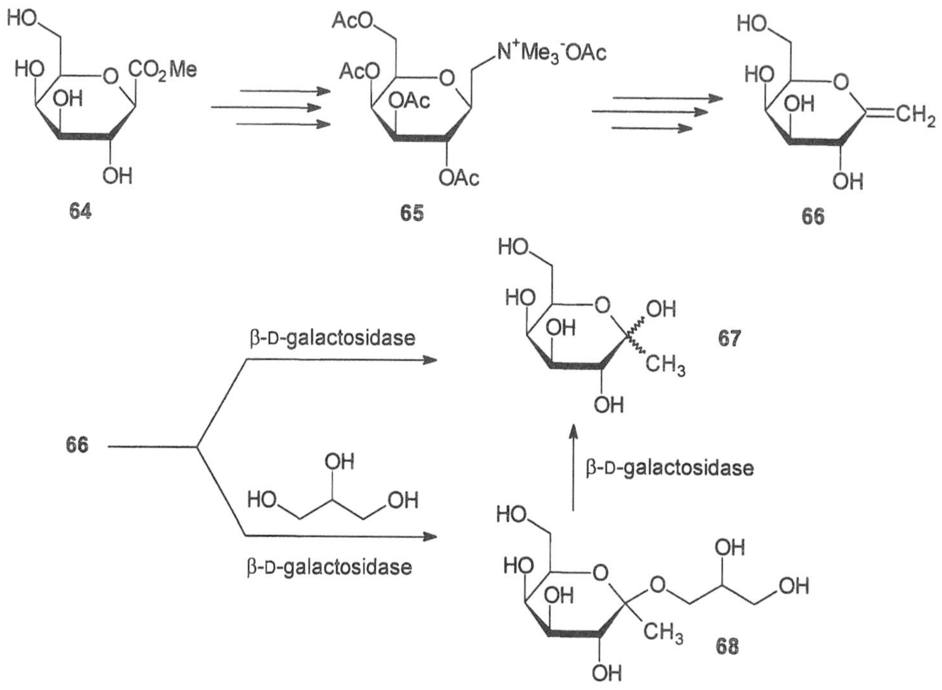

Scheme 10

B.   LABELING

Lehmann and co-workers[15] have also used the *C*-glycoside disaccharide **69** as an affinity label for Immunoglobulin A X24. Irradiation of **69** in the presence of IgA X24 gave labeled immunoglobulin. Presumably the covalent modification of the antibody occurs via a carbene insertion reaction that comes about as a result of diazirine photolysis.

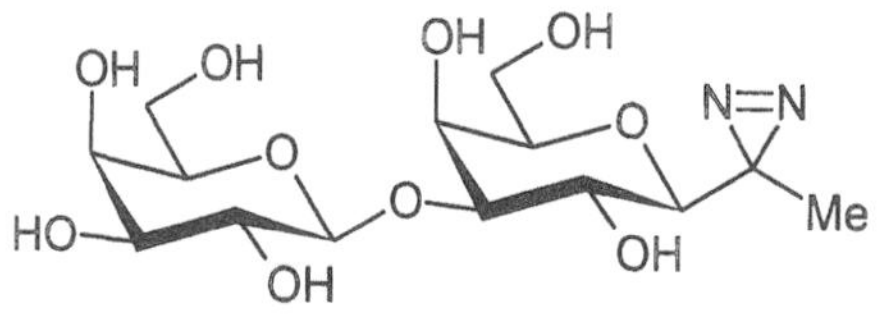

69

Figure 6

Work[16] with **70** has shown that it has a K_i value of 21000 µM for the enzyme β-D-galactosidase from *E. coli*. It was shown that **70** covalently attaches itself to the enzyme probably via a bond at *C*-3 as shown in Figure 7. The inhibitor does not only link itself to the active site but to groups on the protein molecule since bound enzyme lost only 7% of its activity. When **70** was incubated with β-D-galactosidase from *E. coli* and ^{14}C labeled glycerol the enzyme became ^{14}C covalently labeled as determined by SDS electrophoresis on acrylamide gel. The chemical basis for labeling is nucleophilic attack of a suitable group on the protein to the electrophilic *C*-3 of compound **70**.

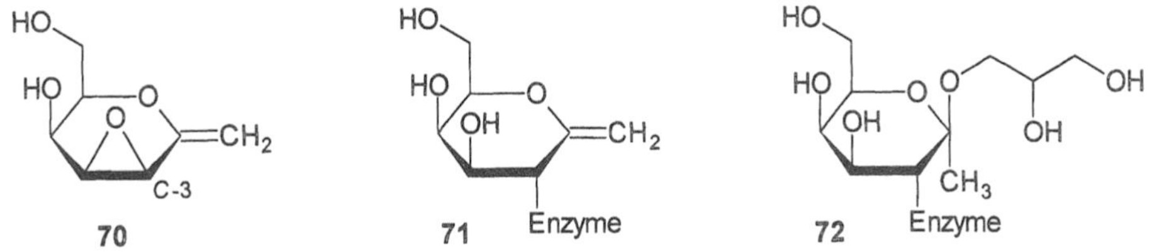

Figure 7

Compounds **73** and **74** were also assayed for their potential as labeling agents, but both were found to be ineffective labeling reagents for β-D-galactosidase from *E. coli*.[17]

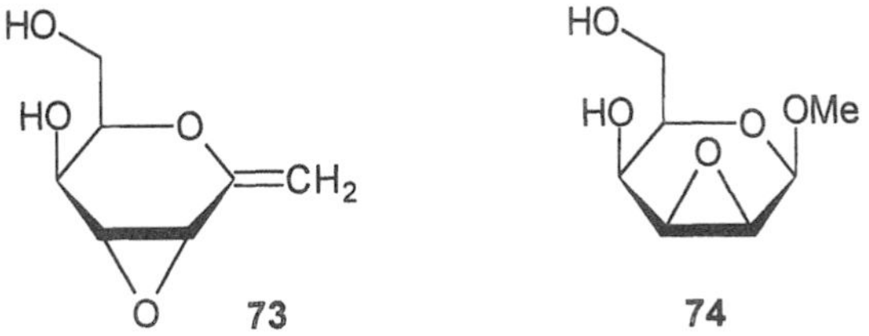

Figure 8

C. ANTI-VIRAL AND ANTI-MICROBIAL AGENTS

It has recently been demonstrated that certain enzymatic inhibitors[18] show a propensity to reduce the replicative ability of the HIV virus. Since *C*-glycosides are prime candidates for enzyme inhibition the application of *C*-glycosidic inhibitors to anti-viral study is a worthy endeavor.

It has been shown that polyvalently linked *O*-linked sialic acids serve to inhibit the attachment of the virus to erythrocytes. This could potentially be a method of combating the virus, but the hydrolysis of the saccharide linkages by neuraminidases (NA) limits its use. Work led by Nagy and Bednarski[19] has shown that if the sialic acid linkages are replaced by more glycosidase-stable carbon linkages (*C*-glycosidic linkages) then viral activity is still inhibited. The *C*-glycoside **76** was prepared from **75** by suitable modification of the allyl group.

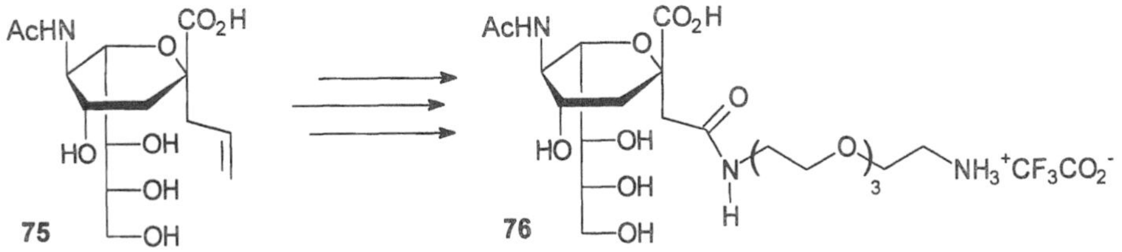

Scheme 11

Attachment of **76** to the carbohydrate monomer **77** was realized via a reductive amination process. Two polymers, one with 5% *C*-glycoside **79** and the other with 30% *C*-glycoside **80** incorporation, were tested. A binding assay showed that both had binding constant in line with similar multivalent *O*-sialoside complexes. Polymer **80** caused a 50% reduction in viral plaque formation when it was present in 100μM concentration while at 500μm concentration an 80% inhibition was observed. Polymer **79** showed only a minimal decrease in plaque reduction, and the *O*-methyl sialoside and *C*-sialoside **76** showed no plaque reduction. Compound **75** weakly inhibited viral neuraminidase whereas polymer **79** or **80** did not inhibit neuraminidase.

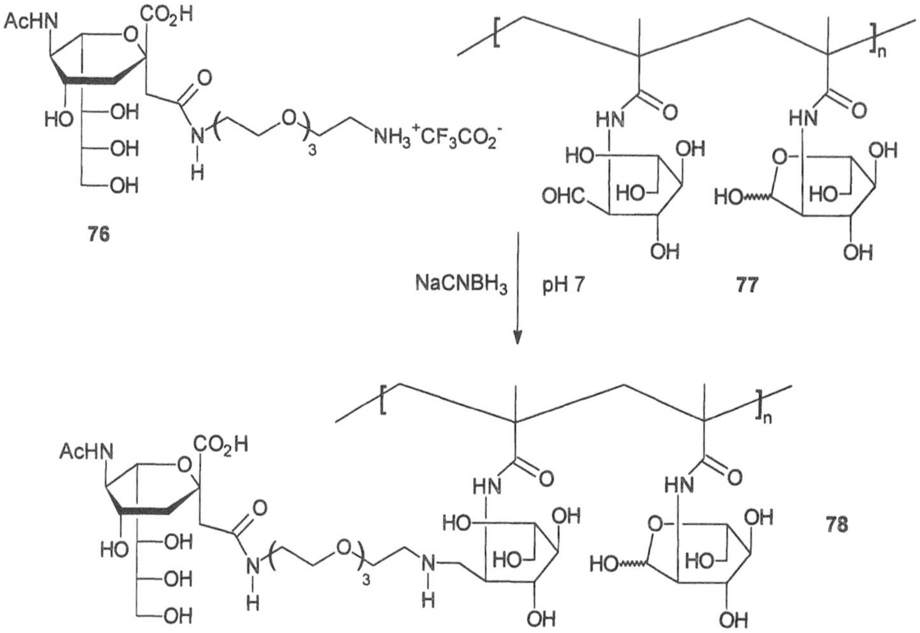

Scheme 12

Work by Bednarski[20] used a sialic acid *C*-glycoside to make polymerized liposomes to determine if they would inhibit the activity of the influenza virus. Synthetic polyvalent sialosides have been observed to inhibit viral adhesion to erythrocytes and these workers have extended this concept to include the use of *C*-glycosidic liposomes which were prepared by standard methods (Scheme 13). Sonication of monomers **83** then provided liposomes.

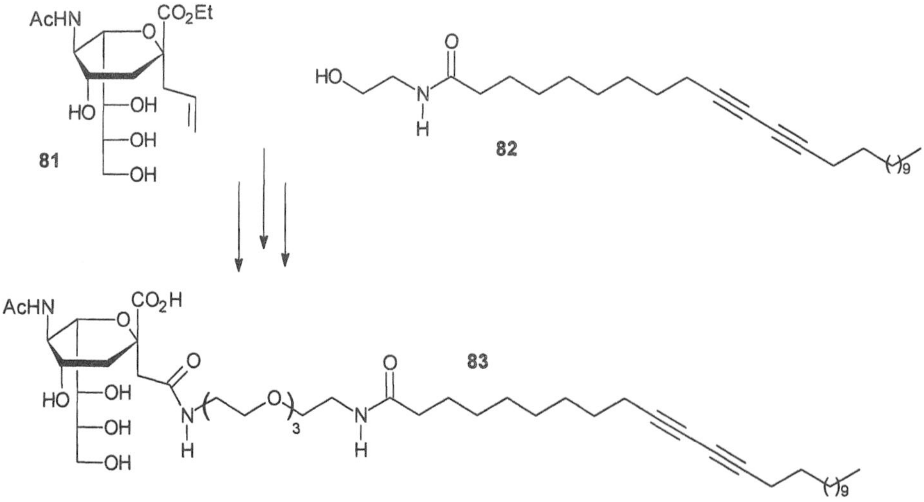

Scheme 13

The six liposome preparations were tested for hemagglutination inhibition and the results are listed in Table 1.

Table 1. Hemagglutination Inhibition (HAI) and Plaque Reduction Assays of Liposome Preparations I-VI.

Entry	Inhibitor	HAI [83], M	[83], mM	Reduction %
1	lip. I (0%, **83**)	0	0.000	0
2	lip. II (1%, **83**)	4.0×10^{-6}	0.003	96
3	lip. III (5%, **83**)	5.7×10^{-7}	0.016	97
4	lip. IV (10%, **83**)	3.3×10^{-7}	0.030	46
5	lip. V (30%, **83**)	8.0×10^{-5}	3.750	0
6	lip. VI (60%, **83**)	1.5×10^{-4}	7.500	0

Reprinted with permission from ref. 20, Copyright 1993 American Chemical Society.

Liposomes III and IV achieved a 50% inhibition of viral binding at concentrations of 5.7×10^{-7} and 33 μM. This is contrasted to the 2000 μM concentration required for the α-*O*-methyl glycoside of sialic acid to achieve the same binding. Liposome II gave impressive results in a standard plaque reduction assay against Madin-Darby canine kidney cells. Liposome II gave 96% inhibition at a sialoside concentration of 3μM. The results show that the capacity of a sialoside to inhibit hemagglutination is not necessarily related to its ability to prevent viral infectivity.

Lipopolysaccharides are endotoxins that make up the cell walls of gram negative bacteria (they consist of linear polysaccharides, one of which LPS A is shown in Figure 9). Lipid A has immunopharmacological properties and is responsible for inducing endotoxic shock in mammals. The main interest in this molecule stems from the observation that it is a powerful immunostimulant, but due to its associated toxicity and the difficulty to obtain it in a pure state Vyplel and co-workers[21] have prepared *C*-glycosidic analogs that would be stable, easily purified, and not undergo unwanted glycosydation reactions at the 1-position during synthesis since this position is somewhat activated.

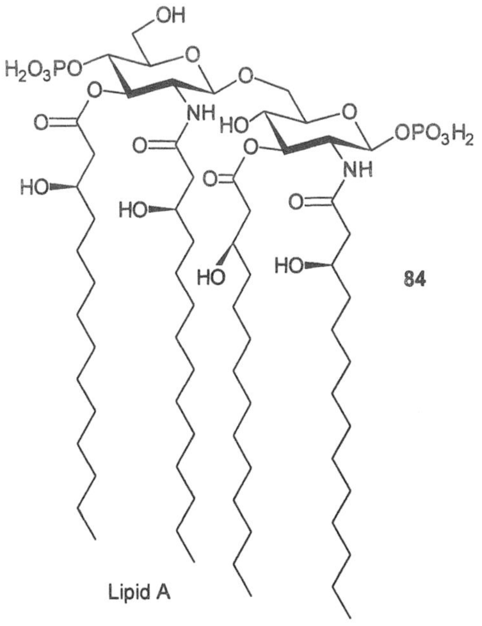

Lipid A

Figure 9

The workers used a Wittig-cyclization approach as outlined in Scheme 14. Both anomers were isolated along with some manno isomer (not shown). The workers then put on the second fatty acid group (either MyrOBn or MyrOMyr), deprotected the anomeric acid and obtained **88** and **89**. Compound **86** could also be manipulated to give **90**. The biological results showed that at least three fatty acid chains are required for immunostimulant activity and that the α-anomer is more active than the β-anomer. The bioisosteric acid *C*-glycosides do not seem to offer any therapeutic advantage over the phosphorylated *O*-glycosides.

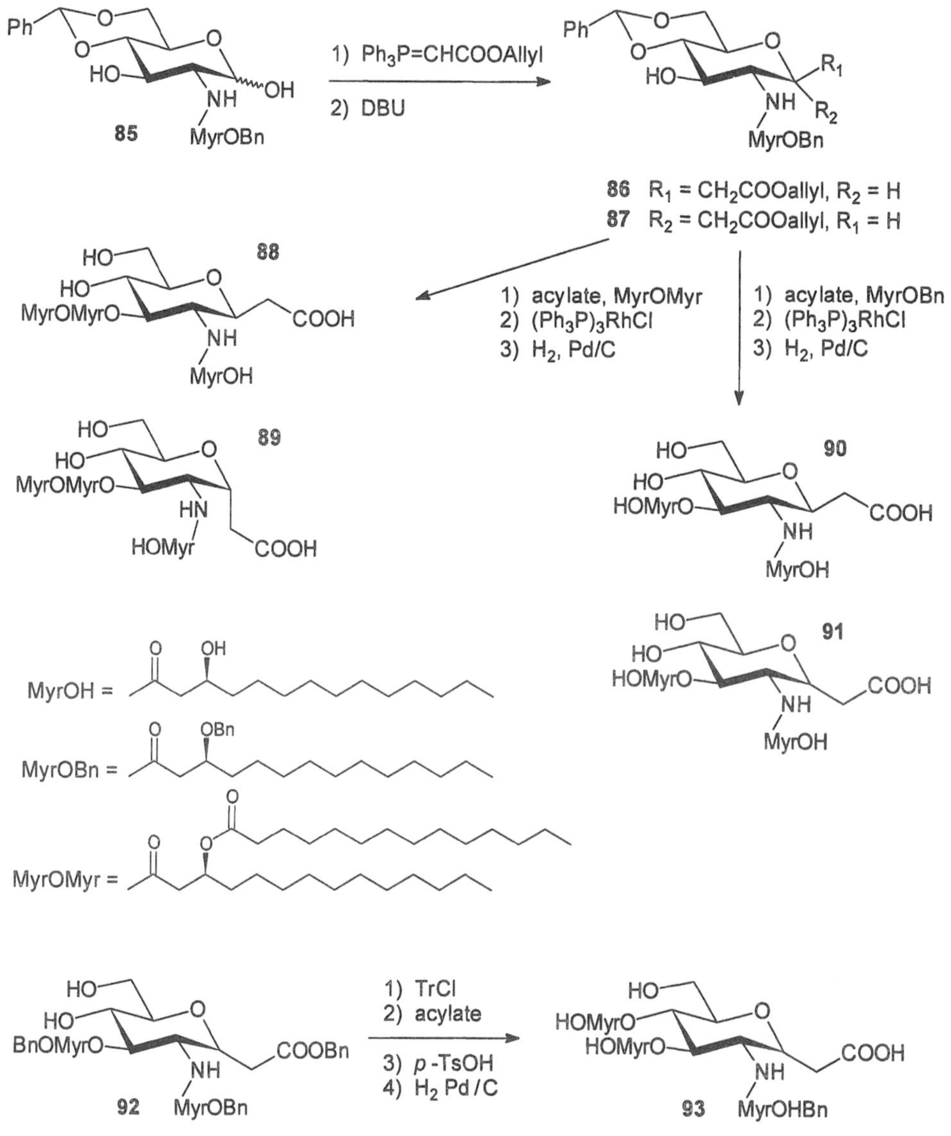
85
1) Ph3P=CHCOOAllyl
2) DBU
86 R1 = CH2COOallyl, R2 = H
87 R2 = CH2COOallyl, R1 = H
1) acylate, MyrOMyr
2) (Ph3P)3RhCl
3) H2, Pd/C
1) acylate, MyrOBn
2) (Ph3P)3RhCl
3) H2, Pd/C
88
89
90
91
MyrOH =
MyrOBn =
MyrOMyr =
1) TrCl
2) acylate
3) p-TsOH
4) H2 Pd/C
92
93

Scheme 14

Vederas and Qiao[22] have incorporated a *C*-glycosidic into a disaccharide analog **101** that combines the structural features of the active portion of the antimicrobial agent moenomycin A and the peptidoglycan **94**. Compound **94** is the natural substrate for the enzyme transglycolase. It is responsible for the formation of the polysaccharide backbone of bacterial peptidoglycan.

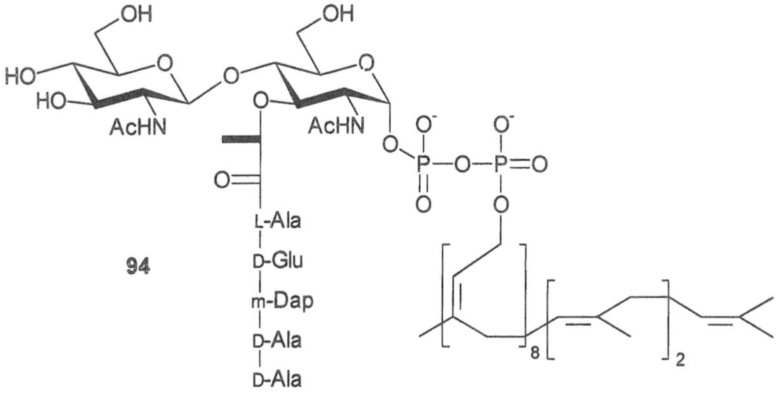

Figure 10

The *C*-glycoside analog was made by careful Wittig olefination followed by mercurio-cyclization to give, after suitable manipulations, **98**. Coupling then gave the disaccharide **100** which was elaborated by standard methods into **101**.

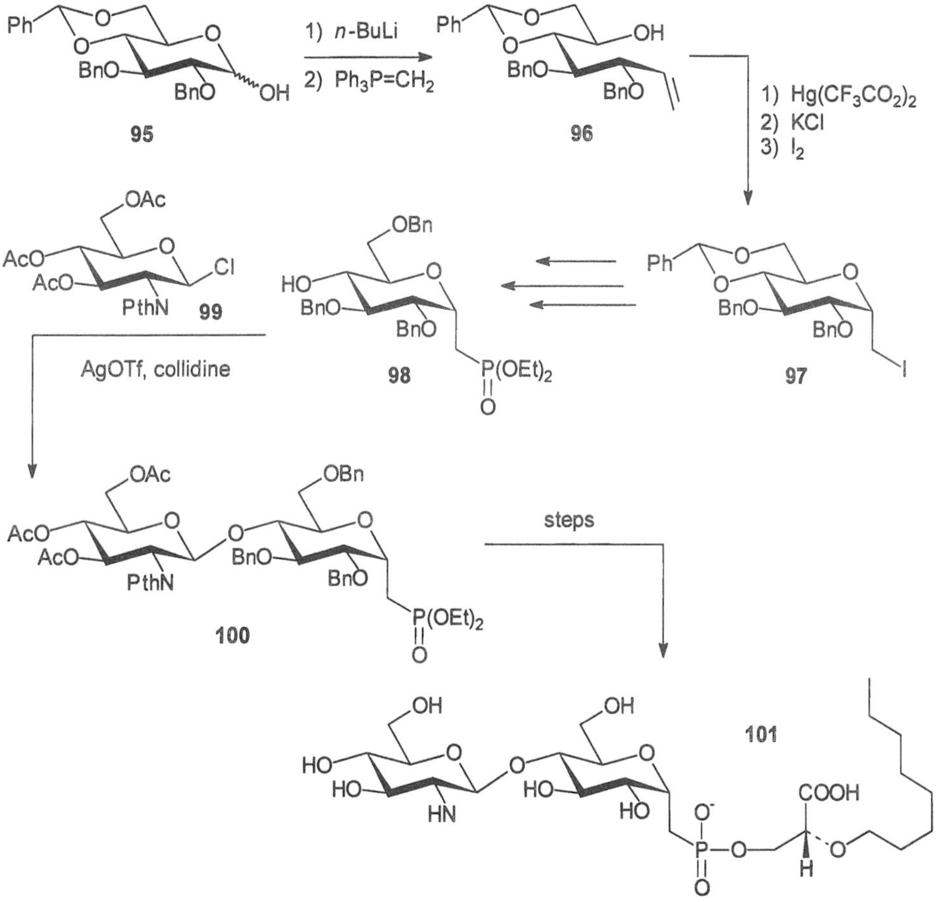

Scheme 15

The chemical characteristics of this analog include a stable non-cleavable C-phosphonate moiety, and a lack of the physiologically active peptide chain. The C-phosphonate also has the added feature that it may "end-cap" growing peptidoglycan polysaccharide chains since glycosylation is now impossible.

D.   ANTI-CANCER ACTIVITY

Curley Jr. and collaborators[23] have prepared C-glycoside analogs of O-glucuronide metabolites of retinoic acid to probe for anti-cancer activity since, unlike the O-glucuronides, the C-glycosides are not susceptible to β-glucuronidase or acid mediated cleavage. The C-glycoside was oxidized to an acid and underwent ring nitration to 103. Reduction to the amine was followed by separation, acylation and deprotection to furnish analog 104. The gluco analog 105 was made along similar lines with omission of the oxidation step. The corresponding benzyl C-glycoside analogs were made in a similar fashion.

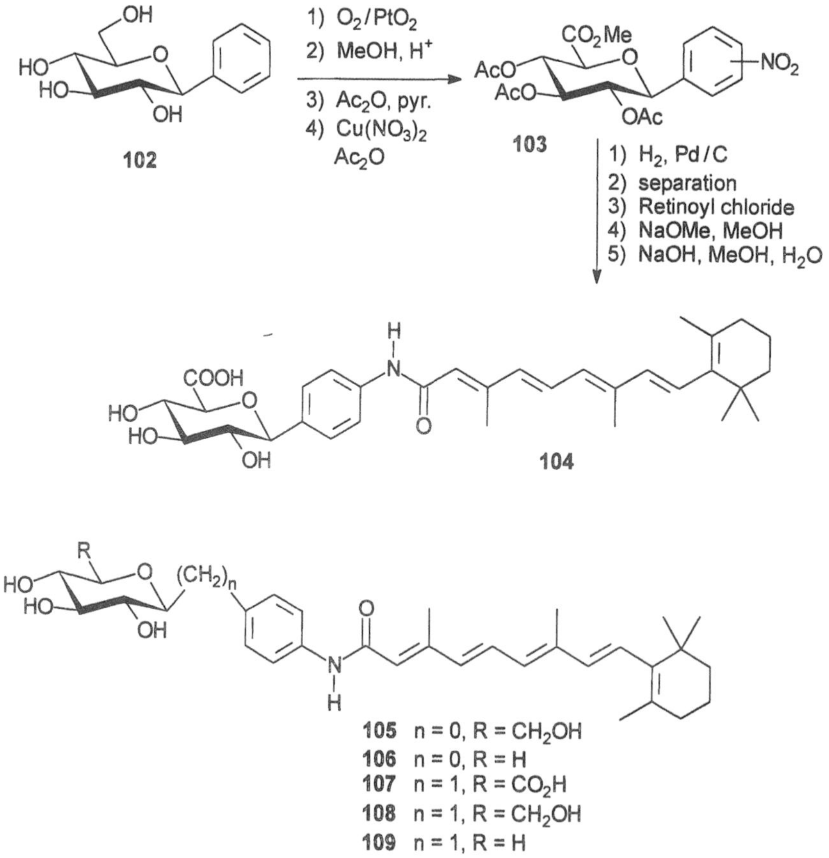

Scheme 16
Reprinted from ref. 23, pg. 307 by courtesy Marcel Dekker Inc.

Compound 105 was stable to acid hydrolysis (0.1N HCl, MeOH) for two hours at 37°C, while the corresponding O-glucuronide underwent partial hydrolysis under these conditions. The C-benzyl compound 107 was a slightly better inhibitor of β-glucuronidase than the C-phenyl compound 104 with an IC_{50} of 267 μM while the corresponding O-glucuronide had IC_{50} of 184.5 μM. The antiproliferative activity of 105-109 against MCF-7 human mammary tumor cells culture model appeared promising but no other data was given. It was also reported that the *C*-glucuronides appear to bind more strongly to the retinoic acid receptors than the corresponding *O*-glucuronides.

Work by Allevi[24] focused on the preparation of the potential anti-cancer analogs of the clinical antitumor agent etoposide (**110**) which is currently in use for treatment of small cell lung cancer and testicular cancer. Several analogs of **110** have been prepared in which the 4 β-*O*-glucosidic moiety has been replaced by other heteroatomic groups such as RS-, RHN-, and RO- in which R can be an alkyl or aryl group.

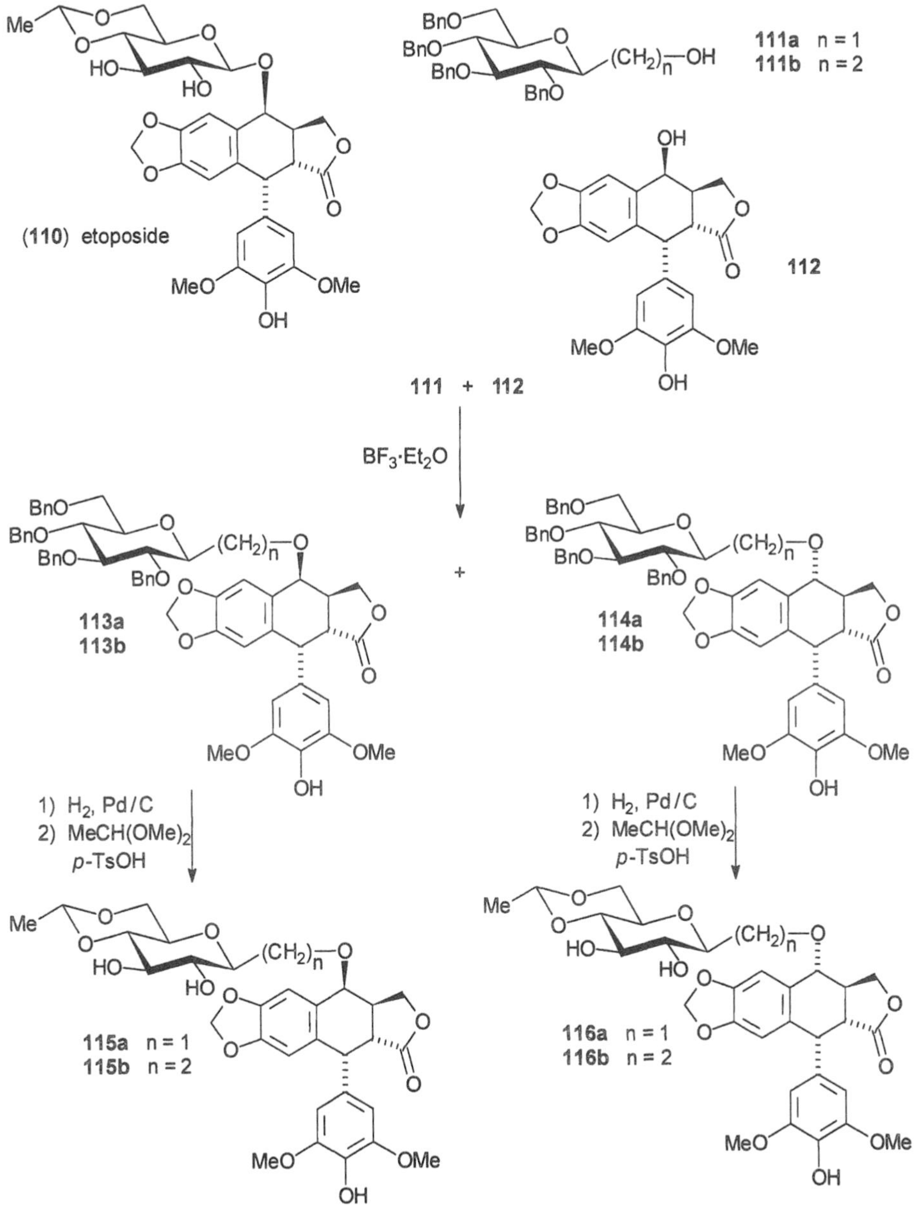

Scheme 17
Reprinted from ref. 24, pg. 212 by courtesy Marcel Dekker Inc.

The workers chose to replace the glycosidic portion with a *C*-glycoside since the stability of the *C*-glycosyl bond would prevent hydrolysis and differing chain lengths could offer control in targeting the active portion of the molecule. The analogs were prepared by coupling of either **111a** or **111b** with **112** under boron trifluoride catalysis (3.0 to 5.0 eq. and 2 h rxn time) to give in the case of compound **111a** compound **114a** (60%) and **113a** (14%). Analogous reaction with the homologue **111b** provided a mixture of **113b** and **114b** in 7% and 45% yield, respectively. Use of less of Lewis acid and longer reaction reversed the ratios of compounds. The sequence was completed by removal of the benzyl groups and ketalization to provide **115** and **116**, Scheme 17. No information on the biological activity of these compounds was reported.

E.   C-DISACCHARIDES

C-Disaccharides have also been the focus of many synthetic studies and an entire chapter of this book has been dedicated to an overview of their preparation. It is thought that not only do *C*-disaccharides have the potential to act as enzyme inhibitors,[1] but that they may also be useful compounds as inhibitors of sucrase since they may offer promise in wieght reduction programs.[25] The carbon analog of sucrose, *C*-sucrose, is shown in Figure 11. This molecule has been made by several groups[26] and may find use as a non-metabolizeable sweetener and could be used to probe structure-activity relationships in sweetness studies. Only recently has the synthetic chemistry become available to quickly and efficiently synthesize *C*-disaccharides. Within the next few years one should expect to see more and more reports dealing with the biological testing and evaluation of this class of compounds.

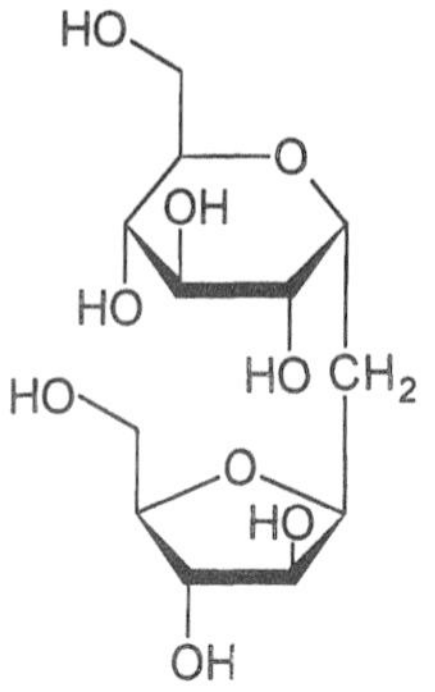

C-sucrose (**117**)
Figure 11

Further interest in these molecules and *C*-glycosides in general has arisen since it has been recently found that certain glycosidase inhibitors may be useful in the treatment of viral infections,[27] cancer[28] and as use as anti-fertility agents.[29]

F.   *C*-PHOSPHONATE ANALOGS

Several analogs of phosphono sugars have been synthesized and evaluated over recent years.[30] Replacing the anomeric oxygen with a carbon atom is feasible since the geometrical similarity with the oxygen compound is retained and the carbon phosphorous bond is now hydrolytically stable.   Wittig chemistry has been the most popular method for the preparation of these compounds since it is usually highly reliable and the available methods for cyclization offer some stereoselectivity. Some of the known *C*-phosphonates are shown in Figure 12. These compounds are of interest since they are isosteric analogs of sugar 1-phosphates and may have the ability to inhibit or regulate metabolic processes.

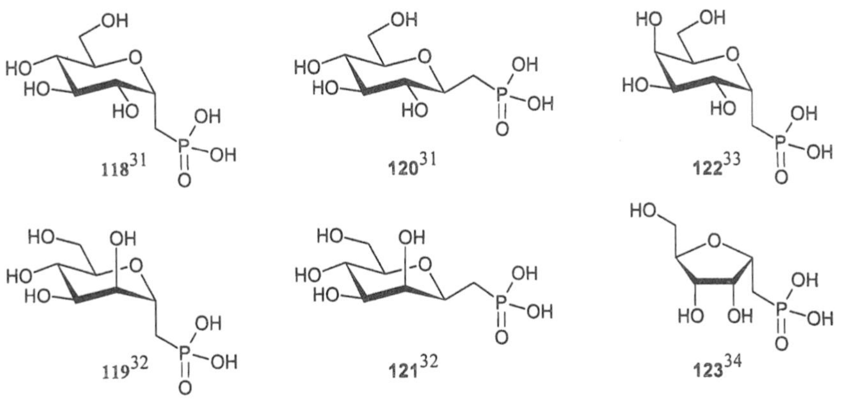

Figure 12

III. OTHER APPLICATIONS

The robust character of the *C*-glycoside linkage coupled with the naturally occurring chirality of sugars makes these compounds useful building blocks in organic chemistry. *C*-Glycosides have been used as starting materials or intermediates in total synthesis and several examples can be found throughout this book as well as in the text of Hanessian.[35]

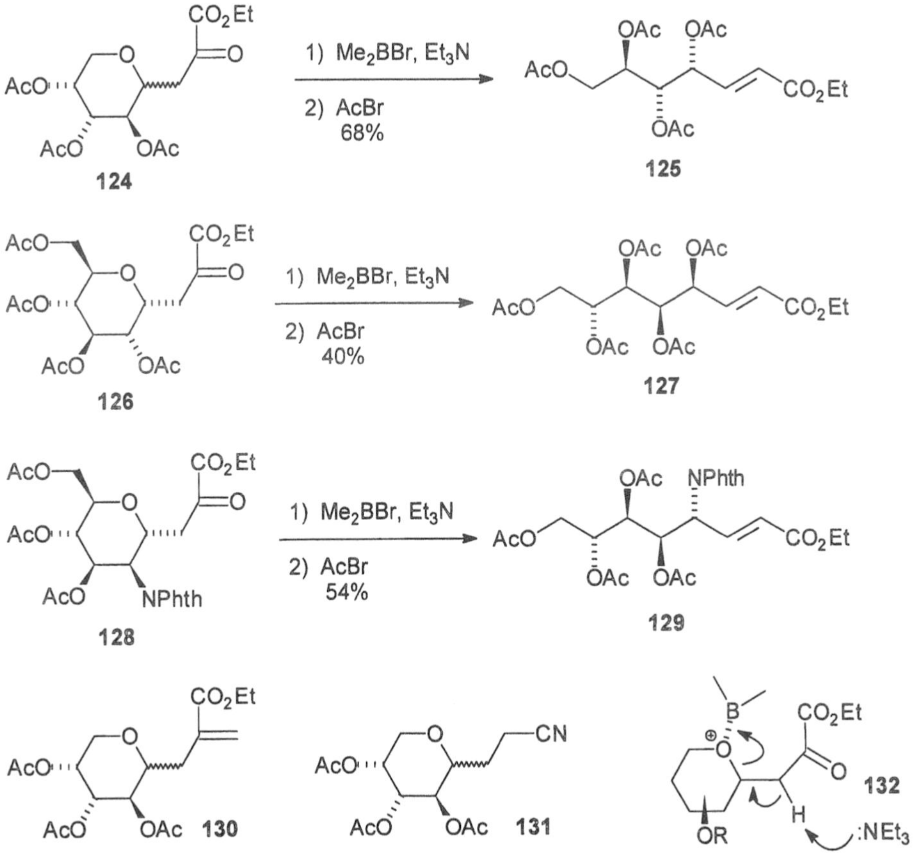

Scheme 18

A good example of the utility of *C*-glycoside is provided in Scheme 18. Abel, Giese, and Linker[36] have found that *C*-glycosides can be ring opened to the open chain derivatives by treatment with Me$_2$BBr followed by acylation to give the *E*-olefins which are useful chemical synthons. The reaction fails with the *C*-glycosides 130 and 131 because no acidic protons are available for the elimination reaction that causes ring opening.

Nonionic surfactants have become increasingly important compounds since they have been used as detergents, solubilizers, and emulsifiers.[37] They are also useful in a biological sense since they are able to solubilize protein and phospholipid components of biological membranes.   Since β-octyl glucoside (133) has been used to solubilize the LDL receptor and aid in its purification,[38] Falck, Mioskowski and co-workers[39] have suggested that the *C*-glycoside 134, which also exhibits interesting detergent properties, may also be a useful compound.

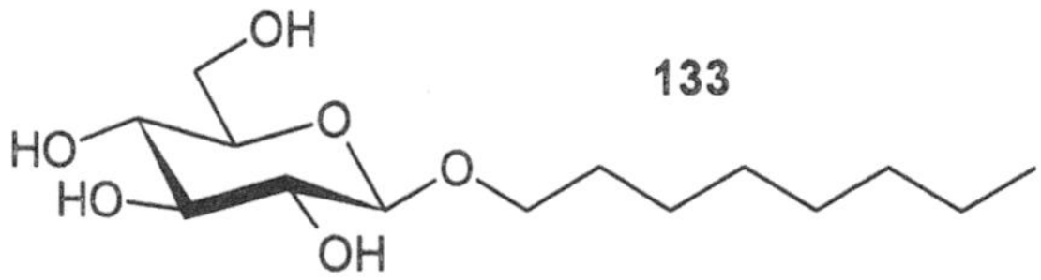

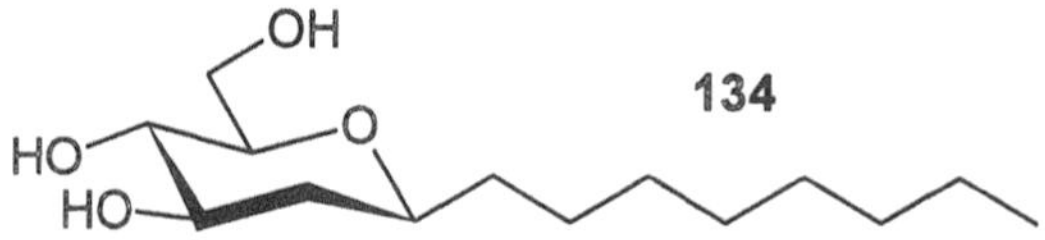

Figure 13

Armstrong[40] has used the *C*-disaccharide 135 as a building block for the assembly of a linear chain such as 135, Figure 14. The workers believe that such a structure may prove useful for polymerization reactions, electron transfer chemistry or molecular recognition events, as well as be water soluble tethers to link bipolymers. Such hydrolytically stable compounds may also find use as rigid frameworks to probe the three dimensional structure of pendant sugars in drugs.

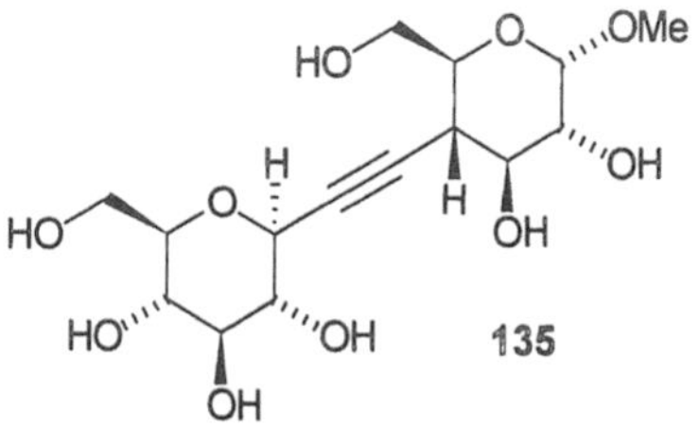

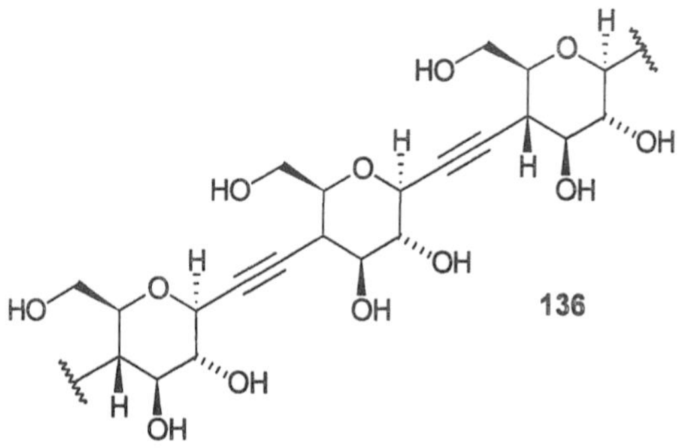

Figure 14

IV. CONCLUSION

This chapter has illustrated some of the emerging applications of *C*-glycoside compounds. From a biological point of view, this class of compounds is starting to be recognized as a source of enzyme inhibitors that can be tailored to fit the requirements of the enzyme. Of particular interest is the ability to modify the aglycone portion in such a way that it can carry "war-heads" to chemically alter functional groups on proteins or other biomolecules. The synthetic knowledge required to make these compounds is now at a level where almost any type of *C*-glycoside should be available to the medicinal chemist. A natural extension of the naturally occurring and often biologically active *C*-glycoside natural products would be to synthesize analogs of these compounds and it is certain that over the next few years reports dealing with this aspect of *C*-glycoside chemistry will start to appear in the literature. We have seen only the beginning of a much larger field that is starting to emerge. Just as *C*-glycoside synthesis was blossoming ten years ago, research today involving biological application of this potentially important class of compounds is just beginning to flourish.

V. REFERENCES

1. (a) Truscheit, E.; Frommer, W.; Junge, B.; Müller, L.; Schmidt, D.D.; Wingender, W. *Angew. Chem. Intl. Ed. Engl.* **1981**, *20*, 744. (b) LaLégerie, P.; Legler, G. and Yon, J.M. *Biochimie* **1982**, *64*, 977. (c) Sinnott, M.L. In *Enzyme Mechanisms*; Page, M.I. and Williams, A., Eds.; The Royal Society of Chemistry; London, 1987, 259. (d) Dale, M.P.; Ensley, H.E.; Kern, K.; Sastry, K.A.R. and Byers, L.D. *Biochemistry* **1985**, *24*, 3530.

2. Liu, P.S. *J. Org. Chem.* **1987**, *52*, 4717.

3. See for example: Winkler, D.A.; Holan, G. *J. Med. Chem.* **1989**, *32*, 2084 and references cited therein.

4. Schmidt, R.R.; Dietrich, H. *Angew. Chem. Intl. Ed. Engl.* **1991**, *30*, 1328.

5. Dietrich, H.; Schmidt, R.R. *Carbohydr. Res.* **1993**, *250*, 161.

6. Lai, W.; Martin, O.R. *Carbohydr. Res.* **1993**, *250*, 185.

7. Dietrich, H.; Schmidt, R.R. *Bioorg. Med. Chem. Lett.* **1994**, *4*, 599.

8. BeMiller, J.N.; Gilson, R.J.; Myers, R.W.; Santoro, M.M. *Carbohydr. Res.* **1993**, *250*, 101.

9. Sheffer-Dee-Noor, S.; Belakhov, V.; Baasov, T. *Bioorg. Med. Chem. Lett.* **1993**, *3*, 1583.

10. Schmidt, R.R. ; Frische, K. *Bioorg. Med. Chem. Lett.* **1993**, *3*, 1747.

11. Vaghefi, M.M.; Bernacki, R.J.; Hennen, W.J.; Robins, R.K. *J. Med. Chem.* **1987**, *30*, 1391 and Vaghefi, M.M.; Bernacki, R.J.; Dalley, N.K.; Wilson, B.E.; Robins, R.K. *J. Med. Chem.* **1987**, *30*, 1383.

12. Shulman, M.L.; Shiyan, S.D.; Khorlin, A.Ya. *Carbohydr. Res.* **1974**, *33*, 229.

13. Luengo, J.I.; Gleason, J.G. *Tetrahedron Lett.* **1992**, *33*, 6911.

14. Brockhaus, M.; Lehmann, J. *Carbohydr. Res.* **1977**, *53*, 21.

15. Kuhn, C.-S.; Glaudemans, C.P.J.; Lehmann, J. *Liebigs Ann. Chem.* **1989**, 357.

16. Lehmann, J. and Schwesinger, B. *Carbohydr. Res.* **1982**, *110*, 181.

17. Lehmann, J.; Schwesinger, B. *Carbohydr. Res.* **1982**, *107*, 43.

18. Gruters, R.A.; Neefjes, J.J.; Tersmette, M.; de Goede, R.E.Y.; Tulp, A.; Huisman, H.G.; Miedema, F.; Ploegh, H.L. *Nature* **1987**, *330*, 74.

19. Nagy, J.O.; Wang, P.; Gilbert, J.H.; Schaefer, M.E.; Hill, T.G.;Callstrom, M.R.; Bednarski, M.D. *J. Med. Chem.* **1992**, *35*, 4501.

20. Spevak, W.; Nagy, J.O.; Charych, D.H.; Schaefer, M.E.; Gilbert, J.H.; Bednarski, M.D. *J. Am. Chem. Soc.* **1993**, *115*, 1146. A related approach using *O*-glycosides had been reported earlier: Kingery-Wood, J.E.; Williams, K.W.; Sigal, G.B.; Whitesides, G.M. *J. Am. Chem. Soc.* **1992**, *114*, 7303.

21. Vyplel, H.; Scholz, D.; Macher, I.; Schindlmaier, K.; Schütze, E. *J. Med. Chem.* **1991**, *34*, 2759.

22. Qiao, L.; Vederas, J.C. *J. Org. Chem.* **1993**, *58*, 3480.

23. Pangiot, M.J.; Humphries, K.A.; Curley, Jr., R.W. *J. Carbohydr. Chem.* **1994**, *13*, 303.

24. Allevi, P.; Anastasia, M.; Ciuffreda, P.; Scala, A. *J. Carbohydr. Chem.* **1993**, *12*, 209.

25. Layer, P.; Carlson, G.L.; DiMagno, E.P. *Gastroenterology* **1985** *88*, 1895.

26. Dyer, U.C.; Kishi, Y. *J. Org. Chem.* **1988**, *53*, 3383 and Carcano, M.; Nicotra, F.; Panza, L.; Russo, G. *J. Chem. Soc., Chem. Commun.* **1989**, 642.

27. Datema, R.; Olofsson, S.; Romero, P.A. *Pharmacol. Ther.* **1987**, *33*, 221.

28. Humphries, M.J.; Matsumoto, K.; White, S.L.; Olden, K. *Cancer Res.* **1986**, *46*, 5215.

29. Wassarman, P.M. *Science* **1987**, *235*, 553.

30. Engel, R. *Chem. Rev.* **1977**, *77*, 349.

31. Nicotra, F.; Ronchetti, F.; Russo, G. *J. Org, Chem.* **1982**, *47*, 4459.

32. Nicotra, F.; Perego, R.; Ronchetti, F.; Russo, G.; Toma, L. *Carbohydr. Res.* **1984**, *131*, 180.

33. Chmielewski, M.; BeMiller, J.; Cerretti, D.P. *Carbohydr. Res.* **1981**, *97*, C1.

34. Nicotra, F.; Panza, L.; Ronchetti, F.; Toma, L. *Tetrahedron Lett.* **1984**, *25*, 5937.

35. Hanessian, S. *Total Synthesis of Natural Products. The Chiron Approach*; Pergamon Press, Oxford, 1983.

36. Abel, S.; Linker, T.; Giese, B. *Synlett* **1991**, 171.

37. Dennis, E.A.; Robson, R.J. *Acc. Chem. Res.* **1983**, *16*, 251.

38. Schneider, W.J.; Basu, S.K.; McPhaul, M.J.; Goldstein, J.L.; Brown, M.S. *Proc. Natl. Acad. Sci. USA* **1979**, *76*, 5577.

39. Ousset, J.B.; Mioskowski, C.; Yang, Y.-L.; Falck, J.R. *Tetrahedron Lett.* **1984**, *25*, 5903.

40. Sutherlin, D.P.; Armstrong, R. W. *Tetrahedron Lett.* **1993**, *34*, 4897 and references cited therein.

Index

-a-

-e-

-p-